Greenhouse Gas Removal Technologies

Energy and Environment Series

Series editors:
Nigel Brandon OBE FREng, *Imperial College London, UK*
Roberto Rinaldi, *Imperial College London, UK*
Vivian Wing-Wah Yam, *University of Hong Kong, Hong Kong*

Titles in the series:
1: Thermochemical Conversion of Biomass to Liquid Fuels and Chemicals
2: Innovations in Fuel Cell Technologies
3: Energy Crops
4: Chemical and Biochemical Catalysis for Next Generation Biofuels
5: Molecular Solar Fuels
6: Catalysts for Alcohol-Fuelled Direct Oxidation Fuel Cells
7: Solid Oxide Fuel Cells: From Materials to System Modeling
8: Solar Energy Conversion: Dynamics of Interfacial Electron and Excitation Transfer
9: Photoelectrochemical Water Splitting: Materials, Processes and Architectures
10: Biological Conversion of Biomass for Fuels and Chemicals: Explorations from Natural Utilization Systems
11: Advanced Concepts in Photovoltaics
12: Materials Challenges: Inorganic Photovoltaic Solar Energy
13: Catalytic Hydrogenation for Biomass Valorization
14: Photocatalysis: Fundamentals and Perspectives
15: Photocatalysis: Applications
16: Unconventional Thin Film Photovoltaics
17: Thermoelectric Materials and Devices
18: X-Ray Free Electron Lasers: Applications in Materials, Chemistry and Biology
19: Lignin Valorization: Emerging Approaches
20: Advances in Photoelectrochemical Water Splitting: Theory, Experiment and Systems Analysis
21: Electrochemical Reduction of Carbon Dioxide: Overcoming the Limitations of Photosynthesis
22: Integrated Solar Fuel Generators
23: Magnesium Batteries: Research and Applications
24: Organic Thermoelectric Materials
25: Electrochemical Methods for Hydrogen Production
26: Carbon Capture and Storage
27: Heterogeneous Catalysis for Energy Applications
28: Carbon Dioxide Electrochemistry: Homogeneous and Heterogeneous Catalysis

29: Thermal Energy Storage
30: Ultrathin Oxide Layers for Solar and Electrocatalytic Systems
31: Greenhouse Gas Removal Technologies

How to obtain future titles on publication:
A standing order plan is available for this series. A standing order will bring delivery of each new volume immediately on publication.

For further information please contact:
Book Sales Department, Royal Society of Chemistry, Thomas Graham House, Science Park, Milton Road, Cambridge, CB4 0WF, UK
Telephone: +44 (0)1223 420066, Fax: +44 (0)1223 420247
Email: booksales@rsc.org
Visit our website at www.rsc.org/books

Greenhouse Gas Removal Technologies

Edited by

Mai Bui
Imperial College London, UK
Email: m.bui@imperial.ac.uk

and

Niall Mac Dowell
Imperial College London, UK
Email: niall@imperial.ac.uk

ROYAL SOCIETY OF CHEMISTRY

Energy and Environment Series No. 31

Print ISBN: 978-1-83916-199-5
PDF ISBN: 978-1-83916-524-5
EPUB ISBN: 978-1-83916-525-2
Print ISSN: 2044-0774
Electronic ISSN: 2044-0782

A catalogue record for this book is available from the British Library

The Royal Society of Chemistry is a charity, registered in England and Wales, Number 207890, and a company incorporated in England by Royal Charter (Registered No. RC000524), registered office: Burlington House, Piccadilly, London W1J 0BA, UK, Telephone: +44 (0) 20 7437 8656.

For further information see our website at www.rsc.org

Printed in the United Kingdom by CPI Group (UK) Ltd, Croydon, CR0 4YY, UK

Preface

Greenhouse gas removal (GGR) technologies are broadly recognised as being indispensable to a 1.5 °C target, and even more so with a growing number of countries committing to net-zero emissions by 2050. GGR approaches are designed to enable the removal of CO_2, or other greenhouse gases, from the atmosphere. GGR development covers a broad range of disciplines, from agricultural practices (*e.g.*, biochar and soil carbon sequestration) to engineering processes (*e.g.*, BECCS, direct air capture). The deployment of GGR at scale will have environmental, economic, and social impacts, as well as affecting competition for land, water and energy usage. There are potential impacts to biodiversity, physical climate feedbacks, *e.g.*, albedo modification, or the release of other greenhouse gases. Therefore, understanding the fundamental science behind GGR technologies will be essential. Also critical will be the development of policies to support GGR deployment, ensuring there is synergy with other decarbonisation efforts. In this context, recognising the specific climate repair value (CRV) that GGR pathways provide and carefully distinguishing CRV from other environmental co-benefits will be key.

This book sets out to provide a resource that is of use to those working in as many of these areas as possible. Our aspiration is that graduate students, researchers, and industry professionals who are joining our community will find this book a useful resource to help them in their work.

Mai Bui and Niall Mac Dowell
London

Energy and Environment Series No. 31
Greenhouse Gas Removal Technologies
Edited by Mai Bui and Niall Mac Dowell

Published by the Royal Society of Chemistry, www.rsc.org

Contents

Chapter 1 Introduction 1
Mai Bui and Niall Mac Dowell

1.1 Where Are We Now? 1
1.2 What To Do? 2
1.3 Conclusions 3
References 4

Chapter 2 Greenhouse Gas Removal: Overview and Current Status of Deployment 6
Maxwell Pisciotta, Justine Davids and Jennifer Wilcox

2.1 Perspective Shift 6
2.2 Removing CO_2 from the Air 9
2.3 Challenges and Opportunities of GGR 10
2.3.1 Bioenergy with Carbon Capture and Storage (BECCS) 14
2.3.2 Direct Air Capture (DAC) 14
2.3.3 Mineral Carbonation 15
2.3.4 Ocean Alkalinity Enhancement 16
2.3.5 Soil Carbon Sequestration 17
2.3.6 Biochar 18
2.3.7 Afforestation and Reforestation 19
2.3.8 Ocean Fertilisation 20
2.3.9 Coastal Blue Carbon 20
2.4 CO_2 Storage 21
2.5 Co-benefits and Societal Considerations 21

Energy and Environment Series No. 31
Greenhouse Gas Removal Technologies
Edited by Mai Bui and Niall Mac Dowell

Published by the Royal Society of Chemistry, www.rsc.org

2.6 Concluding Remarks 22
References 22

Chapter 3 Negative Emissions: The Role and Response of the Climate System 27
Chris D. Jones

3.1 Introduction 28
3.1.1 The Need for Negative Emissions to Achieve Climate Goals 28
3.1.2 The Role of the Earth System in NETs Research 29
3.2 Supply and Demand for Negative Emissions 30
3.3 Earth System Implications for the Demand for Negative Emissions 33
3.3.1 Earth System Model Response to Negative Emissions in Complex Models 34
3.4 Earth System Implications for the Supply of Negative Emissions 39
3.4.1 BECCS and Afforestation 40
3.4.2 Land Management and Restoration 41
3.4.3 Summary 41
3.5 Discussion and Conclusions 42
3.5.1 Co-benefits and Unintended Consequences 42
3.5.2 Other Considerations: The Role of Non-CO_2 Activities 43
Glossary 44
References 45

Chapter 4 Accounting Negative Emissions. How Difficult Could It Be? 57
Andrea Ramírez Ramírez

4.1 Introduction 57
4.2 Carbon Accounting 59
4.3 Storage Time: How Long is Long Enough? 60
4.3.1 Time Horizon and Lifetime of Carbon Storage in Products 61
4.4 The Timing of Emissions and Removals 67
4.4.1 Delayed Emissions 68
4.5 CO_2 Storage and CO_2 Avoidance 71
4.6 Some Final Words 73
References 74

Chapter 5 Bioenergy with Carbon Capture and Storage (BECCS) **80**
Mathilde Fajardy

5.1 Introduction 80
5.1.1 BECCS Principle 80
5.1.2 BECCS Value in Decarbonisation Pathways 81
5.1.3 How Much BECCS is there Today? 83
5.2 Definitions 83
5.2.1 BECCS Value Chain 83
5.2.2 BECCS Carbon Metrics 86
5.2.3 BECCS Energy Metrics 89
5.2.4 BECCS Financial Metrics 90
5.2.5 BECCS Environmental Metrics 90
5.3 BECCS Feedstocks 91
5.3.1 Woody Biomass 92
5.3.2 Energy Crops and Residues 92
5.3.3 Wastes 93
5.4 BECCS Pathways 94
5.4.1 BECCS to Power 94
5.4.2 BECCS to Liquid Fuel 95
5.4.3 BECCS for Hydrogen 98
5.4.4 BECCS in Industry 101
5.4.5 Mitigation *vs.* Removal Potential and System Integration Effects 102
5.5 Deployment Opportunities and Constraints 103
5.5.1 BECCS Global Sustainable Potential and Impacts 103
5.5.2 Low Hanging Fruit 105
5.5.3 Optimal BECCS Deployment at the Global Level 106
5.5.4 NET Co-deployment and Synergies 107
5.6 Conclusion 108
References 108

Chapter 6 Direct Air Capture **115**
Howard Herzog

6.1 Overview 115
6.2 Process Fundamentals 117
6.2.1 Air Handling and Contactor Design 118
6.2.2 Minimum Work and Energy Requirements 120

6.3 Process Design 123
6.3.1 Climeworks 124
6.3.2 Carbon Engineering 127
6.4 Costs 130
6.5 Assessment and Takeaway Messages 135
Acknowledgements 136
References 136

Chapter 7 Geochemical Negative Emission Technologies 138
Olivia Hawrot, James Campbell, Frances Buckingham and Phil Renforth

7.1 Historical Context 139
7.1.1 Introduction 139
7.1.2 *Ex-situ* Mineral Carbonation 140
7.1.3 *In-situ* Carbonation 145
7.1.4 Enhanced Rock Weathering (Surficial) 146
7.1.5 Anthropogenic Alkaline Materials 147
7.2 Resources 150
7.2.1 Chemistry 150
7.2.2 Dissolution Minerals 151
7.2.3 Secondary Minerals 154
7.2.4 Ocean Alkalinity 155
7.3 The Kinetics of Geochemical NETs 157
7.3.1 Gas–Solid Processes 157
7.3.2 Aqueous Processes 158
7.3.3 Summary 163
7.4 Geochemical NET systems 163
7.4.1 *Ex-situ* Mineral Carbonation 163
7.4.2 *In-situ* Mineral Carbonation 172
7.4.3 Anthropogenic Alkaline Material Carbonation 172
7.4.4 Proposals for Increasing Ocean Alkalinity 173
7.4.5 Hybrid Approaches 174
7.5 Risks and Co-benefits 174
7.5.1 Valuable Products 174
7.5.2 Valuable Services 175
7.5.3 Risks 175
7.6 Conclusions 177
Glossary 178
Acknowledgements 183
References 183

Chapter 8 Soil Carbon Sequestration and Biochar **194**
Sylvia H. Vetter, Mohamed Abdalla, Matthias Kuhnert and Pete Smith

8.1 Introduction 195
8.2 Soil Carbon Sequestration (SCS) 197
8.2.1 SCS Management 197
8.2.2 Impacts of SCS Managements 207
8.2.3 SCS as a Greenhouse Gas Removal Technology 212
8.3 Biochar 216
8.3.1 What is Biochar? 216
8.3.2 Biochar as a Greenhouse Gas Removal Technology 218
8.3.3 Impact of Biochar Beyond the Function as GGR 222
8.4 Implementation Frameworks 228
8.4.1 The "4 per 1000" Initiative – Soils for Food Security and Climate 228
8.5 Summary and Conclusions 229
References 231

Chapter 9 Afforestation/Reforestation (AR) **244**
Solene Chiquier and Niall Mac Dowell

9.1 Introduction 244
9.1.1 Definition and Scope of the Study 244
9.1.2 Accounting for CO_2 Sequestration in AR Projects 245
9.1.3 Chapter Contributions 248
9.2 Forest Growth Model 248
9.2.1 Model Scope & Characterisation 248
9.2.2 Above-ground Biomass 251
9.2.3 Afforested Above-ground Biomass 253
9.2.4 Below-ground Biomass 259
9.2.5 Total Biomass 259
9.2.6 Overview 259
9.3 Biogenic C (and CO_2) Sequestration Model 261
9.3.1 Biogenic C Pools 261
9.3.2 Biomass C Pool 262
9.3.3 Dead Organic Matter C Pool 264
9.3.4 Harvested Wood Product C Pool 264
9.3.5 Overview 265

9.4 Risk-adjusted C Sequestration Model 266
9.4.1 Permanence of Biogenic CO_2 Sequestration 266
9.4.2 Wildfire-penalty Coefficient 267
9.4.3 CO_2 Sequestration Potential 268
9.5 Forestry Operations Model 269
9.5.1 Site Establishment 270
9.5.2 Forest Road Construction and Maintenance 270
9.5.3 Annual Forestry Operations 271
9.6 CO_2 and N_2O Balance Model 271
9.6.1 CO_2 and N_2O Emission Contributions 271
9.6.2 CO_2 Removal Efficiency 273
9.6.3 GHG Balance/CO_2 Removal Potential 275
9.7 Cost Balance Model 275
9.7.1 Cost Contributions 275
9.7.2 Cost Balance Over 100 Years 277
9.7.3 Cost Balance Over 1000 Years 278
9.8 Water Footprint Model 280
9.8.1 Water Contributions 280
9.8.2 Water Balance 284
9.9 Conclusion 285
References 286

Chapter 10 Feasibility of Using Biologically-based Processes in the Open Ocean and Coastal Seas for Atmospheric CO_2 Removal 291
Phillip Williamson, Philip W. Boyd, Daniel P. Harrison, Nick Reynard and Ali Mashayek

10.1 Introduction 292
10.2 Biologically-based CO_2 Removal in the Open Ocean 295
10.2.1 Direct Ocean Iron Fertilisation 295
10.2.2 Atmospheric Ocean Iron Fertilisation (Iron Salt Aerosol) 303
10.2.3 Macronutrient Addition 305
10.2.4 Enhanced Upwelling to Increase Productivity 309
10.3 Biologically-based CO_2 Removal in Coastal Seas 311
10.3.1 Coastal Blue Carbon Ecosystems 311
10.3.2 Marine BECCS Using Seaweed (Macroalgae) 322
10.4 Conclusions 326
Acknowledgements and Competing Interests 330
References 330

Chapter 11 Comparison of Technologies and Practices for Removing Carbon Dioxide from the Atmosphere **351**
Sabine Fuss

11.1 Introduction 351
11.2 Readiness Framework for CDR 357
11.3 Comparison of CDR Technologies and Practices 358
11.3.1 Bioenergy with Carbon Capture and Storage (BECCS) 358
11.3.2 Direct Air Carbon Capture and Storage (DACCS) 360
11.3.3 Mineral Carbonation and Alkaline Materials 362
11.3.4 Soil Carbon Sequestration and Biochar 363
11.3.5 Afforestation 365
11.3.6 Marine Biologically-based CDR 366
11.4 Synthesis 368
11.5 Discussion 369
Abbreviations 371
Acknowledgements 371
References 372

Chapter 12 Global Distribution of Negative Emissions **378**
Carlos Pozo, Ángel Galán-Martín, David M. Reiner, Gonzalo Guillén-Gosálbez, Mai Bui and Niall Mac Dowell

12.1 Introduction 379
12.2 Definition of Fair CDR Quotas 380
12.3 Two Possible Pathways: Isolationist and Cooperative 381
12.4 Encouraging Cooperation: United We Stand 384
12.5 Effective Policies to Remove Barriers 385
12.6 Conclusions 386
References 388

Chapter 13 Political and Socio-economic Challenges of Greenhouse Gas Removal Technologies **390**
Zeynep Clulow and David M. Reiner

13.1 Introduction 391
13.2 Economic Challenges 395
13.2.1 Cost Structures 395
13.2.2 Compatibility with Other Priorities 397
13.2.3 Incentives 398

13.3 Stakeholder Views 399
13.3.1 Governments 400
13.3.2 Private Sector Views 402
13.3.3 NGOs 408
13.3.4 Public Engagement 410
13.4 Governance 413
13.4.1 Incentivising GGR 413
13.4.2 Carbon Pricing 414
13.4.3 Accounting 414
13.4.4 Risk Management 415
13.4.5 Side-effects and Broader Implications 416
13.4.6 Procedural Justice 416
13.5 Socio-political Aspects of International Governance 417
13.6 Conclusion 419
References 421

Chapter 14 Priorities for Policy Design **430**
Habiba Ahut Daggash, Mai Bui and Niall Mac Dowell

14.1 Introduction 431
14.1.1 Characterising NETs for Policy Development 432
14.2 Research and Development 434
14.2.1 Technological Feasibility 434
14.2.2 Costs 434
14.2.3 Scalability 437
14.2.4 Measuring, Verification and Reporting Guidelines 437
14.2.5 Environmental Impacts 437
14.3 Demonstration and Scale-up 437
14.4 Financing 438
14.4.1 NET-based Cost–benefit Analysis 439
14.4.2 NET-based Carbon Taxes 441
14.4.3 NET-based Carbon Markets 442
14.4.4 NET-based Subsidies 443
14.5 Niche Markets 444
14.6 Regulation and Governance 445
14.6.1 Recognising NETs 449
14.6.2 Measurement, Reporting and Verification 451
14.6.3 Cross-border Carbon Accounting 453
14.6.4 The Role of Diplomacy 454

14.7 Public Acceptability 455
14.7.1 Moral Hazard 455
14.7.2 Food–Fuel Debate 455
14.7.3 CCS 456
14.8 Conclusion 456
Abbreviations 457
References 458

Chapter 15 Socially Equitable Energy Transitions: Analytical Challenges and Policy Implications **465**
Piera Patrizio and Niall Mac Dowell

15.1 Introduction 466
15.2 Energy Transition Modelling 467
15.2.1 Existing Approaches 467
15.2.2 Key Features of Socially Equitable Energy Transition 469
15.3 Energy Transition Policies 474
15.3.1 Cost-benefit Analysis (CBA) 474
15.3.2 Economic Impact Assessment (EIA) 475
15.3.3 Key Socio-economic Metrics 476
15.4 Conclusions 480
References 481

Subject Index **484**

CHAPTER 1

Introduction

MAI BUI[a,b] AND NIALL MAC DOWELL*[a]

[a] Centre for Environmental Policy, Imperial College London, UK; [b] Centre for Process Systems Engineering, Imperial College London, UK
*Email: niall@imperial.ac.uk

Greenhouse gas removal (GGR) technologies, also known as carbon dioxide removal (CDR), are understood to be essential to all scenarios consistent with a 1.5 °C target. Technologies that provide net removal of CO_2 emissions include nature-based solutions (*e.g.*, afforestation, reforestation and coastal blue carbon), enhanced weathering, biochar, soil carbon sequestration, ocean fertilisation, direct air capture of CO_2 (DAC) and bioenergy with carbon capture and storage (BECCS). Advancing these GGR technologies forward will require scale-up and commercialisation. Importantly, we will need to identify which GGR pathways can be relied upon to be a material option within the portfolio of solutions, and which will be niche players. The key areas of GGR that require further development in understanding include the climate repair value, the permanence of the carbon removal (*e.g.*, one year or 1000 years), the types of services provided (*e.g.*, environmental, energy) by each GGR option, how cost might vary in different regions, and the protocols for monitoring, reporting and verification (MRV).

1.1 Where Are We Now?

Referencing the 2015 Paris Agreement at the beginning of books and research papers has emerged as an almost unspoken convention amongst the community who are concerned with climate change. In the year 2015, the

Energy and Environment Series No. 31
Greenhouse Gas Removal Technologies
Edited by Mai Bui and Niall Mac Dowell

Published by the Royal Society of Chemistry, www.rsc.org

world agreed to the now famous "well below 2 °C" target, with the equally famous aspiration to meet a more stringent 1.5 °C target, and anthropogenic CO_2 emissions of 35.2 Gt per year.[1]

In 2018, the IPCC published their special report on 1.5 °C[2] which, *inter alia*, highlighted the importance of reducing greenhouse gas emissions to net zero by "around 2050", and also firmly established the need to remove CO_2 from the atmosphere, both as part of the transition to net zero and also as being integral to the overarching 1.5 °C aspiration.

By 2019, starting with the UK, an increasing number of corporations and nations pledged to achieve a net zero emissions goal. National pledges to reach net zero cover 61% of global emissions, 68% of the global economy, and 52% of the global population.[3] In sum, 124 countries are committed to, or are actively considering net zero targets. In short, the direction of travel appears clear, and it would appear that net zero targets are becoming the norm. Importantly, both the EU and US appear to be actively considering border carbon adjustment mechanisms (BCAM)[4–6] to address issues surrounding carbon leakage, and China has pledged to peak emissions by 2030 and thereafter decline to net zero by 2060.[7] It would appear difficult to say that we are lacking in ambition.

However, ambition is one thing, results are decidedly another. By 2019, total anthropogenic CO_2 emissions had increased to 36.4 Gt per year[1] – an increase of approximately 3.5% relative to 2015. The year 2020 brought with it the COVID-19 pandemic and an associated reduction in CO_2 emissions to 34.07 Gt per year.[1] This is instructive in that it indicates the extent of reduction that can be achieved *via* behavioural change alone.

On the 6th of August 2021, the IPCC published the contribution of Working Group I to their Sixth Assessment Report.[8] It was widely reported as being a "code red for humanity"[9] with similarly bombastic language having been used by a variety of media outlets, and the current Biden administration[10] emphasising the need for action with POTUS tweeting "*We can't wait to tackle the climate crisis. The signs are unmistakable. The science is undeniable. And the cost of inaction keeps mounting*".[11] However, on the 12th of August 2021, the Biden administration also called on OPEC to boost oil production so as to avoid "harming the ongoing global recovery (from the economic impact of the COVID-19 pandemic)". Similarly, China has announced an additional forty three coal-fired power plants and thirty five million tonnes of coal-dependent ironmaking capacity in the first half of 2021.[12] It would appear that the net zero goals are rapidly in danger of becoming "more honoured in the breach than the observance."

1.2 What To Do?

Given the above, it seems increasingly clear that whilst efforts to mitigate climate change through the now-conventional methods of fuel switching, renewable energy, the use of nuclear power, and deploying CO_2 capture and storage (CCS) technology ought to be accelerated where possible, achieving the aforementioned climate change targets is going to increasingly rely on the ability to remove CO_2, and potentially other greenhouse gases, from the atmosphere at the gigatonne scale.

This practice of greenhouse gas removal (GGR), also known as carbon dioxide removal (CDR), is now extensively used as a shorthand to refer to an increasing portfolio of approaches to achieve this aim. From the so-called nature-based solutions (NBS), such as soil carbon sequestration (SCS), to afforestation and reforestation (AR) to more technology-focused approaches such as biochar (BC), enhanced weathering (EW), direct air CO_2 capture and storage (DACCS) and bioenergy with carbon capture and storage (BECCS). Importantly, this list only refers to some of the land-based approaches, and it is important to recognise that, given that approximately 71% of the earth's surface is water, and that the oceans have absorbed tens of billions of tonnes of CO_2 since the industrial revolution, there is merit in considering approaches to remove CO_2 directly from the oceans as well as from the atmosphere. Many of these approaches involve "fertilising" the oceans in some way, or the restoration of coastal or "blue carbon" ecosystems, or finally the cultivation and harvesting of ocean biomass for use in marine-BECCS approaches.

It is important to recognise that, despite its prominence in current national and international policy discourse, and the arguably near-commercial status of BECCS and DACCS concepts, the field of GGR is very much in its infancy, and there remain many questions that are, as yet unanswered.

1.3 Conclusions

So, how can GGR move forward? This will, inevitably, rest upon the scale-up and commercialisation of the various options so we can quickly discern between which GGR pathways can be relied upon to be a material element of the portfolio of solutions, and which are niche players.

To achieve this, there are several key steps forward, which include the following:

1. We need to quantify the climate repair value provided by each of the GGR pathways. Anthropogenic CO_2 in the atmosphere is a problem because it is such a long-lived greenhouse gas. Thus, in order to fully compensate for the impact of a given CO_2 emission, an equivalent amount of CO_2 will need to be permanently removed from the atmosphere at approximately the same time.
2. We need to decide what "permanent" means in this context. Does it mean one year, or one million? Or perhaps something in between.
3. We need to distinguish between the environmental service of carbon removal and the other services that the various GGR pathways can provide. For example, DACCS is something that one would pursue exclusively for reasons of climate repair. BECCS provides a combination of GGR and energy services. Afforestation, however, can provide a range

of services in addition to carbon management, such as air quality, water management, biodiversity, and ecosystem services, amongst others. All of these services ought to be explicitly valued.

4. The cost of delivering credible GGR will vary around the world. Each nation will have its own supply curve, and it is not obvious why these comparative advantages should not be exploited to deliver cost-optimal GGR in combination with a rigorous monitoring, verification and regulatory regime.
5. In order to write a contract for the provision of GGR services, it will be necessary to know (a) when CO_2 removal begins, (b) at what rate CO_2 is being removed, and (c) when a given store is saturated, and a project can be completed. Achieving these outcomes will require the development of detailed monitoring, reporting and verification (MRV) protocols. These will need to be credible, and internationally recognised.

Finally, we have learned from recent experience during the COVID-19 pandemic that initiating and rapidly scaling new infrastructure is challenging, to say the least, and that best results are achieved by leveraging existing assets and flexing them. In this context, existing emissions trading schemes (ETS) may provide a good option for the commoditisation of GGRs. Understanding how credible GGRs can be integrated with an ETS is an option which ought to be explored with alacrity.

References

1. H. Ritchie and M. Roser, *CO_2 emissions*, https://ourworldindata.org/co2-emissions, accessed 13 November 2021.
2. IPCC, *Global warming of 1.5 °C*, Intergovernmental Panel on Climate Change (IPCC), Switzerland, 2018.
3. R. Black, K. Cullen, B. Fay, T. Hale, J. Lang, S. Mahmood and S. M. Smith, Taking stock: A global assessment of net zero target, Energy & Climate Intelligence Unit and Oxford Net Zero, https://ca1-eci.edcdn.com/reports/ECIU-Oxford_Taking_Stock.pdf?mtime=20210323005817&focal=none, 2021.
4. European Commission, *EU Green Deal (carbon border adjustment mechanism)*, https://ec.europa.eu/info/law/better-regulation/have-your-say/initiatives/12228-EU-Green-Deal-carbon-border-adjustment-mechanism-_en, 2021.
5. K. Taylor, US lawmakers push carbon border tariff similar to EU's CBAM, EURACTIV, https://www.euractiv.com/section/energy-environment/news/us-lawmakers-push-carbon-border-tariff-similar-to-eus-cbam/, 2021.
6. F. Jordans, Kerry says US examining carbon border tax, sees risks, AP News, https://apnews.com/article/europe-environment-and-nature-business-government-and-politics-6a020cd7bb93a639e7445cf4999276a2, 2021.

7. V. Volcovici, *China pledges to achieve CO2 emissions peak before 2030, carbon neutrality before 2060 -Xi*, Reuters, 2020, https://www.reuters.com/article/un-assembly-climatechange-idUSL2N2GJ105.
8. IPCC, *Climate Change 2021: The Physical Science Basis, the Working Group I contribution to the Sixth Assessment Report*, Intergovernmental Panel on Climate Change, 2021, https://www.ipcc.ch/report/sixth-assessment-report-working-group-i/.
9. UN Secretary-General, *Secretary-General's statement on the IPCC Working Group 1 Report on the Physical Science Basis of the Sixth Assessment*, United Nations, 2021, https://www.un.org/sg/en/content/secretary-generals-statement-the-ipcc-working-group-1-report-the-physical-science-basis-of-the-sixth-assessment.
10. The White House, ICYMI: Biden-Harris Administration Leaders Underscore Importance of Global Action to Combat Climate Change on Heels of New IPCC Report, Press Release, Briefing Room, https://www.whitehouse.gov/ostp/news-updates/2021/08/11/icymi-biden-harris-administration-leaders-underscore-importance-of-global-action-to-combat-climate-change-on-heels-of-new-ipcc-report/, 2021.
11. President Joe Biden, *The White House*, Twitter, https://twitter.com/POTUS/status/1424759606845313028, 2021.
12. C. Shepherd, *China Puts Growth Ahead of Climate With Surge in Coal-powered Steel Mills*, Financial Times, 2021, https://www.ft.com/content/c4c79efb-9a4e-4f22-a75a-6b3ea3161bf1.

CHAPTER 2

Greenhouse Gas Removal: Overview and Current Status of Deployment

MAXWELL PISCIOTTA,[a] JUSTINE DAVIDS[b] AND JENNIFER WILCOX*[a]

[a] University of Pennsylvania, Philadelphia, PA, USA; [b] Worcester Polytechnic Institute, Worcester, MA, USA
*Email: jlwilcox@seas.upenn.edu

Greenhouse gas removal (GGR) plays an increasingly important role in meeting the climate goals of keeping the global temperature rise below 2 °C above that of the preindustrial era. There are many GGR methods which can achieve negative emissions, each with their own advantages and drawbacks. This overview chapter provides insight into the challenges of removing CO_2 from the air while highlighting opportunities for the GGR technologies discussed throughout the rest of the book, and considerations facing their future development and deployment.

2.1 Perspective Shift

The potential for climate change as a result of increased carbon dioxide (CO_2) emissions was explored as early as 1856 by Eunice Newton Foote, when she published her work concluding that an atmosphere with higher concentrations of CO_2 could increase the Earth's temperature.[1] Although her experiments had proven that increased CO_2 emissions would increase the

Energy and Environment Series No. 31
Greenhouse Gas Removal Technologies
Edited by Mai Bui and Niall Mac Dowell

Published by the Royal Society of Chemistry, www.rsc.org

Earth's temperature, it was not until years later that The Intergovernmental Panel on Climate Change (IPCC) was established. In 1988, the IPCC was formed with the purpose of assessing climate change based on the latest science.[2] The IPCC has reviewed the scientific evidence that greenhouse gas (GHG) emissions greatly impact the atmosphere, resulting in increased prevalence of weather anomalies and abnormal warming. For this reason, greenhouse gas removal (GGR), which includes negative emissions technologies (NETs) or CO_2 removal (CDR) technologies, are considered to be a part of the portfolio of solutions, along with deep decarbonisation, for reducing and avoiding atmospheric CO_2. Greenhouse gas removal mainly focuses on CO_2 removal (CDR), which is defined as technologies that provide the net removal of CO_2 from the atmosphere (*i.e.*, net negative CO_2 emissions); possible pathways include:

- using biological processes to increase carbon stocks in soil, forests, and/or wetlands;
- producing energy from biomass while simultaneously capturing and storing any CO_2 emissions;
- using chemical processes to capture CO_2 directly from the air and then sequestering it in geologic reservoirs;
- enhancing geologic processes that capture CO_2 from the atmosphere and permanently bind it with minerals.[3]

In 1992, the IPCC released a supplement to its First Assessment Report (AR1), in which it stated that "technologies to capture and sequester CO_2 from fossil fuel combustion deserve investigation.[4]" In this same report, solutions such as afforestation, reforestation, bioenergy, and soil carbon sequestration were proposed. As more time has passed, GGR technologies have continued to appear in the IPCC's publications (see Figure 2.1).

GGR technologies are becoming a focal point due to side effects from climate change being felt around the globe and the evidence that points to emissions reductions alone being insufficient for limiting the increased average surface warming to less than 2 °C.[9] Removing CO_2 (or other GHGs) from air can reduce historical emissions in the atmosphere and reduce the impact of sectors that are more difficult to decarbonise. Additionally, during the transition from fossil fuels to renewable energy sources, it will be important to utilise GGR to offset emissions while fossil fuels remain to serve many energy needs. Lastly, the IPCC predicts that for the likelihood of 66 percent that the average surface warming temperature will have increased between 0.9 and 2.3 °C in 2100 (consistent with the IPCC stringent mitigation scenario, RCP2.6), GGR will be required to achieve a net-negative anthropogenic CO_2 balance around 2070[9,10] (see Figure 2.2). However, the deployment of GGR technologies should not replace emissions reductions: both are needed to meet climate goals.

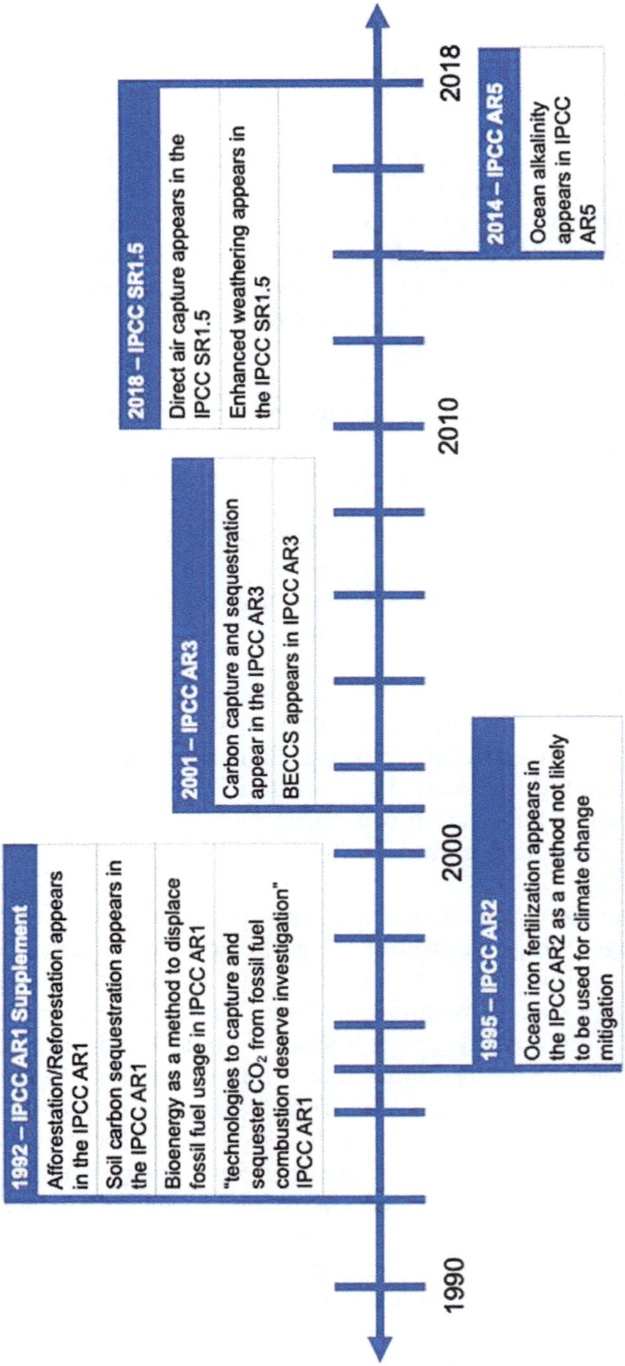

Figure 2.1 Timeline showing when the IPCC publications first mentioned specific negative emissions technologies or ideologies.[4–8]

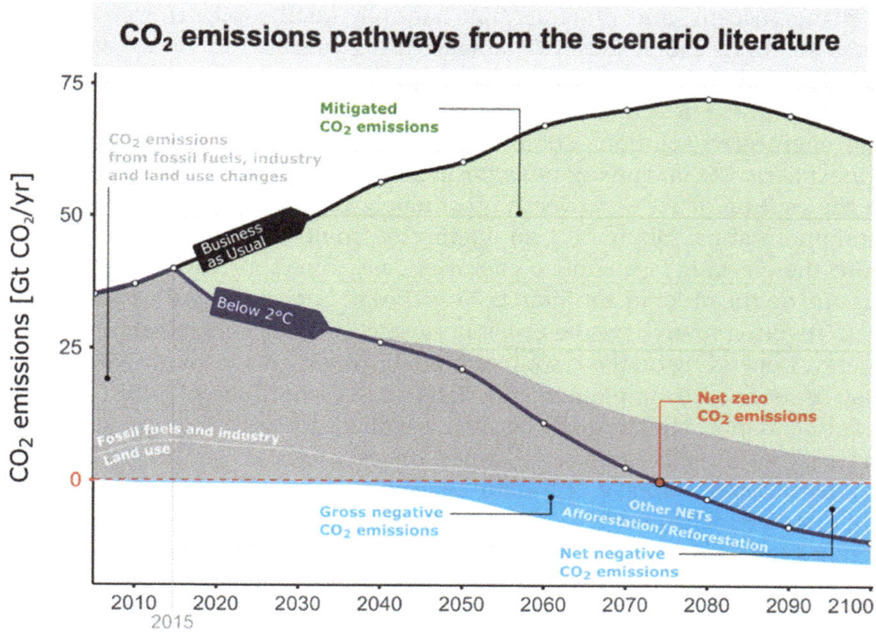

Figure 2.2 Graph showing the historical CO_2 emissions since 2010 and the predicted "Business as Usual" and "Below 2 °C" trajectories. It highlights the importance of emissions reductions, indicated by the mitigated CO_2 emissions, and NETs/GGR in achieving the net-negative anthropogenic CO_2 balance near 2070. Adapted from ref. 17, https://doi.org/10.1088/1748-9326/aabf9f, under the terms of the CC BY 3.0 license https://creativecommons.org/licenses/by/3.0/.

The need for negative CO_2 emissions indicates that emissions reductions alone will not be sufficient in limiting the average surface warming to less than 2 °C. However, emissions reductions should continue to be pursued simultaneously, as relying on GGR to be deployed on a large scale could pose social and environmental risks if they are not deployed equitably.[11] Additionally, many emissions reductions pathways involve less financial and technical investment to implement compared to GGR. Utilising both emissions reductions and GGR emphasises the need to rely on a portfolio of options to achieve net-negative emissions, rather than utilising GGR to justify the continuation of anthropogenic CO_2 emissions.

2.2 Removing CO_2 from the Air

The concentration of CO_2 in the atmosphere is currently around 415 parts per million (ppm).[12] Air is predominantly composed of nitrogen (78%) and oxygen (21%), with the remainder being argon, CO_2, and other gases. The separation of CO_2 from air takes place through natural sinks, which often consist of plants, oceans, and minerals that naturally uptake CO_2. However, engineered solutions can achieve CO_2 uptake with a 30–100% increased efficiency. Due to the amount of CO_2 that needs to be removed from ongoing industrial activity, and the need to achieve net-negative emissions by 2070, both natural and engineered solutions should be leveraged. For example, trees take up atmospheric CO_2 at varying rates throughout their lifetime, which can result in the capture of 1 t CO_2 over approximately 40 years, only requiring arable land and manual labour.[13,14] An engineered solution such as removing CO_2 from the air with regenerative chemicals, *i.e.*, direct air capture (DAC), can capture on the order of 1 million t CO_2 per year, but can be costly to achieve.[15] DAC requires many raw materials (*i.e.*, steel, cement, chemicals) and heat energy. Depending on the resources available in an area, certain GGR options may be more easily implemented at large scale. A portfolio of GGR options is required to meet climate goals because not all locations are equipped with the same physical land features, soil types, or resources.

When the CO_2 uptake of a nature-based GGR approach is considered, it can often be measured in gigatonnes of carbon (Gt C) because it is stored as organic carbon within the system. This uptake can be directly converted to gigatonnes of CO_2 (Gt CO_2) through the ratio of molar masses. The conversion is shown in eqn (2.1).

$$\text{Gt CO}_2 = (\text{Gt C}) \cdot \frac{\text{MW}_{\text{CO}_2}}{\text{MW}_\text{C}} = (\text{Gt C}) \cdot \frac{44\,\text{g mol}^{-1}}{12\,\text{g mol}^{-1}} \cong 3.67 \quad (2.1)$$

2.3 Challenges and Opportunities of GGR

There are many approaches to achieve negative emissions (see Figure 2.3). Some of these GGR pathways rely on biological systems, while others involve a more engineered approach. Some of these GGR approaches include, but are not limited to:

- **Bioenergy with Carbon Capture and Storage (BECCS):** a form of energy production through the conversion of biomass, either thermochemical (*e.g.*, combustion, gasification) or biological (*e.g.*, fermentation). For this process to result in net-negative CO_2 emissions, the emissions associated with the biomass supply chain need to be less than the captured CO_2 from the exhaust gas, which is subsequently stored. Permanent storage is achieved by pairing this process with geological sequestration or mineral carbonation.
- **Direct Air Capture (DAC):** Chemically removing CO_2 from air. For this process to result in net-negative CO_2 emissions, the carbon intensity of the energy flows required for capture should be minimised and the captured CO_2 must be permanently removed from air, which may be achieved by pairing this process with geological sequestration or mineral carbonation.

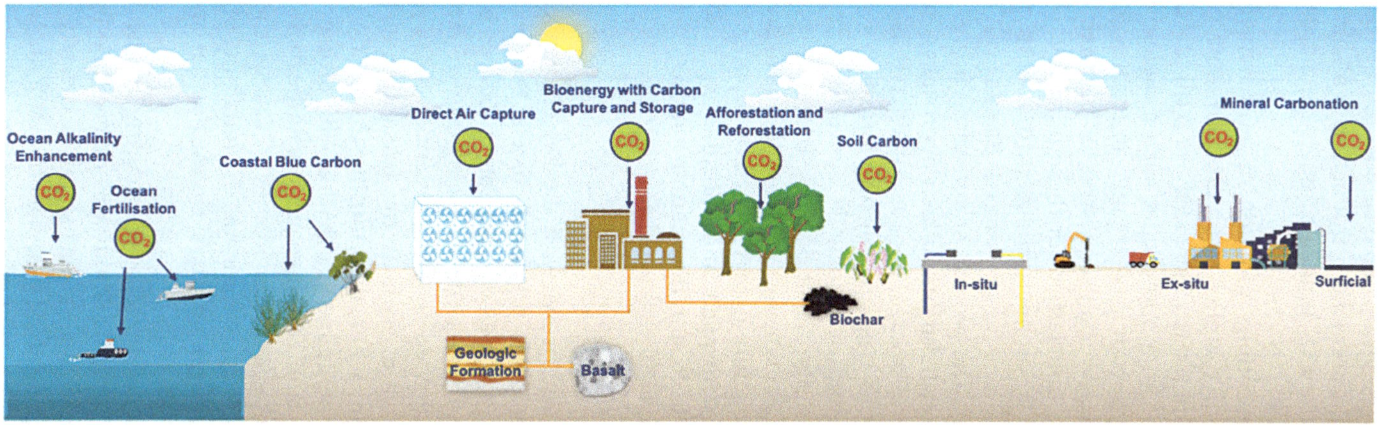

Figure 2.3 Pathways to achieve negative emissions, including, ocean alkalinity enhancement, ocean fertilisation, coastal blue carbon, direct air capture, bioenergy with carbon capture and storage, afforestation and reforestation, soil carbon, and various forms of mineral carbonation. Adapted from ref. 3 with permission from NASEM, Copyright 2019.

- **Mineral Carbonation:** A process in which alkaline-rich minerals form a bond with CO_2 and result in stable carbonate minerals.
- **Ocean Alkalinity Enhancement:** Introduction of alkaline minerals to the ocean to increase the CO_2 uptake through the natural production of stable carbonate minerals.
- **Soil Carbon Sequestration:** Adoption of land management practices that enhance the soil's ability to uptake and store carbon. These include, but are not limited to, decreased or no-till practices, cover crops, and crop rotations. Other approaches such as adaptive multi-paddock grazing and the application of organic amendments have also been proven to increase soil carbon sequestration and their region variability are comparably less understood for these practices.
- **Biochar:** Charcoal-like product produced from burning biomass in reduced oxygen conditions (*i.e.*, pyrolysis). Biochar can be used as a soil amendment through the process of spreading it over crop or pastureland or used in fuel switching in industrial processes and when coupled with CO_2 capture and storage, may be considered as BECCS.
- **Afforestation:** Creating a completely new forest.
- **Reforestation:** Planting trees, usually native, in a previously forested area that has seen a significant decrease in trees.
- **Ocean Fertilization:** Enhancing ocean conditions to promote the growth of photosynthesisers for increased carbon uptake. Ocean fertilisers usually include iron, nitrogen, and phosphorus.
- **Coastal Blue Carbon:** Restoration and/or maintenance of wetlands for the purpose of enhanced CO_2 uptake.

This chapter provides an overview of the key challenges and opportunities of the GGR technologies previously listed. Further technical information can be found in the following chapters focused on the individual technologies. In this overview, GGR technologies were evaluated based on their feasibility, scalability, co-benefits, potential consequences, and permanence of storage. Factors such as the feasibility and scalability of a technology are important when determining if specific GGR approaches will be able to be implemented at a scale that will mitigate climate change, while co-benefits and potential consequences indicate how these deployments could further impact the environment. Lastly, the permanence of storage must be considered as it indicates the reliability of keeping the CO_2 securely out of the atmosphere on a timescale that will contribute to climate change mitigation.

A GGR's feasibility was evaluated by considering its scientific viability and current scale of deployment. Scalability was determined by the known and anticipated technological barriers that stand in the way of deploying the GGR approach on a large-scale. The co-benefits were analysed to determine if the deployment of a GGR technology would provide benefits aside from CO_2 removal, while potential consequences encompass the negative side

Figure 2.4 Subjective evaluation of different GGR technologies using green as a favourable quality/aspect, yellow as mediocre, orange as unfavourable, and red as the most unfavourable. Further elaboration regarding the evaluation of each GGR option is included in the subsequent sections.

effects that can result from large-scale deployment. Lastly, the permanence of storage was evaluated based on how long and how reliably the CO_2 will be stored (covered in more detail in Section 2.4).

2.3.1 Bioenergy with Carbon Capture and Storage (BECCS)

BECCS has been proven to be feasible through many notable projects worldwide.[16] To ensure responsible land use practices, it is imperative that the biomass supply for BECCS is obtained from a sustainable source, such as agricultural or forest residue. Given that there are limited sustainably sourced biomass resources and they are distributed unevenly across the world, the scalability of BECCS can be capacity and locationally limited. If BECCS is deployed on a large-scale, it could reduce dependence on fossil fuel energy sources through combustion for electricity, gasification for hydrogen, and fermentation for bioethanol.[17–19] However, using BECCS to produce carbon-based energy products, such as bioethanol or bio-oil, can result in the release of CO_2 back into the atmosphere when the product is used. The gasification of biomass for hydrogen is a BECCS pathway that allows for near-complete capture of CO_2 in a single step.[20,21] Being able to produce renewable hydrogen for further energy use while potentially resulting in net-negative CO_2 emissions is a vital facet within the GGR portfolio. Additionally, the deployment of sustainable BECCS has the potential to contribute to the protection of degraded and marginal land for further biomass growth, or waste biomass collection.

In addition to the geologic or mineral carbonation methods, storage of CO_2 from BECCS can also come in the form of the by-product biochar.[22] Biochar can store the organic carbon from biomass and then be repurposed as a soil amendment or in building materials.

2.3.2 Direct Air Capture (DAC)

DAC is currently implemented using both liquid solvent and solid sorbent technology. Carbon Engineering, a company based out of British Columbia, has developed a pilot plant that uses liquid solvents.[23] Climeworks, another company focused on DAC, has deployed commercial plants globally with notable projects located in Switzerland and Iceland, using solid sorbents.[24] The main differences between these technologies are the chemistry used to separate the CO_2 from the other atmospheric components and the energy requirements. In addition to the energy required to move air through the system, energy is also needed to regenerate the solvents or solid sorbents to promote process sustainability and longer lifetimes of the capture materials.

Provided it requires energy to operate a DAC facility, the scalability of this technology may be limited by the overall energy demand and how it compares to global or regional supply. However, the scalability of this technology could be improved through deep decarbonisation of the electrical grid and calciner technologies for the liquid solvent case. In the solid sorbent case, the quality of the heat for regeneration is lower (*i.e.*, 100 °C compared to 900 °C for solvents), and can be coupled to waste industrial heat.[24] Although these facilities do not have any biophysical land constraints, as they do not require arable land to operate, it is advantageous that they be located near low-carbon energy resources and reliable geological storage, so they will not be solely reliant on fossil energy, and the CO_2 transportation costs are minimised.[23–26] Not requiring arable land keeps them from competing with land for food production or other nature-based GGR options.

Solvent-based DAC solely acts as a tool for CO_2 removal. However, solid-sorbent based DAC has the additional benefit of resulting in water production.[27] The primary ecological considerations with DAC technology scale-up are the disposal and the environmental impacts of the solvent or sorbent material. Disposal of solvent material is comparable to the disposal of monoethanolamine (MEA), a common adsorbent used in oil refineries, whereas there is more uncertainty with regards to the disposal of solid-sorbent material.[3] Some solid-sorbent materials can result in negative environmental impacts which will only be exacerbated when DAC is scaled-up. Deutz and Bardow (2020) analyse the environmental impacts that solid-sorbent materials can contribute to, such as further pollution, ozone depletion, and human toxicity.[28] This highlights that the choice of solid-sorbent materials for use in DAC should be evaluated and understood to ensure that all components used result in a net-negative emission technology.[28] To ensure CO_2 is not re-released into the atmosphere, it needs to be injected into sedimentary basins or alkaline-mineral deposits for permanent storage. One example of this storage is Climeworks in Iceland, which is a collaborator on the CarbFix2 project where they are storing CO_2 as carbonate minerals by exposing it to alkaline-rich minerals.[29,30]

2.3.3 Mineral Carbonation

Mineral carbonation can be achieved in three ways: *in situ*, surficial, and *ex situ* (see Figure 2.3):

- *In situ* mineral carbonation is the process of exposing alkaline-rich minerals, in the ground at depth, to CO_2.
- Surficial mineral carbonation is achieved through exposing silicate or oxide-rich wastes to CO_2-bearing gas or fluids, such that they form stable carbonates.[31]
- *Ex situ* mineral carbonation is the process of extracting alkaline-rich minerals so they can be exposed to CO_2 in a location separate from where they naturally form.

These processes are rooted in well-understood chemistry and have been proven through different projects.[31–35] However, there are still opportunities to advance the process for more effective CO_2 uptake. A project focused on *in situ* mineral carbonation is CarbFix2, which uses CO_2 captured from the air, and then mixed with water, then injected into alkaline-rich minerals 500 m underground to form solid carbonates[32,36] (see Figure 2.5). An investigation into surficial mineral carbonation is being undertaken by the De Beers Group, the world's leading diamond mining company, in

Figure 2.5 Stable carbonates (white) forming after *in situ* CO_2 injection at the CarbFix2 site. Reproduced from ref. 30 with permission from Carbfix, Copyright 2021, photo credit to Sandra Snæbjörnsdóttir.

which they are experimenting with their mine tailings in Canada and Botswana to determine how the differences in the climate and geology contribute to the CO_2 sequestration potential.[36] *Ex situ* mineral carbonation projects have often used olivine, due to its high reaction rate and abundance.[31] Recently, a looping process with magnesium-rich materials has been proposed, as a method to reduce the amount of extraction and processing that would be required for using mineral feedstock for *ex situ* mineral carbonation.[37]

In order to scale-up mineral carbonation to meet climate goals, considerations regarding mineral extraction, heating, and land use could prove to be limiting.[3,38] By putting the alkaline-rich minerals on agricultural soil to naturally uptake CO_2, they could also replace soil lost from erosion and increase crop yields.[38] Beerling *et al.* (2018) estimated that for basalt ground to particle sizes ranging from 10–30 μm in diameter and applied to agricultural soil, ~0.3 t CO_2 can be captured per tonne of basalt over a decadal timescale.[39] The carbonation rate and optimal particle size can differ depending on the mineral, specific mineralogy, and soil conditions of study. The stable carbonates produced as a result of mineral carbonation will last centuries to millennia, and therefore represent a permanent form of storage.[31]

2.3.4 Ocean Alkalinity Enhancement

Ocean alkalinity enhancement involves the same "permanent" chemical bonding found in mineral carbonation, but rather than the carbonate minerals being produced underground or on the land, the CO_2 is converted to bicarbonate ions that are stored in the ocean. Although this form of carbon capture is an acceleration of a natural process that already occurs on Earth, there is still a lot of technological development needed in this area.

In order to start the process of putting alkaline material in the ocean, the alkalinity first needs to be sourced. A common source of alkaline material that is being considered for this process is lime (CaO). Upscaling the lime industry for this purpose would require greater extraction, grinding and processing, at approximately the same scale of the cement industry today.[40] The creation of lime also needs to be taken into consideration when upscaling ocean alkalinity enhancement due to the emissions-intensive process involved in creating this material. The implementation of ocean alkalinity enhancement for CO_2 sequestration can contribute to the reversal of ocean acidification.[41] However, some of the consequences of this technique include the potential for increased pH in local ecosystems and environmental impact from trace elements in the materials added.[13] Other co-benefits and potential consequences are currently unknown due to the lack of experimentation.

2.3.5 Soil Carbon Sequestration

Many land-use practices used to date have resulted in a loss of soil organic carbon content.[42] This loss can be reversed through well understood, proven prevalent practices within certain regions to enhance soil carbon sequestration.[43] Previously, the reasons for implementation have historically not been for carbon dioxide removal, but rather because of the co-benefits associated with soil carbon sequestration practices. Soil organic carbon is increased as plants uptake CO_2 from the atmosphere and their organic components deposit carbon into the soils. The best practices to enhance the carbon uptake and storage by soil depends on the climate and soil type of each region. Specific field-based research is still needed to determine how these factors affect sequestration.[13] Some of these practices, such as no- or low-till agriculture, may contrast with what landowners have been taught previously, and may be met with skepticism in some contexts.[43] In addition, the strict economic constraint for farmers who depend on their crop yields to earn a living, is yet another barrier to implementation. From the experiences of agricultural operators that have implemented these practices, the scalability is largely dependent on the education and perception the landowners have towards these practices and their willingness to adopt them.[44–46] Taking on these new practices can be inhibited by policies or insurance that do not support or accommodate landowners in engaging in these practices.[3,45,46]

Although soil carbon sequestration practices may require significant education and decision support efforts to facilitate widespread adoption, these practices can increase overall soil health, improving resilience and yeilds.[47,48] Such practices include no- or low-tillage, planting cover crops, and rotating pasturelands. These can improve crop yields, soil resilience, biodiversity, and nutrients, such as nitrogen, phosphorus, and potassium.[3,13,49] In addition to the benefits of thriving soil, there is also evidence that shows these practices can result in a near-term spike of nitrous oxide (N_2O) emissions, another greenhouse gas.[49] N_2O is currently produced when nitrogen fertiliser is used so further research is needed to determine if soil carbon sequestration practices would still require this additive. However, there is evidence that with increased soil organic carbon content, crop yields can improve naturally and the reliance of nitrogen-based fertilisers could decrease.[48] To keep the carbon that is sequestered from being re-released into the atmosphere, the soil carbon practices need to be consistent. Quantifying the amount of carbon sequestered by soil through the implementation of these practices is a challenge that will need to be met through robust measurement, monitoring and verification which may differ depending on location. The data collected by government and private conservation agriculture programs could provide insights into which actions are the most efficient to enhance soil carbon uptake and storage, the required maintenance and monitoring to keep carbon stored, and determining if and/or how land owners will be compensated for adopting these practices.[46] Depending on the climate zone and maintenance, carbon can continue to be sequestered for anywhere from 10–100 years.[13,49] After a certain point, the soil will reach carbon saturation, at which it will not be able to sequester any additional carbon.

2.3.6 Biochar

Biochar is a carbon-rich, charcoal-like substance, often used as a soil amendment, that incorporates and progresses CO_2 storage. Biochar is produced by pyrolysing biomass. This process traps some of the carbon from the biomass into a solid that is resistant to decomposition and applied to soils or used in building materials. Biochar can also be used to co-fire with biomass in power plants or as an oxidising agent in steel production, but these are not pathways that result in negative emissions, as the carbon is released as CO_2 into the atmosphere.[50] There are no further technological advances needed to create biochar on a large scale and can even be a by-product of BECCS when the biomass is pyrolysed. In the pyrolysis process, biomass is heated to temperatures as high as 900 °C in the absence of oxygen to remove volatiles and increase its carbon density. The process results in biochar, bio-oil, and hot gases, which can consist of hydrogen, carbon monoxide, CO_2 and methane.[51,52] These hot gases can be used to heat water to create steam for energy generation. However, the energy that is recovered from this process is lower than that of direct combustion of biomass, contributing to the high energy intensity of biochar.[53] Additionally, for biochar to result in negative emissions, the CO_2 from the pyrolysis needs to be captured and stored. The main feedstock for biochar is biomass, so, similar to BECCS, the biomass used needs to be sustainably sourced to be considered a negative emissions technology. The land area required for the biomass growth as well as the expansion of pyrolysis facilities needed to produce biochar contribute to the limited scalability of this technology. The carbon intensity of the pyrolysis process, thermal energy requirement and therefore biochar, can be reduced by integrating carbon capture and storage. The slow progress of decarbonising industrial heat also contributes to the limited scalability of this approach.

Biochar can be used as a soil amendment and can result in increased water retention, plant growth, and microbial activity while also reducing the N_2O and methane that is released as a gas by soil. However, plants that grow in soil where biochar is used can be more vulnerable to insects, pathogens, and drought.[17] When using biochar as a soil amendment, topsoil disruption should be considered. If tillage is required to apply biochar, the soil carbon can oxidise, reversing any previous soil carbon sequestration.[53] Another consideration when evaluating the potential consequences of upscaling biochar use, due to its dark colour, is that it can also result in a negative albedo[†] effect and increased radiative forces due to biochar particles lost to the environment. Biochar can remain in soil for a few decades to centuries depending on the soil practices that are used on the land.[49,54]

2.3.7 Afforestation and Reforestation

Afforestation is the creation of an entirely new forest while reforestation entails planting typically native trees in a previously forested or depleted forest area. Although planting trees has been done for many years, scaling either of these options to have a significant climate change mitigation impact, faces difficulties when it comes to competition for land use and impacts toward the Earth's albedo. The scalability for reforestation is more favorable than afforestation because reforested areas have a predetermined plot of land and are composed of native plant species that already have a known impact on the climate in which they reside. This is not always the case for afforestation. For example, planting certain trees in tropical regions can result in local cooling and can act as natural water filters,[13] where others can become invasive species or damage the environment they were meant to protect. Depending on the reason for deforestation and how the land has been used since the trees have been removed, can be a determining factor as to whether reforestation will contribute to competition with food production or a gain or loss in the Earth's albedo. Additionally, by planting or replanting trees responsibly, biodiversity can also be increased.[3] The storage of the CO_2 that is absorbed by the trees typically remains stored between 20 and 100 years, assuming that the forest is not disturbed by human activity, drought, or fire. This storage can also be extended depending on the end use of the dead biomass.[13]

[†]Albedo refers to the reflectivity of the Earth. A higher albedo indicates that more light is being reflected off the Earth back into space, while a lowered albedo indicates that more light is being absorbed by the Earth. Increasing dark objects, structures, and areas on the Earth's surface lowers the overall albedo.

2.3.8 Ocean Fertilisation

Ocean fertilisation refers to the addition of nutrients to promote the growth of photosynthesisers in the ocean for greater CO_2 uptake. The fertilisers typically used are iron, nitrogen and phosphorus. The feasibility of fertilisation using iron has been well established from laboratory scale to small scale through many sets of experiments. The main uncertainty regarding the feasibility is on a scale that would impact climate change mitigation, since it has only been considered in models, which show that iron enrichment may not prove to be as influential when it comes to sequestering carbon due to other externalities, such as ocean circulation and gas exchange. Fertilisation using nitrogen and phosphorus have also been tested to determine feasibility. In laboratory and small-scale studies, additions of nitrogen have shown to increase photosynthesis activity, but for phosphorus, the results are inconclusive.[40]

There are few technological barriers in scaling ocean fertilisation because there are no advances needed for deployment. The fertilisers and distribution methods use infrastructure that is already in place.[13] Consequences from large-scale deployment of ocean fertilisation include the growth of toxic diatom species and harmful algal blooms, in addition to "dead-zones," depending on the amount of fertiliser used in specific areas. Researchers have also collected limited data regarding the increased concentrations of methane and NO_x as a result of iron-stimulated blooms decomposing in the subsurface. The CO_2 sequestered through the growth of photosynthesisers in the ocean can directly feed into the carbon cycle when they decompose. It is still unclear how much of the carbon taken up by these photosynthesisers ends up in the deep ocean stock and how much will be re-released when the oceans give off gas to reach equilibrium when the atmospheric CO_2 concentrations are reduced.[40]

2.3.9 Coastal Blue Carbon

Coastal blue carbon is the restoration or maintenance of wetlands for the purpose of enhanced CO_2 uptake. Although coastal wetlands are proven to sequester CO_2 at relatively high rates, there is less known about the restoration and maintenance of them than afforestation and reforestation.[3] Due to the ease of implementation, coastal blue carbon does not rely on technological innovations, energy, or significant water consumption to be scaled up. However, sea level rises as climate change progresses should be considered when evaluating the prospective implementation.

The restoration and maintenance of these wetlands can provide protection to the uplands‡ from flooding and increase water quality, benefiting the habitats that rely on these as primary water sources.[3] Wetland ecosystems

‡Upland refers to land near wetlands at a higher elevation, usually at a farther distance from the body of water.

tend to emit significant quantities of N_2O and methane, so that as these are restored, their emissions are also expected to increase. A portion of the CO_2 sequestered by these wetlands can be re-released if the ecosystem is dug up or destroyed. The portion that remains will be trapped in any soil that is left undisturbed or will enter the ocean, if the ecosystem is to be washed away.[13]

2.4 CO_2 Storage

In order for the removal of CO_2 to have an impact on climate change, the level of storage permanence of each GGR approach must be considered based on the length and reliability of the storage. Some GGR technologies have inherent storage methods within the process of sequestration while others must be coupled to storage to result in net-negative emissions. For instance, BECCS and DAC require coupled storage methods to ensure the CO_2 is not re-released to the atmosphere. Common storage types for these NETs include *in situ* mineral carbonation and geologic storage, both of which result in high levels of permanence. The availability of storage is not a limiting factor in the deployment of BECCS and DAC. This can be shown through an estimate of the mafic and ultramafic rock formations, sedimentary basins, and saline aquifers available for storage.[31] GGR methods that have inherent storage include mineral carbonation, soil carbon sequestration, biochar (so long as it is not burned or oxidised), afforestation/reforestation, ocean fertilisation and coastal blue carbon. Of these methods, the level of permanence of storage ranges (see Figure 2.4). The level of permanence that a storage method has is reliant on how well the type of storage retains the CO_2 over time and the risk of disturbance that would cause carbon storage reversal. For many nature-based GGR options, methods for monitoring and verification are essential in determining the actual CO_2 uptake and subsequent storage permanence. For instance, trees are a viable way to store CO_2, but if a forest fire occurs or if the trees are cut down, carbon is unintentionally released which causes it to lose its permanence of storage. This calls scientists to look into the upper maximum projected limits (theoretical potential of the solution) in comparison to the amount able to be stored based on societal constraints (realistic potential). When looking at societal constraints, the economic or cost-effective potential must also be considered as a factor for implementation. The resources in a given area can help to drive decisions about which GGR options are the most promising for deployment and the additional storage methods, if any, that can be utilised.

2.5 Co-benefits and Societal Considerations

The large-scale deployment of GGR has more considerations than just carbon removal; this also includes evaluating the lifecycle emissions, co-benefits, consequences, methods for ensuring storage permanence, and public acceptance. In order for GGR technologies to result in negative emissions, they must offset any CO_2 or GHG that originates from their lifecycle, such as rock crushing, meeting energy requirements, and transportation. Some pushbacks to the implementation of GGR technologies are related to ethics and ecological issues, while other technologies face barriers relating to education. Some of the hurdles concerning education can be overcome with appropriate funding to provide educational programs. For example, this could include funding for avid no- or low-till practitioners to educate their fellow landowners on how to efficiently adopt these practices and what outcomes they can expect.

Other societal considerations include policies or incentives required to promote the investment in GGR and how the job landscape will change with these emerging technologies. For example, if BECCS plants were to replace coal-fired power plants, how comparable a typical coal plant worker's job training and responsibilities are to those of a BECCS plant worker should be considered. Also, if additional training or education would be needed to adequately prepare coal plant workers to work in BECCS plants, there would need to be a protocol that would determine who would make the investment for this re-training.

2.6 Concluding Remarks

Many GGR technologies can work together in a mutually beneficial way, such as pairing DAC with mineral carbonation or creating biochar from a BECCS process. These synergies are not limited to the technologies alone. Given the different resources available in various locations, such as low-carbon energy infrastructure, geologic storage sites, or waste biomass, cooperation between regional efforts can optimise the development and deployment of GGR. Further work into classifying and quantifying resource availability in different regions can lead to insight where regional efforts can be combined resulting in a strategic and equitable deployment plan for GGR technologies. Investment in any deployment plan will be needed from the public, regional governments, and policymakers to ensure the success of an effort of this magnitude. Cooperation among these entities can assist in determining regional resources, deciding how much of these resources will be devoted to GGR, and who is financially responsible for their deployment or maintenance.

Although some GGR technologies are ready to be deployed today with little or no technological advancements, reaching climate goals will first and foremost require reducing the overall CO_2 emissions produced. These solutions alone will not be successful in mitigating the CO_2 that needs to be removed; it will require a portfolio of techniques.

References

1. E. Foote, Circumstances affecting the Heat of the Sun's Rays, *Am. J. Sci. Arts*, 1856, **XXII**, 382.
2. Union of Concerned Scientists, The IPCC: Who Are They, 2018 [accessed 2020 Aug 12], https://www.ucsusa.org/resources/ipcc-who-are-they.

3. NASEM, Negative Emissions Technologies and Reliable Sequestration, Negative Emissions Technologies and Reliable Sequestration, 2019.
4. IPCC, Climate Change: The IPCC 1990 and 1992 Assessments, 1992 [accessed 2021 Jan 11], https://www.ipcc.ch/site/assets/uploads/2018/05/ipcc_90_92_assessments_far_full_report.pdf.
5. IPCC Working Group II of Second Assessment Report, Climate Change 1995: Impacts, Adaptations and Mitigation of Climate Change: Scientific-Technical Analyses, 1995, pp. 811–812.
6. IPCC, Climate Change 2001: Synthesis Report, 2001, p. 119–120, 131, 143, 296–297, https://www.ipcc.ch/report/ar3/syr/
7. IPCC, Working Group III of the Fifth Assessment Report, *Climate Change 2014: Mitigation of Climate Change*, 2014.
8. IPCC, *Global Warming of 1.5 °C*, 2018, p. 17.
9. IPCC, *Climate Change 2014: Synthesis Report*, 2014.
10. S. Fuss, J. G. Canadell, G. P. Peters, M. Tavoni, R. M. Andrew and P. Ciais, *et al.*, Betting on negative emissions, *Nature Climate Change*, 2014, 4, 850–853.
11. H. Shue, Climate Dreaming: Negative Emissions, Risk Transfer, and Irreversibility, *SSRN Electronic J.*, 2017.
12. Climate Change: Atmospheric Carbon Dioxide | NOAA Climate.gov, [accessed 2020 Aug 24], https://www.climate.gov/news-features/understanding-climate/climate-change-atmospheric-carbon-dioxide
13. A. Azapagic, D. Beerling, C. Cheeseman, G. Henderson, C. Hepburn, J. House, C. Le Quéré, N. Markusson, N. Shah, J. Shepherd, and P. Smith *Greenhouse Gas Removal, Royal Society and Royal Academy of Engineering*, 2018. https://www.raeng.org.uk/publications/reports/greenhouse-gas-removal
14. reThink Wood, Evaluating the Carbon Footprint of Wood Buildings, *GreenSource*, 2015, **8**.
15. R. Socolow, M. Desmond, R. Aines, J. Blackstock, O. Bolland and T. Kaarsberg, *et al.*, Direct Air Capture of CO_2 with Chemicals, A Technology Assessment for the APS Panel on Public Affairs, *The American Physical Society*, 2011. https://www.aps.org/policy/reports/assessments/upload/dac2011.pdf
16. C. Consoli, *Bioenergy and Carbon Capture and Storage*, 2019.
17. S. Fuss, W. F. Lamb, M. W. Callaghan, J. Hilaire, F. Creutzig and T. Amann, *et al.*, Negative emissions-Part 2: Costs, potentials and side effects, *Environ. Res. Lett.*, 2018, **13**, 063002.
18. A. C. C. Chang, H.-F. Chang, F.-J. Lin, K.-H. Lin and C.-H. Chen, Biomass gasification for hydrogen production, *Int. J. Hydrogen Energy*, 2011, **36**, 14252.
19. Ó. J. Sánchez and S. Montoya, Production of bioethanol from biomass: An overview, in *Biofuel Technologies: Recent Developments*, Springer-Verlag Berlin Heidelberg, 2013, [cited 2021 Jan 13], pp. 397–441.
20. Lawrence Livermore National Laboratory, Getting to Neutral, Options for negative carbon emissions in California, Livermore, CA, 2020, https://gs.llnl.gov/sites/gs/files/2021-08/getting_to_neutral.pdf

21. E. Larson, C. Greig, J. Jenkins, E. Mayfield, A. Pascale and C. Zhang, *et al.*, *Net-Zero America: Potential Pathways, Infrastructure, and Impacts*, Princeton, NJ, 2020.
22. D. Wang, P. Jiang, H. Zhang and W. Yuan, Biochar production and applications in agro and forestry systems: A review, *Science Total Environ.*, 2020, **723**, 1.
23. D. W. Keith, G. Holmes, D. St Angelo and K. Heidel, A Process for Capturing CO_2 from the Atmosphere, *Joule*, 2018, **2**, 1573.
24. C. Beuttler, L. Charles and J. Wurzbacher, The Role of Direct Air Capture in Mitigation of Anthropogenic Greenhouse Gas Emissions, *Front. Clim.*, 2019, **1**, 1.
25. N. McQueen, M. J. Desmond, R. H. Socolow, P. Psarras and J. Wilcox, Natural Gas versus Electricity for Solvent-Based Direct Air Capture, *Front. Clim.*, 2020.
26. H. Pilorgé, B. Kolosz, G. C. Wu and J. Freeman, in *Global Mapping of CDR Opportunities*, 2021, ed. Wilcox J., Kolosz B. and Freeman J., CDR Primer.
27. K. Lebling, N. McQueen, M. Pisciotta and J. Wilcox, Direct Air Capture: Definition, Cost, & Considerations World Resources Institute, World Resources Institute, 2021 [accessed 2021 Jan 27], https://www.wri.org/blog/2021/01/direct-air-capture-definition-cost-considerations
28. S. Deutz, and A. Bardow, How (carbon) negative is direct air capture? Life cycle assessment of an industrial temperature-vacuum swing adsorption process, ChemRxiv, 2020, https://chemrxiv.org/engage/chemrxiv/article-details/60c75474f96a005d1c288673
29. Climeworks, Climeworks' Response to the IPCC report, 2018, [accessed 2020 Aug 12], https://www.climeworks.com/news/climeworks-response-to-the-ipcc-report
30. CarbFix, Where does it work?, 2021 [accessed 2021 Jan 22], https://www.carbfix.com/atlas/
31. P. Kelemen, S. M. Benson, H. Pilorgé, P. Psarras and J. Wilcox, An Overview of the Status and Challenges of CO_2 Storage in Minerals and Geological Formations, *Front. Climate*, 2019, **1**, 1.
32. S.Ó. Snæbjörnsdóttir, B. Sigfússon, C. Marieni, D. Goldberg, S. R. Gislason and E. H. Oelkers, Carbon dioxide storage through mineral carbonation, *Nat. Rev. Earth Environ.*, 2020, **1**, 90.
33. J. L. Hamilton, S. A. Wilson, B. Morgan and A. Harrison, Accelerating Mineral Carbonation in Ultramafic Mine Tailings via Direct CO_2 Reaction and Heap Leaching with Potential for Base Metal Enrichment and Recovery, *Economic Geology*, 2020, 115(2), 303–323.
34. S. Hariharan and M. Mazzotti, Kinetics of Flue Gas CO_2 Mineralization Processes using Partially Dehydroxylated Lizardite, *Chemical Engineering Journal*, 2017, **324**, 397–413.
35. P. Renforth, P. A. E. Pogge von Strandmann and G. M. Henderson, The dissolution of olivine added to soil: Implications for enhanced weathering, *Applied Geochemistry*, 2015, **61**, 109.

36. T. Ormsby, *De Beers Group Carbon Capture Project Receives Research Funding from Natural Resources Canada's Clean Growth Program*, 2019, [accessed 2020 Aug 12]. https://www.debeersgroup.com/media/company-news/2019/carbon-capture-project-receives-research-funding
37. N. McQueen, P. Kelemen, G. Dipple, P. Renforth and J. Wilcox, Ambient weathering of magnesium oxide for CO_2 removal from air, *Nat. Commun.*, 2020, **11**, 1.
38. L. T. Bach, S. J. Gill, R. E. M. Rickaby, S. Gore and P. Renforth, CO_2 Removal With Enhanced Weathering and Ocean Alkalinity Enhancement: Potential Risks and Co-benefits for Marine Pelagic Ecosystems, *Front. Clim.*, 2019, **1**, 7.
39. D. J. Beerling, J. R. Leake, S. P. Long, J. D. Scholes, J. Ton and P. N. Nelson, *et al.*, Farming with crops and rocks to address global climate, food and soil security, *Nat. Plants*, 2018, **4**, 138–147.
40. GESAMP Working Group 41, High Level Review of a Wide Range of Proposed Marine Geoengineering Techniques, 2019, www.imo.org.
41. P. Renforth and G. Henderson, Assessing ocean alkalinity for carbon sequestration, *Rev. Geophys.*, 2017, **55**, 636.
42. J. Sanderman, T. Hengl and G. J. Fiske, Soil carbon debt of 12,000 years of human land use, *Proc. Natl. Acad. Sci. U. S. Am.*, 2017, **114**, 9575.
43. P. Byck, *Soil Carbon Cowboys: During the Drought*, 2020, [accessed 2020 Aug 12], https://www.youtube.com/watch?v=w0cGRKwP9W8&feature=emb_logo
44. N. P. Dumbrell, M. E. Kragt and F. L. Gibson, What carbon farming activities are farmers likely to adopt? A best-worst scaling survey, *Land Use Policy*, 2016, **54**, 29–37.
45. K. Tang, M. E. Kragt, A. Hailu and C. Ma, Carbon farming economics: What have we learned?, *J. Environ. Manage.*, 2016, **172**, 49.
46. A. J. Sykes, M. Macleod, V. Eory, R. M. Rees, F. Payen and V. Myrgiotis, *et al.*, Characterising the biophysical, economic and social impacts of soil carbon sequestration as a greenhouse gas removal technology, *Global Change Biology*, Blackwell Publishing Ltd, 2020, vol. 26, pp. 1085–1108.
47. R. Lal, Societal value of soil carbon, *J. Soil Water Conserv.*, 2014, **69**, 186A.
48. E. E. Oldfield, M. A. Bradford and S. A. Wood, Global meta-analysis of the relationship between soil organic matter and crop yields, *Soil*, 2019, **5**, 15.
49. P. Smith, Soil carbon sequestration and biochar as negative emission technologies, *Global Change Biol.*, 2016, **22**, 1315.
50. H.-P. Schmidt, C. Kammann and N. Hagemann, Certification of the carbon sink potential of biochar (Version 1), [accessed 2021 Jan 13], www.european-biochar.org.
51. S. Zafar, *Biomass Pyrolysis*, Alt Energy Magazine, New York, NY, 2009, Springer, New York, [accessed 2021 Jan 13], https://www.altenergymag.com/article/2009/02/biomass-pyrolysis/502/

52. A. V. Bridgwater, Pyrolysis of Solid Biomass: Basics, Processes and Products, in *Energy from Organic Materials (Biomass)*, Springer, New York, 2019. pp. 1221–1250.
53. NASEM, Climate intervention: Carbon dioxide removal and reliable sequestration, Climate Intervention: Carbon Dioxide Removal and Reliable Sequestration, 2015.
54. A. Bergman and A. Rinberg, in *The Case for Carbon Dioxide Removal: From Science to Justice*, ed. Wilcox J., Kolosz B. and Freeman J., CDR Primer, 2021.

CHAPTER 3

Negative Emissions: The Role and Response of the Climate System

CHRIS D. JONES

Met Office Hadley Centre, FitzRoy Road, Exeter EX1 3PB, UK
Email: chris.d.jones@metoffice.gov.uk

Our climate is changing and the role of human activity in this is unequivocal. There is now a well-defined relationship between the amount of carbon we emit as CO_2 and the global temperature rise. This allows us to quantify a global carbon budget that is consistent with the goals of the Paris Agreement to limit warming to well below 2 °C above pre-industrial levels. Almost all scenarios of future socioeconomic activity and carbon emissions that meet the Paris targets rely on society developing and deploying techniques to directly remove some of the emitted carbon. These negative emissions technologies (NETs) and greenhouse gas removal (GGR) are receiving much scientific attention in terms of their feasibility, costs, limitations, effectiveness, and unintended consequences. But there are considerable uncertainties in our knowledge surrounding how much CO_2 removal, *i.e.*, negative emissions, would be required to achieve targets, and also how much is feasible. Feedback between climate and the carbon cycle represent the main processes which determine the remaining carbon budgets, and our imperfect knowledge of them represents a key gap in being able to inform mitigation policy more quantitatively. This chapter aims to highlight the role of the physical Earth System in this scientific debate – showing that it is a major control of the amount of negative emissions

Energy and Environment Series No. 31
Greenhouse Gas Removal Technologies
Edited by Mai Bui and Niall Mac Dowell

Published by the Royal Society of Chemistry, www.rsc.org

which are required to achieve climate goals (*i.e.*, the "demand" for negative emissions), and also that many NETs and GGR are affected by the climate which therefore regulates the potential supply of negative emissions.

3.1 Introduction

3.1.1 The Need for Negative Emissions to Achieve Climate Goals

In 2013, the IPCC Fifth Assessment Report[1] presented a new development in the understanding of the climate's response to emissions of carbon dioxide (CO_2). Namely the direct, and almost linear, relationship between global warming and the total accumulated amount of CO_2 emitted. Twenty years previously, it had been noted that the saturation of the radiative effect of CO_2 in the atmosphere could balance declining carbon sinks leading to an approximate insensitivity of global warming to the timing, or pathway, of CO_2 emissions.[2] And in 2009, a number of different studies quantified more closely how the degree of global warming was directly linked to the total carbon emissions rather than to the pathway of the emissions.[3–5] Prior to this discovery, many "stabilisation pathways" had been proposed, ranging from idealised conceptual pathways to more complex assumptions about rates of decrease of energy use.[6,7] As a result, constraints from physical sciences (such as climate and carbon cycle feedbacks) became intricately mixed with constraints from technological and socio-economic communities (*e.g.*, infrastructure lock-in, public acceptance, behavioural change and maximal rates of decarbonisation).

This new relationship between emissions and warming was termed the Transient Response to cumulative Carbon Emissions (TCRE) and was used by the IPCC to define a total carbon emission compatible with a climate target. This framing of the mitigation challenge simplified how it was treated: the Earth system determines the total carbon emissions compatible with attaining a climate goal, but then social sciences and available technology could help determine the optimal method of achieving this. In other words, we can now quantify a "carbon budget" which can be "spent" (*i.e.*, emitted) in order to avoid dangerous interference with the climate system. The timing of how we spend (emit) this does not strongly affect the eventual warming, and so it is a societal choice whether to spend the budget quickly in the near term or save more for future generations.

Using this TCRE relationship, IPCC AR5 assessed a total carbon budget of 790 PgC[†] or 2900 Gt CO_2 in order to likely (66% chance) remain below 2 °C

[†]A note on units. Peta-grammes (Pg = 10^{15}g) is the SI unit of mass, but is the same as giga-tonnes (Gt = 10^9 metric tonnes). Both are frequently used in expressing carbon emissions. It is also required to state whether the amount is expressing the mass of carbon (C) or of CO_2, which differ by the ratio of their molecular weights: 44/12. So, 1 PgC = 1 Gt C = 3.67 Gt CO_2. Throughout this paper we use the units of Gt CO_2 which are most frequently used in policy discussions of carbon emissions and budgets.

above the pre-industrial level, of which about 2310 Gt CO_2 has been emitted over the 1870–2018 period.[8] Staying within this finite quota of cumulative CO_2 emissions in a cost-effective way will require strong mitigation action across all sectors. The longer mitigation is delayed, the greater the effort that will be needed. Some proposed technologies that could help us stay within our carbon budget include the deliberate removal of CO_2 from the atmosphere by human intervention, *i.e.*, so-called "negative emissions", or carbon dioxide removal (CDR). For example, the use of bioenergy with carbon capture and storage (BECCS) or direct capture of CO_2 from the atmosphere itself (DAC).

BECCS is the negative emissions technology most widely included in Integrated Assessment models to generate future scenarios, and the vast majority of such scenarios that limit warming to 1.5 °C or 2 °C rely on negative CO_2 emissions.[9] Even scenarios which do not require globally net-negative CO_2 emissions can have substantial use of negative emissions by using CDR, offsetting some portion of anthropogenic activity. Hence, it is vital for us to be able to determine how many negative emissions may be required and whether or not technology can deliver them.

3.1.2 The Role of the Earth System in NETs Research

Negative emissions technologies (NETs) affect the climate. This is by design as they are intentional actions which aim to slow, or even reverse, the rate of anthropogenic climate change by removing CO_2 or other greenhouse gases from the atmosphere. But they are also *affected by* climate – this is partly because the way they work is affected by changes in climate and partly because their effectiveness in reducing CO_2 concentrations is affected by climate–carbon cycle interactions. For example, the productivity of biofuels on land can increase or decrease due to changes in climate or CO_2 levels. Similarly, the response of natural carbon sinks determines how much of the negative emission acts to reduce the CO_2 in the atmosphere. The world's requirement for NETs also depends on climate, and the rate it is changing – the impacts at certain levels of warming determine society's demand to limit the harmful effects of climate change, and carbon cycle feedbacks affect the cumulative emissions compatible with such climate targets.

However, a large uncertainty in carbon cycle feedbacks, which can be seen in the substantial spread in carbon cycle modelling results,[10] means that the uncertainty in the remaining carbon budget to achieve 2 °C is very large – in fact possibly larger than the remaining budget itself.[11] This large uncertainty reduces the usefulness of this simplifying concept to policy makers.

Previous analysis showed that the future fractions of emissions absorbed by land and ocean, and remaining in the atmosphere would be quite different under high or low emissions scenarios.[12,126] We know that the near-constancy of the airborne fraction in recent decades is a direct result of the near-exponential increase in emissions,[13] and so deviation from an exponential increase will cause deviation from this constant airborne fraction. As the policy focus has now moved towards low climate targets, with the intent that future emissions stop their continued exponential increase and begin to decline, it is therefore inevitable that the future airborne fraction will depart from its historical value of approximately 50%. But by how much it changes, and how it responds to negative emissions is not well understood yet and is vitally important to determine the degree to which NETs are required.

It is the complex response of land and ocean carbon sinks to CO_2 that drives their future behaviour under low CO_2 pathways where climate change is limited, and CO_2 levels stabilise, or peak and decline. If natural carbon sinks are caused by an increasing level of CO_2, then reducing the amount of CO_2 in the atmosphere will reduce these carbon sinks again. Specifically, vegetation productivity will decrease when CO_2 decreases, and ocean CO_2 uptake will also decrease with decreasing CO_2. If the sinks weaken enough, then any carbon stored may begin to be emitted again (*i.e.*, "outgassing"). In other words, reduction of atmospheric CO_2 leads to a redistribution across natural carbon stocks and therefore weakens the effect of the negative emissions on atmospheric CO_2 levels.[14,15]

In this way, the response of the Earth system directly affects both the potential supply of, and the demand for, negative emissions. Other sectors also have direct effects on the supply and demand of NETs. From technology to societal acceptance through to international governance all must be taken into account.[16,17] The cross-disciplinary nature of the issue is clear and has led to calls for sustained cross-cutting research.[18]

In this chapter, I assess the latest knowledge in aspects associated with the physical and biogeochemical Earth System. In Section 3.2, I develop a narrative presenting the cross-disciplinary issue of NETs as a matter of matching supply and demand. In Section 3.3, I examine implications of the Earth system for the demand for NETs – how do we know how much CO_2 removal is required to achieve climate goals? Finally, in Section 3.4, I examine implications of the Earth system for their supply. I conclude with some recommendations for ensuring future research enables these complimentary approaches to be joined up.

3.2 Supply and Demand for Negative Emissions

There is a fundamental difference between the questions "how much NETs can we deliver?" (*i.e.*, the "supply") and "how much NETs do we need?" (*i.e.*, the "demand"), which is why studies typically address one or the other. Jones and Albanito[19] published one of the few studies which explicitly addresses both aspects and assesses both the efficiency of biomass energy in removing CO_2 and also secondary impacts of it in terms of shifting land-use emissions to other regions.

Extensive literature on NETs has grown rapidly in recent years,[20–22] but studies rarely explicitly acknowledge which side of the equation is being assessed. For example, the extensive review of Smith *et al.*[23] simply chose a level of CO_2 removal and looked at costs and constraints on different NETs to achieve this. Conversely, Griscom *et al.*[24] looking at nature-based solutions, quantified the potential of different activities to remove CO_2. These are both "supply" side assessments. On the "demand" side, van Vuuren *et al.*[25] quantified the amount of CO_2 removal to achieve the RCP2.6[‡] scenario and both Kriegler *et al.*[26] and van Vuuren *et al.*[27] show the possibility of still achieving 2-degrees without the need for NETs.

Both aspects clearly need to be considered – it is not enough to know how much NETs we need without knowing if this is possible, nor is it enough to quantify what is possible without consideration of what is required. Disciplines spanning technology, physical sciences and social sciences contribute to both aspects. Different NETs also bring different potential co-benefits and trade-offs. Here I develop a narrative of NETs expressed in terms of supply *versus* demand, and show analysis of how physical Earth system science contributes to both. This framing enables research to be targeted to address the question of whether or not the requirement for NETs to achieve climate targets can actually be met. To date, global scale requirement is hard to compare with bottom-up regional scale studies on feasibility.

For example, assessment of the requirement for NETs requires cross-disciplinary input that could be crudely described along the lines:

(i) Climate science and impacts assessments quantify how changes in global climate will affect people, societies and ecosystems;[28–31]
(ii) Society and policymakers make a subjective value choice of what can be considered "dangerous" and sets a target of, say, 2 °C (or 1.5 °C) above the pre-industrial level;[32]
(iii) Carbon cycle science can then help determine a carbon budget to achieve this, but does not distinguish net or gross positive/negative emissions nor the temporal pathway; it simply quantifies the available budget coherent with a given target;[5,11,33,34]
(iv) Technology and infrastructure lock-in impose constraints on the maximum rate of decarbonisation and therefore determines the magnitude of NET required to close the gap.[35–37]

Equally on the supply side, multiple disciplines must provide input, although there is less of an obvious chain of thinking. Here there are many proposed actions which can contribute to a global portfolio of negative emissions. These can be thought of as wedges which individually are not enough but combined can add up to give the required degree of action – much in the same way that Pacala and Socolow[38] suggested a wedge-based approach for thinking about technologies to stabilise CO_2

‡One of the Representative Concentration Pathways – see Glossary for definition.

concentrations below double pre-industrial levels. Issues include availability of land on which to deploy the NET (*e.g.*, for biofuel);[39] subsequent impacts on food prices;[40] impacts on water resource or stream flow[41] or potentially counterproductive changes in biophysical properties of the land surface.[42,43]

This is summarised in Figure 3.1 which shows how aspects of different research disciplines such as physical earth system sciences, technology and social sciences contribute to understanding the demand and supply of NETs.

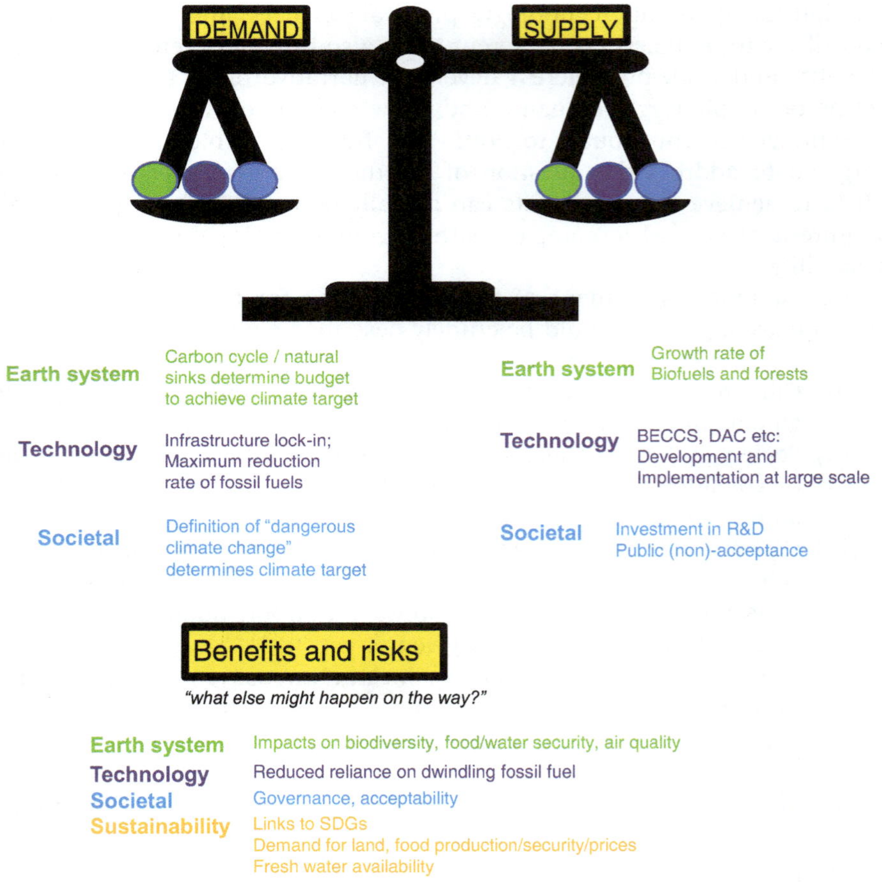

Figure 3.1 Schematic figure outlining how the Earth system, and other research sectors influence both the "demand" and "supply" side of the equation for NETs. Literature is often not clear which side it is addressing, but both sides need to be quantitative to enable comparison and balance. The figure is not exhaustive but illustrates how the research community can combine to more clearly address the issues that determine our ability for supply to meet demand.

Explicitly recognising whether studies address one side or the other of the supply/demand equation enables us to address the question whether supply can meet the demand. We must also check there are no unintended consequences along the way: What are the risks, co-benefits or trade-offs of achieving our target through NETs activities? For this, it is important to link back to issues of impacts and sustainability – such as through the Sustainable Development Goals (SDGs) and the need that any pathway is sustainable in that context. Although helpful in framing the problem, we must eventually consider much more than the global-mean-temperature. In fact, the globally averaged surface air temperature is a very crude metric of climate change and hides many of the aspects of our weather and environment which directly affect society. If actions taken to control it bring other unwanted effects – an obvious example being compromising food security to grow bioenergy crops – then we must take this into account.

3.3 Earth System Implications for the Demand for Negative Emissions

The role of the Earth system in the demand for NETs is perhaps easier to address at a global level than its implications for supply (Section 3.4). CO_2 is well mixed in the atmosphere and so the location of carbon removal or storage is not important in determining its long-term effect on climate. Similarly, impacts of climate on ecosystems and the carbon cycle are well modelled at global scales – much better than detailed land-management practices for example.[44,45] Earth system modelling also enables the inclusion of other processes, often far away geographically from the NET activity, but which can be central to the effectiveness of any CO_2 removal – for example thawing of permafrost,[46] productivity response of forests[47] or changes in the ocean circulation.[48]

As outlined in Section 3.1.1, it has been well established that a global temperature increase is near-linear with cumulative CO_2 emissions.[4,5,33] This relationship – the transient climate response to cumulative carbon (CO_2) emissions – TCRE – allows climate goals to be associated with global total carbon budgets, and is due to the diminishing radiative forcing[§] from CO_2 in the atmosphere and the diminishing efficiency of ocean heat uptake.[49–51] In other words, the Earth system sensitivity tells us a carbon budget compatible with achieving a climate target. This concept of near linearity of climate change to cumulative carbon emissions holds for measures other than just global average temperature, such as regional climate[52] or extremes.[53,54]

Deployment of a small number of NETs may just reduce the anthropogenic total and slow down the increase due to our emissions. However, if we need to strongly reduce the atmospheric CO_2 through the use of globally net

---

[§]See the Glossary.

negative emissions (*i.e.*, CO_2 removal by NETs exceed any residual fossil fuel emission), then we need to understand to what extent changes to the Earth system are reversible.[55] In such cases, the global temperature may reach a peak value before declining – so-called "overshoot" behaviour.[56] The IPCC Special Report on 1.5 degrees[57] concluded with high confidence that such overshoot trajectories "result in higher impacts and associated challenges compared to pathways that limit global warming to 1.5 °C with no or limited overshoot". While not all components of the system are fully reversible, studies suggest that any asymmetry of global temperature to negative emissions is likely small.[127,128] The extent and duration of overshoot affects the risks and impacts likely to be experienced[29] and the negative emissions required to achieve it.[58] Sea-level in particular responds to the integrated amount of heat taken up by the oceans and may be irreversible on the timescales of the next few centuries.[59–61] Earth system model simulations under the protocol defined by CDRMIP (Carbon dioxide removal model intercomparison project[62]) are designed to quantify the removal potential of different NETs, their long-term effectiveness and the reversibility – or otherwise – of the Earth system in response.

Understanding this relationship is central to knowing the requirement for an amount of CO_2 removal. Specifically of relevance to the issue of NETs is the symmetry of the carbon cycle response to the removal of CO_2. Is this simply a reverse of the response to the initial emissions? To some extent, the existence of TCRE as a pseudo-invariant, pathway-independent metric of Earth system sensitivity implies that we can expect the carbon cycle response to CO_2 removal to be symmetrical. Although ocean heat and carbon uptake can exhibit path dependence for higher levels of overshoot, leading to deviation from the TCRE relationship,[63] it is expected that to a large extent, removal of CO_2 will help to reduce global temperatures.

As we have injected more and more CO_2 into the atmosphere, this has stimulated carbon sinks and changed the climate. If we remove the CO_2, we would expect the impact on climate to reduce and the carbon sinks to reduce also. This leads to the notion of a "rebound effect"[64] – negative emissions are subject to the same airborne fraction as positive emissions, and so if we reduce the CO_2 in the atmosphere, the natural sinks will weaken or even reverse.[14] Approximately speaking, if we add 2 tonnes of CO_2 to the atmosphere only 1 tonne will remain there due to natural carbon sinks. Likewise, we must remove 2 tonnes of CO_2 to reduce the atmospheric burden by 1 tonne due to the weakening of the natural sinks. Earth system models can tell us quantitatively how the complex dynamics of the land and ocean carbon cycles respond to removal of CO_2 from the atmosphere.

3.3.1 Earth System Model Response to Negative Emissions in Complex Models

We know that nature responds to both increases in CO_2 and potential decreases too: the IPCC Special Report on 1.5 degrees[11] concluded that overshoot trajectories "result in higher impacts and associated challenges compared to pathways that limit global warming to 1.5 °C with no or limited overshoot". The degree and duration of overshoot affects the emissions pathway required to achieve it[58] and may lead to irreversible changes in sea-level,[59–61] ocean circulation,[60] and marine ecosystems.[61,65,66]

This response directly determines the magnitude of any CO_2 removal required to achieve climate targets. The natural carbon cycle acts as a buffer to CO_2 emissions from human activity: approximately half of our CO_2 emissions are absorbed by land and ocean carbon sinks, thus slowing the rate of CO_2 increase in the atmosphere.[67] The same processes that act to slow down the CO_2 growth rate will also respond to negative emissions, but in reverse: as CO_2 is removed from the atmosphere, the CO_2 fertilisation effect on vegetation will weaken and the oceans will reduce their uptake to equilibrate with the lower amount of CO_2 in the atmosphere. Some of these responses are very rapid – such as the leaf-level photosynthesis response to ambient CO_2 concentration, while others occur on longer timescales such as changes in vegetation structure and deep ocean carbon storage. Others, such as geological weathering and sediments on the ocean floor lead to changes in CO_2 over thousands to millions of years.[68]

In order to understand this complex behaviour, we must understand, quantitatively, how different sources and sinks of carbon can vary in sign (*i.e.*, positive or negative) and combine to determine whether the amount in the atmosphere will increase or decrease (Figure 3.2). Anthropogenic activity can be split into sources (*i.e.* emissions) – such as burning fossil fuel or deforestation. These dominate the global total to date, in the region of 37 and 6 Gt CO_2 per year, respectively.[8] Alongside these are negative emissions from human activity – such as the techniques discussed above, *i.e.* from BECCS or direct air capture. While small now, these are projected to grow in the future. The sign of the combination of these therefore may change in the future from positive (net emission) to negative (net removal). Natural ecosystems – both for land and ocean – are currently carbon sinks[8] (*i.e.*, negative in Figure 3.2) but may become positive if CO_2 is removed in large quantities in future.[69] Ultimately what we care about is the amount of CO_2 in the atmosphere and whether this is increasing or decreasing – in order to determine this, we need to be able to quantitatively combine the net anthropogenic and the net natural terms. Only complex, process-based Earth system models are equipped to help us understand this fine balance, and how this dynamic system will evolve in the future.

Jones *et al.*[69] analysed the centennial scale response of the carbon cycle in several CMIP5[¶] models in response to a "peak and decline" CO_2 pathway (RCP2.6 inherently includes large-scale BECCS deployment from 2020 onwards, van Vuuren *et al.*[25]). Here, I repeat that analysis using the most up to date Earth system models which took part in CMIP6.[70] In this case, there are

---

¶Coupled Model Intercomparison Project 5 – see the Glossary.

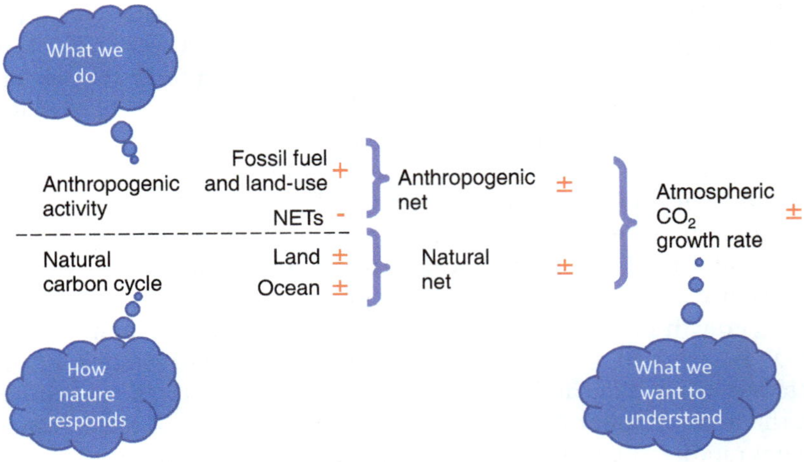

Figure 3.2 Schematic representation of how anthropogenic and natural carbon sinks and sources combine in a complex interplay to determine whether the amount of CO_2 in the atmosphere will increase or decrease. The individual terms of "what we do" (fossil fuel use, deforestation, negative emissions) can be positive or negative, as can the individual terms of natural carbon cycle response (both land and ocean). These may combine to give either positive or negative net behaviour of human and natural systems. It is the end combination of all of these which is what we want to understand – in what way will the amount of CO_2 in the atmosphere change?

still only 4 models which performed the simulations up to the year 2300 that enable us to look into the long-term dynamics of the carbon cycle system and how it responds to negative emissions. I use here results from the SSP5-3.4-OS‖ overshoot scenario (O'Neill *et al.*[71]) which explores the implications of a peak and decline in forcing during the 21st century. In it, CO_2 peaks at 571 ppm in the 2060s and reverts below 500 ppm by 2100. To achieve this, it has strong negative emissions of CO_2 exceeding those in SSP1-2.6 and SSP1-1.9 from 2070 onwards and reaching −20 Gt CO_2 per year by 2100. This aggressive mitigation and negative emissions bring the CO_2 concentration back down to similar levels to the SSP1-2.6 scenario during the 23rd century.[71,72]

Figure 3.3 shows how the natural carbon sinks respond to the peak and decline in atmospheric CO_2. The profile is very similar to that seen before, but by using a scenario with a larger overshoot, the magnitude of changes is larger, and there are different balances of sink and source behaviour further in the future as CO_2 concentrations stabilise. Land and ocean sinks behave in a similar manner when the changes in CO_2 are strong, with both halves of the natural carbon cycle showing strong increases in sink behaviour during the 21st century. Both peak at approximately 14 Gt CO_2 per year in the

‖One of the Shared Socioeconomic Pathways – see the Glossary.

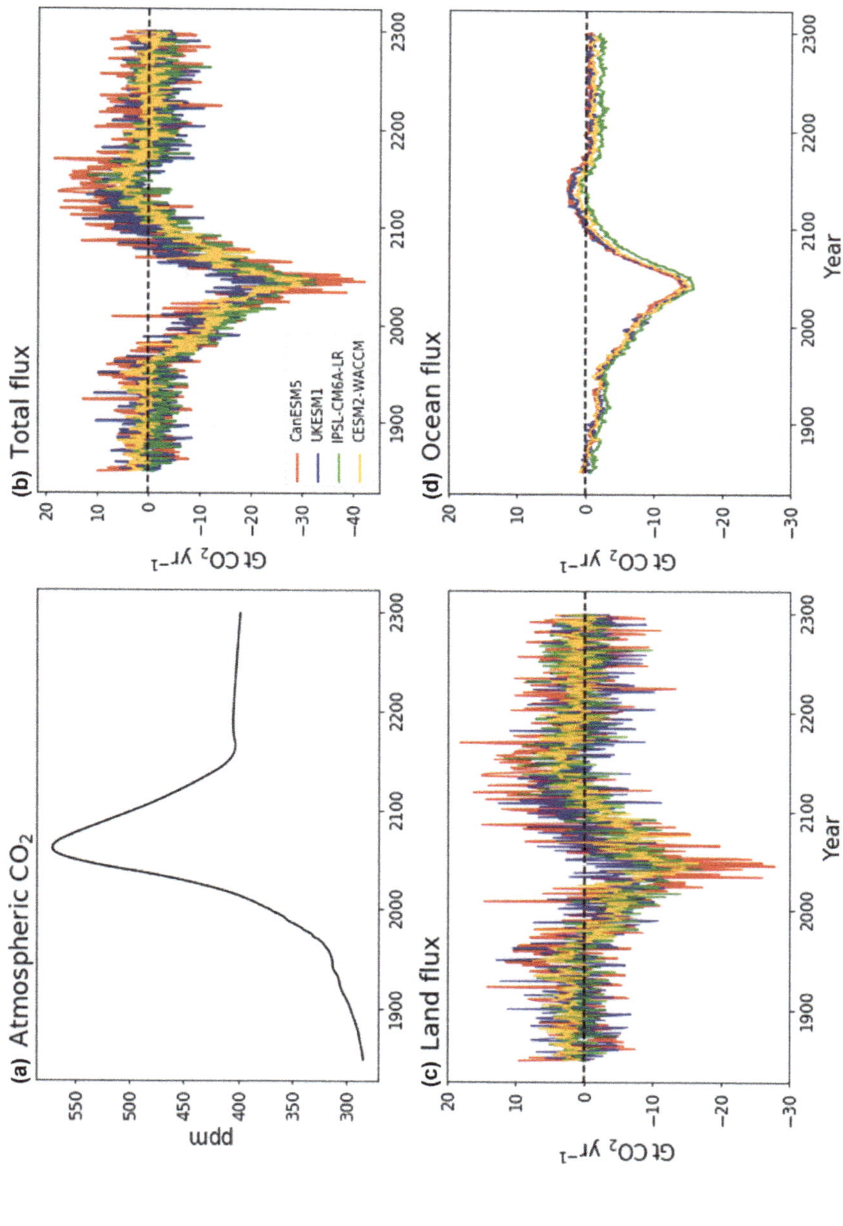

Figure 3.3 Response of natural carbon sinks to the peak and decline atmospheric CO_2 concentration in SSP5-3.4-OS. (a) Atmospheric CO_2 concentration profile; (b–d) CMIP6 ESM simulated results of carbon flux (PgC per year): (b) Total (land + ocean) flux; (c) land component; (d) ocean component. Sign convention is taken such that negative flux implies a removal of carbon from the atmosphere. CMIP6 data is available from the Earth System Grid Federation (https://esgf-node.llnl.gov/projects/cmip6/).

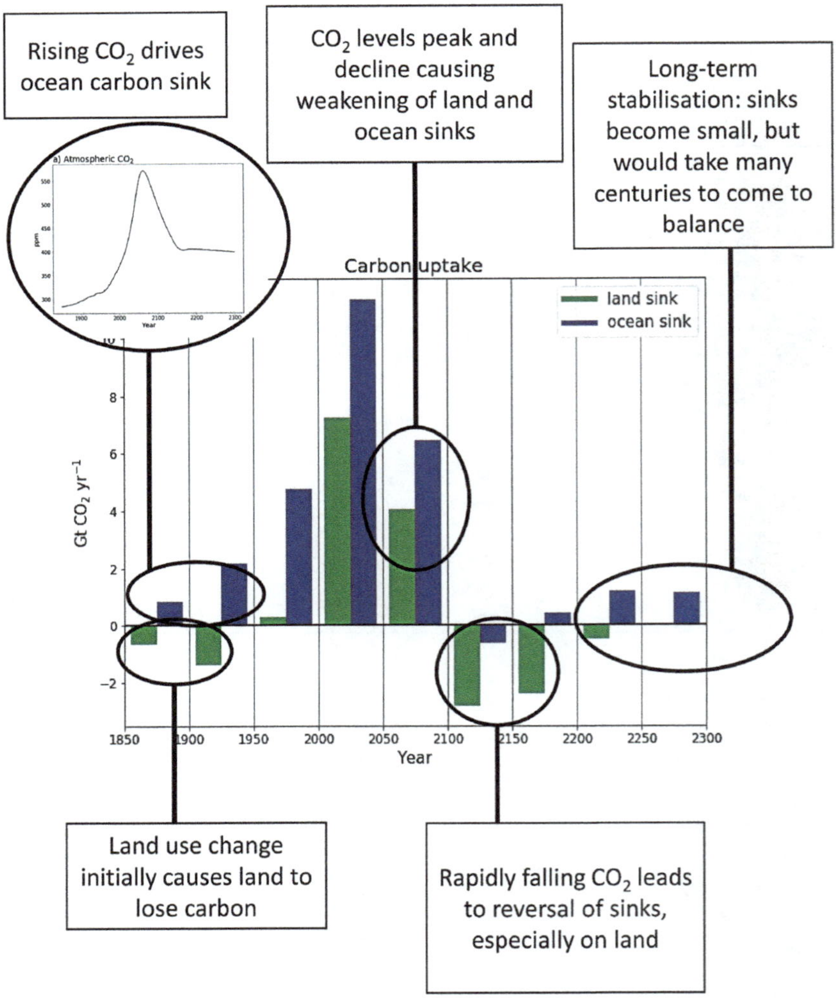

Figure 3.4 50-year average behaviour of natural carbon sinks (land: green; ocean: blue). The inset shows the CO_2 concentrations as depicted in Figure 3.3, panel (a). Initially, the rise in CO_2 since the pre-industrial era drives an ocean carbon sink, but on land the effect of land-use change (deforestation) outweighs this and the land loses carbon. Strong sink behaviour is then seen during the faster rise in atmospheric CO_2 during the late 20th and early 21st century, followed by reduced sink behaviour as CO_2 concentration peaks and declines. Sinks saturate and become sources after 2100 before the ocean finally returns to being a sink again after 2150.

middle of the century when the CO_2 growth rate is at its fastest. As expected, land sinks have much greater year-to-year variability, but on average their magnitude is similar to the ocean sinks. The total sink therefore peaks at about 29 Gt CO_2 per year before declining to zero in the first half of the 22nd century. In contrast to previous analysis using CMIP5 models, the 4 Earth system models (ESMs**) used here (CanESM5,[73] CESM2-WACCM,[74] IPSL-CM6-LR,[75] UKESM1-0-LL[76]) show strong agreement with CanESM5 having the biggest magnitude of response of land sinks and IPSL-CM6-LR the strongest response of ocean sinks.

The interesting behaviour of the system is when the CO_2 concentration in the atmosphere decreases, causing a decrease in the disequilibrium between atmosphere and land/ocean stores, which therefore decreases the sink strength. A feature of the SSP5-3.4-OS scenario is a small minimum in CO_2 around 2150 with a small increase again before stabilising by 2300. This has the effect of driving an outgassing from both land and ocean sinks which is more prominent than seen under RCP2.6.[69]

Figure 3.4 shows the land and ocean sinks aggregated into 50-year periods. From 2100–2150, both land and ocean become net sources of CO_2 with the land showing the strongest response – a sustained source of more than 2 Gt CO_2 per year for the whole 22nd century. This represents more than 250 Gt CO_2 of release of CO_2 from land ecosystems during this period. The ocean responds in a more muted way displaying a source for just the first half of the 22nd century and reverting to a sink thereafter.

The conclusion is that the dynamics of natural carbon sinks play a strong role in modulating and generally offsetting the effectiveness of negative emissions in reducing the amount of CO_2 in the atmosphere. For every tonne of CO_2 removed *via* negative emissions, natural sinks weaken and lead to only a fraction of the removal being effective.

This behaviour is well known and is represented in integrated assessment models (IAMs), which are used to create scenarios used by IPCC. But often the uncertainty in these processes, especially on longer timescales, is not fully appreciated. This means that policies and proposed pathways to meet climate goals must be adaptive and take into account adjusted information on these feedbacks and sensitivities.[77] In order to chart an "optimal" pathway, early and rapid emission reductions now will lead to reduced demand for negative emissions later.[78] Progress in modelling since IPCC AR5 has helped reduce the model disagreement in airborne fraction and therefore has reduced uncertainty in the demand for NETs.[79]

3.4 Earth System Implications for the Supply of Negative Emissions

The Earth system is central in determining the feasibility and implications of different NETs, especially those requiring large areas of land and/or biological activity of ecosystems. Nature-based solutions (NBS) are approaches that work with and enhance nature to address societal challenges.[80] Those

**See the Glossary.

activities described in the literature as "nature-based solutions", or sometimes "natural climate solutions", are clearly highly dependent on the climate and environment within which nature resides, as are NETs which rely on land vegetation to remove CO_2 from the atmosphere. I do not attempt here to repeat previous studies but mention key aspects with direct relevance to the Earth system, or where the role of climate is important in determining their feasibility.

Griscom *et al.* (2017) discussed numerous NBSs and found that the most likely to contribute meaningful quantities over the next decade or so is reforestation, although avoided deforestation and management of forests also have potentially strong contributions. Smith *et al.* (2015)[43] discussed four main classes of negative emissions, outlining costs and limitations which may affect the potential supply for each one. Direct air capture (DAC) and enhanced weathering (EW) are less strongly affected by weather and climate, but biofuel (whether with, or without, carbon capture and storage) and afforestation are strongly affected by their environment and any changes in this.

3.4.1 BECCS and Afforestation

For both afforestation and BECCS, as major users of land area, there are some common Earth system limitations. These land vegetation-based options require much greater areas of land than direct air capture or enhanced weathering, but estimates of the amount of land required per tonne of CO_2 removed vary. Harper *et al.*[39] demonstrate that the land use and carbon stored prior to land being utilised for biofuel has a large impact on the net CO_2 benefit of BECCS deployment. This builds on earlier modelling by Hughes *et al.*[81] who calculated that the pay-back time of converting primary vegetation to biofuels (in this case, miscanthus) varied markedly regionally. Harper *et al.*[39] describe how the balance of land required for biofuel and food crops is achieved in IAMs through assumptions of increased agricultural intensity. But uncertainties in these assumptions are not routinely analysed. van Vuuren *et al.*[82] show how uncertainties in socio-economic assumptions outweigh uncertainties in Earth system modelling for determining mitigation costs in low emissions scenarios. Much of the carbon sequestered in IAM scenarios with major BECCS deployment is in the form of crop residues, the governance and sustainability of which have not always been well understood.[83]

The effectiveness of BECCS in meeting supply depends on the productivity of the biofuel grown (Hughes *et al.*,[81]), but this itself is potentially vulnerable to climate change and especially water scarcity.[84] The permanence of these activities must also be considered. We know that forests are undergoing substantial changes around the world,[47,85] and their ability to continue to act as a carbon sink is not guaranteed to be maintained. Tropical forests may already have reached an observable tendency for their carbon sink to saturate,[86] and as CO_2 concentrations in the atmosphere stabilise or even decline, their ability to absorb carbon will be dramatically reduced (see Section 3.3.1).

Issues of NETs require consideration of more than just carbon. When looking at the full greenhouse gas (GHG) balance of forest bioenergy, the net GHG reduction potential is limited,[87] and for BECCS, consideration of other greenhouse gases may be enough to change potential net negative activity into a net positive climate forcing.[88] As well as the biogeochemistry, changes in land cover have major impacts on biophysical properties such as albedo. We have known for many years that reforestation of northern high latitudes may have a counter-productive warming effect.[42] All current Earth system models agree that high latitude tree cover leads to warming of the climate, but the magnitude of this effect remains uncertain.[89]

Not all CO_2 absorbed is permanently sequestered,[19] meaning that the potential supply has to be adjusted for the efficiency (and permanence) of the storage.

3.4.2 Land Management and Restoration

Given that a large fraction of current anthropogenic greenhouse gas emissions comes from agriculture[90] and our use of the land to produce food and fibre, there are numerous techniques which could reduce our emissions. I will not discuss here techniques to reduce emissions – such as changes in grazing or fertiliser management – as these are not greenhouse gas removal techniques. But I will mention a couple of activities that might remove GHGs from the atmosphere.

Farming practices that are deliberately altered to increase the organic carbon content of the soil could contribute to CO_2 removal without compromising food production or security. The addition of biochar to soil offers a potential to sequester carbon and also potentially increase the fertility of the soil.[91,92] Fuss *et al.*[22] reviewed multiple studies on soil carbon sequestration and estimated substantial technical feasibility, but issues remain over the possibility to monitor and verify any changes in soil organic carbon;[93] this is discussed in Chapter 8. Although global Earth system models do not capture the process complexity required to assess these issues reliably, initial studies imply that under future climate projections significant organic input of carbon to soil is certainly possible.[94] Regardless of the technique, all activities which sequester organic material into soil are subject to possible loss in the future if the land-use is abandoned or returned to previous usage.

3.4.3 Summary

Many of these constraints on supply are very detailed and require specific knowledge of the technique in question and the location proposed for its deployment. As such, they are not easy to study collectively but rather require many smaller studies, which can be synthesised as in Smith *et al.*[23] and Griscom *et al.*[24] For example, the FAB-GGR[††] research programme in the UK explored UK-specific actions and implications, however, these would not necessarily apply in other regions or countries. This project highlighted important societal and governance aspects of NETs feasibility that are not well understood and vary regionally,[95] highlighting the need for social science research to inform the debate around the implementation of NETs.[96]

A final constraint on the supply side is the need for storage of the CO_2,[97] which could be a limiting factor in the feasibility of delivering large amounts of NETs. Whether this is by sequestration in geological stores or enhancing the biomass of forests or dissolving CO_2 in the oceans, all storage options require consideration of capacity and permanence.[98]

3.5 Discussion and Conclusions

The discussion in this chapter has been laid out to emphasise that we must consider both the demand for NETs and the potential supply in order to ascertain if the two sides of this equation are compatible. Intrinsic in the literature is the assumption that they are broadly compatible – or else NETs would not hold such a prominent place in climate scenarios and mitigation policies. If either NETs were not required, or if they were deemed to be impossible, then their inclusion in integrated assessment models for example would have been redundant long ago. So, given that there is ample evidence that both the supply and demand are potentially significant, I argue that more effort is required now to explicitly balance both sides of the equation.

3.5.1 Co-benefits and Unintended Consequences

In Figure 3.1, I introduced the idea that there are other considerations aside from meeting the magnitude of supply and demand. In particular, there are specific co-benefits or unintended consequences/trade-offs which must be taken into account. In a way, severe trade-offs are included in the assessments of supply – as they are seen to limit how much NETs can be safely achieved.

A major issue with land-based NETs is the demand for land, and possible competition for food production. Even if the demand for NETs can be met, the subsequent impacts on food prices[40] could be severe, but could be mitigated if they are foreseen. Other studies have found impacts on water resource or reduced stream flow[41] in areas of new tree plantations. In general, Chausson *et al.* (2020) found that major perturbations – *i.e.*, newly created ecosystems such as afforestation – led to trade-offs, whereas interventions in natural ecosystems were more likely to be beneficial.

[††]Feasibility of Afforestation and Biomass energy with carbon capture and storage for Greenhouse Gas Removal (FAB-GGR), https://fab-ggr.org/.

Some biofuel crops are known for being strong emitters of isoprene or other biogenic volatile compounds (BVOCs) which affect ozone formation and production of organic aerosols.[99,100] This brings the risk that bioenergy production may have consequences for air quality.[101,102] But emissions vary significantly by species, so the choice of biofuel is an important consideration.[103] Furthermore, burning of biofuels, such as bioethanol, may change the chemical composition of exhaust gases and affect local air quality although whether in beneficial or detrimental ways is not conclusive.[104]

Co-benefits are harder to include in quantification of supply-side estimates, and are generally listed as additional reasons – sometimes the primary reason – for investing in NETs or a nature-based GGR solution. Griscom *et al.* (2017) summarise the co-benefits of nature-based solutions and almost all have multiple benefits across air quality, biodiversity, water quality and soil fertility.

3.5.2 Other Considerations: The Role of Non-CO_2 Activities

So far, I have discussed primarily the removal of CO_2 – in other words negative emissions of CO_2. But of course, climate change is the result of multiple greenhouse gases also including methane and nitrous oxide as primary contributors to radiative forcing since pre-industrial times.[105] Methane contributes approximately 0.5 W m^{-2} and N_2O approximately 0.2 W m^{-2} of radiative forcing compared to that of CO_2 (1.8 W m^{-2}). It is therefore pertinent to ask if the removal of other greenhouse gases (greenhouse gas removal: "GGR" rather than "CDR") could play a role in a portfolio of climate mitigation strategies. There is also significant interest in, and associated huge literature on, the use of solar radiation management (SRM) to control climate. While this chapter does not intend to cover this in detail it is worth mentioning for completeness as SRM and CDR are often discussed hand-in-hand as geoengineering activities,[106] although the label itself and its possible governance are contentious.[107]

While large amounts of deliberate aerosol injection would be required to offset an increasing greenhouse gas forcing of unmitigated emissions, thus leading to many decades or even centuries committed to SRM,[108] recent literature argues that such studies are poor choices to assess implications of SRM and new scenarios of SRM as the short-term "sticking plaster" response could be viable in order to buy time for CO_2 emissions reductions to be effective.[109,110] Thus, while it seems much less likely that SRM may be deployed as the sole response to climate change, proponents still argue that its consequences as a small-magnitude, finite lifetime application may be less severe than those of overshooting climate targets. SRM may also interact with the supply and demand for NETs because it directly affects ecosystems[111] and natural carbon sinks such as permafrost[112] and fire.[113] However, while Keith *et al.*[114] argued that SRM brings therefore a double advantage of reducing climate change and increasing carbon sinks, Plazzotta *et al.*[115] show the impact is likely both small (a modest 150 Gt CO_2 after 50 years of SRM deployment) and is not sustained if the SRM deployment is terminated.

Regarding removal of non-CO_2 greenhouse gases, the literature is still in its infancy. Boucher and Folberth[116] discussed the issue a decade ago, but only recently has there been interest in technological advances which may be able to deliver meaningful amounts of "negative methane emissions". As with CO_2, methane mitigation (*i.e.*, reducing emissions) is the primary way of avoiding future climate change[117] and due to the shorter lifetime of CH_4, it can be effective more quickly than reducing CO_2 emissions. Jones *et al.*[118] explored in an idealised scenario the potential climate benefit of reducing CH_4 at 1% or 2% per year (up to 4 times the rate of reduction supposed in the RCP2.6 scenario). The climate benefit, around 0.5 °C by 2100, is potentially an important contribution to meeting the Paris Agreement climate goals, but would require anthropogenic removal of CH_4 and global net-negative CH_4 emissions by around 2060. Jackson *et al.*[119] discussed the use of zeolites and other technologies as a means towards atmospheric restoration. This paves the way for a new generation of climate models to answer these new questions – while it is increasingly common for climate models ("earth system models") to represent atmospheric chemistry,[120] it is much less common to be able to simulate the full interactive methane cycle including scenarios of anthropogenic emissions and natural feedbacks such as from wetlands.[121] Previous studies found a dual benefit of reduced climate change and also improved air quality through reductions in surface ozone when CH_4 emissions were reduced to tackle climate change.[122] Therefore, direct CH_4 removal may bring associated co-benefits to climate and air quality are not yet available.[129] Jackson et al. (2021)[130] therefore recommend a renewed research agenda into the feasibility and implications of methane removal.

An overall conclusion is that global temperature is only a crude metric of how our climate is changing. While it is natural that global policy targets are couched in such terms, it should be remembered that these warming levels have different impacts depending on the rate and method with which we approach or exceed them. Similarly, the concept of carbon budgets is a robust simplifying idea which aides societal planning, but hides the complexity of how a global total should be fairly achieved at national levels, either for total budgets[123] or for negative emissions specifically.[124] As with all mitigation options, NETs have other, sometimes direct effects, on things we care about. Not all "2-degree worlds" are the same, and the path taken to arrive at any global temperature is as important as the end destination. There is a need, therefore, for ongoing nuanced discussion on the role of NETs, the urgent need for emissions reduction, and how these relate to sustainability, social equity and fairness.

Glossary

CMIP. The Coupled Model Intercomparison Project (CMIP) is a coordinated activity of the World Climate Research Programme. It facilitates climate research by organising shared protocols and experimental design for research groups to perform numerical simulations with their Earth system models. It has undergone multiple generations of experiments with CMIP5[125] being used to underpin research contributing to the IPCC's Fifth Assessment Reports, and CMIP6[70] contributing to the IPCC's Sixth Assessment Reports.

ESM. Earth system models (ESMs) are a class of numerical model used for making climate projections. They are the most complex of such models and incorporate physical, dynamical and biogeochemical processes associated with land surface, atmospheric and oceanic processes.

Radiative forcing. The energy imbalance at the top of the Earth's atmosphere due to changes in atmospheric composition – such as changes in greenhouse gases or aerosols. Commonly expressed in units of $W m^{-2}$.

RCPs. Representative Concentration Pathways (RCPs) are a set of scenarios of future greenhouse gas and aerosol emissions and concentrations that were used by climate models to investigate different future climates in response to different societal choices.[25] They are labelled by the approximate radiative forcing by the end of the 21st century, *e.g.*, RCP2.6 represents a scenario where global radiative forcing is 2.6 $W m^{-2}$ in 2100. There are 4 main RCPs: 2.6, 4.5, 6.0 and 8.5 spanning from aggressive mitigation (RCP2.6) to very high emissions (RCP8.5).

SSPs. Shared Socioeconomic Pathways (SSPs[71]) are a set of scenarios of future greenhouse gas and aerosol emissions, and other activity such as land-use/land-use change. They supersede RCP scenarios and are widely used by climate models participating in CMIP6.

References

1. IPCC, *AR5 - Summary for Policymakers*, 2013.
2. K. Caldeira and J. F. Kasting, *Nature*, 1993, **366**, 251–253.
3. M. R. Allen, D. J. Frame, C. Huntingford, C. D. Jones, J. A. Lowe, M. Meinshausen and N. Meinshausen, *Nature*, 2009, **458**, 1163–1166.
4. M. Meinshausen, N. Meinshausen, W. Hare, S. C. B. Raper, K. Frieler, R. Knutti, D. J. Frame and M. R. Allen, *Nature*, 2009, **458**, 1158–1162.
5. H. D. Matthews, N. P. Gillett, P. A. Stott and K. Zickfeld, *Nature*, 2009, **459**, 829–U3.
6. I. G. Enting, T. M. L. Wigley and M. Heimann, *Future Emissions and Concentrations of Carbon Dioxide: Key Ocean/Atmosphere/Land Analyses*, CSIRO, [Aspendale, Vic.], 1994, DOI: 10.4225/08/585eba6ad9286.
7. T. M. L. Wigley, R. Richels and J. A. Edmonds, *Nature*, 1996, **379**, 240–243.
8. P. Friedlingstein, M. W. Jones, M. O'Sullivan, R. M. Andrew, J. Hauck, G. P. Peters, W. Peters, J. Pongratz, S. Sitch, C. Le Quéré, D. C. E. Bakker, J. G. Canadell, P. Ciais, R. B. Jackson, P. Anthoni, L. Barbero, A. Bastos, V. Bastrikov, M. Becker, L. Bopp, E. Buitenhuis, N. Chandra, F. Chevallier, L. P. Chini, K. I. Currie, R. A. Feely, M. Gehlen, D. Gilfillan, T. Gkritzalis, D. S. Goll, N. Gruber, S. Gutekunst, I. Harris, V. Haverd, R. A. Houghton, G. Hurtt, T. Ilyina, A. K. Jain, E. Joetzjer, J. O. Kaplan,

E. Kato, K. Klein Goldewijk, J. I. Korsbakken, P. Landschützer, S. K. Lauvset, N. Lefèvre, A. Lenton, S. Lienert, D. Lombardozzi, G. Marland, P. C. McGuire, J. R. Melton, N. Metzl, D. R. Munro, J. E. M. S. Nabel, S.-I. Nakaoka, C. Neill, A. M. Omar, T. Ono, A. Peregon, D. Pierrot, B. Poulter, G. Rehder, L. Resplandy, E. Robertson, C. Rödenbeck, R. Séférian, J. Schwinger, N. Smith, P. P. Tans, H. Tian, B. Tilbrook, F. N. Tubiello, G. R. van der Werf, A. J. Wiltshire and S. Zaehle, *Earth Syst. Sci. Data*, 2019, **11**, 1783–1838.

9. S. Fuss, J. G. Canadell, G. P. Peters, M. Tavoni, R. M. Andrew, P. Ciais, R. B. Jackson, C. D. Jones, F. Kraxner, N. Nakicenovic, C. Le Quere, M. R. Raupach, A. Sharifi, P. Smith and Y. Yamagata, *Nat. Clim. Change*, 2014, **4**, 850–853.
10. P. Friedlingstein, M. Meinshausen, V. K. Arora, C. D. Jones, A. Anav, S. K. Liddicoat and R. Knutti, *J. Clim.*, 2014, **27**, 511–526.
11. IPCC, SPECIAL REPORT Global Warming of 1.5 °C Chapter 2 Mitigation Pathways Compatible with 1.5 °C in the Context of Sustainable Development, 2018.
12. C. Jones, E. Robertson, V. Arora, P. Friedlingstein, E. Shevliakova, L. Bopp, V. Brovkin, T. Hajima, E. Kato, M. Kawamiya, S. Liddicoat, K. Lindsay, C. H. Reick, C. Roelandt, J. Segschneider and J. Tjiputra, *J. Clim.*, 2013, **26**, 4398–4413.
13. M. R. Raupach, *Earth Syst. Dyn.*, 2013, **4**, 31–49.
14. L. Cao and K. Caldeira, *Environ. Res. Lett.*, 2010, **5**, 024011.
15. H. D. Matthews and A. J. Weaver, *Nat. Geosci.*, 2010, **3**, 142.
16. S. Fuss, J. G. Canadell, G. P. Peters, M. Tavoni, R. M. Andrew, P. Ciais, R. B. Jackson, C. D. Jones, F. Kraxner, N. Nakicenovic, C. Le Quéré, M. R. Raupach, A. Sharifi, P. Smith and Y. Yamagata, *Nat. Clim. Change*, 2014, **4**(10), 850–853.
17. S. Fuss, C. D. Jones, F. Kraxner, G. P. Peters, P. Smith, M. Tavoni, D. P. van Vuuren, J. G. Canadell, R. B. Jackson, J. Milne, J. R. Moreira, N. Nakicenovic, A. Sharifi and Y. Yamagata, *Environ. Res. Lett.*, 2016, **11**, 115007.
18. S. Fuss, J. G. Canadell, P. Ciais, R. B. Jackson, C. D. Jones, A. Lyngfelt, G. P. Peters and D. P. van Vuuren, *One Earth*, 2020, **3**, 145–149.
19. M. B. Jones and F. Albanito, *Global Change Biol.*, 2020, **26**, 5358–5364.
20. G. F. Nemet, M. W. Callaghan, F. Creutzig, S. Fuss, J. Hartmann, J. Hilaire, W. F. Lamb, J. C. Minx, S. Rogers and P. Smith, *Environ. Res. Lett.*, 2018, **13**, 063003.
21. J. C. Minx, W. F. Lamb, M. W. Callaghan, S. Fuss, J. Hilaire, F. Creutzig, T. Amann, T. Beringer, W. De Oliveira Garcia, J. Hartmann, T. Khanna, D. Lenzi, G. Luderer, G. F. Nemet, J. Rogelj, P. Smith, J. L. Vicente Vicente, J. Wilcox and M. Del Mar Zamora Dominguez, *Environ. Res. Lett.*, 2018, **13**, 063001.
22. S. Fuss, W. F. Lamb, M. W. Callaghan, J. Hilaire, F. Creutzig, T. Amann, T. Beringer, W. De Oliveira Garcia, J. Hartmann, T. Khanna, G. Luderer,

G. F. Nemet, J. Rogelj, P. Smith, J. V. Vicente, J. Wilcox, M. Del Mar Zamora Dominguez and J. C. Minx, *Environ. Res. Lett.*, 2018, **13**, 063002.

23. P. Smith, S. J. Davis, F. Creutzig, S. Fuss, J. Minx, B. Gabrielle, E. Kato, R. B. Jackson, A. Cowie, E. Kriegler, D. P. van Vuuren, J. Rogelj, P. Ciais, J. Milne, J. G. Canadell, D. McCollum, G. Peters, R. Andrew, V. Krey, G. Shrestha, P. Friedlingstein, T. Gasser, A. Grubler, W. K. Heidug, M. Jonas, C. D. Jones, F. Kraxner, E. Littleton, J. Lowe, J. R. Moreira, N. Nakicenovic, M. Obersteiner, A. Patwardhan, M. Rogner, E. Rubin, A. Sharifi, A. Torvanger, Y. Yamagata, J. Edmonds and Y. Cho, *Nat. Clim. Change*, 2016, **6**, 42–50.
24. B. W. Griscom, J. Adams, P. W. Ellis, R. A. Houghton, G. Lomax, D. A. Miteva, W. H. Schlesinger, D. Shoch, J. V. Siikamäki, P. Smith, P. Woodbury, C. Zganjar, A. Blackman, J. Campari, R. T. Conant, C. Delgado, P. Elias, T. Gopalakrishna, M. R. Hamsik, M. Herrero, J. Kiesecker, E. Landis, L. Laestadius, S. M. Leavitt, S. Minnemeyer, S. Polasky, P. Potapov, F. E. Putz, J. Sanderman, M. Silvius, E. Wollenberg and J. Fargione, *Proc. Natl. Acad. Sci. U. S. A.*, 2017, **114**(44), 11645–11650.
25. D. P. van Vuuren, J. Edmonds, M. Kainuma, K. Riahi, A. Thomson, K. Hibbard, G. C. Hurtt, T. Kram, V. Krey, J. F. Lamarque, T. Masui, M. Meinshausen, N. Nakicenovic, S. J. Smith and S. K. Rose, *Clim. Change*, 2011, **109**, 5–31.
26. E. Kriegler, G. Luderer, N. Bauer, L. Baumstark, S. Fujimori, A. Popp, J. Rogelj, J. Strefer and D. P. van Vuuren, *Philos. Trans. R. Soc. A Math. Phys. Eng. Sci.*, 2018, **376**(2119), 20160457.
27. D. P. van Vuuren, E. Stehfest, D. E. H. J. Gernaat, M. Van Den Berg, D. L. Bijl, H. S. De Boer, V. Daioglou, J. C. Doelman, O. Y. Edelenbosch, M. Harmsen, A. F. Hof and M. A. E. Van Sluisveld, *Nat. Clim. Change*, 2018, **8**, 391–397.
28. IPCC, *Climate Change 2014: Impacts, Adaptation, and Vulnerability. Summaries, Frequently Asked Questions, and Cross-Chapter Boxes*, 2014, https://www.ipcc.ch/site/assets/uploads/2018/03/WGIIAR5-Integration Brochure_FINAL-1.pdf.
29. O. Hoegh - Guldberg, D. Jacob, M. Taylor, M. Bindi, S. Brown, I. Camilloni, A. Diedhiou, R. Djalante, K. Ebi, F. Engelbrecht, J. Guiot, Y. Hijioka, S. Mehrotra, A. Payne, S. I. Seneviratne, A. Thomas, R. Warren and G. Zhou, in Global Warming of 1.5 °C an IPCC special report on the impacts of global warming of 1.5 °C above pre-industrial levels and related global greenhouse gas emission pathways, in the context of strengthening the global response to the threat of climate change, 2018.
30. R. A. Betts, L. Alfieri, C. Bradshaw, J. Caesar, L. Feyen, P. Friedlingstein, L. Gohar, A. Koutroulis, K. Lewis, C. Morfopoulos, L. Papadimitriou, K. J. Richardson, I. Tsanis and K. Wyser, *Philos. Trans. Ser. A, Math. Phys. Eng. Sci*, 2018, **376**(2119), 20160452.

31. IPCC, M. Oppenheimer, B. Glavovic, J. Hinkel, R. van de Wal, A. K. Magnan, A. Abd-Elgawad, R. Cai, M. Cifuentes-Jara, R. M. DeConto, T. Ghosh, J. Hay, F. Isla, B. Marzeion, B. Meyssignac and Z. Sebesvari, in *IPCC Special Report on the Ocean and Cryosphere in a Changing Climate*, 2019.
32. M. R. Allen, O. P. Dube, W. Solecki, F. Aragón–Durand, W. Cramer, S. Humphreys, M. Kainuma, J. Kala, N. Mahowald, Y. Mulugetta, R. Perez, M. Wairiu and K. Zickfeld, in *Global warming of 1.5 °C. An IPCC Special Report on the impacts of global warming of 1.5 °C above pre-industrial levels and related global greenhouse gas emission pathways, in the context of strengthening the global response to the threat of climate change,* 2018.
33. M. R. Allen, D. J. Frame, C. Huntingford, C. D. Jones, J. A. Lowe, M. Meinshausen and N. Meinshausen, *Nature*, 2009, **458**, 1163–1166.
34. K. Zickfeld, A. H. MacDougall and H. D. Matthews, *Environ. Res. Lett.*, 2016, **11**, 55006.
35. S. J. Davis, K. Caldeira and H. D. Matthews, *Science*, 2010, **329**, 1330–1333.
36. D. Tong, Q. Zhang, Y. Zheng, K. Caldeira, C. Shearer, C. Hong, Y. Qin and S. J. Davis, *Nature*, 2019, **572**, 373–377.
37. C. J. Smith, P. M. Forster, M. Allen, J. Fuglestvedt, R. J. Millar, J. Rogelj and K. Zickfeld, *Nat. Commun*, 2019, **10**, 101.
38. S. Pacala and R. Socolow, *Science*, 2004, **305**, 968–972.
39. A. B. Harper, T. Powell, P. M. Cox, J. House, C. Huntingford, T. M. Lenton, S. Sitch, E. Burke, S. E. Chadburn, W. J. Collins, E. Comyn-Platt, V. Daioglou, J. C. Doelman, G. Hayman, E. Robertson, D. van Vuuren, A. Wiltshire, C. P. Webber, A. Bastos, L. Boysen, P. Ciais, N. Devaraju, A. K. Jain, A. Krause, B. Poulter and S. Shu, *Nat. Commun.*, 2018, **9**, 2938.
40. J. Fuhrman, H. McJeon, P. Patel, S. C. Doney, W. M. Shobe and A. F. Clarens, *Nat. Clim. Change*, 2020, **10**, 920–927.
41. R. B. Jackson, E. G. Jobbágy, R. Avissar, S. B. Roy, D. J. Barrett, C. W. Cook, K. A. Farley, D. C. le Maitre, B. A. McCarl and B. C. Murray, *Science*, 2005, **310**(5756), 1944–1947.
42. R. A. Betts, *Nature*, 2000, **408**, 187–190.
43. P. Smith, S. J. Davis, F. Creutzig, S. Fuss, J. Minx, B. Gabrielle, E. Kato, R. B. Jackson, A. Cowie, E. Kriegler, D. P. van Vuuren, J. Rogelj, P. Ciais, J. Milne, J. G. Canadell, D. McCollum, G. Peters, R. Andrew, V. Krey, G. Shrestha, P. Friedlingstein, T. Gasser, A. Grübler, W. K. Heidug, M. Jonas, C. D. Jones, F. Kraxner, E. Littleton, J. Lowe, J. R. Moreira, N. Nakicenovic, M. Obersteiner, A. Patwardhan, M. Rogner, E. Rubin, A. Sharifi, A. Torvanger, Y. Yamagata, J. Edmonds and C. Yongsung, *Nat. Clim. Change*, 2016, **6**, 42–50.
44. J. Pongratz, H. Dolman, A. Don, K. H. Erb, R. Fuchs, M. Herold, C. Jones, T. Kuemmerle, S. Luyssaert, P. Meyfroidt and K. Naudts, *Glob. Chang. Biol.*, 2018, **24**, 1470–1487.

45. K. H. Erb, S. Luyssaert, P. Meyfroidt, J. Pongratz, A. Don, S. Kloster, T. Kuemmerle, T. Fetzel, R. Fuchs, M. Herold, H. Haberl, C. D. Jones, E. Marín-Spiotta, I. McCallum, E. Robertson, V. Seufert, S. Fritz, A. Valade, A. Wiltshire and A. J. Dolman, *Glob. Chang. Biol.*, 2017, **23**(2), 512–533.
46. E. J. Burke, I. P. Hartley and C. D. Jones, *Cryosphere*, 2012, **6**, 1063–1076.
47. R. J. W. Brienen, O. L. Phillips, T. R. Feldpausch, E. Gloor, T. R. Baker, J. Lloyd, G. Lopez-Gonzalez, A. Monteagudo-Mendoza, Y. Malhi, S. L. Lewis, R. Vásquez Martinez, M. Alexiades, E. Álvarez Dávila, P. Alvarez-Loayza, A. Andrade, L. E. O. C. Aragaõ, A. Araujo-Murakami, E. J. M. M. Arets, L. Arroyo, G. A. Aymard, O. S. Bánki, C. Baraloto, J. Barroso, D. Bonal, R. G. A. Boot, J. L. C. Camargo, C. V. Castilho, V. Chama, K. J. Chao, J. Chave, J. A. Comiskey, F. Cornejo Valverde, L. Da Costa, E. A. De Oliveira, A. Di Fiore, T. L. Erwin, S. Fauset, M. Forsthofer, D. R. Galbraith, E. S. Grahame, N. Groot, B. Hérault, N. Higuchi, E. N. Honorio Coronado, H. Keeling, T. J. Killeen, W. F. Laurance, S. Laurance, J. Licona, W. E. Magnussen, B. S. Marimon, B. H. Marimon-Junior, C. Mendoza, D. A. Neill, E. M. Nogueira, P. Núñez, N. C. Pallqui Camacho, A. Parada, G. Pardo-Molina, J. Peacock, M. Penã-Claros, G. C. Pickavance, N. C. A. Pitman, L. Poorter, A. Prieto, C. A. Quesada, F. Ramírez, H. Ramírez-Angulo, Z. Restrepo, A. Roopsind, A. Rudas, R. P. Salomaõ, M. Schwarz, N. Silva, J. E. Silva-Espejo, M. Silveira, J. Stropp, J. Talbot, H. Ter Steege, J. Teran-Aguilar, J. Terborgh, R. Thomas-Caesar, M. Toledo, M. Torello-Raventos, R. K. Umetsu, G. M. F. Van Der Heijden, P. Van Der Hout, I. C. Guimarães Vieira, S. A. Vieira, E. Vilanova, V. A. Vos and R. J. Zagt, *Nature*, 2015, 519, 344–348.
48. A. J. Watson, U. Schuster, J. D. Shutler, T. Holding, I. G. C. Ashton, P. Landschützer, D. K. Woolf and L. Goddijn-Murphy, *Nat. Commun.*, 2020, **11**, 4422.
49. P. Goodwin, R. G. Williams and A. Ridgwell, *Nat. Geosci.*, 2015, **8**, 29–34.
50. A. H. MacDougall, *Curr. Clim. Change Rep.*, 2016, **2**, 39–47.
51. D. Ehlert, K. Zickfeld, M. Eby and N. Gillett, *J. Clim.*, 2017, **30**, 2921–2935.
52. M. Leduc, H. D. Matthews and R. de Elía, *Nat. Clim. Change*, 2016, **6**, 474–478.
53. L. J. Harrington, D. J. Frame, E. M. Fischer, E. Hawkins, M. Joshi and C. D. Jones, *Environ. Res. Lett.*, 2016, **11**, 055007.
54. S. I. Seneviratne, M. G. Donat, A. J. Pitman, R. Knutti and R. L. Wilby, *Nature*, 2016, **529**, 477–483.
55. O. Boucher, P. R. Halloran, E. J. Burke, M. Doutriaux-Boucher, C. D. Jones, J. Lowe, M. A. Ringer, E. Robertson and P. Wu, *Environ. Res. Lett.*, 2012, 7, 024013.
56. D. P. van Vuuren, J. Edmonds, M. Kainuma, K. Riahi, A. Thomson, K. Hibbard, G. C. Hurtt, T. Kram, V. Krey, J. F. Lamarque, T. Masui,

M. Meinshausen, N. Nakicenovic, S. J. Smith and S. K. Rose, *Clim. Change*, 2011, **109**, 5–31.
57. V. Masson-Delmotte, P. Zhai, H.-O. Pörtner, D. Roberts, J. Skea, P. R. Shukla, A. Pirani, W. Moufouma-Okia, C. Péan, R. Pidcock, S. Connors, J. B. R. Matthews, Y. Chen, X. Zhou, M. I. Gomis, E. Lonnoy, T. Maycock, M. Tignor and T. Waterfield, IPCC Special Report 1.5 - Summary for Policymakers, 2018.
58. K. Akimoto, F. Sano and T. Tomoda, *Mitigation Adapt. Strategies Global Change*, 2018, **23**, 839–852.
59. K. B. Tokarska and K. Zickfeld, *Environ. Res. Lett.*, 2015, **10**, 094013.
60. J. B. Palter, T. L. Frölicher, D. Paynter and J. G. John, *Earth Syst. Dyn.*, 2018, **9**, 817–828.
61. K. Tachiiri, D. Silva Herran, X. Su and M. Kawamiya, *Environ. Res. Lett.*, 2019, **14**, 124063.
62. D. P. Keller, A. Lenton, V. Scott, N. E. Vaughan, N. Bauer, D. Ji, C. D. Jones, B. Kravitz, H. Muri and K. Zickfeld, *Geosci. Model Dev.*, 2018, **11**, 1133–1160.
63. K. B. Tokarska, K. Zickfeld and J. Rogelj, *Earth's Future*, 2019, 7, 1283–1295.
64. P. Ciais, C. Sabine, G. Bala, L. Bopp, V. Brovkin, J. Canadell, A. Chhabra, R. DeFries, J. Galloway, M. Heimann, C. Jones, C. Le Quéré, R. B. Myneni, S. Piao and P. Thornton, in *Climate Change 2013 - The Physical Science Basis*, ed. Intergovernmental Panel on Climate Change, Cambridge University Press, Cambridge, 2013, pp. 465–570.
65. S. Mathesius, M. Hofmann, K. Caldeira and H. J. Schellnhuber, *Nat. Clim. Change*, 2015, **5**, 1107–1113.
66. X. Li, K. Zickfeld, S. Mathesius, K. Kohfeld and J. B. R. Matthews, *Geophys. Res. Lett.*, 2020, **47**, e2020GL088507.
67. C. Le Quéré, R. M. Andrew, P. Friedlingstein, S. Sitch, J. Hauck, J. Pongratz, P. A. Pickers, J. I. Korsbakken, G. P. Peters, J. G. Canadell, A. Arneth, V. K. Arora, L. Barbero, A. Bastos, L. Bopp, F. Chevallier, L. P. Chini, P. Ciais, S. C. Doney, T. Gkritzalis, D. S. Goll, I. Harris, V. Haverd, F. M. Hoffman, M. Hoppema, R. A. Houghton, G. Hurtt, T. Ilyina, A. K. Jain, T. Johannessen, C. D. Jones, E. Kato, R. F. Keeling, K. K. Goldewijk, P. Landschützer, N. Lefèvre, S. Lienert, Z. Liu, D. Lombardozzi, N. Metzl, D. R. Munro, J. E. M. S. Nabel, S. Nakaoka, C. Neill, A. Olsen, T. Ono, P. Patra, A. Peregon, W. Peters, P. Peylin, B. Pfeil, D. Pierrot, B. Poulter, G. Rehder, L. Resplandy, E. Robertson, M. Rocher, C. Rödenbeck, U. Schuster, J. Schwinger, R. Séférian, I. Skjelvan, T. Steinhoff, A. Sutton, P. P. Tans, H. Tian, B. Tilbrook, F. N. Tubiello, I. T. van der Laan-Luijkx, G. R. van der Werf, N. Viovy, A. P. Walker, A. J. Wiltshire, R. Wright, S. Zaehle and B. Zheng, *Earth Syst. Sci. Data*, 2018, **10**, 2141–2194.
68. F. Joos, R. Roth, J. S. Fuglestvedt, G. P. Peters, I. G. Enting, W. von Bloh, V. Brovkin, E. J. Burke, M. Eby, N. R. Edwards, T. Friedrich, T. L. Frolicher, P. R. Halloran, P. B. Holden, C. Jones, T. Kleinen,

F. T. Mackenzie, K. Matsumoto, M. Meinshausen, G. K. Plattner, A. Reisinger, J. Segschneider, G. Shaffer, M. Steinacher, K. Strassmann, K. Tanaka, A. Timmermann and A. J. Weaver, *Atmos. Chem. Phys.*, 2013, **13**, 2793–2825.

69. C. D. Jones, P. Ciais, S. J. Davis, P. Friedlingstein, T. Gasser, G. P. Peters, J. Rogelj, D. P. van Vuuren, J. G. Canadell, A. Cowie, R. B. Jackson, M. Jonas, E. Kriegler, E. Littleton, J. A. Lowe, J. Milne, G. Shrestha, P. Smith, A. Torvanger and A. Wiltshire, *Environ. Res. Lett.*, 2016, **11**, 095012.
70. V. Eyring, S. Bony, G. A. Meehl, C. A. Senior, B. Stevens, R. J. Stouffer and K. E. Taylor, *Geosci. Model Dev.*, 2016, **9**, 1937–1958.
71. B. C. O'Neill, C. Tebaldi, D. P. van Vuuren, V. Eyring, P. Friedlingstein, G. Hurtt, R. Knutti, E. Kriegler, J.-F. Lamarque, J. Lowe, G. A. Meehl, R. Moss, K. Riahi and B. M. Sanderson, *Geosci. Model Dev.*, 2016, **9**, 3461–3482.
72. M. Meinshausen, Z. R. J. Nicholls, J. Lewis, M. J. Gidden, E. Vogel, M. Freund, U. Beyerle, C. Gessner, A. Nauels, N. Bauer, J. G. Canadell, J. S. Daniel, A. John, P. B. Krummel, G. Luderer, N. Meinshausen, S. A. Montzka, P. J. Rayner, S. Reimann, S. J. Smith, M. van den Berg, G. J. M. Velders, M. K. Vollmer and R. H. J. Wang, *Geosci. Model Dev.*, 2020, **13**, 3571–3605.
73. N. C. Swart, J. N. S. Cole, V. V. Kharin, M. Lazare, J. F. Scinocca, N. P. Gillett, J. Anstey, V. Arora, J. R. Christian, S. Hanna, Y. Jiao, W. G. Lee, F. Majaess, O. A. Saenko, C. Seiler, C. Seinen, A. Shao, M. Sigmond, L. Solheim, K. von Salzen, D. Yang and B. Winter, *Geosci. Model Dev.*, 2019, **12**, 4823–4873.
74. G. Danabasoglu, J.-F. Lamarque, J. Bacmeister, D. A. Bailey, A. K. DuVivier, J. Edwards, L. K. Emmons, J. Fasullo, R. Garcia, A. Gettelman, C. Hannay, M. M. Holland, W. G. Large, P. H. Lauritzen, D. M. Lawrence, J. T. M. Lenaerts, K. Lindsay, W. H. Lipscomb, M. J. Mills, R. Neale, K. W. Oleson, B. Otto-Bliesner, A. S. Phillips, W. Sacks, S. Tilmes, L. van Kampenhout, M. Vertenstein, A. Bertini, J. Dennis, C. Deser, C. Fischer, B. Fox-Kemper, J. E. Kay, D. Kinnison, P. J. Kushner, V. E. Larson, M. C. Long, S. Mickelson, J. K. Moore, E. Nienhouse, L. Polvani, P. J. Rasch and W. G. Strand, *J. Adv. Model. Earth Syst.*, 2020, **12**, e2019MS001916.
75. F. Hourdin, C. Rio, J. Grandpeix, J. Madeleine, F. Cheruy, N. Rochetin, A. Jam, I. Musat, A. Idelkadi, L. Fairhead, M. Foujols, L. Mellul, A. Traore, J. Dufresne, O. Boucher, M. Lefebvre, E. Millour, E. Vignon, J. Jouhaud, F. B. Diallo, F. Lott, G. Gastineau, A. Caubel, Y. Meurdesoif and J. Ghattas, *J. Adv. Model. Earth Syst.*, 2020, **12**, e2019MS001892.
76. A. A. Sellar, C. G. Jones, J. Mulcahy, Y. Tang, A. Yool, A. Wiltshire, F. M. O'Connor, M. Stringer, R. Hill, J. Palmieri, S. Woodward, L. de Mora, T. Kuhlbrodt, S. Rumbold, D. I. Kelley, R. Ellis, C. E. Johnson, J. Walton, N. L. Abraham, M. B. Andrews, T. Andrews, A. T. Archibald, S. Berthou, E. Burke, E. Blockley, K. Carslaw, M. Dalvi,

J. Edwards, G. A. Folberth, N. Gedney, P. T. Griffiths, A. B. Harper, M. A. Hendry, A. J. Hewitt, B. Johnson, A. Jones, C. D. Jones, J. Keeble, S. Liddicoat, O. Morgenstern, R. J. Parker, V. Predoi, E. Robertson, A. Siahaan, R. S. Smith, R. Swaminathan, M. T. Woodhouse, G. Zeng and M. Zerroukat, *J. Adv. Model. Earth Syst.*, 2019, **11**, 4513–4558.

77. P. Goodwin, S. Brown, I. D. Haigh, R. J. Nicholls and J. M. Matter, *Earth's Future*, 2018, **6**, 601–615.
78. M. Schaeffer, L. Gohar, E. Kriegler, J. Lowe, K. Riahi and D. van Vuuren, *Technol. Forecast. Soc. Change*, 2015, **90**, 257–268.
79. C. D. Jones and P. Friedlingstein, *Environ. Res. Lett.*, 2020, **15**, 074019.
80. A. Chausson, B. Turner, D. Seddon, N. Chabaneix, C. A. J. Girardin, V. Kapos, I. Key, D. Roe, A. Smith, S. Woroniecki and N. Seddon, *Global Change Biol.*, 2020, gcb.15310.
81. J. K. Hughes, A. J. Lloyd, C. Huntingford, J. W. Finch and R. J. Harding, *GCB Bioenergy*, 2010, **2**, 79–88.
82. D. P. van Vuuren, K.-I. van der Wijst, S. Marsman, M. van den Berg, A. F. Hof and C. D. Jones, *Nat. Clim. Change*, 2020, **10**, 329–334.
83. N. E. Vaughan, C. Gough, S. Mander, E. W. Littleton, A. Welfle, D. E. H. J. Gernaat and D. P. van Vuuren, *Environ. Res. Lett.*, 2018, **13**, 044014.
84. R. Séférian, M. Rocher, C. Guivarch and J. Colin, *Environ. Res. Lett.*, 2018, **13**, 054011.
85. N. G. McDowell, C. D. Allen, K. Anderson-Teixeira, B. H. Aukema, B. Bond-Lamberty, L. Chini, J. S. Clark, M. Dietze, C. Grossiord, A. Hanbury-Brown, G. C. Hurtt, R. B. Jackson, D. J. Johnson, L. Kueppers, J. W. Lichstein, K. Ogle, B. Poulter, T. A. M. Pugh, R. Seidl, M. G. Turner, M. Uriarte, A. P. Walker and C. Xu, *Science*, 2020, **368**(6494), eaaz9463.
86. W. Hubau, S. L. Lewis, O. L. Phillips, K. Affum-Baffoe, H. Beeckman, A. Cuní-Sanchez, A. K. Daniels, C. E. N. Ewango, S. Fauset, J. M. Mukinzi, D. Sheil, B. Sonké, M. J. P. Sullivan, T. C. H. Sunderland, H. Taedoumg, S. C. Thomas, L. J. T. White, K. A. Abernethy, S. Adu-Bredu, C. A. Amani, T. R. Baker, L. F. Banin, F. Baya, S. K. Begne, A. C. Bennett, F. Benedet, R. Bitariho, Y. E. Bocko, P. Boeckx, P. Boundja, R. J. W. Brienen, T. Brncic, E. Chezeaux, G. B. Chuyong, C. J. Clark, M. Collins, J. A. Comiskey, D. A. Coomes, G. C. Dargie, T. de Haulleville, M. N. D. Kamdem, J.-L. Doucet, A. Esquivel-Muelbert, T. R. Feldpausch, A. Fofanah, E. G. Foli, M. Gilpin, E. Gloor, C. Gonmadje, S. Gourlet-Fleury, J. S. Hall, A. C. Hamilton, D. J. Harris, T. B. Hart, M. B. N. Hockemba, A. Hladik, S. A. Ifo, K. J. Jeffery, T. Jucker, E. K. Yakusu, E. Kearsley, D. Kenfack, A. Koch, M. E. Leal, A. Levesley, J. A. Lindsell, J. Lisingo, G. Lopez-Gonzalez, J. C. Lovett, J.-R. Makana, Y. Malhi, A. R. Marshall, J. Martin, E. H. Martin, F. M. Mbayu, V. P. Medjibe, V. Mihindou, E. T. A. Mitchard, S. Moore, P. K. T. Munishi, N. N. Bengone, L. Ojo, F. E. Ondo, K. S.-H. Peh, G. C. Pickavance, A. D. Poulsen, J. R. Poulsen, L. Qie, J. Reitsma,

F. Rovero, M. D. Swaine, J. Talbot, J. Taplin, D. M. Taylor, D. W. Thomas, B. Toirambe, J. T. Mukendi, D. Tuagben, P. M. Umunay, G. M. F. van der Heijden, H. Verbeeck, J. Vleminckx, S. Willcock, H. Wöll, J. T. Woods and L. Zemagho, *Nature*, 2020, **579**, 80–87.

87. M. Röder, E. Thiffault, C. Martínez-Alonso, F. Senez-Gagnon, L. Paradis and P. Thornley, *Biomass Bioenergy*, 2019, **121**, 99–114.
88. S. V. Hanssen, V. Daioglou, Z. J. N. Steinmann, J. C. Doelman, D. P. van Vuuren and M. A. J. Huijbregts, *Nat. Clim. Change*, 2020, 1–7.
89. L. R. Boysen, V. Brovkin, J. Pongratz, D. M. Lawrence, P. Lawrence, N. Vuichard, P. Peylin, S. Liddicoat, T. Hajima, Y. Zhang, M. Rocher, C. Delire, R. Séférian, V. K. Arora, L. Nieradzik, P. Anthoni, W. Thiery, M. M. Laguë, D. Lawrence and M.-H. Lo, *Biogeosciences*, 2020, **17**, 5615–5638.
90. P. Smith, M. Bustamenta, T. Krug, G.-J. Nabuurs, M. Molodovskaya Canada, M. Bustamante, H. Ahammad, H. Clark, H. Dong, E. A. Elsiddig, H. Haberl, R. Harper, J. House, M. Jafari, O. Masera, C. Mbow, N. H. Ravindranath, C. W. Rice, C. Robledo Abad, Y. Sokona, E. Farahani, S. Kadner, K. Seyboth, A. Adler, I. Baum, S. Brunner, P. Eickemeier, B. Kriemann, J. Savolainen, S. Schlömer, C. von Stechow, T. Zwickel and J. Minx, *IPCC Fifth Assessment Report: Agriculture, Forestry and Other Land Use (AFOLU)*, 2014.
91. A. El-Naggar, S. S. Lee, J. Rinklebe, M. Farooq, H. Song, A. K. Sarmah, A. R. Zimmerman, M. Ahmad, S. M. Shaheen and Y. S. Ok, *Geoderma*, 2019, **337**, 536–554.
92. Y. Ding, Y. Liu, S. Liu, Z. Li, X. Tan, X. Huang, G. Zeng, L. Zhou and B. Zheng, *Agron. Sustainability Dev.*, 2016, **36**, 36.
93. P. Smith, J. Soussana, D. Angers, L. Schipper, C. Chenu, D. P. Rasse, N. H. Batjes, F. Egmond, S. McNeill, M. Kuhnert, C. Arias-Navarro, J. E. Olesen, N. Chirinda, D. Fornara, E. Wollenberg, J. Álvaro-Fuentes, A. Sanz-Cobena and K. Klumpp, *Global Change Biol.*, 2020, **26**, 219–241.
94. A. Ito, T. Hajima, D. M. Lawrence, V. Brovkin, C. Delire, B. Guenet, C. D. Jones, S. Malyshev, S. Materia, S. P. McDermid, D. Peano, J. Pongratz, E. Robertson, E. Shevliakova, N. Vuichard, D. Wårlind, A. Wiltshire and T. Ziehn, *Environ. Res. Lett.*, 2020, **15**(12), 124061.
95. J. Forster, N. E. Vaughan, C. Gough, I. Lorenzoni and J. Chilvers, *Global Environ. Change*, 2020, **63**, 102073.
96. C. Gough and S. Mander, *Curr. Sustainability Energy Rep.*, 2019, **6**, 116–123.
97. T. Gasser, C. Guivarch, K. Tachiiri, C. D. Jones and P. Ciais, *Nat. Commun.*, 2015, **6**, 7958.
98. S. Benson, P. Cook, J. Anderson, S. Bachu, H. Bashir Nimir, B. Basu, J. Bradshaw, G. Deguchi, J. Gale, G. von Goerne, W. Heidug, S. Holloway, R. Kamal, D. Keith, P. Lloyd, P. Rocha, B. Senior, J. Thomson, T. Torp, T. Wildenborg, M. Wilson, F. Zarlenga and D. Zhou, in *Carbon Dioxide Capture and Storage an IPCC Special Report*,

ed. B. Metz, O. Davidson, H. de Coninck, M. Loos and L. Meyer, Cambridge University Press, UK, 2005, p. 431.
99. N. Carslaw, D. Creasey, D. Harrison, D. Heard, M. Hunter, P. Jacobs, M. Jenkin, J. Lee, A. Lewis, M. Pilling, S. Saunders and P. Seakins, *Atmos. Environ.*, 2001, **35**, 4725–4737.
100. W. J. Collins, R. G. Derwent, C. E. Johnson and D. S. Stevenson, *Clim. Change*, 2002, **52**, 453–479.
101. F. Pacifico, G. A. Folberth, C. D. Jones, S. P. Harrison and W. J. Collins, *J. Geophys. Res. Atmos.*, 2012, **117**, D22302.
102. K. Ashworth, G. Folberth, C. N. Hewitt and O. Wild, *Atmos. Chem. Phys.*, 2012, **12**, 919–939.
103. E. C. Morrison, J. Drewer and M. R. Heal, *GCB Bioenergy*, 2016, **8**, 211–225.
104. N. Scovronick, D. França, M. Alonso, C. Almeida, K. Longo, S. Freitas, B. Rudorff and P. Wilkinson, *Int. J. Environ. Res. Public Health*, 2016, **13**, 695.
105. G. Myhre, D. Shindell, F.-M. Bréon, W. D. Collins, J. Fuglestvedt, J. Huang, D. Koch, J.-F. Lamarque, D. Lee, B. Mendoza, T. Nakajima, A. Robock, G. Stephens, T. Takemura and H. Zhan, in *Climate Change 2013: The Physical Science Basis. Contribution of Working Group I to the Fifth Assessment Report of the Intergovernmental Panel on Climate Change*, 2013.
106. N. E. Vaughan and T. M. Lenton, *Clim. Change*, 2011, **109**, 745–790.
107. J. L. Reynolds, *WIREs Clim. Change*, 2021, **12**, e690.
108. O. Boucher, J. A. Lowe and C. D. Jones, *Clim. Change*, 2009, **92**, 261–273.
109. D. W. Keith and D. G. MacMartin, *Nat. Clim. Change*, 2015, **5**, 201–206.
110. M. Sugiyama, Y. Arino, T. Kosugi, A. Kurosawa and S. Watanabe, *Clim. Policy*, 2017, 1–9.
111. L. M. Russell, P. J. Rasch, G. M. Mace, R. B. Jackson, J. Shepherd, P. Liss, M. Leinen, D. Schimel, N. E. Vaughan, A. C. Janetos, P. W. Boyd, R. J. Norby, K. Caldeira, J. Merikanto, P. Artaxo, J. Melillo and M. G. Morgan, *Ambio*, 2012, **41**, 350–369.
112. H. Lee, A. Ekici, J. Tjiputra, H. Muri, S. E. Chadburn, D. M. Lawrence and J. Schwinger, *Earth's Future*, 2019, 7, 605–614.
113. C. Burton, R. A. Betts, C. D. Jones and K. Williams, *Geophys. Res. Lett.*, 2018, **45**, 3644–3652.
114. D. W. Keith, G. Wagner and C. L. Zabel, *Nat. Clim. Change*, 2017, 7, 617–619.
115. M. Plazzotta, R. Séférian and H. Douville, *Earth's Future*, 2019, 7, 664–676.
116. O. Boucher and G. A. Folberth, *Atmos. Environ.*, 2010, **44**, 3343–3345.
117. E. G. Nisbet, R. E. Fisher, D. Lowry, J. L. France, G. Allen, S. Bakkaloglu, T. J. Broderick, M. Cain, M. Coleman, J. Fernandez, G. Forster, P. T. Griffiths, C. P. Iverach, B. F. J. Kelly, M. R. Manning, P. B. R. Nisbet-Jones, J. A. Pyle, A. Townsend-Small, A. al-Shalaan, N. Warwick and G. Zazzeri, *Rev. Geophys.*, 2020, 58.

118. A. Jones, J. M. Haywood and C. D. Jones, *Atmos. Sci. Lett.*, 2018, **19**, e821.
119. R. B. Jackson, E. I. Solomon, J. G. Canadell, M. Cargnello and C. B. Field, *Nat. Sustain.*, 2019.
120. W. J. Collins, J.-F. Lamarque, M. Schulz, O. Boucher, V. Eyring, M. I. Hegglin, A. Maycock, G. Myhre, M. Prather, D. Shindell and S. J. Smith, *Geosci. Model Dev.*, 2017, **10**, 585–607.
121. N. Gedney, C. Huntingford, E. Comyn-Platt and A. Wiltshire, *Environ. Res. Lett.*, 2019, **14**, 084027.
122. D. Shindell, J. C. I. Kuylenstierna, E. Vignati, R. van Dingenen, M. Amann, Z. Klimont, S. C. Anenberg, N. Muller, G. Janssens-Maenhout, F. Raes, J. Schwartz, G. Faluvegi, L. Pozzoli, K. Kupiainen, L. Hoglund-Isaksson, L. Emberson, D. Streets, V. Ramanathan, K. Hicks, N. T. K. Oanh, G. Milly, M. Williams, V. Demkine and D. Fowler, *Science*, 2012, **335**, 183–189.
123. H. D. Matthews, K. B. Tokarska, Z. R. J. Nicholls, J. Rogelj, J. G. Canadell, P. Friedlingstein, T. L. Frölicher, P. M. Forster, N. P. Gillett, T. Ilyina, R. B. Jackson, C. D. Jones, C. Koven, R. Knutti, A. H. MacDougall, M. Meinshausen, N. Mengis, R. Séférian and K. Zickfeld, *Nat. Geosci.*, 2020, **13**, 769–779.
124. C. Pozo, Á. Galán-Martín, D. M. Reiner, N. Mac Dowell and G. Guillén-Gosálbez, *Nat. Clim. Change*, 2020, **10**, 640–646.
125. K. E. Taylor, R. J. Stouffer and G. A. Meehl, *Bull. Am. Meteorol. Soc.*, 2012, **93**, 485–498.
126. IPCC, Summary for Policymakers, in *Climate Change 2021: The Physical Science Basis. Contribution of Working Group I to the Sixth Assessment Report of the Intergovernmental Panel on Climate Change*, ed. V. Masson-Delmotte, P. Zhai, A. Pirani, S. L. Connors, C. Péan, S. Berger, N. Caud, Y. Chen, L. Goldfarb, M. I. Gomis, M. Huang, K. Leitzell, E. Lonnoy, J. B. R. Matthews, T. K. Maycock, T. Waterfield, O. Yelekçi, R. Yu and B. Zhou, Cambridge University Press, Cambridge, 2021, pp. 3–32.
127. K. Zickfeld, D. Azevedo, S. Mathesius and H. D. Matthews, Asymmetry in the climate–carbon cycle response to positive and negative CO_2 emissions, *Nat. Clim. Change*, 2021, **11**(7), 613–617.
128. J. G. Canadell, P. M. S. Monteiro, M. H. Costa, L. Cotrim da Cunha, P. M. Cox, A. V. Eliseev, S. Henson, M. Ishii, S. Jaccard, C. Koven, A. Lohila, P. K. Patra, S. Piao, J. Rogelj, S. Syampungani, S. Zaehle and K. Zickfeld, Global Carbon and other Biogeochemical Cycles and Feedbacks, in *Climate Change 2021: The Physical Science Basis. Contribution of Working Group I to the Sixth Assessment Report of the Intergovernmental Panel on Climate Change*, ed. V. Masson-Delmotte, P. Zhai, A. Pirani, S. L. Connors, C. Péan, S. Berger, N. Caud, Y. Chen, L. Goldfarb, M. I. Gomis, M. Huang, K. Leitzell, E. Lonnoy, J. B. R. Matthews, T. K. Maycock, T. Waterfield, O. Yelekçi, R. Yu and B. Zhou, Cambridge University Press, Cambridge, 2021, pp. 673–816.

129. R. B. Jackson, S. Abernethy, J. G. Canadell, M. Cargnello, S. J. Davis, S. Féron, S. Fuss, A. J. Heyer, C. Hong, C. D. Jones, M. H. Damon, F. M. O'Connor, M. Pisciotta, H. M. Rhoda, R. de Richter, E. I. Solomon, J. L. Wilcox and K. Zickfeld, Atmospheric methane removal: a research agenda, *Philos. Trans. R. Soc. A*, 2021, **379**, 20200454.
130. S. Abernethy, F. M. O'Connor, C. D. Jones and R. B. Jackson, Methane removal and the proportional reductions in surface temperature and ozone, *Philos. Trans. R. Soc. A*, 2021, **379**, 20210104.

CHAPTER 4

Accounting Negative Emissions. How Difficult Could It Be?

ANDREA RAMÍREZ RAMÍREZ

Section Energy & Industry, Faculty of Technology, Policy and Management, Delft University of Technology, The Netherlands
Email: c.a.ramirezramirez@tudelft.nl

Negative emissions will play a key role in our ability to reach ambitious climate targets. Although carbon accounting appears a simple task, the lack of agreed definition, accounting methodologies and assumptions will hinder decision makers from appropriately and efficiently incentivising, deploying, monitoring and verifying those emissions. This chapter highlights the importance of appropriately defining and accounting emissions by exploring four key challenges for carbon accounting of negative emissions, namely: assessing permanence, accounting for the temporal distribution of emissions and removals, choosing time horizons, and conflating CO_2 avoidance and CO_2 negativity.

4.1 Introduction

In the last decade, the need for more ambitious climate targets has intensified as the impacts of climate change become clearer and more evident. The Intergovernmental Panel on Climate Change Special Report on Global Warming of 1.5 °C (SR1.5) showed the importance of limiting global warming to 1.5 °C compared to pre-industrial levels and emphasised that negative emissions may already be needed by mid-century.[1] In 2020,

Energy and Environment Series No. 31
Greenhouse Gas Removal Technologies
Edited by Mai Bui and Niall Mac Dowell

Published by the Royal Society of Chemistry, www.rsc.org

pledges from business and countries to reduce their carbon emissions to zero by 2050 doubled, compared to 2019.[2] More than 76 national governments or regions and over 800 cities have set or are considering setting net-zero emissions targets.[3] As it is very unlikely that all sectors will be able to reduce their emissions to zero, such net zero emission targets heavily rely on the existence of offsetting through negative emissions. According to Daggash *et al.*[4] about 50 megatonnes of CO_2 per year of negative emissions will be needed in the UK by 2050 to offset hard-to-decarbonise sectors, such as aviation and shipping. In the EU, as part of the European's Commission's vision for a climate neutral EU, emissions covered by the European trade system (ETS) are expected to reach net zero by 2045, with substantial net negative levels achieved by 2050.[5] In Japan, as part of its strategy to achieve carbon neutrality by 2050, the government "recognises the key role of negative emissions technologies to capture and store carbon dioxide given that even with energy saving, electrification, power source decarbonization and hydrogen fuel initiatives, zero use of fossil fuels is probably unrealistic".[6]

Negative emissions are also a key component of companies' mitigation strategies. Microsoft, for instance, contracted 1.3 million greenhouse gas (GHG) emission offsets for 2021 as part of its strategy to become carbon negative by 2030, and wants to remove its entire historical CO_2 emissions by mid-century.[7] The Musk foundation has teamed up with XPrize to set up a 100 million dollar 4-year competition to develop negative emissions solutions that take CO_2 directly from the atmosphere or oceans and store it durably.[8] Amazon co-founded the Climate Pledge where, as of April 2021, more than 100 companies have committed to achieving net-zero carbon emissions by 2040, and as part of the pledge they commit to "neutralise any remaining emissions with additional, quantifiable, real, permanent and socially beneficial offsets to achieve net-zero annual carbon emissions...".[9]

In literature, negative emissions are often needed to offset the carbon debt that results from overshooting the carbon budget, which is most likely to already occur in this decade.[10,11] Moreover, negative emissions appear to be a key dominant factor determining whether, and if so when, a country could reach net-zero emissions.[12] Oberstein *et al.*[13] argue that the timing of negative emissions technologies needs to be urgently revisited to start deploying them as soon as possible, and that climate targets should be reformulated to include targets of early action in carbon dioxide removal (CDR) technologies. Although CDR has a vital role in meeting ambitious climate targets, several studies have already warned that the lack of agreed definition, accounting methodologies and assumptions will hinder decision makers to appropriately and efficiently incentivise, deploy, monitor and verify those emissions.[14,15] Lack of standardisation and transparency has resulted in a conflation of concepts and approaches that increase susceptibility to greenwashing. It also limits companies' ability to invest in negative emission solutions and hinders public trust and acceptability. A better understanding of carbon accounting and the challenges of obtaining carbon negativity is therefore needed to make better decisions concerning carbon strategies, while steering them away from choices that have a lesser potential impact or can indeed result in an increase of carbon dioxide in the atmosphere.

4.2 Carbon Accounting

Carbon accounting addresses "the measuring, collation, assessment and communication of GHG emissions emitted by a source or sequestered in a sink and the monetary valuation of GHG emissions (as assets and liabilities) to provide this information to internal or external audiences".[16] Carbon accounting is at the core of any CO_2 mitigation strategy and plays a key role for identifying risks and liabilities, conducting cost-benefits analysis, and developing benchmarking metrics.[17]

Schaltegger and Csutora[18] identified the core functions of carbon accounting according to whether they are used for unveiling past and current problems (through creating transparency about past and current activities or by forecasting future impacts) or for designing strategies for improvement (through identifying reduction potentials, evaluating reduction measures or supporting the implementation of reduction measures). Interestingly, carbon accounting of negative emissions is used for both categories. Accounting methods can roughly be divided into methods using financial terms (*e.g.*, carbon cost accounting) and non-financial terms (*e.g.*, life cycle assessment).[17] Without disregarding the importance of economic and social approaches, this chapter focuses on non-financial aspects as understanding and addressing the challenges of accounting net negative physical flows will be needed to further address the challenges imposed when accounting for the impact in financial and social terms.

The first step in any accounting methodology is to define what one wants to account for. Tanzer and Ramírez[14] conducted a literature review of how negative emissions are defined in the literature and examined the importance of system boundary selection. Furthermore, they defined four key criteria as "minimum qualifications" for determining whether a technology results in negative emissions, namely:

1. Physical greenhouse gases are removed from the atmosphere.
2. The removed gases are stored out of the atmosphere in a manner intended to be permanent.
3. Upstream and downstream greenhouse gas emissions associated with the removal and storage process, such as biomass origin, energy use, gas fate, and co-product fate, are comprehensively estimated and included in the emission balance.
4. The total quantity of atmospheric greenhouse gases removed and permanently stored is greater than the total quantity of greenhouse gases emitted to the atmosphere.

In this chapter, we will attain to these criteria and explore some of the key challenges that they impose for carbon accounting, namely assessing permanence, accounting for the temporal distribution of emissions and removals, choosing time horizons, and conflating CO_2 avoidance and CO_2 negativity.

4.3 Storage Time: How Long is Long Enough?

Defining negative emissions requires for the CO_2 to be physically removed, and for the carbon to be stored in a manner intended to be permanent. But how long is long enough? Time is a relative concept and therefore, when answering these questions, it is important to remember that negative emissions systems are (to be) developed in the context of climate change, and therefore permanence also needs to be defined in the context of climate change. Permanence is defined as the length of time that the carbon will remain *stored* after having been fixed in, for instance, vegetation, soil, a geological reservoir, or a product. Here, two terms need to be distinguished that are used in the context of carbon permanence: carbon sequestration and carbon storage. *Carbon sequestration*, or uptake, refers to the active process of removing CO_2 from the atmosphere, while *carbon storage* refers to the process of keeping the CO_2 from the atmosphere.

Within carbon storage, we can distinguish between temporal and permanent storage. Temporal carbon storage is, by definition, reversible after a given period of time. This length of time is referred to as the accounting time horizon (TH). For practical purposes, it provides the time beyond which impacts caused by releasing the CO_2 back into the atmosphere are not considered. The time horizon juggles the benefits of temporal mitigation (*i.e.*, delaying radiative forcing which decreases the cumulative impact) with those of permanent actions.[19] The setting of finite THs is an implicit form of discounting for long-term impacts. It is, therefore, a value-loaded choice on the importance of present impacts in comparison to future impacts.

Brandão and Levasseur[20] relate this choice to two different worldviews, one that values sooner emission reduction options compared to later ones, *versus* one that requires managing carbon now as well as in the future without any distinction. The benefits and drawbacks of temporal storage can, in fact, be linked to these viewpoints. Those favouring sooner emission reduction often name environmental and economic arguments such as, buying time for technological progress and learning; postponing climate change; the potential of some temporary sequestration may become permanent over time; incentivising the transition to longer, circular products.[21–23] Those opposing temporal storage highlight the main argument that it 'pushes' impacts further into the future, ignoring long-term climate impacts affecting future generations, thereby violating the principle of inter-generational justice.[19]

4.3.1 Time Horizon and Lifetime of Carbon Storage in Products

Determining the time horizon (TH) plays a key role in determining the impact of temporal carbon storage. The first aspect to consider when addressing THs is the fact that there is not a direct correlation between the THs used in literature and the atmospheric lifetime of carbon dioxide. A single lifetime for CO_2 cannot be defined because of the different uptake rates by removal processes. The IPCC indicates that while more than 50% of the CO_2 emitted is currently removed from the atmosphere within a century, about 20% of emitted CO_2 stays in the atmosphere for many millennia. Due to the lack of a single lifetime, several time horizons are used in the literature to distinguish between temporal and permanent storage. For instance, according to the UNFCC[24] and the Publicly Available Specification (PAS) 2050,[25] carbon storage longer than 100 years can be considered permanent. In the Product Environmental Footprint (PEF) guidelines,[26] carbon storage is considered permanent if the carbon is only expected to be reemitted more than three centuries after its uptake; emissions expected to occur within three centuries (delayed emissions or temporary storage) are modelled as if emitted now. The International Reference Life Cycle Data (ILCD) handbook[27] differentiates temporary storage from permanent storage if guaranteed for over 10 000 years. THs included in life cycle assessments are 20, 50, 100 and 500 years, although 100 years is the most common TH used as this is often applied for calculating Global Warming Potential (GWP) impacts. As discussed previously, the choice of a TH is subjective and is driven by how long-term impacts are considered (or discounted). It is important to note here that while the impact of delaying an emission can be positive for a given metric (GWP, *i.e.*, cumulative radiative forcing over a given time frame), it can be negative for another metric (instantaneous temperature increase at a given time in the future). This is because there is no direct linear relation between the cumulative radiative forcing and physical endpoints in the system (*e.g.*, temperature change or sea-level rise). Currently, consensus is yet to be reached on which metric to use.

Carbon can be stored in natural and anthropogenic sinks. Natural sinks are the ocean, the forest, and soils; anthropogenic sinks are, for instance, products and geological reservoirs. The focus of this chapter is on anthropogenic sinks. For geological reservoirs, there is agreement that if the geological location is properly chosen, CO_2 could remain stored for thousands of years. For man-made products, the release time depends on the average lifetime of the product (including end-of-life). Table 4.1 shows an overview of the average lifetimes reported in the literature. Depending on the product, lifetimes can vary from weeks (short-lived chemicals or plastics) to centuries (*e.g.*, for some infrastructure). Below, each product category is briefly discussed.

Table 4.1 Overview of the average lifetime of products reported in the literature. SD is standard deviation.

Product category	Average lifetime [years]	Reference
Wood-based products	Household furniture: 30	28
	Sawn wood products: 35	29
	Wood-based panels: 25	
	Paper and paperboard: 2	
	Furniture: 13	30
Building materials: steel	Transportation: 20	31
	Machinery: 30	
	Construction: 75	
	Products: 15	
	Pipes and tubes: 19	30
	Fabricated metal products: 11	
	Gas and water pipelines: 30 (user) *vs.* 50 (designed lifetime) -China	32
Building materials: concrete (infrastructure)	Total life cycle: 100 (based on 40 years of service life and recycling span of 60 years after demolition)	33
	Total life cycle: 100 (based on 70 years of service life and recycling span of 30 years after demolition)	34
	Buildings: 39 to 73 (service life)	30
	Roads: 22 to 100 (service life)	
	Buildings: 50	35
	Residential buildings: 34 to 100 (depending on the country)	36
	Data for China.	32
	Buildings: 30 user time, 50 designed service lifetime	
	Highway and first-class roads: 10 user lifetime, 20 (designed)	
	Second-class roads: 8 (user) to 15 (designed)	
	Third-class and fourth-class roads: 3–10	
	Bridge: 80 years use life + 30 years after end of life	37
	Residential building: 50–60 (US); >100 (Europe); 30–40 (Japan); 25 (China)	38
	Buildings (Italy): 900 (Historical); 40 (buildings constructed after 1980)	39
Plastics	In:	40
	Packaging: 0.5 (SD: 0,1)	
	Transportation: 13 (SD: 3)	
	Building and construction 35 (SD: 7)	
	Electrical/electronic 8 (SD: 2)	
	Industrial Machinery: 20 (SD: 3)	
	Textiles: 5 (SD: 1,5)	
	Packaging: 0.6 (use phase)	41
	Consumer products: 3 (use phase)	
	Packaging: <1	42
	High impact polystyrene in electrical and electronic equipment: 10	
Fuels and chemicals	Fuels: from weeks to months	
	Automotive fuel (butanol): <1 year	35
	Textile coating agents: 10	42
	Urea: 1 week	43
	Chemical intermediates: <10 years	44
	Methanol: <1 (when used for short-lived plastics and fuels) and 20 (long-life insulation)	35

4.3.1.1 Plastics

As indicated in Table 4.1, the lifetime of plastic is dependent on its application, varying between very short times for packaging applications (weeks to months) to longer lifetime for plastic used in machinery and appliances (years to decades). Currently, plastic has four fates: recycling, thermal pyrolysis, containment in sanitary landfills, and left uncontained in open dumps. There is an increasing interest in plastic as a potential carbon sink either through storage of non-biodegradable plastic in landfills or through physically keeping the carbon stored in a value chain through circularity. The Joint Research Center,[45] for instance, published guidelines for conducting life cycle assessments of plastic articles. It indicates that if the material is not degraded within 100 years from deposition in a landfill, it can be assumed to undergo no further degradation and can be considered permanently stored in the landfill body. The report, however, cautions that a) the situation may be different for unmanaged landfills, and b) that over an infinite TH (instead of 100 years), additional material degradation and carbon mineralisation may take place, making the storage of CO_2 dependent on the TH selected, and on the *in situ* managing conditions of the landfill. Eriksson and Finvedden[46] assume in their study that plastic will degrade in a landfill by 3% during the first century due to chemical reactions, whereas 70% of the degraded carbon will form methane with a yield of 0.025 kg CH_4 per kg plastic. Part of this methane is captured in a (managed) landfill but part of it is emitted into the atmosphere.

One of the most attractive characteristics of plastics, from a carbon storage point of view, is the possibility for recycling which could, in principle, keep the carbon from being emitted back into the atmosphere. In theory, it is possible to design value chains that allow us to keep track of plastics, reuse them and recycle them (either physically or chemically). Doing so for many cycles can significantly extend the plastic's lifetime. There are, however, two main caveats that need to be considered. First, circularity requires much more than technological breakthroughs. Although technically possible, most plastics are not recycled today. Geyer *et al.*[40] presented a global first analysis of all mass-produced plastics. They focused on the most prevalent resins and fibres[†] and estimated that the global cumulative production in the period 1950–2015 was about 7800 Mt, of which 3900 Mt were produced in the period 2002–2015. By the end of 2015, all plastic waste ever generated from primary plastics had reached about 5800 Mt, 700 Mt of which were polyester, polyamide, and acrylic (PP&A) fibres. The authors calculated that historically 600 Mt have been recycled (corresponding to 9% of all plastic produced between 1950 and 2015) and that of this, only 10% has been recycled more than once. 60% of all plastics ever produced are

[†]The authors focused on the following plastics: high density polyethylene (PE), low-density and linear low-density PE, polypropylene (PP), polystyrene (PS), polyvinylchloride (PVC), polyethylene terephthalate (PET), and PUR resins; and polyester, polyamide, and acrylic (PP&A) fibres.

accumulating in landfills or the natural environment. Furthermore, they expect that if current global waste management trends stay as they are now by 2050, 9000 Mt of plastic waste will have been recycled (once), 12 000 Mt will be incinerated, and 12 000 Mt discarded in landfills or the natural environment. A similar picture was drawn by Vora *et al.*[41] Here, the authors indicate that only 10% of polyester worldwide is mechanically recycled more than once and that a large portion of the collected plastic waste in the United States (92%) and in Europe (69%) ends up in landfills or waste incineration plans. Crippa *et al.*[47] reports that in 2017, a total of 2580 kt of plastic were recycled in the European Union while 7920 Kt were incinerated and 3830 kt were landfilled. Using current trends, Vora *et al.*[41] expect a 44% recycling rate by 2050, which is slightly higher than the one reported by Geyer *et al.*[40] Although the environmental and economic benefits of plastic recycling have been recognised for several decades, there are factors inhibiting recycling at every state of the product life cycle. A study conducted by d'Ambrières[48] provides an overview of current plastic recycling worldwide. The study indicates that sustainable recycles only appear when the very large number of actors in the ecosystem, at every stage of the life cycle, are aligned. Furthermore, Vora *et al.*[41] point out that even in a case with a 100% recycling, a purely circular plastic processing facility cannot be achieved due to material and efficiency losses in the refinery requiring additional makeup material.

The second challenge is that circularity, as negative emissions, is a systems concept. The potential of using circular chains as CO_2 sinks can only be assessed from a systems point of view, that is, such potential needs to consider not only the carbon that is contained in the plastic, but also the carbon that is added in the processes of reaching full circularity. For instance, it has to not only account for the carbon emissions due to capturing and processing carbon from the atmosphere, but also account for CO_2 emissions from the fuels used when transporting the plastic product and the plastic waste, any carbon released when providing energy for mechanical or chemical recycling processes, the new carbon added as make-up, *etc.* Basically, one could consider the (negative) carbon stored in a circular chain as a credit that decreases over time if fossil-based carbon is released in each cycle to keep the circular chain ongoing. In this case, the credit can deplete over time and, as with any other temporal storage solutions, the circular plastic chain could then turn from sink to neutral to a net source of CO_2 to the atmosphere. Circularity also requires developing monitoring and verification mechanisms that keep custody of the CO_2 through the different uses and cycles. This kind of system is still to be developed and implemented.

4.3.1.2 Building Materials: Wood Products

Carbon stored in harvested wood products (HWPs) is increasingly being discussed as an option for climate change mitigation. In 2011, at the UNFCCC Conference in Durban, carbon storage in HWPs was recognised as a measure to mitigate climate change by agreeing on the primary carbon accounting and reporting rules.[49] Of particular significance is the potential benefit of replacing steel with wood products, given the large amount of carbon inherently stored in timber. The longer lifetime of HWP is in construction products (see Table 4.1). Note that the studies in Table 4.1 do not (explicitly) consider reusing or recycling in their assumptions. Churkina *et al.*,[50] for instance, indicate that while mineral-based materials (such as concrete or steel) have substantial embodied carbon emissions with minimal carbon storage capacities, timber stores a considerable quantity of carbon with a relatively small ratio of carbon emissions to material volume. According to the authors, the storage capacity of timber is about 0.48 tonne carbon per tonne timber, compared to 0.004 tonne carbon per tonne steel. The authors estimate that the construction of timber buildings for new urban dwellers could store 0.01–0.68 Gt C per year depending on the scenario. Johnston and Radeloff[29] estimate the global HWP pool as a net sink of about 335 Mt $CO_{2,eq}$ per year in 2015. Although significant, the authors note that this amount of carbon stored annually in HWPs is less than 1% of global emissions and warn that economic shock can easily turn a HWP pool from a carbon sink into a carbon source. The latest is an inherent characteristic of temporal storage, as at some point in time the carbon will be released back into the atmosphere. A similar argument is made by Pompini *et al.*[51] when indicating that if forests are not sustainably managed, then timber-based construction is an inefficient form of displacing carbon from forests to cities that minimises the carbon-storage potential of forests. Cardellini *et al.*[52] found that the benefits of HWPs is dependent on the LCA methodology used. The authors indicated that whilst using a static LCA accounting approach results in a positive climate substitution effect, this is not necessarily the case when using dynamic approaches, which are influenced by the way the analysis is performed. Peñaloza *et al.*[37] highlights the need for extending the lifetime of biobased alternatives, as biobased products usually have shorter service lives than non-biobased alternatives, and thus must be replaced more frequently. However, as with other products, the reduction in climate change that was achieved by extending the service life was less pronounced with a long-term time horizon.

4.3.1.3 Building Materials: Steel and Concrete

Steel and concrete are two important products for negative emission scenarios because of their significant contribution to global carbon emissions and the long carbon storage time in the products. In the case of steel, carbon is incorporated in the blast furnace by adding coking coal to the molten iron. The amount of carbon in steel does not change over time. Fresh cement, on the other hand, contains no carbon, but this changes over time because calcium oxide in cement materials is not stable.[50] Atmospheric CO_2 reacts with cement hydrated products in the presence of water, slowly increasing the amount of carbon in cement. In both cases, after the lifetime of the product, the carbon stays in the product (or waste). The carbon stored can therefore be considered permanent regardless of the lifetime of the product. For the purposes of carbon storage, however, it is important to note that the carbon storage capacity of steel and cement is low (0.004 tonne carbon per tonne steel and up to 0.12 tonne carbon per tonne cement[50]).

Besides their long lifetime, steel and concrete are interesting for negative emissions because their potential as storage sinks could be enhanced through mineral carbonation. In the case of steel, the amount of carbon stored can be increased by carbonation of hydraulic steel slags.[‡] In this process, the minerals in slags react with CO_2 at elevated/moderated pressure and temperature resulting in carbonated products which are thermodynamically stable. In Europe, 76% of the steel slags are already recycled as aggregates in construction,[53] making carbonation particularly interesting for steel slags that are either landfilled or used for low-value application. Pullin *et al.*[54] estimate the CO_2 capture potential of direct carbonation of steel slags at 296–337 kg of CO_2 per tonne. The authors also studied passive carbonation of slags[§] (*i.e.*, enhanced weathering), and although it has a large theoretical potential (422 to 584 kg CO_2 per tonne product), the authors found that this potential is not reached in practice as the mineralogy, particle size and emplacement conditions of the slag can be limiting factors.

Mineral carbonation also plays a key role for CO_2 storage in concrete. A main difference with steel is that the amount of CO_2 sequestered by concrete occurs gradually throughout its service lifetime and increases after the lifetime of the building has been reached. This is referred to as the sponge effect.[55] After end of life (*e.g.*, demolition), if the concrete is exposed to air, the contact surface between the concrete and atmospheric CO_2 increases, then so does the carbonation percentage. Although all the CO_2 released from limestone during calcination could be theoretically bound back, in reality, less than half of the CO_2 is bound during the structure's lifetime and subsequent demolition since the carbonation rate is low,[56] and less than 1% of demolished concrete is reported to be recycled in an exposed environment.[55] Nonetheless, Cao *et al.*[55] estimated the cumulative uptake of CO_2 in the global cement cycle from 2015 to 2100 to sum up to 116 Gt CO_2. Interestingly, the authors point out that the net CO_2 emission balance in the cement cycle is a nonlinear amortisation function over a specific timeframe, and therefore despite the long lifetime of buildings, the choice of TH for evaluating decarbonisation strategies in this system is of high relevance. This point was also highlighted in Tanzer and Ramírez (2021).[57] In this study, net negativity was only reached after end-of life by a combination of accelerated carbonation (injecting captured CO_2 from a bio-source or the atmosphere) and natural carbonation (natural contact of concrete with CO_2 in the atmosphere).

‡Stainless steel (SS) slag and basic oxygen furnace (BOF) slags.

§Passive carbonation happens at atmospheric pressure and temperatures by natural contact of the slag with atmospheric CO_2.

4.3.1.4 *Chemicals and Fuels*

The expected lifetime reported for chemicals and fuels is significantly shorter than for other products. Although there are limited studies on the lifespan of chemicals, most of these studies focus on plastics, which we have treated in this chapter as a separate category. Other chemicals often studied are fuels and urea but the average lifetime of a fuel (from the refinery to end-of-life) is in the order of months and in the case of urea, CO_2 is released within a week of application to the soil. These are very limited storage times, and therefore not relevant for the discussion on permanence.

The lifetime of (other) chemicals is dependent on the product and the application. It is, however, difficult to assess the carbon stock in chemical products, particularly for longer value chain chemicals. Although interest in this topic has increased as chemical companies are trying to quantity scope 3 CO_2 emissions,[¶] the amount of public information available to date is very limited. In terms of CO_2 permanence, the most common approach in LCA of chemical products is to assume that all carbon contained in a product will be eventually emitted to the atmosphere as CO_2, and any benefits (or impacts) from temporal storage are not included in the assessment. This approach is driven by the limited lifetime of most chemical products, lack of data, and the uncertainties around verifying that the carbon is indeed stored. There is some discussion in the literature on the potential of circularity to extend the lifetime of chemicals, but a real understanding of the (theoretical) potential of this option would require first a systematic quantification of its potential. Note as well that the caveats named for circularity of plastics also apply here.

4.4 The Timing of Emissions and Removals

As discussed previously, the timing of when emissions are taken from the atmosphere, are stored and when the system results in negative emissions may not be the same, which is especially the case for CDR solutions based on natural systems (*e.g.*, bioenergy with carbon capture and storage) or mineral carbonation (*e.g.*, concrete weathering). Challenges with accounting for a synchronised carbon emissions and removals has extensively been discussed in the field of bioenergy (see *e.g.*, Lohmann[58]). A term commonly used in bioenergy is the time to carbon debt repayment, which has been defined as the time between biomass harvest and when overall carbon balance of bioenergy use offsets the loss of carbon stored in biomass at the time of harvest.[59] Leemans *et al.*[60] was one of the first papers describing the transition period where increased deployment of (forest) bioenergy increases CO_2 emissions to the atmosphere followed by an extended period with reduced emissions. Lamers and Junginger[61] show how that the carbon payback time

[¶]That is the total lifetime of sold products including waste disposal, this includes downstream transportation and distribution, processing of sold products, use of sold products, end of life treatments, downstream leased assets, franchises and investments.

of apparently comparable forest bioenergy scenarios can vary by up to 200 years. In Tanzer and Ramírez,[57] the plausibility of reaching negativity in the cement industry was explored by including different levels of bioenergy, carbon capture and storage as well as natural and induced carbonation during product use and after end-of-life. As both biomass and concrete reuptake CO_2 over time, the timing of CO_2 emissions and removals was explicitly accounted for. In their results, they report that, for the cases where bioenergy was used, higher net CO_2 emissions were shown than their fossil counterparts for 15–25 years after concrete production, until sufficient atmospheric CO_2 was reabsorbed by sustainably replanted biomass. Their results also indicate that the net CO_2 footprint did not decrease below that of fossil energy until approximately halfway through the biomass rotation and furthermore, that carbon negativity was achieved after 40 years or more after the biomass was used in the kiln and the resulting CO_2 emissions were stored underground.

Currently, it is possible to find in the literature studies showing that the climate impact of forest bioenergy is instantaneous as well as studies showing that it will take more than 1000 years before it can contribute to climate mitigation.[62] These large differences are common causes of discussion, both in academia and policy making. Bentsen[62] concluded that the outcome of carbon debt studies lies in the assumptions, and methodological rather than ecosystem and management related assumptions determine the findings. Similar challenges can be expected for negative emissions systems, particularly to those based on natural (*e.g.*, ocean, afforestation) or hybrid solutions (bioenergy with carbon capture and storage) and will, alike to the debate on the benefits of bioenergy, create confusion and dispute about their potential climate benefit.

4.4.1 Delayed Emissions

The timing issue as discussed above is related to the fact that, in a counterfactual where the biomass would have not been used, carbon uptake by the biomass would have still taken place. A different aspect of timing is related to the temporal permanence of carbon (as discussed in Section 4.3) and whether there are climate benefits from it. In this chapter, we advise considering the TH to be as broad as possible. Nonetheless, as the debate has not reached consensus on the benefits (or disadvantages) of temporal storage, we present here the main approaches currently being used to estimate the climate impact of delayed emissions. Those methods can basically be categorised into (a) methods derived from static life cycle assessment under a particular time horizon and (b) methods based on dynamic life cycle assessment. It is important to stress that, to date, no agreement has been reached in the scientific community regarding the correct approach to estimate the benefits or impacts of CO_2 temporal storage. For a complete overview, we refer to Brandão *et al.*[63]

4.4.1.1 Tonne-year Approaches

Tonne-year accounting approaches were introduced in the special report on Land Use, Land-Use Change and Forestry by the IPCC,[64] driven by the need for assessing the role of forestry and wood products in climate mitigation. They are based on estimating the cumulative radiate forcing over a period of time.[19] In these approaches, a time horizon (TH) is defined, which allows comparison of the area under two carbon emission pulse curves,[‖] one for the baseline scenario and one for the scenario being studied. There are several ways of calculating tonne-years. Below, three common methods are briefly discussed.

- *Moura-Costa Method:* is based on the concept of absolute global warming potential (AGWP), which is defined as the cumulative radiative forcing potential of a unit of CO_2 over a defined TH.[65] The method calculates tonne-years by assuming that the carbon stored is fixed. The impact is estimated by considering the number of years the CO_2 is stored out of the atmosphere and converting it in terms of impact on global warming. It does so by estimating the cumulative radiative forcing caused by the pulse emission of 1 t CO_2 integrated over a given period (see Figure 4.1). In this approach, an equivalent time is calculated as the point at which the area of the rectangle in Figure 4.1 is equal to the area under the atmospheric C decay curve over the TH, corresponding to 100 years in the figure.[64] Note that each year in the TH is accounted equally (*i.e.*, the impact of storage in year 1 is the same as that in any other year) by assuming a linear relation between the residence of CO_2 in the atmosphere

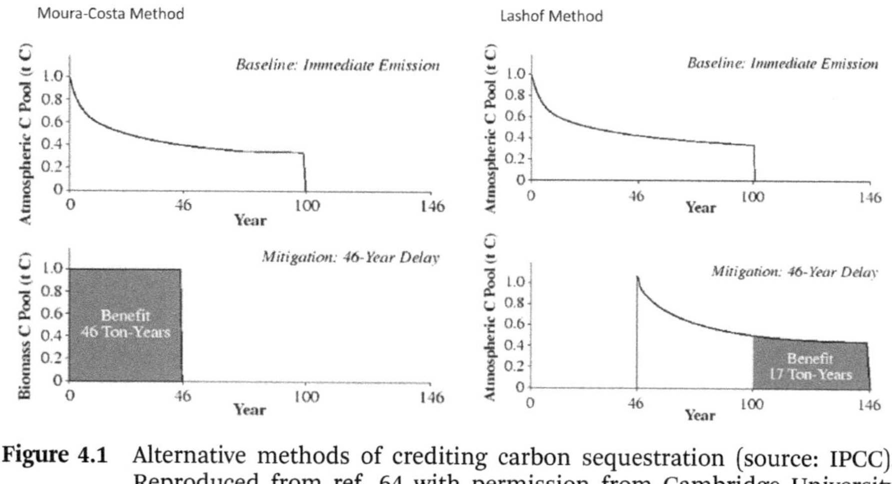

Figure 4.1 Alternative methods of crediting carbon sequestration (source: IPCC). Reproduced from ref. 64 with permission from Cambridge University Press, Copyright 2000.

‖A pulse is an instantaneous emission of CO_2.

and its radiative forcing effect over the TH. Storing one tonne CO_2 for 52.4 years is therefore equivalent to *avoiding* the impact of one tonne CO_2 for 100 years. This results in an equivalence factor of 0.019 (= 1/52.4), which can be used to quantify the benefits of temporal storage.[63]

- *Lashof Method*: calculates tonne-years by assuming that carbon in the atmosphere is removed by natural processes. It considers carbon storage as a *delayed* emission.[66] Based on this method, storing carbon for x number of years is equivalent to delaying a CO_2 emission until year x. The pulse emission then happens from year x onwards (in Figure 4.1, x is 46 years). This approach, therefore, pushes part of the impact beyond the end of the TH. The portion that is outside the TH represents the benefit of the carbon storage (*i.e.*, the area of the tail of the second curve that goes over the TH is equivalent to decreasing the TH by a period equal to the time for which the release is delayed) – corresponds to 17 years in Figure 4.1.
- *PAS 2050-method*: is a simplified method developed by the Publicly Available Specification (PAS) 2050 : 2008 to quantify the carbon storage. The method discounts carbon storage from the total emissions for the life of the product through a discount coefficient. A weighting factor (FW) is applied to the GHG emissions, aiming to reflect the number of years between the formation of the product and the single release of emissions during a 100-year quantification period. FW is calculated using eqn (4.1), where i is each year in which emission occur and x is the proportion of total emissions occurring in any year i. FW reflects the weighted average time that the emission is present in the atmosphere during the TH (100 years).

$$\mathrm{FW} = \frac{\sum_{i=1}^{100} x_i^{*}(100 - i)}{100} \tag{4.1}$$

4.4.1.2 Dynamic Life Cycle Assessment

This method was developed by Levasseur *et al.*[19] driven by the need to better understand the role of forest as a climate mitigation option. Contrary to tonne-year approaches, where the characterisation factors of gasses for global warming impact are considered constant over time, here dynamic characterisation factors (DCF) are used, *i.e.*, AGWP for every time step which are calculated using eqn (4.2). The method requires to (a) select a time zero (the time when the first emission occurs), and (b) divide the TH into one-year time steps for which the impact on radiative forcing is calculated. The cumulative impact is then calculated at a given time t by summing the instantaneous impact on radiative forcing since time zero, eqn (4.3). This approach combines instantaneous and cumulative impacts on the GWP category within a defined TH. Note that the impact of emissions after the TH are not considered. Finally, the results can be expressed relative to the equivalent value for 1 kg CO_2 emitted at time zero which allows comparison of the results with those from static life cycle assessments.

$$\mathrm{DCF}(t) = \int_{t-1}^{t} a_i[C_i(t)]\mathrm{d}t \tag{4.2}$$

$$\mathrm{GWI}(t) = \sum_i \mathrm{GWI}(t) = \sum_i \sum_{j-0}^{t} [g_i]_j{}^*[\mathrm{DCF}_i]_{t-j} \tag{4.3}$$

where, i is the GHG gas; a is the instantaneous radiative forcing per unit mass of GHG in the atmosphere; $C(t)$ is the time-dependent GHG load following a unit mass pulse-emission; g_i is the inventory result for GHG_i at time j.

Although there is increasing consensus of the benefits of using dynamic life cycle approaches for carbon accounting, there are several challenges that hinder their use, including: complexity of dynamic LCAs, limited availability of life cycle inventory data that locate and differentiate activities and flows in time, lack of accessible and transparent software, and lack of knowledge about the methodology by practitioners.[67]

4.5 CO_2 Storage and CO_2 Avoidance

As discussed in Section 4.1, a key condition of negative emissions is that it should result in a *physical* removal of CO_2 emissions from the atmosphere. This seems an obvious condition but a common problem in literature is the conflation of negative emissions and avoided emissions, which can result in assigning negativity to virtual carbon pools (non-physical removals). This point was broadly discussed in Tanzer and Ramírez.[14] To illustrate the problem for calculating negative emission, four simplified examples are listed below, for the purpose of simplicity, all examples consider capturing CO_2 from the atmosphere *via* direct air capture (DAC).

- CO_2 is captured from the atmosphere and stored in a geological formation. Energy is used for the extraction of CO_2, for its compression and for its storage. CO_2 storage is the only service provided by this system. The net CO_2 stored is the difference between the CO_2 captured from the atmosphere minus the CO_2 emitted during the capture, transport and geological injection (*e.g.*, due to carbon intensity of the energy used). In this case the net CO_2 stored is equal to the CO_2 avoided by the system.
- CO_2 is captured from the atmosphere and used as feedstock in a process (together with other feedstocks, *e.g.*, hydrogen) to produce a fuel such as methanol. There is no carbon storage in this option as discussed in Section 4.3. The climate benefit of this option is estimated by assuming that the CO_2-based fuel (*i.e.*, synthetic fuel) *substitutes* a fossil-based carbon intensive fuel. It can therefore be argued that

consequently, emissions will not be produced that would have been emitted otherwise, *i.e.*, avoided CO_2 emissions. In the calculation of net CO_2 emissions, many studies subtract the avoided emissions of the net carbon balance of the technology. In other words, they mix a physical burden with a non-physical (virtual) burden. Mathematically this can result in a negative number that is often wrongly interpreted as negative emissions, especially by non-LCA practitioners.

- CO_2 is captured from the atmosphere and used for enhanced weathering of concrete. The resulting concrete has more strength** than the conventional concrete. The CO_2-concrete then gets credits for permanently storing CO_2 in the product and for using less concrete (assuming that the CO_2-concrete is stronger and therefore a lower amount is required for the same service). Like the case above, the net CO_2 emissions are often calculated by adding the avoided emissions from requiring less carbon intensive materials to the net emissions stored in the concrete. The result therefore mixes physical flows of carbon with virtual flows.
- CO_2 is captured from the atmosphere and used as a feedstock for chemical production. As part of the process, two products are generated: the main (intended) product A and a by-product B. In the reference system, the two products are produced in separated processes (A′ and B′), each having their own value chain and emissions. In the new alternative situation, one process produces both product A and product B, and it can be assumed that product B′ is substituted and the avoided emissions are allocated to product A. If the reference way of producing B′ is more carbon intensive than producing B′ *via* the CO_2-based route, this can easily lead mathematically to a negative value for the life cycle results.

Although it is not the focus of this chapter to address in-depth substitution in life cycle accounting, we will briefly discuss here some key challenges which are impacting carbon accounting of negative emission systems. It is important to note that although substitution is a valid method within LCA, it remains controversial, mostly due to three assumptions. First, a common approach when estimating the impacts from substituting a reference product is that it is often assumed that if the product is technically equivalent to the reference product (*i.e.*, it has the same performance/provides the same service), substitution happens on a 100% basis (one-to-one substitution), which is not necessarily the case. Zink *et al.*[68] indicate, for instance, that full (complete) displacement requires special conditions, and it is also likely that low or even zero displacement would occur. In this case, a CO_2-based product can be added to the market without displacing any of the existing fossil-based product. Valdenbo *et al.*[69] conclude that whether a

**Please note that this is case, formulation and application specific.

product would be displaced or not is not necessarily driven by physical or engineering relationships but rather by the dynamics of demand and supply, which are generally omitted in carbon accounting calculations (as the study is done in a function unit basis, *e.g.*, per tonne of product). A similar finding is reported by Arehart *et al.*[67] and Howard *et al.*[70] In both studies, the authors looked at the way the benefits of substituting traditional fossil-based products in construction are calculated. They indicate that the substitutability is not only influenced by the technical properties of the novel products (compared to, *e.g.*, traditional building materials), but also depends on the price of the product, acceptance of building materials and codes, regulations of new codes, and knowledge among stakeholders. Assuming 100% substitution is only one possible scenario, and therefore carbon accounting needs to examine the impacts under several substitutability scenarios.

Second, the benefits of substitution (*i.e.*, displacing fossil carbon-based product resulting in avoided emissions) are due to the existing economic structure that is dominated by fossil carbon intensive products. As we transition to an economy less reliant on fossil-based products, the fossil footprint of products will decrease in the future and the benefits of substitution will diminish.[71] Many carbon accounting exercises are not dynamic, and therefore assume constant substitution factors, thereby likely overestimating the potential benefits from substitution. A final caveat on substitution is its permanence, *i.e.*, how permanent are the substitution effects? While some authors argue that the substitution benefits are permanent, others point out that there is no evidence showing this has (or will) be the case.[67]

As discussed in Section 4.1, negativity implies that the net physical removal of CO_2 from the atmosphere is greater than the physical emissions. However, a consequence of substitution in carbon accounting is that by including non-physical emissions (virtual flows) the final value can mathematically be negative even when the physical *removals* are less than the physical *emissions*. While avoiding emissions is at the basis of most mitigation measures by incentivising the (future) displacement of carbon intensive products, avoided emissions are less relevant for estimating negative emissions. Negative emissions centre around emissions that are physically withdrawn from the atmosphere. Avoided emissions centre around the relative change in emissions because of the deployment of a product or process. A product that stores CO_2 can result in both negative emissions and avoided emissions, but it can also provide one without the other. It is therefore important to avoid any conflation in terms (or values) and report those separately.

4.6 Some Final Words

Carbon accounting appears as a simple task, how difficult could it be? Compared to, for instance, complex challenges such as developing new carbon capture technologies or monitoring carbon flows in underground storage locations, carbon accounting appears indeed deceitfully straightforward. The current confusion on terminology and on the implications of assumptions that drive carbon accounting practices (time horizons, methods to assess climate benefits or impacts from temporal storage, mixing real and virtual mass flows) are already polarising the public debate, demonstrating that accounting negative emissions is not as straightforward as it seems to be.

The CDR community needs to further build on the lessons gained from the public and academic debates on carbon sinks and bioenergy. Carton *et al.*[72] make a compelling case for avoiding a reductionist approach to developing negative emissions. While the authors recognised that the scale at which negative emissions are expected to be needed surpasses past proposals on carbon sinks, failing to learn from past experiences and rediscovering the obstacles to overcome will mean deployment and incentivisation of carbon removal will not only cost money but also valuable time. Time that society does not have given the high stakes and the speed at which mitigation targets need to be achieved.

In terms of carbon accounting, a first simple but effective step is to always consider several scenarios (*e.g.*, include time differences in carbon assumptions, as well as THs) when conducting assessments, and to report all assumptions and results transparently and consistently to decision makers, so that they can have a better understanding of the implications of the choices they make. Second, international standardised guidelines on thresholds for permanent carbon storage need to be agreed upon. Such guidelines need to be open and explicit on the implications of choosing a given time horizon in terms of trade-offs between incentivising/accelerating mitigation options in the near-term *vs* shifting the consequences of pulse emissions further in time. Furthermore, a better understanding of what circularity can or not mean for negativity is urgently needed as the two concepts are increasingly and too casually being intertwined in policy and industry arenas.

Given the increasing importance of negative emissions for reaching ambitious climate targets, the importance of properly defining and accounting emissions should not be underestimated as it is crucial for incentivizing, developing, monitoring, and verifying negative emissions so that it can collectively meet the demands of policymakers and society.

References

1. IPCC, Summary for Policymakers, in *Global Warming of 1.5 °C. An IPCC Special Report on the Impacts of Global Warming of 1.5 °C Above Pre-industrial Levels and Related Global Greenhouse Gas Emission Pathways, in the Context of Strengthening the Global Response to the Threat of Climate Change, Sustainable Development, and Efforts to Eradicate Poverty*, ed. V. Masson-Delmotte, P. Zhai, H. O. Pörtner, D. Roberts, J. Skea, P. R. Shukla, A. Pirani, W. Moufouma-Okia, C. Péan, R. Pidcock,

S. Connors, J. B. R. Matthews, Y. Chen, X. Zhou, M. I. Gomis, E. Lonnoy, T. Maycock, M. Tignor and T. Waterfield, *World Meteorological Organization, Geneva, Switzerland*, 2018, 32 pp.
2. UNFCC, Commitments to net zero double in less than a year, Press release, September 21, 2020, available at https://unfccc.int/news/commitments-to-net-zero-double-in-less-than-a-year.
3. New climate, https://newclimate.org/ambitiousactions, (accessed May 2021).
4. H. A. Daggash, C. F. Heuberger and N. Mac Dowell, The role and value of negative emissions technologies in decarbonising the UK energy system, *Int. J. Greenhouse Gas Control*, 2019, **81**, 181.
5. W. Rickels, A. Proelß, O. Geden, J. Burhenne and M. Fridahl, Integrating carbon dioxide removal into European emissions trading, *Front. Clim.*, 2021, **3**, 690023.
6. Ministry of Economy, Trade and Industry, METI, https://www.meti.go.jp/english/policy/energy_environment/global_warming/roadmap/innovation/jctgi.html (accessed May 2021).
7. Microsoft, https://blogs.microsoft.com/blog/2020/01/16/microsoft-will-be-carbon-negative-by-2030/, (accessed June 2021).
8. Xprize, https://www.xprize.org/prizes/elonmusk (accessed May 2021).
9. Amazon, https://www.aboutamazon.com/news/sustainability/the-climate-pledge-celebrates-surpassing-100-signatories, (accessed July 2021).
10. J. Bednar, M. Obersteiner, A. Baklanov, M. Thomson, F. Wagner, O. Geden, M. Allen and J. Hall, Operationalizing the net-negative carbon economy, *Nature*, 2021, **596**, 377–383.
11. N. Höhne, M. den Elzen, J. Rogeji, B. Metz, T. Fransen, T. Kuramochi, A. Olhoff, J. Alcamo, H. Winkler, S. Fuzz, M. Schaeffer, R. Scheffer, G. Peters, S. Maxwell and N. I. Dubash, Emissions: world has four times the work or one-third of the time, *Nature*, 2020, **579**, 25.
12. E. van Soest, M. den Elzen and D. van Vuuren, Net-zero emission targets for major emitting countries consistent with the Paris Agreement, *Nat. Commun.*, 2021, **12**, 2140.
13. M. Obersteiner, J. Bednar, F. Wagner, T. Gasser, P. Ciais, N. Forsell, S. Frank, P. Havlik, H. Valin, I. A. Janssens, J. Peñuelas and G. Schmidt-Traub, How to spend a dwindling greenhouse gas budget, *Nat. Clim. Change*, 2018, **8**, 7.
14. S. Tanzer and A. Ramirez, When are negative emissions negative emissions?, *Energy Environ. Sci.*, 2019, **12**, 1210.
15. M. Fajardy, P. Patrizio, H. Daggash and N. Mac Dowell, Negative emissions: priorities for research and policy design, *Front. Clim.*, 2019, **1**, 6.
16. K. Stechemesser and E. Guenthe, Carbon accounting: a systematic literature review, *J. Cleaner Prod.*, 2012, **36**, 17.
17. C. Gulluscio, P. Puntillo, V. Luciani and D. Huisingh, Climate Change Accounting and Reporting: A Systematic Literature Review, *Sustainability*, 2020, **12**, 5455.
18. S. Schaltegger and M. Csutora, Carbon accounting for sustainability and management. Status quo and challenges, *J. Cleaner Prod.*, 2012, **36**, 1.

19. A. Levasseur, M. Brandão, P. Lesage, M. Margni, D. Pennington, R. Cliff and R. Samson, Valuing temporary carbon storage, *Nat. Clim. Change*, 2012, **2**, 6.
20. M. Brandão and A. Levasseur. *Assessing Temporary Carbon Storage in Life Cycle Assessment and Carbon Footprinting: Outcomes of an Expert Workshop*, Publications Office of the European Union, Luxembourg, 2011.
21. G. Guest, F. Cherubini and A. Strømman, Global warming potential of carbon dioxide emissions from biomass stored in the anthroposphere and use for bioenergy at end of life, *J. Ind. Ecol.*, 2012, **17**, 20.
22. H. Herzog, K. Caldeira and J. Reilly, An issue of permanence: assessing the effectiveness of temporary carbon storage, *Clim. Change*, 2003, **59**(3), 293.
23. G. Marland, K. Fruit and R. Sedjo, Accounting for sequestered carbon: the question of permanence, *Environ. Sci. Policy*, 2001, **4**, 259.
24. UNFCCC. *United Nations Framework Convention on Climate Change*, UNEP/IUC, Switzerland, 1992, p. 30.
25. BSI, The guide to PAS 2050:11, How to carbon footprint your products, identify hotspots and reduce emissions in your supply chain, 2012.
26. European Commission, PEFCR Guidance document - Guidance for the development of Product Environmental Footprint Category Rules (PEFCRs), Version 6.3, Publications Office of the European Union, 2018.
27. European Commission International Reference Life Cycle Data System (ILCD), Handbook - General guide for Life Cycle Assessment - Detailed guidance, First edition, EUR 24708 EN, Luxembourg, Publications Office of the European Union, 2010.
28. J. E. Smith, L. S. Heath, K. E. Skog and R. A. Birdsey, *Methods for Calculating Forest Ecosystem and Harvested Carbon with Standard Estimates for Forest Types of the United States. General Technical Report NE-343. Newtown Square, Pennsylvania: U.S. Department of Agriculture, Forest Service, Northeastern Research Station*, Newtown Square, Pennsylvania, 2006, p. 216.
29. C. Johnston and V. Radeloff, Global mitigation potential of carbon stored in harvested wood products, *PNAS*, 2019, **115**, 14526.
30. S. Murakami, M. Oguchi, T. Tasaki, I. Daigo and S. Hashimoto, Lifespam of commodities, part I. The creation of a database and its review, *J. Indus. Ecol.*, 2010, **4**(4), 598.
31. S. Pauliuk, R. L. Milford, D. B. Müller and J. M. Allwood, The steel scrap age, *Environ. Sci. Technol.*, 2013, **47**, 3448.
32. C. Huang, J. Han and W. Chen, Changing patterns and determinants of infrastructures' material stocks in Chinese cities, *Resour., Conserv. Recycl.*, 2016, **123**, 47.
33. K.-H. Yang, E.-A. Seo and S.-H. Tae, Carbonation and CO_2 uptake of concrete, *Environ. Impact Assess. Rev.*, 2014, **46**, 43.
34. I. Galan, C. Andrade, P. Mora and M. San Juan, Sequestration of CO_2 by concrete carbonation, *Environ. Sci. Technol.*, 2010, **44**, 3181.
35. D. Peñaloza, F. Røyne, G. Sandin, M. Svanström and M. Erlandsson, The influence of system boundaries and baseline in climate impact assessment of forest products, *Int. J. Life Cycle Assess.*, 2019, **24**, 160.

36. S. Deetman, S. Marinova, E. van der Voet, D. van Vuuren, O. Edelenbosch and R. Heijungs, Modelling global material stocks and flows for residential and service sector buildings towards 2050, *J. Cleaner Prod.*, 2020, **245**, 118658.
37. D. Peñaloza, M. Erlandsson and A. Pousette, Climate impacts from road bridges: effects of introducing concrete carbonation and biogenic carbon storage in wood, *Struct. Infrastruct. Eng.*, 2018, **14**, 56.
38. E. G. Hertwich, S. Ali, L. Ciacci, T. Fishman, N. Heeren, E. Masanet, F. Nojavan Asghari, E. Olivetti, S. Pauliuk, Q. Tu and P. Wolfram, *Environ. Res. Lett.*, 2019, **14**, 043004.
39. A. Miatto, H. Schandl, L. Forlin, F. Ronzani, P. Borin, A. Giordano and H. Tanikawa, A spatial analysis of material stock accumulation and demolition waste potential of buildings: A case study of Padua, *Resour., Conserv. Recycl.*, 2018, **142**, 245.
40. R. Geyer, J. R. Jambeck and K. L. Law, Production, use and fate of all plastics ever made, *Sci. Adv.*, 2017, **3**, e1700782.
41. N. Vora, P. R. Christensen, J. Demarteau, N. R. Baral, J. D. Keasling, B. A. Helms and C. D. Scown, Leveling the cost and carbon footprint of circular polymers that are chemically recycled to monomer, *Sci. Adv.*, 2021, 7, eabf0187.
42. L. Li, Modeling the fate of chemicals in products, Springer Theses 2020, Singapore, 10.1007/978-981-15-0579-9.
43. J. Tierling, CO_2 emission after urea application. Lan trial at Yara research center, Hanninghof (unpublished data), YARA, 2016.
44. IEA, *Putting CO_2 to use. Creating value from emissions*, International Energy Agency, Paris, 2019.
45. Joint Research Center, Comparative life cycle assessment of alternative feedstock for plastic production, Joint Research Center, European Union, Brussels, 2020.
46. O. Eriksson and G. Finnveden, Plastic waste as a fuel- CO_2 -neutral or not?, *Energy Environ. Sci.*, 2009, **2**, 907.
47. M. Crippa, B. De Wilde, R. Koopmans, J. Leyssens, J. Muncke, A. C. Ritschkoff, K. Van Doorsselaer, C. Velis and M. A. Wagner. Circular economy for plastics – Insights from research and innovation to inform policy and funding decisions, ed. M. De Smet and M. Linder, European Commission, Brussels, 2019.
48. W. d'Ambrières, Plastics recycling worldwide: current overview and desirable changes, Field Actions Science Reports, The journal of field actions 2019, Special Issue 19, 12.
49. United Nations, Land use, land-use change and forestry. Report of the Conference of the Parties serving as the meeting of the Parties to the Kyoto Protocol on its seventh session, Durban, South Africa: Decision 2/CMP.7, 11–21, 2012.
50. G. Churkina, A. Organschi, C. P. O. Reyer, A. Ruff, K. Vinke, Z. Liu, B. K. Reck, T. E. Graedel and H. J. Schellnhuber, Buildings as a global carbon sink, *Nat. Sustainability*, 2020, **3**, 269.

51. F. Pompini, J. Hart, J. Arehart and B. D'amico, Buildings as a global carbon sink? A reality check on feasibility limits, *One Earth*, 2020, **3**, 157–161.
52. G. Cardellini, C. Mutel, E. Vial and B. Muys, Temporalis, a generic method and tool for dynamic life cycle assessment, *Sci. Total Environ*, 2018, **645**, 585.
53. M. Quaghebeur, P. Nielsen, L. Horckmans and D. van Mechelen, Accelerated carbonation of steel slag compacts: development of high-strength construction materials, *Front. Energy Res.*, 2015, **3**, 52.
54. H. Pullin, A. W. Bray, I. T. Burke, D. D. Muir, D. J. Sapsford, W. M. Mayes and P. Renforth, Atmospheric carbon capture performance of legacy iron and steel waste, *Environ. Sci. Technol.*, 2019, **53**, 9502.
55. Z. Cao, R. J. Myers, R. C. Lupton, H. Duan, R. Sacchi, N. Zhou, T. R. Miller, J. M. Cullen, Q. Ge and G. Liu, The sponge effect and carbon emission mitigation potentials of the global cement cycle, *Nat. Commun.*, 2020, **11**, 3777.
56. J. Skocek, M. Zajac and M. Ben Haha, Carbon Capture and Utilization by mineralization of cement pastes derived from recycled concrete, *Sci. Rep.*, 2020, **10**, 5614.
57. S. Tanzer, K. Blok and A. Ramirez, Can bioenergy with carbon capture and storage result in carbon negative carbon steel?, *Int. J. Greenhouse Gas Control*, 2020, **100**, 10314.
58. L. Lohmann, Toward a different debate in environmental accounting: The cases of carbon and cost–benefit, *Account., Organ. Soc.*, 2009, **34**(3–4), 499.
59. S. R. Mitchell, M. E. Harmon and K. E. O'Connell, Carbon debt and carbon sequestration parity in forest bioenergy production, *GCB Bioenergy*, 2012, **4**(6), 818.
60. R. Leemans, A. van Amstel, C. Battjes, E. Kreileman and S. Toet, The land cover and carbon cycle consequences of large-scale utilizations of biomass as an energy source, *Global Environ. Change*, 1996, **6**, 335.
61. P. Lamers and M. Junginger, The 'debt' is in the detail: a synthesis of recent temporal forest carbon analyses on woody biomass for energy, *Biofuels, Bioprod. Biorefin.*, 2013, **7**, 373.
62. N. Bentsen, Carbon debt and payback time- lost in the forest?, *Renewable Sustainable Energy Rev.*, 2017, **13**, 1211.
63. M. Brandão, A. Levasseur, M. Kirschbaum, B. Weidema, A. Cowie, S. Jorgensen, M. Hauschild, D. Pennington and K. Chomkhamsri, Key issues and options in accounting for carbon sequestration and temporary storage in life cycle assessment and carbon footprinting, *Int. J. Life Cycle Assess.*, 2013, **18**, 230.
64. IPCC, *Land Use, Land-use Change and Forestry*, ed. R. T. Watson, I. R. Noble, B. Bolin, N. H. Ravindranath, D. J. Verardo and D. J. Dokken, Cambridge University Press, UK. Cambridge, 2000, p. 375.
65. P. Moura Costa and C. Wilson, An equivalence factor between CO_2 avoided emissions and sequestration-description and application in forestry, *Mitigation Adapt. Strategies Global Change*, 2000, **5**, 51.

66. P. M. Fearnside, D. A. Lashof and P. Moura-Costa, Accounting for time in mitigating global warming through land-use change and forestry, *Mitigation Adapt. Strategies Global Change*, 2000, **5**, 239.
67. J. H. Arehart, J. Hart, F. Pomponi and B. D'Amico, Carbon sequestration and storage in the built environment, *Sustainable production and consumption*, 2021, **27**, 1047.
68. T. Zink, R. Geyer and R. Startz, A market-based framework for quantifying displaced production from recycling or reuse, *J. Ind. Ecol.*, 2016, **20**, 719.
69. C. Valdenbo, S. Hellweg and T. F. Astrup, Let's Be Clear(er) about substitution: a reporting framework to account for product displacement in life cycle assessment, *J. Ind. Ecol.*, 2016, **21**, 1078.
70. C. Howard, C. Dymond, V. Griess, D. Tolkien-Spurr and G. C. van Kooten, Wood product carbon substitution benefits: a critical review of assumptions, *Carbon Balance Manage.*, 2021, **16**, 9.
71. M. E. Harmon, Have product substitution carbon benefits been overestimated? A sensitivity analysis of key assumptions, *Environ. Res. Lett.*, 2019, **14**, 065008.
72. W. Carton, A. Asiyanbi, S. Beck, H. Buck and J. F. Lund, Negative emissions and the long history of carbon removal, *Wiley Interdiscip. Rev.: Clim. Change*, 2020, **11**(6), e671.

CHAPTER 5

Bioenergy with Carbon Capture and Storage (BECCS)

MATHILDE FAJARDY

University of Cambridge, Judge Business School, Trumpington Road, Cambridge, UK
Email: m.fajardy@jbs.cam.ac.uk

Bioenergy with carbon capture and storage or BECCS has been a cornerstone technology of emissions trajectories compatible with a 1.5 to 2 °C target since the IPCC fifth assessment report in 2014, owing to its theoretically large carbon removal potential and mitigation value to the energy system. Sustainability concerns around the scale up of bioenergy production, the diversity of potential bioenergy pathways, the emergence of alternative technology-based and/or natural-based carbon removal methods, and local opposition to bioenergy and CCS deployment have however cast uncertainty and confusion over BECCS future as a mitigation and removal technology. This chapter aims at providing key tools to better understand the technology, clarify important metrics to measure its performance, compare potential feedstocks and application in the energy sector, and appreciate its current state of deployment and future prospects.

5.1 Introduction

5.1.1 BECCS Principle

Biomass with carbon capture and storage, or BECCS, refers to the combination of natural sequestration of atmospheric carbon by biomass during its

Energy and Environment Series No. 31
Greenhouse Gas Removal Technologies
Edited by Mai Bui and Niall Mac Dowell

Published by the Royal Society of Chemistry, www.rsc.org

growth with the capture and geological storage of CO_2 emitted upon conversion of this biomass to energy.[1–4] If more CO_2 is captured and permanently sequestered than is emitted over the entire BECCS value chain, this system results in a net removal of CO_2 from the atmosphere. As such, BECCS is considered part of a group of technologies referred to as Carbon Dioxide Removal (CDR) methods as well as Negative Emissions Technology (NETs), and has been consistently featured in economy-wide emissions and energy use projections consistent with stringent climate change targets.[5]

Biomass is a very versatile natural resource which results in a great diversity of ways in which a bio-feedstock can be converted into a useful energy vector to the energy system. This means that BECCS is not a single technology but rather a family of technologies. Figure 5.1 illustrates the BECCS value chain and a selection of bio-feedstock and energy conversion pathways which can form a BECCS configuration.

5.1.2 BECCS Value in Decarbonisation Pathways

BECCS has been a predominant technology in most Integrated Assessment Models (IAMs) compatible with a "well-below" 2 °C target. In the International Panel for Climate Change (IPCC)'s fifth assessment report, in 2014, BECCS was featured in 73% of the scenarios consistent with a 2 °C target, with up to 870 EJ per year deployed.[6] The 2019 IPCC 1.5 °C report re-asserted BECCS's importance as a key provider of CO_2 removal with between 149 Gt (P2 scenario, "middle-of-the-road") and 1189 Gt (P4 scenario, "fossil intensive") of cumulative removal provided, depending on levels and rate of emissions mitigation in other sectors. Only one category of scenarios assuming drastic demand-side reductions and behavioural changes (P1 or "low energy demand") featured a limited amount of CDR (190 Gt, provided by afforestation and land use management or AFOLU) and no amount of BECCS.[7,8]

A first reason for BECCS's predominance in these models lies in the double facetted nature of the technology.[9] As a CDR method, BECCS can provide scalable carbon removal, and as an energy technology, BECCS also delivers a low carbon energy valuable in mitigating emissions of multiple sectors (*e.g.*, heat, transport, power, industry). Regional energy system modelling studies have shown that firm low carbon power provided by BECCS was very valuable in a decarbonised electricity grid.[10,11] More recently, BECCS *via* biohydrogen was found to be key in helping decarbonise hard-to-reach sectors such as heating and transport.[12] Finally, the use of biomass and CCS in industrial applications could also help decarbonise high emitting sectors such as cement and steel with low availability of low carbon alternatives.[13]

Others factors have also been highlighted in the literature, such as the use of high discount rates (3–5%) which favour delayed action strategies such as removal, the absence of scalable removal alternatives and grid penetration constraints on variable renewable energy.[9]

As a result, in a 1.5 °C and net-zero by mid-century scenario, bioenergy deployment is typically limited by the global availability of sustainable bio-

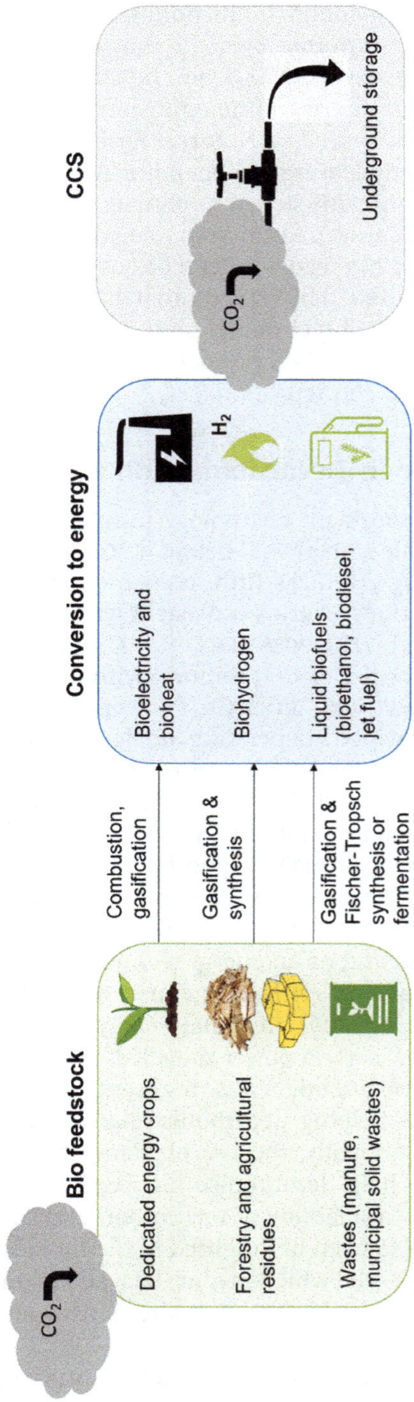

Figure 5.1 Overview of BECCS value chain.

feedstock, which is assumed to be between 100 and 400 EJ per year by the end of the century.[14,15]

5.1.3 How Much BECCS is there Today?

While bioenergy has been used throughout human history, and first CCS deployment dates back to the seventies, the negative emissions potential of BECCS was only first highlighted in the nineties.[16,17] The first commercial-scale BECCS plant started operating in 2017 in the Decatur corn-based bioethanol plant in Illinois, which captures around 1 Mt of CO_2 per year, and stores the CO_2 in a sandstone reservoir underneath the plant.[18]

In early 2019, the DRAX thermal power plant in the UK, which has converted four of its six 660 MW boilers to dedicated biomass combustion, started a BECCS pilot project with the company C-capture to capture 1 t CO_2 per day.[19] No storage site or utilisation of this captured CO_2 has however been identified yet. In 2020 the 50 MW Mikawa biomass plant in Japan started capturing 500 t of CO_2 per day. Other bioenergy and carbon utilisation projects include four bioethanol plants in the US and Canada which currently utilise the CO_2 for enhanced oil recovery (EOR).[20] BECCS projects in planning include the integration of CCS in a North Dakota ethanol plant, and a BECCS cluster in Norway composed of a Nolcem cement plant firing 30% biomass and a waste-to-energy plant. Figure 5.2 provides an overview of the operating and planned BECCS projects across the world in 2020.

In 2020, projects including capture from both fossil and biogenic sources captured about 30 Mt of CO_2 per year.[21] Even when considering all CCS projects, these values are still orders of magnitude below the gigaton-scale requirement of carbon dioxide removal forecasted by climate models. A drastic ramp up of BECCS and CCS deployment is therefore required.

5.2 Definitions

This section summarises key BECCS and bioenergy concepts required to understand the differences between BECCS feedstocks, pathways and regional opportunities and constraints to its deployment.

5.2.1 BECCS Value Chain

BECCS value chain is composed of four main stages: biomass production, biomass processing and transport, biomass conversion to energy, and capture, transport and geological storage of the biogenic CO_2 emitted upon biomass conversion.

5.2.1.1 Biomass Production

The biomass feedstock used in a BECCS value chain is either the main product (dedicated energy crop, harvested wood from forestry), or a by-

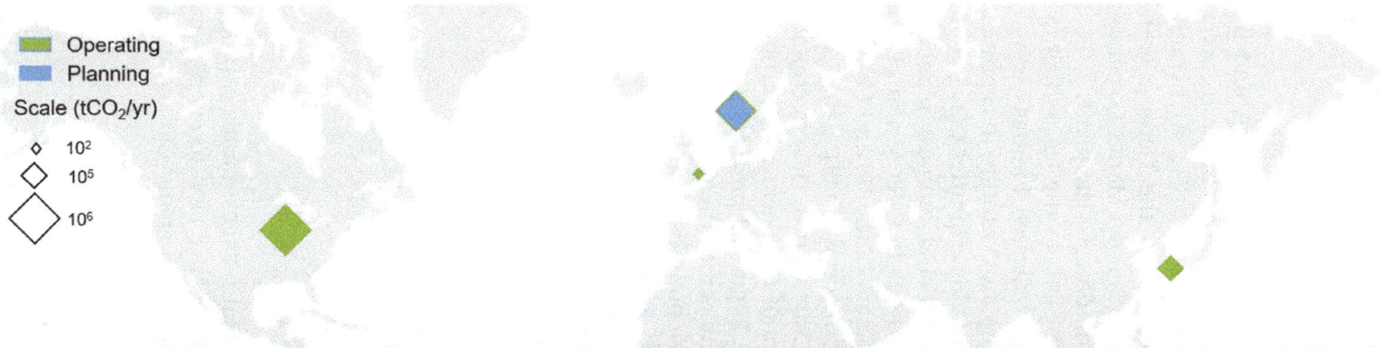

BECCS/BECCU	Technology	Application
Decatur plant (1Mt CO2/yr)	Bioethanol + CO2 capture	Storage
Kansas Arkalon (0.2 Mt CO2/yr)	Bioethanol + CO2 capture	EOR
Bonanza CCS (0.1 Mt CO2/yr)	Bioethanol + CO2 capture	EOR
Husky Energy (250 t CO2/day)	Bioethanol + CO2 capture	EOR
Farnsworth (0.2 Mt CO2/yr)	Bioethanol + CO2 capture	EOR
DRAX and C-capture (1t CO2/day)	Bioelectricity plant + post-combustion amine CO2 capture	Storage tbc
Mikawa power station (500 t CO2/day)	Bioelectricity plant + post-combustion CO2 capture	Storage tbc
Klemetsrud plant (0.4 Mt wastes/yr)	Waste-to-energy	Storage
Norcem cement plant (0.4 Mt CO2/yr)	Cement plant with 30% biomass	Storage

Figure 5.2 Map of BECCS projects and table of BECCS and BECCU projects in planning in 2020.

product (forestry residues, agriculture residues, municipal solid wastes, waste wood from industries, manure) of an activity. When biomass is a main product, a parcel of land is prepared (*e.g.*, with machinery, seeds or plantings), actively managed (*e.g.*, with machinery, fertilisers, irrigation) and harvested according to the crop growth cycle, which is defined by an establishment phase to reach nominal yield (*e.g.* 2 years for miscanthus, 5–6 years for willow), the harvest rotation (*e.g.*, every year for miscanthus, every three years for willow), and the total lifetime of the crop (*e.g.* 15–18 years for miscanthus, 20 for willow). To these activities are associated costs, energy, GHG emissions and environmental impacts. These costs, energy requirements and emissions can be incurred directly: these include direct use of fuel in machinery, direct emissions from the conversion of the land to a crop or a managed forest (also referred to as direct land use change), and direct N_2O emissions from the application of nitrogen fertiliser. Indirect energy requirements and emissions include the GHG emissions from the production of diesel, seeds, machinery and fertilisers, GHG emissions from the re-allocation of an activity if the land converted had a former activity (*i.e.*, indirect land use change).

In the case of wastes, emissions, energy and costs associated with the main activity are typically attributed to the main-product. However, costs, emissions and energy can be attributed to the waste if their collection competes with existing uses of the wastes (*e.g.*, agriculture residues can be used for soil nutrition or fodder for cattle).

5.2.1.2 *Biomass Processing and Transport*

Before being converted to energy, each biomass conversion pathway may require different levels of quality (ash content, moisture content) and format (finely ground biomass, chopped biomass) for the incoming feedstock, resulting in a different value chain energy cost and associated emissions. For biomass direct combustion, biomass needs to be finely ground with a low moisture content (around 10–15 wt%[22,23]), which means drying biomass is generally required (typical moisture content is 50% for woody biomass, and 30–40% for perennial grasses[24]). For other pathways such as biomass gasification, pyrolysis or fermentation, lower quality biomass can be used. Biomass can be supplied to the conversion plant in various forms, such as bales, chips, pellets or briquettes. As transport costs (financial, CO_2 or energy wise) are typically lower for biomass feedstock in a denser form,[25] the choice of the biomass format is a trade-off between biomass processing and biomass transport cost, as well as supply chain dry mass loss (which can be decreased with processing such as pelleting or torrefaction). A pelleting process involves initial size reduction of the harvested biomass (chopping, chipping), drying to a maximum moisture content of 10–15%, and further milling (grinding) before going through the die extruder. Processing operations require energy both in the form of fuel (drying) and electrical power (size reduction, drying, grinding, pelleting, cooling).[26,27]

5.2.1.3 Biomass Conversion

Depending on the final energy product and the initial biomass feedstock, biomass can be converted to energy through different processes. Processes differ by their energy efficiencies, capture efficiencies and costs. Most common BECCS pathways include BECCS power, hydrogen, biodiesel, bioethanol and use in industrial applications. Biomass conversion pathways are detailed in Section 5.3.

5.2.1.4 CO_2 Capture and Storage

CCS can be applied to all conversion pathways with different technologies: post-combustion when biomass is combusted in air, oxy-combustion when biomass is combusted in oxygen, a chemical loop integrated to biomass combustion in a chemical looping reactor, and pre-combustion capture when biomass is gasified. For a full review of the CCS technology see Bui *et al.* 2018.[28]

5.2.2 BECCS Carbon Metrics

5.2.2.1 Carbon Stored, Carbon Emitted and Carbon Removed

As BECCS is primarily deployed for carbon removal, a first key measure of the performance of a BECCS project is the net carbon removal potential, *i.e.*, how much CO_2 is removed when accounting for the life cycle greenhouse gas emissions which occur at different levels of the value chain. To this end, it is imperative to distinguish between different carbon flows along the value chain:

- the carbon *stored* in a BECCS process is the amount of physically stored CO_2, which can be both biogenic and fossil (*e.g.*, if biomass is co-fired with coal in the BECCS plant);
- the carbon *emitted* is the summation of all fossil CO_2 equivalent (*i.e.*, all greenhouse gas emissions converted into CO_2) emissions during biomass production, processing and transport, conversion and CCS stages;
- the carbon *removed* is the net balance between the biogenic carbon stored and the carbon emitted.

When the biogenic carbon stored is greater that the carbon emitted, the BECCS process is considered "carbon negative". Key factors which may decrease the amount of carbon stored by unit of biomass is the CO_2 capture rate during biomass conversion, which can vary from 90–95% when biomass is converted to electricity or hydrogen, to around 15% when biomass is converted to bioethanol with partial CO_2 capture (see Section 5.3). This can result in a potentially large amount of biogenic CO_2 being emitted back to the atmosphere. Another factor of CO_2 stored reduction is the potential biomass loss along the value chain (for example during transport and processing).[29]

Key factors impacting the amount of fossil carbon emitted are usually within the biomass supply chain. Direct greenhouse gas emissions resulting from the conversion of a carbon rich land (*e.g.*, prairies, forest) to a bio-crop, or to the harvesting of trees which would have otherwise sequestered CO_2, is referred to as *direct land use change*, and has been identified as a potential large source of emissions.[30–32] When the land was managed (other crop, pasture) before being converted to bioenergy production, the reallocation of this activity can also cause land use change elsewhere. This is referred to as *indirect land use change*, and could represent an even larger source of emissions, though more uncertain.[33,34] Another potential high source of production emissions is the use of nitrogen fertiliser which can result in both high direct N_2O emissions and indirect greenhouse gas emissions from fertiliser manufacturing. Depending on the carbon intensity of energy sources, biomass transport and processing can also release large amounts of emissions.[14,35]

5.2.2.2 Carbon Efficiency

The carbon removal potential can also be understood in terms of carbon efficiency: out of the initial carbon stored in the biomass by photosynthesis (or biomass carbon content), how much CO_2 is removed from the atmosphere. In Smith and Torn, for example, owing to the physical biogenic carbon losses and fossil carbon emissions along the value chain, 2.11 t CO_2 must be captured in the biomass to yield 1 ton of net CO_2 removal (which means an efficiency of 47%).[36] In Fajardy *et al.*, the carbon efficiency of BECCS is found between 2% and 70% depending on the configuration of the BECCS value chain.[35]

5.2.2.3 Carbon Avoided

Another key metric often used to define the decarbonisation performance of a mitigation technology is avoided carbon. This may happen when BECCS energy displaces another source of energy of higher carbon intensity. The carbon avoided is then measured as a difference between the carbon intensity of BECCS energy (*e.g.*, biofuel, bioelectricity, biohydrogen) and that of the displaced energy source, multiplied by the amount of energy displaced. This metric is less straightforward to calculate for it is defined relative to a counterfactual (for example a biofuel yields much more carbon avoided if it displaces gasoline than electric cars). It is therefore very dependent on energy systems integration effects and technology interdependency.[37] Figure 5.3 from the UK Committee on Climate Change (CCC) illustrates how carbon removal ('CO_2 sequestered') and avoided ('CO_2 displaced') potential may vary as a function of biomass uses and counterfactual considered.

Another element worth noting is that while carbon removal is a physical metric, carbon avoided is also expected to decrease with time, as the carbon intensity of the incumbent energy system decreases with mitigation. It is

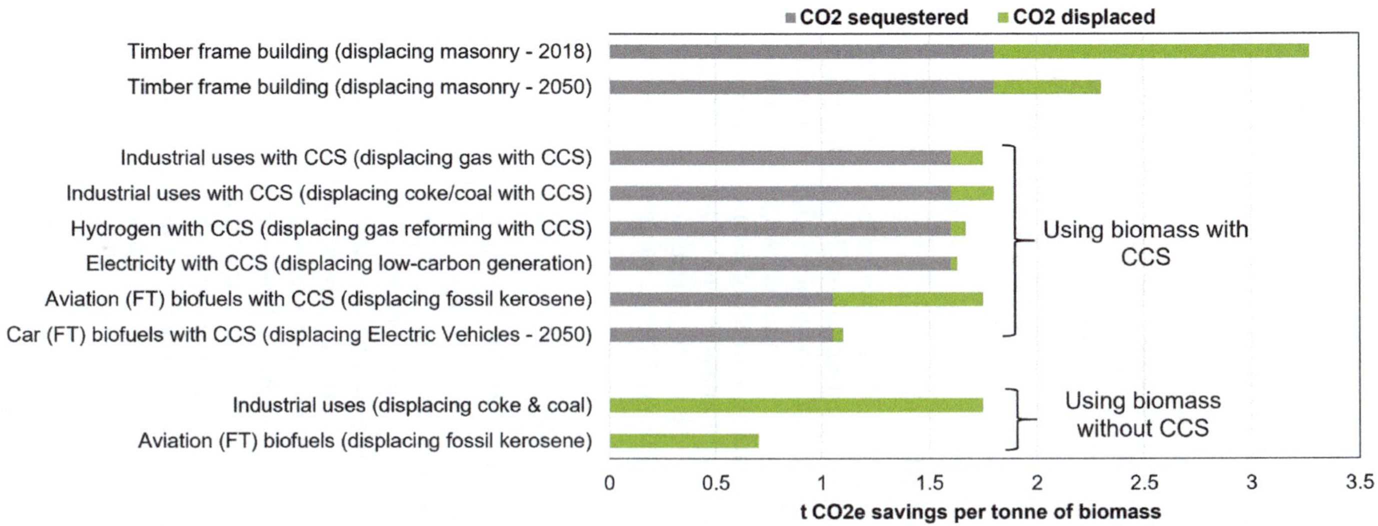

Figure 5.3 Illustration of CO_2 removal and CO_2 avoidance potentials of different biomass utilisation pathways. Figure adapted from CCC (2019).[38]

however a useful metric to quantify the value of a BECCS energy vector used to decarbonise the energy system.

5.2.2.4 *Carbon Breakeven Time*

Carbon negativity can also be defined in time. Emissions at the beginning of a BECCS project (from the establishment of a bio-crop or the harvesting of a forest) can be understood as an initial carbon debt.[30] Over the lifetime of a BECCS project, the CO_2 removed by BECCS may pay back this debt. The time required for a BECCS system to reach carbon negativity can be defined as the *carbon breakeven time*.[29,39] In Fajardy *et al.*,[35] the BECCS carbon breakeven time was calculated to be 1 year for a BECCS system using local agricultural residues, and over fifty years for one using perennial grasses grown on a former tropical forest.[35] This notion has also been defined in the context of avoided carbon. For example, in Fargione *et al.*,[40] the carbon breaktime of biofuels is defined as the time required for a biofuel system to deliver net carbon avoided relative to gasoline. In some systems with a high initial carbon debt from land use change, carbon breakeven time was found to vary between 0 (when growing prairie biomass on marginal land) to 423 years (when producing biodiesel from palm deforesting peatland forests).[40] The definition of a time horizon when performing a life cycle assessment over a BECCS value chain is therefore crucial, as it will directly impact the value of the initial carbon debt relative to the BECCS annual removal potential. In Hanssen *et al.*,[41] the authors showed that BECCS net removal potential increased from 2.5 to 40 Gt CO_2 per year when BECCS LCA was performed over 80 years (thus decreasing land use change emissions relatively to the amount of CO_2 removed) as compared to 30 years, suggesting the importance of deploying energy crops earlier in the century.[41]

5.2.3 BECCS Energy Metrics

Another key performance metric to measure BECCS contribution to decarbonising the energy system is its *energy efficiency*. Per unit of biomass, different biomass pathways can yield different amounts of useful energy to the energy system. For example, typical conversions of biomass to electricity with CCS is 30% while biomass to fuel is found around 40% and biomass to hydrogen 50%. Another important element is the BECCS net energy balance, which measures BECCS net energy production when accounting for BECCS life cycle energy use. Energy can be used at the biomass production level in the form of diesel and chemical inputs, in the form of electricity or fuel for processing (*e.g.*, drying, pelleting), in the form of fuel for transport, *etc.* Akin to BECCS carbon balance or removal potential, BECCS net energy balance can be found to be negative or positive depending on the energy loss (during conversion) and energy uses along the value chain.[42] BECCS energy performance can also be defined in terms of Energy Return on Investment or EROI, which is useful to compare the energy performance of different BECCS pathways. Similar to carbon balance, the BECCS EROI and energy balance can also be calculated as a function of time, though the BECCS initial 'energy debt' is likely to be significantly lower than the BECCS carbon debt when land use change is involved.

5.2.4 BECCS Financial Metrics

BECCS cost is key to compare different BECCS pathways, but also compare BECCS with alternative mitigation and removal methods. However, BECCS cost can take multiple forms and definitions. In Fuss *et al.*,[1] a literature review of BECCS cost suggests a cost range of £12–315 per t CO_2 (2018 GBP), but the detailed database provided by the authors suggest that costs in individual studies follow different boundaries and terminology. In terms of boundaries, while some studies consider all the life cycle costs and emissions of a BECCS process, some only focus on the BECCS plant and exclude life cycle emissions associated with producing and supplying the biomass. In terms of terminology, some costs are ***avoided costs*** and can be defined as a cost differential between the BECCS technology and a relevant counterfactual (*e.g.*, a coal power plant) divided by the ***carbon avoided*** by the BECCS process relative to the counterfactual (see Section 5.4). BECCS cost is also often presented as the cost of **CO_2 capture**, and calculates the total cost of capturing CO_2 at the plant level. This can be different from the **cost of CO_2 removal**, which calculates the BECCS total cost divided by the BECCS total net CO_2 removal, thereby factoring upstream emissions. It is therefore crucial to handle the BECCS costs value in the right context before performing a comparative cost assessment of BECCS processes or carbon removal methods.

Finally, BECCS is a unique carbon removal method in the way that it also yields a marketable energy product. A first useful cost metric to measure the competitivity of a BECCS energy product is the cost per unit of energy produced: for example, the cost per L of bioethanol and cost per MWh of bioelectricity. Given that this energy product can translate into a revenue stream for the BECCS system, BECCS ***net cost of removal***, which measures the BECCS cost of removal minus revenues associated with the energy product, is another key metric to accurately compare the BECCS financial performance with other carbon removal methods. Importantly, this metric also represents the appropriate ***carbon price***, or ***negative emissions credit (NEC)***, required to support a BECCS system. Figure 5.4 represents these different metrics for a BECCS electricity system in different regions of the world using different feedstocks.

5.2.5 BECCS Environmental Metrics

The topic which has raised the most concerns around BECCS deployment is BECCS potential negative impacts on the environment. Key concerns include BECCS land footprint and land use change, water footprint (both from

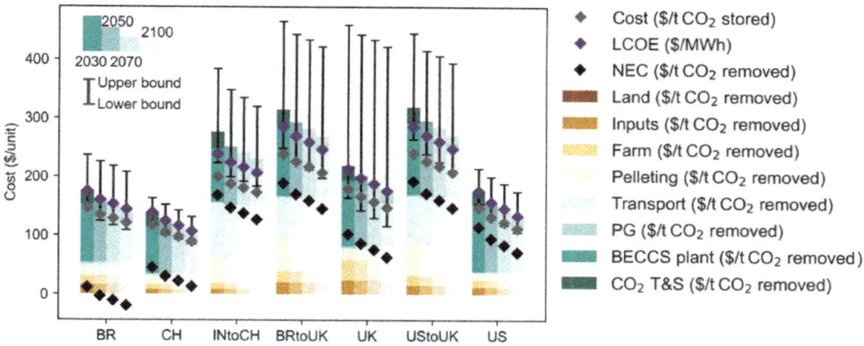

Figure 5.4 Illustration of BECCS total cost of removal (bars), cost of capture, *i.e.*, without accounting for life cycle emissions (grey diamond), levelised cost of electricity (purple diamond) and required negative emissions credit (black diamond) for different BECCS value chains across the world. Reproduced from ref. 43 with permission from Elsevier, Copyright © 2020 The Authors.

biomass production and carbon capture and storage), biodiversity loss from land conversion to bio-crop, food prices increase from the conversion of food crop with bio-crop, water eutrophication from the use of nitrogen fertiliser.[44–48] Apart from the CCS water footprint, which can be substantial,[46] these are mostly related to the biomass production phase and are key metrics to compare the sustainability of BECCS feedstocks. For example, an irrigated and fertilised crop-based biomass is likely to have a much higher environmental impact than residues collected from a forest and field.[14,47,49]

Given the importance of these impacts, attempts to monetise environmental impacts and more widely the ecosystem services have been explored,[50] but uncertainties remain as to whether large scale deployment of BECCS can be done without significant pressure on planetary boundaries.[41,51]

Now that key BECCS concepts have been defined, the next sections explore in detail the main BECCS feedstocks and pathways; as well as regional opportunities and constraints to BECCS deployment.

5.3 BECCS Feedstocks

A range of bio-feedstocks have been studied for bioenergy and BECCS applications, all of which differ by sustainable potential, side-effects and suitability for bioenergy conversion pathways. This range includes harvested wood and residues from forestry, energy crops and residues, waste wood, municipal solid wastes (MSW) and wet biomass (*e.g.*, manure). Figure 5.5 represents 2018 global bioenergy use by biomass type. Out of the 31 EJ of modern bioenergy consumption, the majority (22 EJ) is in the form of solid fuels, while 1 EJ is in the form of biogas, and 4 EJ in the form of liquid biofuels,[52] which suggests that dense solid fuels remain the most cost

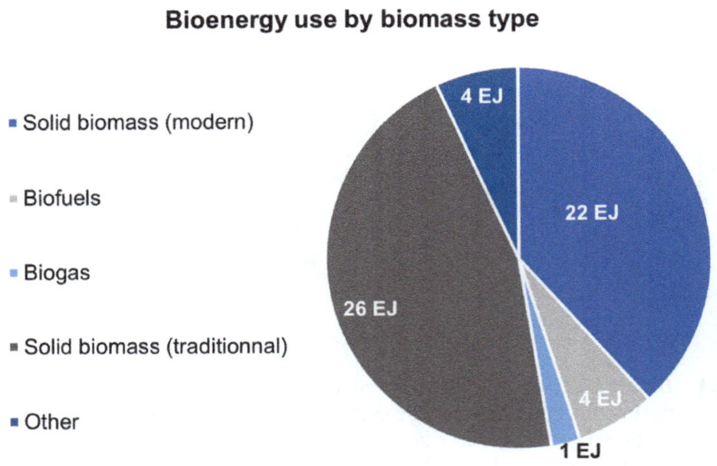

Figure 5.5 Global bioenergy use by biomass type. Data from IEA statistics (2019).[52]

effective format to transport and store bioenergy. Recent literature has investigated the potential of marine biomass and algae for BECCS,[53,54] but these are discussed in Chapter 10.

5.3.1 Woody Biomass

With a high carbon content, woody biomass offers the closest resemblance to coal when dried, which means that it can be used for most of the cellulosic biomass conversion pathways, including direct combustion to heat and power, gasification and synthesis to hydrogen or methanol, and pyrolysis. A key driver of sustainable potential is the sustainability of forestry management, with significant controversy around the carbon debt associated with the harvesting of a tree which would have otherwise continued to sequester carbon both above and below ground. In Creutzig *et al.*,[14] the global potential for woody biomass in 2050 is projected to be between 25 and 50 EJ per year.

Forestry residues are typically less controversial as long as the harvesting of these residues is beneficial to the forest growth. In this context, important synergies could be obtained when co-deploying afforestation and BECCS, depending on the amount of sustainable forest management done with afforestation. The topic of afforestation is explored in detail in Chapter 9.

5.3.2 Energy Crops and Residues

Dedicated energy crops typically present the largest potential, as well as the largest uncertainty. The two main drivers of sustainable potential are biomass yield, available land with low land use change and biodiversity impacts. In Creutzig *et al.*,[14] a review of the literature yields a range between 25 to 625 EJ per year of potential, with a higher confidence towards the lower end of the range. In a recent modelling exercise, authors quantified the energy crop potential between 149 and 245 EJ per year as a function of conservation policies impacting which parcels of land are available.[55] The most commonly assessed type of biomass is lignocellulosic biomass such as perennial grasses (*e.g.*, miscanthus, switchgrass) and short rotation coppice (*e.g.*, poplar, willow) which can follow the same conversion processes as woody biomass. The reason for the focus on this type of biomass is to avoid the use of first-generation crops such as sugar and starch crops, which can be converted to bioethanol through hydrolysis and fermentation, and oil crops to biodiesel through extraction and esterification, owing to competition with food products. As with all crops, however, sustainability risks associated with energy crops involve land use change from natural land conversion to agricultural land,[34,40,56] water eutrophication and direct N_2O and indirect GHG emissions from nitrogen fertiliser application, irrigation, *etc.*

The potential for agricultural residues, such as corn stover and wheat straw, which is also categorised as a lignocellulosic second-generation crop, has been quantified between 30 and 125 EJ per year.[14] Key drivers of sustainable potential are the fraction of residues which can be sustainably used for bioenergy without competing with existing use, such as soil protection and fertilisation (to avoid water run-off and soil degradation) or fodder[57].

Energy crops and agricultural residues typically have a lower moisture content (<30–40%)[58] than woody biomass, which requires less drying energy, but a higher ash content and lower carbon content,[24] which results in an overall lower heating value[59] than woody biomass.

5.3.3 Wastes

The interest for waste biomass such as waste wood and municipal solid waste has received a boost in recent years, to avoid potential negative environmental impacts associated with forestry and agricultural biomass. In Haberl *et al.*,[60] 10 EJ per year of waste biomass is assessed to be the amount available by mid-century. In a recent report by Lawrence Livermore National Laboratory (LLNL) BECCS potential in California is solely assessed from waste and residues availability.[61]

However, the higher ash content and higher moisture content of some of these, MSW for example (70%), and/or the high content in trace elements could represent a limitation for the use of these biomass in large scale biomass conversation applications. In Pour *et al.*,[62] the potential for large scale (500 MW) fluidised bed reactors combined with carbon capture and storage using municipal solid wastes has been evaluated, and showed to be cost-effective as well as sustainable.

Other types of wastes, such as manure, can be converted to biogas through anaerobic digestion, and potentially hydrogen when combined with synthesis. The smaller scale of these plants however makes them difficult to reconcile with the cost-effective scale of carbon capture and storage.

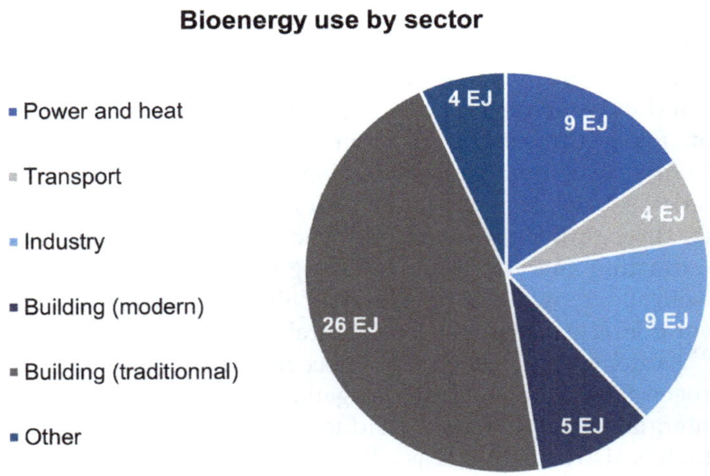

Figure 5.6 Global bioenergy use by sector. Data from IEA statistics (2020).[52]

5.4 BECCS Pathways

We have seen in the previous section that sustainable biomass is a constrained resource. As bioenergy can contribute to the decarbonisation of multiple sectors of the economy, bioenergy deployment is typically limited by feedstock availability. Given that constraint, this raises the question of the best use of biomass for both removal and mitigation. In this context, it is interesting to compare BECCS pathways in terms of CO_2 removal potential – how much CO_2 is physically removed from the atmosphere per unit of biomass, CO_2 avoided – how much emissions are avoided by energy or product substitution per unit of biomass, and cost per ton of CO_2 avoided and removed. This section presents different BECCS pathways and discusses how these pathways compare in terms of these metrics. In 2018, out of 56 EJ of bioenergy consumption, 26 EJ is referred to as traditional use in buildings for cooking and heating and developing countries.[52] Among modern uses (31 EJ), the largest share of bioenergy consumption is dedicated to the production of heat and power (9 EJ), and industrial processes (9 EJ), followed by heating in buildings (5 EJ) and transport (4 EJ).[52] This means that there is an inherently large diversity in potential BECCS processes.

Figure 5.6 presents the current uses of bioenergy by sector.

5.4.1 BECCS to Power

Lignocellulosic biomass can be converted to BECCS to power *via* combustion in pulverised combustion (PC) plants, fluidised bed combustion (FBC) plants, chemical looping combustion plants (CLC) and combined heat and power (CHP) plants to produce heat in addition to power, or *via* gasification in an integrated combined cycle (IGCC).

The energy efficiency of BECCS power plants is lower than coal power plants owing to the 9–10% efficiency (in % points) penalty incurred by CO_2 capture and compression,[63] and the biomass fuel which tends to be of lower heating quality than coal (from both a heating value and ash content perspective). Overall, the energy efficiency of a BECCS pulverised combustion plant is found to range between 17%[64] and 37%[65] in the literature, with potential improvements to 38–42% provided *via* process design enhancement (*e.g.*, heat recovery) and/or the use of advanced solvents.[65,66] Average energy efficiency values are typically calculated around 25–29% depending on the CO_2 capture technology and feedstock.[35,66] IGCC plants are expected to yield higher efficiencies (35–43%[65,66] and 50% with design improvements[65]) but the feasibility of large-scale biomass gasification is yet to be proven at scale.

While the energy efficiency of bioelectricity-CCS is relatively low, the carbon efficiency, however, is one of the highest. Upon biomass conversion to electricity, 90–95% of the emitted CO_2 can be captured and sent for storage.[65,67] Potential capture rate improvements to 96 and 99% could be reached with advanced absorption processes, incurring marginal cost increase.[67–69]

Finally, the cost of a BECCS electricity plant can be found within £69–226 per t CO_2[†] in the Fuss *et al.* review.[70] The wide range can be explained by the use of different cost metrics (cost of carbon avoided, captured or removed), the systems boundaries (whether the balance includes biomass life cycle emissions) and key cost assumptions such as feedstock cost. Table 5.1 summarises the cost of BECCS electricity processes across the literature. When available, the cost of CO_2 removal has been calculated between £49 and £212 per ton of CO_2 removed[43,65,71–74] for biomass combustion plants. As a BECCS plant also produces energy, accounting for the revenues associated with energy production gives insights into the carbon price, or negative emissions credit (NEC), required for the system to be cost effective. In Keller *et al.*,[73] for example, accounting for electricity revenues of a BECCS plant in Japan reduces the cost from £176–212 per ton of CO_2 removed to a NEC of £128–155 per ton of CO_2 removed. In other multi-country studies, the NEC is also found to be highly region specific depending on the cost of feedstock, BECCS CAPEX and price of electricity. While the BECCS cost is found to be £164 per ton of CO_2 removed in the UK, £124 per ton of CO_2 removed in Brazil and £106 per ton of CO_2 removed in China, the corresponding NEC is only £76 per t CO_2 in the UK, £22 per t CO_2 in China and no NEC is required in Brazil.[43]

5.4.2 BECCS to Liquid Fuel

Biomass can be converted to multiple liquid fuels including ethanol, diesel, methanol, jet fuel, *etc.* Ethanol can be produced from the fermentation of grains (*i.e.*, first-generation) and ligno-cellulosic biomass (*i.e.*, second generation).

[†]All cost data is in 2018 GBP. The BEIS GBP deflator rates are used for year on year conversions, and average annual exchange rates from US dollars to GBP are used.

Table 5.1 Review on costs of biomass combustion and CCS technologies.

Technology (Source)	Misc.	£ per t CO_2 captured	£ per t CO_2 avoided	£ per t CO_2 removed	NEC required (£ per t CO_2)	Comments
PC/FBC-CCS (Fuss 2018[1])	69–226					Review – various definitions of CO_2 included
BECCS power (McGlashan 2012[75])		179–222				Life cycle cost of CO_2 captured – but no upstream emissions considered
BECCS power (McLaren 2012[76])		59–210				Single source for BECCS, unclear whether it is carbon removed/captured/avoided, supply chain emissions?
CHP-CCS (Arasto 2014[77])	27					Cost of CO_2 abatement – unclear if upstream emissions are considered
PC-post CCS (Al-Qayim 2015[78])		136.25				Avoided as compared to coal – no upstream emissions considered
PC-post CCS (Bhave 2018[79])		110–225	56–118			Avoided as compared to coal – no upstream emissions considered
PC-post CCS (Yi 2018[74])				207		Life cycle cost of CO_2 removal – including all life cycle emissions; low CAPEX
PC-post CCS (Keller 2019[73])				212	155	CO_2 removal, upstream emissions considered
PC-post CCS (Yan 2020[71])				50		CO_2 removed = CO_2 captured – feedstock/process life cycle emissions
PC-post CCS (Emenike 2020[66])			85–121			Avoided as compared to biomass – includes upstream emissions
PC-post CCS China (Fajardy 2020[43])				106	22	
PC-post CCS Brazil (Fajardy 2020[43])				124	0	
PC-post CCS UK (Fajardy 2020[43])				164	76	
PC-oxy CCS (Abanades 2011[80])			30–61			Avoided as compared to coal – no upstream emissions considered

PC – oxy CCS (Al-Qayim 2015[78])	114				Avoided as compared to coal – no upstream emissions considered
PC-oxy CCS (Bhave 2018[79])	105–215	54–115			Avoided as compared to coal – no upstream emissions considered
PC-oxy CCS (Emenike 2020[66])		105–144			Avoided as compared to biomass – includes upstream emissions
CFB-CCS (IEAGHG 2011[65])			54–64		CO_2 removed = CO_2 captured – all upstream/downstream emissions
MSW CFB CCS (Pour 2018[72])	60–75		128		CO_2 removed = CO_2 captured – feedstock/process life cycle emissions (use MSW)
CLC (Bhave 2018[79])	75–218	30–95			Avoided as compared to coal – no upstream emissions considered
CLC (Keller 2019[73])			176	128	CO_2 removal, upstream emissions considered
CLC (Yan 2020[71])			49		CO_2 removed = CO_2 captured – feedstock/process life cycle emissions

The fuel efficiency of this process is found within the 24–44%[‡ 65,86,87] range (including CO_2 capture and compression), with potential co-production of electricity from an auxiliary boiler combusting lignin residual and biogas from waste water.[87] The capture efficiency of bioethanol conversion is relatively lower than BECCS power in a base configuration. From the fermentation stage, only 10 to 15% of the initial carbon is released as a high purity carbon stream which can be captured, while 19 to 30% of the carbon remains in the ethanol.[65,86–88] However, further capture of the emissions from the auxiliary boiler combusting solid and gas residues can increase the capture efficiency to 74%.[87] Given the high purity of the CO_2 stream (99%), the cost of CO_2 capture is relatively low, between £15 and 25 per ton of CO_2, and £30–37 per t CO_2 if auxiliary emissions are also captured. The cost of removal is however dependent on the process upstream emissions. When factoring high biomass upstream emissions, some configurations where only fermentation emissions are captured are found not to be carbon negative, which leads to an infinite cost of removal.[87,89]

Biodiesel and jet fuel can be produced through gasification and Fischer–Tropsch (FT) synthesis of the biofeedstock. The process energy and capture efficiencies are overall higher than the biomass to ethanol process. The overall process energy efficiency (including CO_2 captured and compression) is found to be around 42–43%,[65,87,90] and the capture efficiency around 51 to 56%[65,87,90] (23–33% of the carbon is in the FT fuel), with a potential increase to 64% if auxiliary emissions from the combustion of the process co-products (char from gasification and purge from FT synthesis) are captured.[87] Though untried with CCS, FT plants are deployed worldwide, mostly using coal gasification and natural gas reforming to yield syngas. Akin to the ethanol process, the cost of capture is relatively low, between £15 and 30 per t CO_2.[87,90,91] The costs for biomass gasification and biofuel processes with CCS across the literature are summarised in Table 5.2 and 5.3.

5.4.3 BECCS for Hydrogen

Biomass can be converted to hydrogen both *via* biological (fermentation) and thermal (gasification) pathways. Though fermentation has the benefit of operating in milder temperature conditions, gasification has so far led to the highest hydrogen yields.[93] After gasification, the gas is stripped of its impurities (ash and contaminants), goes through a water-gas-shift reaction, and hydrogen is separated *via* pressure swing adsorption. The process energy efficiency (in hydrogen terms) can reach 37 to 54%,[87,94–96] and the capture efficiency 87 to 94%.[87,95] While there is no BHCCS plant operating at scale yet, coal gasification to hydrogen represents 20–25% of hydrogen production today, and is therefore a mature technology. The cost evidence on BHCCS is more scarce compared to other bioenergy conversion pathways. In Hannula *et al.*,[87] costs are assessed between £16 and 20 per ton of CO_2 captured (Table 5.4). Given the process high

‡All energy efficiencies are in % lower heating value (LHV).

Table 5.2 Review on costs of biomass gasification and CCS technologies.

Technology (Source)	Misc.	£ per t CO_2 captured	£ per t CO_2 avoided	£ per t CO_2 removed	Comments
BIGCC-CCS (range Fuss 2018[1])	24–60				Review – various definitions of CO_2 included
Black liquor IGCC (Mollersten 2006[81])				22–61	Does not explicitly account for upstream emissions but uses black liquor considered as a waste
Black liquor IGCC (McLaren 2012[76])		37.8			Single source for BECCS, unclear whether it is carbon removed/captured/ avoided, supply chain emissions?
Black liquor IGCC (Arasto 2014[77])	44.72				Cost of CO_2 abatement – unclear if upstream emissions are considered
Black liquor IGCC (Onarheim 2015[82])			18–58	17–23	Review of other studies – unclear if feedstock life cycle emissions are included
Black liquor IGCC (Onarheim 2015[82])					Does not specify the counterfactual used for the avoided cost calculation
bio-SNG (Carbo 2011[83])			56–112		Includes supply chain emissions but cost of CO_2 avoided
bio-SNG (Arasto 2014[77])	71.552				Review – not certain on definition of CO_2 "abated"
BIGCC (IEAGHG 2011[65])				47–58	
BIGCC (Ranjan 2011[84])			128–340		Takes into account revenues from the electricity
BIGCC (Sanchez 2016[85])			49–60		Lower values correspond to displacing coal, and higher values to displacing natural gas – no upstream emissions considered
BIGCC (Bhave 2018[79])		122–250	48–105		Avoided as compared to coal – no upstream emissions considered
BIGCC (Emenike 2020[66])			108–149		Avoided as compared to biomass – includes upstream emissions

Table 5.3 Review of the costs of biofuel + CCS technologies.

Source	Misc.	£ per t CO_2 captured	£ per t CO_2 avoided	£ per t CO_2 removed	Comments
Bioethenol-CCS (Fuss 2018[1])	15–138				Review – various definitions of CO_2 included
Bioethanol CCS (Laude 2011[89])				37–112	Does not include feedstock upstream. According to carbon balance, the scenario is actually not carbon negative.
Bioethanol CCS (all) (Laude 2011[89])				169–187	Does not include feedstock upstream emissions. According to carbon balance, the scenario is only slightly carbon negative
Bioethanol CCS (IEAGHG 2011[65])				192–317	CO_2 removed = CO_2 captured – all upstream/downstream emissions
Bioethanol CCS (Arasto 2014[77])	268				Cost of CO_2 abatement – unclear if upstream or downstream emissions are considered
Bioethanol CCS (Moreira 2016[92])		23			Incremental cost as compared to the base case without CCS. Cost per ton of CO_2 physically captured – no accounting for feedstock upstream emissions or fuel downstream emissions
(corn)-bioethanol CCS (Hannula 2021[87])		18		carbon positive	Cost of CO_2 captured, study includes supply chain emissions, 0% capture efficiency
(lignocellulosic)-bioethanol CCS (Hannula 2021[87])		19			Cost of CO_2 captured, study includes supply chain emissions, 10% capture efficiency
(lignocellulosic)-bioethanol CCS capture + (Hannula 2021[87])		37			Cost of CO_2, study includes supply chain emissions, 74% capture efficiency (auxiliary emissions captured)
FT diesel CCS (IEAGHG 2011[65])				59–76	CO_2 removed = CO_2 captured – all upstream/downstream emissions
FT diesel CCS + (Johnson 2014[91])		15–30			Cost per ton of CO_2 captured at the plant; adapted from total cost per total amount of negative emissions – including feedstock upstream emissions, includes capture of auxiliary emissions
FT diesel (Arasto 2014[77])	36				Cost of CO_2 abatement – unclear if upstream emissions are considered
FT diesel (Hannula 2021[87])		15			Cost of CO_2 captured, study includes supply chain emissions, 52% capture efficiency
FT diesel capture + (Hannula 2021[87])		25			Cost of CO_2 captured, study includes supply chain emissions, 64% capture efficiency (auxiliary emissions captured)

Table 5.4 Cost of biohydrogen + CCS pathways for different configurations.

Technology (Source)	£ per t CO_2 captured	£ per t CO_2 removed	Comments
BHCCS (Hannula 2021[87])	16		Cost of CO_2 captured, includes supply chain emissions, 87% capture efficiency
BHCCS + (Hannula 2021[87])	20		Cost of CO_2 captured, includes supply chain emissions, 94% capture efficiency (auxiliary emissions captured)

energy and carbon efficiency, and the production of a valuable decarbonised energy vector in the form of hydrogen, BHCCS could therefore have both a high removal and avoided potential.

5.4.4 BECCS in Industry

More recently, the use of biomass and CCS has also been investigated in the industrial sector. First, applying CCS in plants which already combust biomass, such as pulp and paper plants, may offer important removal potential at low cost.[97] In Onarheim *et al.*,[97] different rates of capture can be achieved depending on which emissions sources are captured in the plant, between the recovery boiler, the multi-fuel boiler and the lime kiln. The carbon avoided cost of integrating CCS on a pulp and paper plant is assessed between £41–53 per t CO_2 for a kraft pulp mill, and £57–71 per t CO_2 for an integrated pulp and board mill when compared with a plant without CCS, and the negative emissions credit required between £48–56 per t CO_2 for the kraft mill and £56–64 per t CO_2 for the integrated mill.

Another upcoming BECCS application is the integration of biomass fuel and CCS in the iron and steel industry. In steelmaking processes using a blast furnace, both the coal used in the furnace, and the coke and sinter can be partially replaced by biomass in the form of charcoal.[98] Bio-syngas (*e.g.*, *via* wood pellet gasification) can also substitute natural gas as the reducing agent in processes using direct reduction of iron. While this has been demonstrated with coal-based syngas, this has yet to be demonstrated with bio-syngas. Biomass can also be used to provide the thermal energy of both processes. As for CCS integration, post-combustion capture can be applied to the furnace which is responsible for 60% of the plant emissions. CCS can also be applied to additional CO_2 sources (*e.g.*, the pellet, coke and sinter plants), but the CO_2 abatement benefits might be relatively low compared to the incurred cost.[13] In terms of removal potential, Tanzer *et al.*[13] showed that when accounting for full upstream emissions of biomass, processes with CCS and biomass integration could just about reach carbon neutrality, which would effectively lead to infinite costs of removal.

5.4.5 Mitigation *vs.* Removal Potential and System Integration Effects

Overall, BECCS pathways differ by the range and quality of biomass which can be used, their process energy and capture efficiencies, and cost of removal. Another key difference is the mitigation value for the energy sector. BECCS power and BHCCS for example both have similar carbon removal potentials, but yield very different energy vectors. In an energy system, the value of a given energy vector can be measured by the marginal cost of abatement of this technology. On the one hand, firm low carbon power can be very valuable in balancing a system with an increasing share of intermittent renewables,[10] on the other hand, hydrogen is a very versatile vector which can provide a low carbon alternative in multiple sectors (*e.g.*, heating, transport) as well as provide seasonal energy storage.[99] BECCS biofuel pathways have a relatively lower carbon removal potential owing to their low capture rates, but can provide valuable low carbon fuels for the transport sector. Muratori *et al.*[100] show that BECCS in IAMs is deployed over 50% as liquid biofuels, as opposed to bioelectricity. Finally, BECCS in steel-making provides the lowest carbon removal potential with processes which just about reach carbon neutrality, but given the low availability of mitigation alternatives, BECCS integration in the steel-making industry could be very valuable in mitigating emissions from a hard-to-decarbonise sector.[13,98] Multi-sector energy systems modelling at the regional level is therefore required to fully capture the potential of each pathway, and maximise both the mitigation and removal potential of biomass.[101,102] Figure 5.7 represents ranges of carbon removal and carbon avoided potential for different pathways as a function of different counterfactuals.

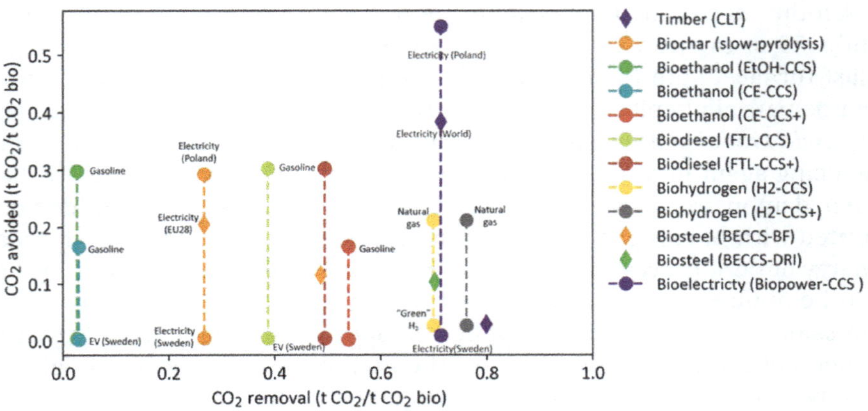

Figure 5.7 Illustration of CO_2 avoided *vs.* CO_2 removal of different BECCS pathways considering different counterfactuals. Reproduced from ref. 103, https://doi.org/10.1016/j.isci.2021.102765, under the terms of the CC BY 4.0 license https://creativecommons.org/licenses/by/4.0/.

5.5 Deployment Opportunities and Constraints

5.5.1 BECCS Global Sustainable Potential and Impacts

Given the high deployment of BECCS in 1.5 °C and 2 °C compatible scenarios, there is a strong imperative to constantly revisit BECCS global potential, as a function of geophysical, sustainability, financial and infrastructure constraints.[104]

5.5.1.1 Bio-geophysical Constraints

CO_2 storage is generally not considered as a global bottleneck, with over 11 700 Gt CO_2 identified available storage with a 75–100% confidence, and 14 000 Gt CO_2 with a confidence of 56% and higher. However, regional assessments may substantially vary in capacity, injectivity, reliability and costs.[20] Taking the UK as an example, while the UK boasts over 70 Gt of CO_2 storage with an injectivity of 1.4 Gt CO_2 per year, storage sites with a total cost below £20 per t CO_2 reach a total capacity of 47 Gt CO_2 and injectivity of 0.95 Gt CO_2 per year.[105,106]

Conversely, global technical and/or sustainable bioenergy potential is a key bottleneck to BECCS deployment. While bioenergy potential assessment by 2050 falls within a broad range (60–1500 EJ per year[1]) there seems to be a higher number of studies pointing to the lower end of the range, around 100 EJ per yr.[107] These ranges typically account for forestry biomass, forestry and agricultural residues, and energy dedicated crops, the latter being the most prone to variability in the literature, owing to land availability and yield improvement assumptions.[108] In a recent assessment of residues and energy crop potential, residues potential remains relatively constant (around 70 EJ per year in 2050 and 80 EJ per year in 2100), but energy crops increase from 46–122 EJ per year in 2050 to 68–276 EJ per year in 2100, assuming significant natural land conversion to bioenergy production.[108]

5.5.1.2 Bioenergy Potential Converted to Removal Potential

The conversion of bioenergy potential into removal potential *via* BECCS is not straightforward. Given the large differences in conversion efficiencies between pathways (for example 90% for bioelectricity and biohydrogen, 55% for biodiesel, 15% for bioethanol), a key factor to translate EJ of bioenergy to removal is the BECCS pathway. In Heck *et al.*,[109] BECCS hydrogen delivers 5 to 20 times more global removal than BECCS *via* biodiesel. In Hanssen *et al.*,[41] BECCS ethanol is found to never deliver removal when accounting for full life cycle emissions, while BECCS biodiesel delivers 8 times less removal than BECCS electricity. Another key factor is the life cycle emissions associated with the BECCS value chain which may further reduce the CO_2 removal potential.[29,36] In the same study, the removal potential of BECCS electricity increases from 2.5 to 40 Gt CO_2 per year when BECCS LCA is performed over 80 years (thus decreasing land use changes emissions relative to the amount of CO_2 removed) as compared to 30 years.[41] However, the authors argued that BECCS deployment in the IPCC 'middle-of-the-road' (SSP2 or 'P1') and 'fossil-fuel deployment' (SSP5 or 'P4') 1.5C scenarios[110] could be approached at 88–94% with BECCS electricity even accounting for the full life cycle emissions of the process, providing the early deployment of energy crops (to ensure a long-time horizon to compensate for life cycle emissions).

Other key factors include biomass energy density (typically 17–21 GJ per dry ton for the higher heating value[111]) and carbon content (47–55%[24]). Assuming that 1 EJ of biomass yields 0.02–0.05 Gt CO_2 of removal,[15] a 100 EJ constraint would mean a global BECCS potential between 2 and 5 Gt CO_2 per year, while a more optimistic range (300–350 EJ per year) would increase the BECCS potential to 6–18 Gt CO_2 per year.

5.5.1.3 Environmental–Economic Constraints

Accounting for environmental impacts can further reduce BECCS potential. In Heck *et al.*,[109] BECCS hydrogen potential reduces from 20–23 Gt CO_2 per year to 0.20 Gt CO_2 per year when considering nitrogen, water use, biodiversity loss and land use change planetary boundaries. Seferian *et al.*[112] quantified that considering water constraints would reduce BECCS deployment by 90% relative to the level of deployment modelled in 1.5 °C scenarios. The 2019 IPCC special report on land use warned of BECCS potential detrimental to ecosystems, and identified 5 Gt CO_2 per year as being an upper limit for sustainable BECCS deployment.[113]

Other constraints to BECCS deployment include the ability to ramp up the feedstock supply chain. In Baik *et al.*,[114] the lack of spatial correlation between CO_2 storage sites and biomass feedstock production sites decreases BECCS near-term removal potential in the US from 370 Mt CO_2 to 100 Mt CO_2.

Finally, BECCS financial competitivity with other generation sources is another constraint.[104] While a BECCS system benefits from potential revenues associated with the generation of low carbon energy, this revenue is generally not enough for a BECCS system to breakeven financially. Depending on BECCS cost and the whole sale price of energy, a revenue associated with the service of carbon removal is required.[115]

5.5.1.4 Social and Policy Constraints

Two last dimensions which could both drive and hinder BECCS deployment are the social acceptability of BECCS deployment at the regional level, and the governance of BECCS at all scales.

Social acceptability, sometimes referred to as the social license to operate, can put a severe halt on BECCS deployment.[116] In particular, strong oppositions to biomass (*e.g.*, in the Netherlands) or to CCS (*e.g.*, in Germany[117]) could significantly limit BECCS deployment potential. Social engagement is key in this context to understand the drivers of social acceptance.[118] Putting forward the economic value around BECCS deployment, for example in the form of job retention or creation, could also be a way to change the perspective around a specific BECCS project at the local level.[119]

International governance is a second key enabler of BECCS. The multiplication of 'net-zero' commitments by mid-century, taken at the macro-region (EU), country, state and city level is a first key step towards this goal, along with the deployment of regional green recovery plans (*e.g.*, the European Green Deal). However, international cooperation in establishing clear regional CDR targets, international biomass sustainable certification, and carbon accounting frameworks over multi-actor, multi-region BECCS value chains, is required to unlock BECCS deployment.[120,121] Furthermore, active collaboration in trading CO_2 flows, in the form of biomass, gaseous or liquefied CO_2, or a negative emissions credit, is key to ensure that the most suitable regions for the deployment of BECCS deploy more than they require, thus providing sustainable and low-cost CDR for less endowed regions.[43]

5.5.2 Low Hanging Fruit

Given these hurdles, it is worth considering potential low hanging fruits to BECCS deployment. Around 31 EJ of modern bioenergy is used today, including 9 EJ of heat and power, 9 EJ in industrial applications and 4 EJ for transport biofuel. If CCS were applied to these processes, close to 1 Gt of CO_2 could be removed from the atmosphere[§]. In particular, the potential for CCS retrofitting on waste-to-energy plants[72] and bioethanol plants[92] has been studied. In industrial applications, retrofitting CCS to pulp and paper plants could also deliver large-scale bio-CCS applications in Nordic countries, with costs of avoided carbon emissions below £41–71 per ton.[122]

One of the common limitations to CCS retrofitting to existing biomass plants is the low scale of current biomass applications. As of June 2019, the world average nameplate capacity of bioelectricity plants is 25 MW, with 84% of the plants with a capacity below 50 MW, and 95% of the plants with a capacity below 100 MW. For waste-to-energy plants, the average nameplate capacity is even lower, at 15 MW, with 92% of the plants below 50 MW, and 99% below 100 MW.[122] Given the cost of CO_2 transport and storage, the low scale of these biomass applications might be difficult to reconcile with a cost-effective scale of CCS. Strategies to share the cost of the CCS infrastructure, such as the creation of five CCS clusters around several biogenic

---

[§]Assuming an average biomass heating value of 18 GJ per ton (dry),[59] a biomass carbon content of 50%,[24] an average carbon efficiency of 75% for heat, power and industrial applications, and 25% for biofuel applications.

and fossil CO_2 emissions sites in the UK, will be needed to lower the cost of CO_2 transport and storage. To this effect, spatial optimisation accounting for both biomass and CCS supply chains and the network is key to fully assess the economic and technical feasibility, as well as the sustainability of such measures. Another limitation is the seasonality of some of the bioeconomy facilities, such as the sugarcane-based bioethanol industry in Brazil. Tagomori *et al.*[123] explored the potential for BECCS and CCS infrastructure deployment in Brazil through exploiting both large fossil power plants and bioethanol plants to lower the cost of CCS infrastructure through increased utilisation.

Conversely, the world average capacity of coal power plants is 840 MW. Fuel switching from coal to biomass and CCS retrofitting on existing coal plants could therefore constitute a more scale-effective way toward BECCS deployment. In Beagle *et al.*[124] for example, the conversion of coal plants to BECCS plants in the US could provide 357 Mt of removal per year, though spatial optimisation and consideration of co-location of biomass production sites and CO_2 storage sites constitute a limitation to BECCS potential.[114]

5.5.3 Optimal BECCS Deployment at the Global Level

In global modelling exercises, BECCS is deployed according to cost-minimising optimisation. In IAMs, cost-optimal deployment depends strongly on regional bio-geophysical availability (bioresources, CO_2 storage), and assumptions as to BECCS regional environmental and economic performance, but also on systems dynamics, regional structure of the economy, regional energy system, regional mitigation technology and policies, *etc.* Peters *et al.*[128] show that across four different IAMS (IMAGE, MERGE-ETL, WITCH and POLES), 44 to 66% of global BECCS deployment to meet a 2 °C target is provided by China, USA, India, Russia, Europe and Brazil. There is however a high variation between IAMs outcomes in the regional shares of deployment. For example, China provides 21–22% of global removal *via* BECCS in MERGE and IMAGE, but only 5% in WITCH and POLES. In a recent study using the EPPA model, a general equilibrium model of the world economy, BECCS does not get deployed in China and India at all because of regional land availability constraints due to large population expansion and competition with food production.[126] Figure 5.8 represents the total and regional removal *via* BECCS in 2100 in different modelling exercises.

When only considering the BECCS supply chain without systems integration considerations, cost minimising deployment of BECCS between Brazil, China, Europe, India and USA under land and water constraints leads to Brazil providing 34–44% of BECCS deployment under optimistic assumptions as to CO_2 storage availability in Brazil. Overall, factors such as high biomass yield, low cost of inputs, high electricity wholesale price, low carbon intensity of electricity, and high biomass potential close to storage sites are key drivers of regional BECCS deployment. The availability of geophysical resources is, however, a key constraint. When conservative

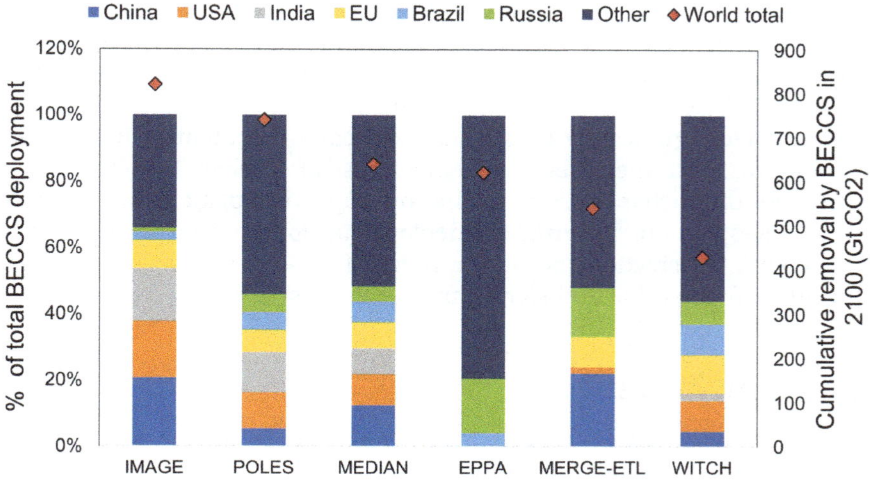

Figure 5.8 Cost-optimal level of removal by BECCS in different regions of the world in 2 °C IAMs scenarios. Data from Peters *et al.* (2017)[125] and Fajardy *et al.* (2020).[43]

estimates are used for CO_2 storage availability in Brazil (only 1 Gt CO_2 quantified in the Santos Basin), China becomes the main provider of CO_2 removal.[43]

Finally, when other objectives are considered in the optimisation, such as water use, land use change, biochemical flows or biodiversity loss, the qualitative picture of BECCS optimal deployment can also change greatly. In Heck *et al.*,[109] moving optimisation from minimising biodiversity loss to minimising water use moves BECCS deployment to tropical and equatorial areas, such as Brazil, southern US, southern China and equatorial Africa, causing substantially higher land use change and biodiversity loss. The same trade-offs are also observed in another study where water minimising BECCS deployment led to 300 Mha extra land use change, as compared to the land minimising scenario.[127] This points to the need for multi-optimisation accounting for the monetisation of BECCS environmental impacts.[50]

5.5.4 NET Co-deployment and Synergies

Most IAMs consider BECCS and afforestation within their CDR methods portfolio.[8] Recent studies also added Direct Air Capture (DAC) to the portfolio.[127] There is, however, a paucity of systematic studies exploring the co-deployment of a broad range of CDR methods to fully capture the integration effects of these technologies.[128] In the context of BECCS in particular, potential synergies could be derived with afforestation[129] and biochar.[130] On afforestation, a forestry modelling exercise showed that forest carbon sequestration and bioenergy expansion policies could be complementary in the short to medium term, bringing about cost reductions and increased carbon sequestration and/or mitigation. This study showed that these synergies could be limited in time, with complementarity turning into competition when demand for woody biomass becomes too important.[129] In a study by Buss *et al.*,[118] the authors showed that using residual wood ash from a BECCS process could significantly enrich the fixed carbon content of biochar and therefore decrease the carbon abatement cost of biochar. However, the effectiveness of biochar nutrient cycling on improving plant growth requires further investigation.[130] Further co-deployment optimisation studies modelling potential interactions between a range of CDR methods are required to fully quantify the value and significance of these synergies.

5.6 Conclusion

BECCS is a suite of technologies which, when deployed carefully, can contribute to both CO_2 emissions mitigation – by replacing fossil energy – and carbon removal. As demonstrated in this chapter, clear definitions and metrics to measure the performance, cost and resource efficiency of BECCS mitigation and removal potential, are required to accurately compare BECCS with alternative mitigation and removal technologies. Given the potential limitations of sustainable biomass supply, these metrics are also important in identifying the best use of biomass for both mitigation removal, depending on regional energy systems and bio-geophysical resources. Finally, BECCS deployment in tandem with alternative carbon removal methods could bring about substantial synergies. Co-deployment of a portfolio of carbon removal methods and their interaction with the energy system, in different regional contexts, should be a research priority to better define BECCS deployment strategies in the next few decades.

References

1. S. Fuss, W. F. Lamb, M. W. Callaghan, J. Hilaire, F. Creutzig, T. Amann, T. Beringer, W. D. O. Garcia, J. Hartmann, T. Khanna, G. Luderer, G. F. Nemet, J. Rogelj, P. Smith, J. L. V. Vicente, J. Wilcox, M. Z. Dominguez and J. C. Minx, *Environ. Res. Lett.*, 2018, **13**, 063002.
2. J. Kemper, *Int. J. Greenhouse Gas Control*, 2015, **40**, 401–430.
3. A. Mishra, M. Kumar, K. Medhi and I. S. Thakur, *Biomass energy with carbon capture and storage (BECCS)*, ed. R. Kataki, A. Pandey, S. K. Khanal and D. Pant, Elsevier, 2020, ch. 17, pp. 399–427.
4. *Bioenergy with Carbon Capture and Storage*, ed. J. C. M. Pires and A. L. D. C. Gonçalves, Academic Press, 2019, DOI: 10.1016/C2017-0-04413-0.
5. J. Rogelj, A. Popp, K. V. Calvin, G. Luderer, J. Emmerling, D. Gernaat, S. Fujimori, J. Strefler, T. Hasegawa, G. Marangoni, V. Krey, E. Kriegler, K. Riahi, D. P. van Vuuren, J. Doelman, L. Drouet, J. Edmonds, O. Fricko, M. Harmsen, P. Havlík, F. Humpenöder, E. Stehfest and M. Tavoni, *Nat. Clim. Change*, 2018, **8**, 325–332.

6. IPCC, *Climate Change 2014, Mitigation of Climate Change. Contribution of Working Group III to the Fifth Assessment Report of the Intergovernmental Panel on Climate Change*, 2014.
7. D. Huppmann, J. Rogelj, V. Krey, E. Kriegler and K. Riahi, *Nat. Clim. Change,* 2018, **8**, 1027–1030.
8. J. Rogelj, D. Shindell, K. Jiang, S. Fifita, P. Forster, V. Ginzburg, C. Handa, H. Kheshgi, S. Kobayashi, E. Kriegler, L. Mundaca, R. Séférian and M. V. Vilariño, *Special Report on the impacts of global warming of 1.5 °C*, Intergovernmental Panel on Climate Change, Geneva, 2018.
9. A. C. Köberle, *Curr. Sustainable Renewable Energy,* 2019, **6**, 107–115.
10. H. A. Daggash, C. F. Heuberger and N. Mac Dowell, *Int. J. Greenhouse Gas Control*, 2019, **81**, 181–198.
11. V. Johansson, M. Lehtveer and L. Göransson, *Energy*, 2019, **168**, 532–541.
12. Energy Systems Catapult (ESC), *Innovating to Net Zero*, 2020.
13. S. E. Tanzer and A. Ramirez, *Int. J. Greenhouse. Gas Control.,* 2020, **100**, 103104.
14. F. Creutzig, N. H. Ravindranath, G. Berndes, S. Bolwig, R. Bright, F. Cherubini, H. Chum, E. Corbera, M. Delucchi, A. Faaij, J. Fargione, H. Haberl, G. Heath, O. Lucon, R. Plevin, A. Popp, C. Robledo-Abad, S. Rose, P. Smith, A. Stromman, S. Suh and O. Masera, *GCB Bioenergy*, 2015, 7, 916–944.
15. S. Fuss, W. F. Lamb, M. W. Callaghan, J. Hilaire, F. Creutzig, T. Amann, T. Beringer, W. de Oliveira Garcia, J. Hartmann, T. Khanna, G. Luderer, G. F. Nemet, J. Rogelj, P. Smith, J. L. V. Vicente, J. Wilcox, M. del Mar Zamora Dominguez and J. C. Minx, *Environ. Res. Lett.*, 2018, **13**, 063002.
16. R. H. Williams, Fuel Decarbonization for Fuel Cell Applications and Sequestration of the Separated CO_2 CEES Report *295*, 1996.
17. H. J. Herzog and E. M. Drake, *Annu. Rev. Energy Environ.*, 1996, **21**, 145–166.
18. S. Gollakota and S. McDonald, *Energy Procedia*, 2014, **63**, 5986–5993.
19. F. Cherubini, G. P. Peters, T. Berntsen, A. H. Strømman and E. Hertwich, *GCB Bioenergy*, 2011, **3**, 413–426.
20. Global CCS Institute, *Bioenergy and Carbon Capture adn Storage: 2019 Perspective*, 2019.
21. Global CCS Institute, *The Global Status of CCS*, 2018.
22. T. Gebreegziabher, A. O. Oyedun and C. W. Hui, *Energy*, 2013, **53**, 67–73.
23. H. T. Luk, T. Y. G. Lam, A. O. Oyedun, T. Gebreegziabher and C. W. Hui, *Energy*, 2013, **63**, 205–215.
24. S. V. Vassilev, D. Baxter, L. K. Andersen and C. G. Vassileva, *Fuel*, 2010, **89**, 913–933.
25. X. Lu, M. R. Withers, N. Seifkar, R. P. Field, S. R. H. Barrett and H. J. Herzog, *Bioresour. Technol.*, 2015, **183**, 1–9.
26. I. Obernberger and G. Thek, *The Pellet Hanbook, the Production and Thermal Utilisation of Biomass Pellets*, 2010.

27. D. J. Murphy, C. A. S. Hall, M. Dale and C. Cleveland, *Sustainability*, 2011, **3**, 1888–1907.
28. M. Bui, C. S. Adjiman, A. Bardow, E. J. Anthony, A. Boston, S. Brown, P. S. Fennell, S. Fuss, A. Galindo, L. A. Hackett, J. P. Hallett, H. J. Herzog, G. Jackson, J. Kemper, S. Krevor, G. C. Maitland, M. Matuszewski, I. S. Metcalfe, C. Petit, G. Puxty, J. Reimer, D. M. Reiner, E. S. Rubin, S. A. Scott, N. Shah, B. Smit, J. P. M. Trusler, P. Webley, J. Wilcox and N. Mac Dowell, *Energy Environ. Sci.*, 2018, **11**, 1062–1176.
29. M. Fajardy and N. Mac Dowell, *Energy Environ. Sci.*, 2017, **10**, 1389–1426.
30. P. H. Joseph Fargione, J. Hill, D. Tilman and S. Polasky, *Science*, 2008, **319**, 1235–1239.
31. J. Daystar, C. Reeb, R. Gonzalez, R. Venditti and S. S. Kelley, *Fuel Process. Technol.*, 2015, **138**, 164–174.
32. A. Don, B. Osborne, A. Hastings, U. Skiba, M. S. Carter, J. Drewer, H. Flessa, A. Freibauer, N. Hyvönen, M. B. Jones, G. J. Lanigan, Ü. Mander, A. Monti, S. N. Djomo, J. Valentine, K. Walter, W. Zegada-Lizarazu and T. Zenone, *GCB Bioenergy*, 2012, **4**, 372–391.
33. T. Searchinger, R. Heimlich, R. A. Houghton, F. Dong, A. Elobeid, J. Fabiosa, S. Tokgoz, D. Hayes and T. Yu, *Science*, 2008, **423**, 1238–1241.
34. R. J. Plevin, M. O'hare, A. D. Jones, M. S. Torn and H. K. Gibbs, *Environ. Sci. Technol.*, 2010, **44**, 8015–8021.
35. M. Fajardy and N. Mac Dowell, *Energy Environ. Sci.*, 2017, **11**, 1389–1426.
36. L. J. Smith and M. S. Torn, *Clim. Change*, 2013, **118**, 89–103.
37. S. Pye, P.-H. Li, I. Keppo and B. O'Gallachoir, *Energy Strategies Rev.*, 2019, **24**, 314–330.
38. Committee on Climate Change (CCC), *Net Zero - The UK' s Contribution to Stopping Global Warming*, 2019.
39. S. E. Tanzer and A. Ramirez, *Energy Environ. Sci.*, 2019, **12**, 1210–1218.
40. J. Fargione, J. Hill, D. Tilman, S. Polasky and P. Hawthorne, *Science*, 2008, **319**, 1235–1237.
41. S. V. Hanssen, V. Daioglou, V. Steinmann, Z. J. N. Doelman, J. C. Van Vuuren and M. A. J. Huijbregts, *Nat. Clim. Change*, 2020, **10**, 1023–1029.
42. M. Fajardy and N. Mac Dowell, *Energy Environ. Sci.*, 2018, **11**, 1581–1594.
43. M. Fajardy and N. Mac Dowell, *One Earth*, 2020, **3**, 214–225.
44. M. M. Núñez-Regueiro, S. F. Siddiqui and R. J. Fletcher Jr, *Conserv. Biol.* 2019, **35**(1), 77–87.
45. P. Smith, S. J. Davis, F. Creutzig, S. Fuss, J. Minx, B. Gabrielle, E. Kato, R. B. Jackson, A. Cowie, E. Kriegler, D. P. van Vuuren, J. Rogelj, P. Ciais, J. Milne, J. G. Canadell, D. McCollum, G. Peters, R. Andrew, V. Krey, G. Shrestha, P. Friedlingstein, T. Gasser, A. Grubler, W. K. Heidug, M. Jonas, C. D. Jones, F. Kraxner, E. Littleton, J. Lowe, J. R. Moreira, N. Nakicenovic, M. Obersteiner, A. Patwardhan, M. Rogner, E. Rubin, A. Sharifi, A. Torvanger, Y. Yamagata, J. Edmonds and C. Yongsung, *Nat. Clim. Chang.*, 2016, **6**, 42–50.

46. L. Rosa, D. L. Sanchez, G. Realmonte, D. Baldocchi and P. D'Odorico, *Renewable Sustainable Energy Rev.*, 2020, 110511.
47. M. M. Mekonnen and A. Y. Hoekstra, *The Green, Blue and Grey Water Footprint of Crops and Derived Crop Products. Volume 2: Appendices*, 2010, vol. 2.
48. A. B. Harper, T. Powell, P. M. Cox, J. House, C. Huntingford, T. M. Lenton, S. Sitch, E. Burke, S. E. Chadburn, W. J. Collins, E. Comyn-Platt, V. Daioglou, J. C. Doelman, G. Hayman, E. Robertson, D. van Vuuren, A. Wiltshire, C. P. Webber, A. Bastos, L. Boysen, P. Ciais, N. Devaraju, A. K. Jain, A. Krause, B. Poulter and S. Shu, *Nat. Commun.*, 2018, **9**, 2938.
49. M. Fajardy, S. Chiquier and N. Mac Dowell, *Energy Environ. Sci.*, 2018, **11**, 3408–3430.
50. S. K. Mishra, M. C. Negri, J. Kozak, J. F. Cacho, J. Quinn, S. Secchi and H. Ssegane, *GCB Bioenergy*, 2019, **11**, 748–762.
51. V. Heck, D. Gerten, W. Lucht and A. Popp, *Nat. Clim. Change*, 2018, **8**, 151–155.
52. IEA Statistics, IEA Statistics, https://www.iea.org/statistics/.
53. C. M. Beal, I. Archibald, M. E. Huntley, C. H. Greene and Z. I. Johnson, *Earth's Future*, 2018, **6**, 524–542.
54. A. J. Melara, U. Singh and L. M. Colosi, *Energy Convers. Manag.*, 2020, **224**, 113300.
55. W. Wu, T. Hasegawa, S. Fujimori, K. Takahashi and K. Oshiro, *Renew. Energy*, 2020, **162**, 308–321.
56. T. W. Hertel, A. A. Golub, A. D. Jones, M. O'Hare, R. J. Plevin and D. M. Kammen, *Bioscience*, 2010, **60**, 223–231.
57. D. J. Garcia, B. M. Lovett and F. You, *J. Cleaner Prod.*, 2019, **228**, 941–955.
58. I. Lewandowski, J. C. Clifton-Brown, J. M. O. Scurlock and W. Huisman, *Biomass Bioenergy*, 2000, **19**, 209–227.
59. D. R. Nhuchhen and P. A. Salam, *Fuel*, 2012, **99**, 55–63.
60. H. Haberl, T. Beringer, S. C. Bhattacharya, K. H. Erb and M. Hoogwijk, *Curr. Opin. Environ. Sustain.*, 2010, **2**, 394–403.
61. Lawrence Livermore National Laboratory (LLNL), *Getting to Neutral: Options for Negative Carbon Emissions in California*, 2020.
62. N. Pour, P. A. Webley and P. J. Cook, *Int. J. Greenhouse Gas Control*, 2018, **68**, 1–15.
63. M. Bui, M. Fajardy and N. Mac Dowell, *Appl. Energy*, 2017, **195**, 289–302.
64. J. Hetland, P. Yowargana, S. Leduc and F. Kraxner, *Int. J. Greenhouse Gas Control*, 2016, **49**, 330–342.
65. IEAGHG, *Potential for Biomass and Carbon Dioxide Capture and Storage*, 2011.
66. O. Emenike, S. Michailos, K. N. Finney, K. J. Hughes, D. Ingham and M. Pourkashanian, *Sustainability Energy Technol. Assess.*, 2020, **40**, 100743.
67. P. Feron, A. Cousins, K. Jiang, R. Zhai, S. Shwe Hla, R. Thiruvenkatachari and K. Burnard, *Int. J. Greenhouse Gas Control*, 2019, **87**, 188–202.

68. T. Hirata, T. Tsujiuchi, T. Kamijo, S. Kishimoto, M. Inui, S. Kawasaki, Y. J. Lin, Y. Nakagami and T. Nojo, *Int. J. Greenhouse Gas Control*, 2020, **111**, 103473.
69. K. Jiang, P. Feron, A. Cousins, R. Zhai and K. Li, *Environ. Sci. Technol.*, 2020, **54**(4), 2429–2438.
70. S. Fuss, W. F. Lamb, M. W. Callaghan, J. Hilaire, F. Creutzig, T. Amann, T. Beringer, W. de Oliveira Garcia, J. Hartmann, T. Khanna, G. Luderer, G. F. Nemet, J. Rogelj, P. Smith, J. L. V. Vicente, J. Wilcox, M. del Mar Zamora Dominguez and J. C. Minx, *Environ. Res. Lett.*, 2018, **13**(6), 063002.
71. L. Yan, Z. Wang, Y. Cao and B. He, *Renewable Energy*, 2020, **147**, 1188–1198.
72. N. Pour, P. A. Webley and P. J. Cook, *Int. J. Greenhouse Gas Control*, 2018, **68**, 1–15.
73. M. Keller, K. Kaibe, H. Hatano and J. Otomo, *Int. J. Greenhouse Gas Control*, 2019, **83**, 69–82.
74. Q. Yi, Y. Zhao, Y. Huang, G. Wei, Y. Hao, J. Feng, U. Mohamed, M. Pourkashanian, W. Nimmo and W. Li, *Appl. Energy*, 2018, **225**, 258–272.
75. N. McGlashan, N. Shah, B. Caldecott and M. Workman, *Process Saf. Environ. Prot.*, 2012, **90**, 501–510.
76. D. McLaren, *Process Saf. Environ. Prot.*, 2012, **90**, 489–500.
77. A. Arasto, K. Onarheim, E. Tsupari and J. Kärki, *Energy Procedia*, 2014, **63**, 6756–6769.
78. K. Al-qayim, W. Nimmo and M. Pourkashanian, *Int. J. Greenhouse Gas Control*, 2015, **43**, 82–92.
79. A. Bhave, R. H. S. Taylor, P. Fennell, W. R. Livingston, N. Shah, N. Mac Dowell, J. Dennis, M. Kraft, M. Pourkashanian, M. Insa, J. Jones, N. Burdett, A. Bauen, C. Beal, A. Smallbone and J. Akroyd, *Appl. Energy*, 2017, **190**, 481–489.
80. J. C. Abanades, M. Alonso and N. Rodríguez, *Ind. Eng. Chem. Res.*, 2011, **50**, 6972–6981.
81. K. Möllersten, L. Gao and J. Yan, *Mitigation Adapt. Strategies Global Change*, 2006, **11**, 1129–1150.
82. K. Onarheim, A. Mathisen and A. Arasto, *Int. J. Greenhouse Gas Control*, 2015, **36**, 93–105.
83. M. C. Carbo, R. Smit, B. Van Der Drift and D. Jansen, *Energy Procedia*, 2011, **4**, 2950–2954.
84. M. Ranjan and H. J. Herzog, *Energy Procedia*, 2011, **4**, 2869–2876.
85. D. L. Sanchez and D. S. Callaway, *Appl. Energy*, 2016, **170**, 437–444.
86. D. Humbird, R. E. Davis, L. Tao, C. M. Kinchin, D. D. Hsu, A. Aden, P. Schoen, J. Lukas, B. Olthof, M. Worley, D. Sexton and D. Dudgeon, *Process Design and Economics for Biochemical Conversion of Lignocellulosic Biomass to Ethanol*, 2011.
87. I. Hannula and K. Melin, *Biorefineries with CCS: Report for IEAGHG*, 2021.

88. X. Lu and H. Herzog, *Biomass to Liquid Fuels Pathways: A Techno-Economic Environmental Evaluation*, 2015.
89. A. Laude, O. Ricci, G. Bureau, J. Royer-Adnot and A. Fabbri, *Int. J. Greenhouse Gas Control*, 2011, **5**, 1220–1231.
90. G. Liu, E. D. Larson, R. H. Williams, T. G. Kreutz and X. Guo, *Energy Fuels*, 2011, **25**, 415–437.
91. N. Johnson, N. Parker and J. Ogden, *Energy Procedia*, 2014, **63**, 6770–6791.
92. J. R. Moreira, V. Romeiro, S. Fuss, F. Kraxner and S. A. Pacca, *Appl. Energy*, 2016, **179**, 55–63.
93. H. Balat and E. Kırtay, *Int. J. Hydrogen Energy*, 2010, **35**, 7416–7426.
94. P. Parthasarathy and K. S. Narayanan, *Renew. Energy*, 2014, **66**, 570–579.
95. IEAGHG, *CO_2 Capture at Coal Based Power and Hydrogen Plants*, 2014.
96. C. Koroneos, A. Dompros and G. Roumbas, *Chem. Eng. Process.*, 2008, **47**, 1261–1268.
97. K. Onarheim, S. Santos, P. Kangas and V. Hankalin, *Int. J. Greenhouse Gas Control*, 2017, **66**, 60–75.
98. H. Mandova, P. Patrizio, S. Leduc, J. Kjärstad, C. Wang, E. Wetterlund, F. Kraxner and G. William, *J. Cleaner Prod.*, 2019, **218**, 118–129.
99. A. Anca-Couce, C. Hochenauer and R. Scharler, *Renewable Sustainability Energy Rev.*, 2021, **135**, 110237.
100. M. Muratori, H. Kheshgi, B. Mignone, L. Clarke, H. McJeon and J. Edmonds, *Int. J. Greenhouse Gas Control*, 2017, **57**, 34–41.
101. A. Poluzzi, G. Guandalini and M. C. Romano, *Biomass Bioenergy*, 2020, **142**, 105618.
102. D. Thrän, M. Bauschmann, N. Dahmen, B. Erlach, K. Heinbach, B. Hirschl, J. Hildebrand, I. Rau, S. Majer, K. Oehmichen, P. Schweizer-Ries and C. Hennig, *Biomass Bioenergy*, 2020, **142**, 105769.
103. P. Patrizio, M. Fajardy, M. Bui and N. Mac Dowell, *iScience*, 2021, 102765.
104. C. S. Galik, *Nat. Clim. Change*, 2019, **10**, 2–3.
105. M. Bentham, J. Lowndes and A. Green, *Energy Procedia*, 2014, **63**, 5103–5113.
106. Energy Technologies Institute (ETI), *A Picture of CO_2 Storage in the UK: Learnings from the ETI's UKSAP and Derived Projects*, 2013.
107. F. Creutzig, N. H. Ravindranath, G. Berndes, S. Bolwig, R. Bright, F. Cherubini, H. Chum, E. Corbera, M. Delucchi, A. Faaij, J. Fargione, H. Haberl, G. Heath, O. Lucon, R. Plevin, A. Popp, C. Robledo-Abad, S. Rose, P. Smith, A. Stromman, S. Suh and O. Masera, *GCB Bioenergy*, 2015, 7, 916–944.
108. V. Daioglou, J. C. Doelman, B. Wicke, A. Faaij and D. P. van Vuuren, *Glob. Environ. Change*, 2019, **54**, 88–101.
109. V. Heck, D. Gerten, W. Lucht and A. Popp, *Nat. Clim. Change*, 2018, **8**, 151–155.
110. J. Rogelj, G. Luderer, R. C. Pietzcker, E. Kriegler, M. Schaeffer, V. Krey and K. Riahi, *Nat. Clim. Change*, 2015, **5**, 519–527.

111. International Labor Organisation, ILOSTAT, https://www.ilo.org/ilostat/faces/wcnav_defaultSelection?_afrLoop=1325411423028499.
112. R. Séférian, M. Rocher, C. Guivarch and J. Colin, *Environ. Res. Lett.*, 2018, **13**, 054011.
113. IPCC, *Climate change and land*, 2019.
114. E. Baik, D. L. Sanchez, P. A. Turner, K. J. Mach, C. B. Field and S. M. Benson, *Proc. Natl. Acad. Sci.*, 2018, 201720338.
115. H. A. Daggash and N. Mac Dowell, *Joule*, 2019, **3**, 1239–1251.
116. M. Fridahl and M. Lehtveer, *Energy Res. Soc. Sci.*, 2018, **42**, 155–165.
117. A. Linzenich, K. Arning, J. Offermann-van Heek and M. Ziefle, *Energy Res. Soc. Sci.*, 2019, **48**, 205–218.
118. E. Cox, E. Spence and N. Pidgeon, *Nat. Clim. Change*, 2020, **10**, 744–749.
119. P. Patrizio, S. Leduc, F. Kraxner, S. Fuss, G. Kindermann, S. Mesfun, K. Spokas, A. Mendoza, N. Mac Dowell, E. Wetterlund, J. Lundgren, E. Dotzauer, P. Yowargana and M. Obersteiner, *Joule*, 2018, 1–16.
120. M. Fajardy, P. Patrizio, H. A. Daggash and N. Mac Dowell, *Front. Clim.*, 2019, **1**, 6.
121. S. Fuss, J. G. Canadell, P. Ciais, R. B. Jackson, C. D. Jones, A. Lyngfelt, G. P. Peters and D. P. Van Vuuren, *One Earth*, 2020, **3**, 145–149.
122. Global Energy Observatory, Google, *KTH Royal Institute of Technology in Stockholm, Enipedia and World Resources Institute (WRI)*, 2018.
123. I. S. Tagomori, F. M. Carvalho, F. T. F. da Silva, P. R. Paulo, P. R. R. Rochedo, A. Szklo and R. Schaeffer, *Int. J. Greenhouse Gas Control*, 2018, **68**, 112–127.
124. E. Beagle and E. Belmont, *BECCS Negative Emission Potential Data*, Texas Data Repository, 2019, V1, DOI: 10.18738/T8/ZORXGL.
125. G. P. Peters and O. Geden, *Nat. Clim. Change*, 2017, 7, 619–621.
126. M. Fajardy, J. Morris, A. Gurgel, H. Herzog, N. Mac Dowell and S. Paltsev, *The Economics of Bioenergy with Carbon Capture and Storage (BECCS) Deployment in a 1.5 °C or 2 °C World*, 2020.
127. J. Fuhrman, H. McJeon, P. Patel, S. C. Doney, W. M. Shobe and A. F. Clarens, *Nat. Clim. Change.*, 2020, **10**, 920–927.
128. R. R. Tan, S. Bandyopadhyay and D. C. Y. Foo, *Resour., Conserv. Recycl.*, 2020, **153**, 104540.
129. J. S. Baker, C. M. Wade, B. L. Sohngen, S. Ohrel and A. A. Fawcett, *Energy Policy*, 2019, **126**, 391–401.
130. W. Buss, S. Jansson, C. Wurzer and O. Mašek, *ACS Sustainability Chem. Eng.*, 2019, 7, 4204–4209.

CHAPTER 6

Direct Air Capture

HOWARD HERZOG

Massachusetts Institute of Technology, Cambridge, MA 02139, USA
Email: hjherzog@mit.edu

Direct air capture (DAC) is technically feasible today, with commercial units already in operation. However, understanding DAC's economic feasibility is required to understand whether DAC will be a significant negative emissions technology. Several start-up companies are trying to commercialise DAC using processes based on chemical absorption or adsorption. Today, there is only one DAC installation generating negative emissions, with a selling price of \$1200 per metric ton of CO_2 (t CO_2). This chapter explores the process fundamentals that apply to all types of DAC processes, looks at the process design for the two major types of DAC processes being developed today, and assesses estimates for DAC costs going forward. The assessment suggests that the low range of cost estimates in the literature, \$100–300 per t CO_2, will not be reached anytime soon, if at all. Reasonable expectations for DAC costs in 2030 are in the range of \$600–1000 per net t CO_2 removed.

6.1 Overview

Direct Air Capture (DAC) can be defined as "chemical processes that capture CO_2 from ambient air and concentrate it, so that it can be injected into a storage reservoir".[1] This invokes images of machines sucking carbon dioxide (CO_2) out of the air, allowing the world to continue business as usual, emitting CO_2 from the chimneys of our houses, the tailpipes of our cars, and the smokestacks of our industries. No wonder DAC is such a seductive concept.

Energy and Environment Series No. 31
Greenhouse Gas Removal Technologies
Edited by Mai Bui and Niall Mac Dowell

Published by the Royal Society of Chemistry, www.rsc.org

This chapter will examine the various chemical processes that capture CO_2 from the air. The topic of injecting CO_2 into storage reservoirs is shared by many CO_2 capture and storage (CCS) technologies, including bioenergy with CCS (BECCS) (see Chapter 5). It should be pointed out that some DAC concepts call for utilisation of the captured CO_2, such as conversion into transport fuels. However, this does not result in negative emissions, since CO_2 is released back to the atmosphere when the fuels are burned. If the captured CO_2 is used to make long-lasting products, such as building materials, then it may qualify as negative emissions. See Chapter 4 for a discussion of the appropriate way to account for negative emissions.

"The question for DAC is not whether we can suck CO_2 out of the air, but whether we can do it economically on a large scale."[2] We know we can remove CO_2 from the air because it has been done commercially for over 80 years in cryogenic oxygen plants. If the CO_2 were not removed from the air fed to these plants, it would form a solid under cryogenic conditions and upset the process, such as clogging heat exchangers. Submarines and space craft are other examples where CO_2 needs to be removed from the air. In these later cases, the reason is to provide breathable air for the human inhabitants.[3]

In the above examples, the goal was to remove the CO_2 from the air. For DAC, this is only a first step, as the CO_2 not only needs to be removed from the air, but also needs to be "captured". This requires the removed CO_2 to be collected at high concentration, pressurised, transported, and stored. In addition, for DAC to be effective as a climate change mitigation technology, it has to operate on a gigatonne (billions of metric tons) scale, which is six orders of magnitude greater than today's commercial DAC operations.

One of the earliest papers proposing DAC as an option for addressing climate change concerns was published in 1999.[4] Since then, the literature has proliferated with papers on DAC, as well as major reports including those from the American Physical Society[5] and the National Academies of Sciences.[1] A number of start-up companies have been formed that are trying to commercialise DAC technology. Two of these, Climeworks and Carbon Engineering, will be discussed later in this chapter.

While the specifics of the different processes for DAC that have been proposed or are under development differ, almost all are based on chemical absorption or adsorption (see Figure 6.1). These processes all have the following three steps in common:

- Step 1. Air is contacted with a sorbent. The CO_2 is scrubbed from the air by chemically binding with the sorbent.
- Step 2. The CO_2 is released from the sorbent and the sorbent is reused to capture more CO_2 from the air. Releasing the CO_2 involves elevating temperatures (temperature-swing), decreasing pressures (pressure-swing), and/or chemical processing.
- Step 3. The CO_2 that is released needs to be pressurised and transported for storage.

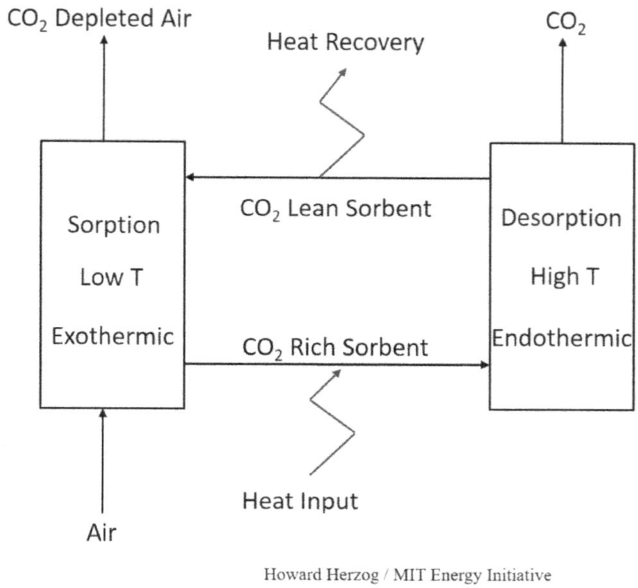

Figure 6.1 Schematic of generic sorption/desorption DAC processes.

It should be noted that some companies only want to make CO_2-enriched air to feed microalgae or greenhouses. These processes are not covered in this chapter because they do not lead to negative emissions. To achieve negative emissions, the CO_2 removed from the air must be kept out of the atmosphere for long periods of time, probably millennia or more. It is also important to note that alternate processes to sorption/desorption processes, such as electrochemical processes, have been suggested. However, they are not reviewed here because they are in a very early stage of development and there is very little information in the open literature.

The next section explores process fundamentals that apply to all types of DAC processes. Section 6.3 looks at process design considerations for the two major types of DAC processes being worked on today and includes commercial examples. Section 6.4 is an analysis of DAC costs and the chapter concludes with an overall assessment and a summary of the take-away messages in Section 6.5.

6.2 Process Fundamentals

A major difference between DAC processes compared to CCS processes at power plants or industrial facilities is the concentration of CO_2 in the feed stream. Today, the CO_2 concentration in air is about 410 parts per million (ppm) or 0.041%. Typical CO_2 concentrations range from 3–5% in gas turbine exhausts, 10–15% in coal-fired boiler flue gases, and 15–30% in cement plant exhausts. This low concentration of CO_2 in air makes the capture process more difficult and costlier compared to capture from power plants or industrial facilities.

The following simple example can help one visualise the impact of concentration on CO_2 capture. Imagine having a bowl on your desk containing 400 red marbles and 3600 blue marbles. The red marbles represent CO_2 molecules, which have a concentration of 10%. Your job is to remove the red marbles. Now imagine a much bigger bowl containing one million marbles, which 400 are red (0.04%) and the other 999 600 are blue. Once again, your task is to remove the red marbles. The latter task is much more difficult than the former, just as DAC is much more difficult than CO_2 capture from flue gases.

This section explores two important implications of the low concentration of CO_2 in air. The first is the large amount of air that must be treated in DAC processes. In the above analogy, though both bowls contain 400 red marbles, the total number of marbles is much larger in the bowl that corresponds to DAC. The second is the amount of energy that is required by DAC processes. This corresponds to the increased time and difficulty it takes to separate out the red marbles from the larger bowl. The analyses in these sections apply to all DAC processes, regardless of design or technology differences.

6.2.1 Air Handling and Contactor Design

What volume of air must be processed to capture one metric ton (tonne) of CO_2 (t CO_2)? To answer this question, we must first know how much CO_2 is in a given volume of air. Since air acts like an ideal gas, we can calculate this using the ideal gas law. Taking an atmospheric concentration of 410 ppm CO_2, the concentration of CO_2 in the air at 25 °C and atmospheric pressure is 0.74 grams per cubic meter ($g\,m^{-3}$). If we can capture all the CO_2 in a given volume of air, we would need to process 1.36 million m^3 of air to capture one tonne of CO_2. However, real DAC processes do not capture all of the CO_2 in the air, so even more air must be processed. For a DAC process that captures 75% of the CO_2 in the air, it must process 1.81 million m^3 of air to capture one tonne of CO_2.

The first step in DAC processes is to contact the air with a sorbent in what is generically called a contactor. Contactors are filled with packing to maximise the surface area for air-sorbent contacting. The CO_2 is removed from the air by chemically binding to the sorbent. The contactor has two dimensions, the cross-sectional area (perpendicular to the air flow) and the dimension parallel to the air flow, which will be referred to as the depth in this chapter (see Figure 6.2). DAC contactors are characterised by very large cross-sectional areas and relatively shallow depths.

We can calculate the required DAC contactor cross-sectional area as follows:

$$A = \frac{V_{\text{air}} \cdot r}{v \cdot \text{cf} \cdot \mu \cdot \text{spy}}$$

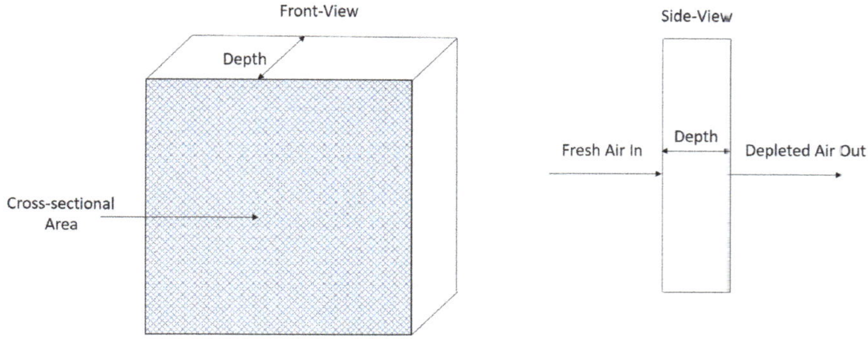

Figure 6.2 Generic schematic of an air contactor for DAC processes defining the cross-sectional area and the depth.

where A is the cross-sectional area in square meters (m^2), V_{air} is the volume of air containing one tonne of CO_2 (1.36 million m^3 per t CO_2 as calculated above for 410 ppm CO_2 concentrations), r is the capture rate in tonnes CO_2 per year, v is the linear velocity (*i.e.*, air speed) entering the contactor in meters per second ($m\,s^{-1}$), cf is the capture fraction, μ is the utilisation fraction (*i.e.*, what fraction of the time the process is running), and spy is the conversion factor for the number of seconds per year (31.536 million).

As an example, Carbon Engineering's design of a million tonne per year DAC plant[6] calls for a linear velocity of 1.4 $m\,s^{-1}$, a capture fraction of 0.745, and a utilisation fraction of 0.9; this requires about 46 000 m^2 of cross-sectional area. If the contactor was 10 m high (about the height of a 3-story building), it would need to be 4.6 kilometres (km) long. In reality it would not be a single unit, but broken down into multiple smaller units.

Figure 6.3 shows how the contactor cross-sectional area varies with linear velocities and capture fraction. Linear velocities for these types of contactors are almost always in the 1–5 $m\,s^{-1}$ range. If the linear velocity is too fast it will result in unacceptably high pressure drops through the contactor (see the discussion below). For a given sorbent, the capture fraction is a function of contactor depth, with capture fraction increasing with depth. While a larger capture fraction is desired in theory (*e.g.*, it reduces the cross-sectional area and the amount of air that must be processed), it also results in larger capital costs associated with increased depth and larger operating costs due to increased pressure drops. Therefore, increasing the capture fraction will result in some cost savings and some cost increases. The optimal capture fraction for a DAC process is determined by trading off these costs.

Fans are used to move the air through the air-sorbent contactor. As the air moves through the contactor, a pressure drop is inevitable. At a minimum, the fan must provide enough power to overcome this pressure drop. The amount of power required depends on both the volume of air going through the contactor and the pressure drop in the contactor. For the Carbon Engineering design, they estimate a pressure drop of 68 Pascals (Pa) and an

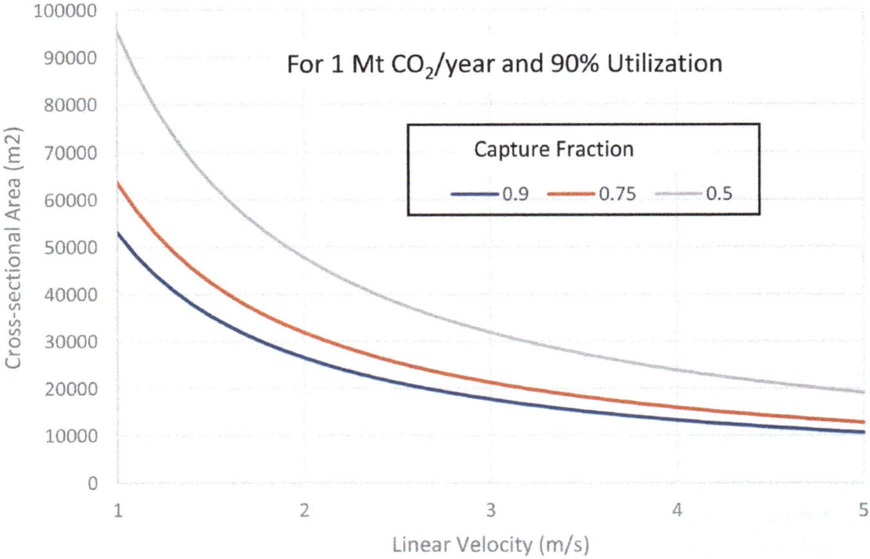

Figure 6.3 Required cross-sectional area of the air contactors for a million t CO_2 per year plant running 90% of the time as a function of linear velocity of the input air stream and the capture fraction.

energy requirement for the fan of 61 kilowatt-hours (kW h) per t CO_2.[6] Because of the large air flow, even an extremely small pressure drop like 68 Pa results in a fairly significant energy use. As a result, a requirement for DAC processes is to have low pressure drops through the contactor.

As discussed above, for large DAC plants (*e.g.*, a million tonnes per year), multiple contactors will be required. A good analogy is a wind farm that is made up of multiple wind turbines. Just as the spacing of wind turbines is important to make sure one turbine is not in the "wind shadow" of another turbine, so too is the spacing of the air contactors at a DAC facility. One wants to avoid feeding an air contactor with the CO_2-depleted air exhaust from another contactor. While models exist to design the proper spacing for wind farms, they do not yet exist (at least in the published literature) for DAC facilities. The land required to space out air contactors for a million t CO_2 per year DAC facility have been estimated to be in the range of 1–7 km^2.[1] Since one must purchase or pay for access to the land, as well as connecting all these air contactors together to gather the CO_2, provide power, *etc.*, costs will increase with larger land footprints.

6.2.2 Minimum Work and Energy Requirements

All DAC processes will require significant energy to remove CO_2 from the air. In this section, we will use the concept of "minimum work" to help understand the energy requirements of a DAC process. The minimum work calculation assumes that you have a reversible process that goes from the initial state to the final state. Minimum work tells us the absolute minimum amount of work needed to go between the end states, regardless of the specific process. Figure 6.4 is a schematic showing the end states for all carbon capture processes, including DAC. The minimum work is simply the difference in exergy (also called availability) between the output and input streams. Details of the minimum work calculation can be found in many texts and papers, including the APS report[5] and House *et al.*[7]

The CO_2 concentration of the inlet stream has a major impact on minimum work, with the minimum work for a DAC process being about three times as much as the carbon capture at a coal-fired power plant (Table 6.1). On the other hand, changing the capture fraction of a DAC process has only a small impact on minimum work.

Most carbon capture processes, including DAC processes, work in two steps, sorption followed by desorption (see Figure 6.1). The sorption process is exothermic, meaning it releases energy. The desorption is endothermic, meaning it requires energy. In practice, capture processes do not recover the energy released by sorption because that process takes place at low temperatures and desorption takes place at much higher temperatures. Therefore, a more meaningful way to assess minimum work is to look just at the desorption step (see Figure 6.5). Depending on the exact sorbent, the minimum work may increase significantly from the values shown in Table 6.1.

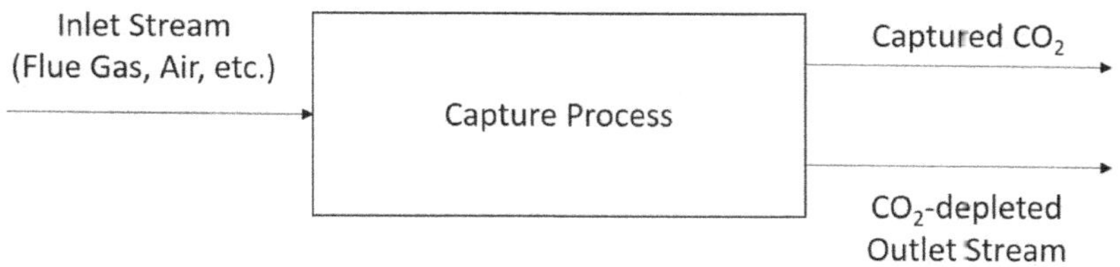

Figure 6.4 Schematic for carbon capture processes, including DAC, showing the inlet and outlet streams used in calculating the minimum work.

Table 6.1 Minimum work for several different carbon capture processes, assuming all inlet and outlet streams are at atmospheric pressure and 25 °C. The carbon capture processes are characterised by the CO_2 concentration of the inlet stream and the fraction of CO_2 removed from the inlet stream (*i.e.*, capture fraction). The calculations assume that the captured CO_2 outlet stream is 100% CO_2.

Inlet stream	CO_2 concentration	Capture fraction	Minimum work kW h per t CO_2
Air	410 ppm	0.9	133
Air	410 ppm	0.75	130
Air	410 ppm	0.5	126
Coal flue gas	11%	0.9	45.2
Natural gas flue gas	4%	0.9	61.5

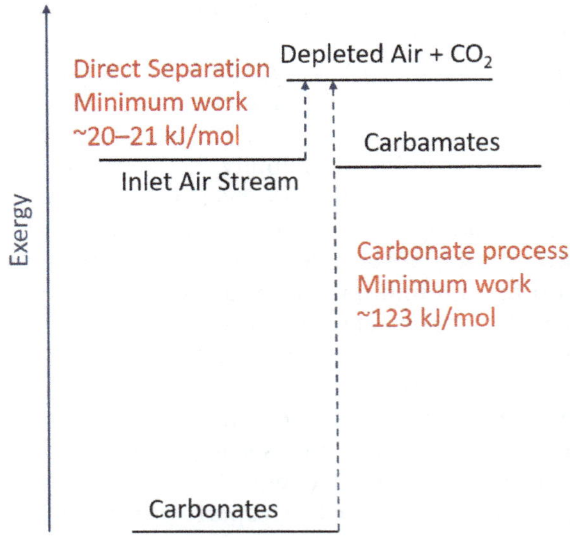

Figure 6.5 Schematic showing the differences in exergy between the streams in Figure 6.4 plus carbonates (formed by sorption of CO_2 by hydroxides) and carbamates (formed by sorption of CO_2 by amines). Based on material from ref. 8.

Using amines as a sorbent will increase the minimum work about 10%, while using hydroxides will increase the minimum work by 500–600%.[8]

The amount of work required by all real processes will be more than the minimum work. Actual work can be related to minimum work by a second law efficiency:

$$\mu_{2law} = W_{min}/W_{act}$$

where μ_{2law} is the second law efficiency, W_{min} is the minimum work for the process, and W_{act} is the actual work of the process. The second law efficiency is not a value that can be calculated rigorously like the minimum work, but it can be estimated empirically.[7]

Using data from MIT's The Future of Coal study,[9] the real work required to capture 90% of the CO_2 from a supercritical coal-fired power plant flue gas (11% CO_2 concentration), excluding compression, is 200 kW h per t CO_2. Comparing with the minimum work of 45.2 kW h per t CO_2 from Table 6.1 yields a second law efficiency of 0.23. Empirically, it has been shown that the more dilute the concentration in the inlet stream, the smaller the second law efficiency.[7] Since the concentration of CO_2 in air is two orders of magnitude below the concentration of CO_2 in a coal-fired power plant flue gas, one can expect the second law efficiencies for DAC to be significantly less than 0.23. If the second law efficiency was 0.1 for a DAC process, it would require about 1300 kW h per t CO_2, which is consistent with the energy requirements for the DAC processes discussed in the next section.

The carbon intensity of the energy used to operate the DAC process is a key factor in determining the amount of negative emissions generated. If the required 1300 kW h per t CO_2 of work came from a coal-fired power plant, you get no negative emissions, but actually add CO_2 to the atmosphere. This is because on a life-cycle basis the coal plant would emit about 1.2 t CO_2 for every t CO_2 captured from the air. For a gas plant, those emissions would be about 0.6 t CO_2 for every t CO_2 captured from the air. Therefore, you can get negative emissions using natural gas as an energy source, but the negative emissions are only 0.4 t CO_2 for every t CO_2 removed from the atmosphere. This increases costs for the negative emissions by 250%, making it impractical (see Section 6.4). As a result of the large energy requirement for DAC, it is imperative to use very low (*e.g.*, <0.1 t CO_2 emitted per t CO_2 removed from the atmosphere) or zero carbon energy sources for the DAC processes. This does not eliminate fossil fuels as an energy source, but does require the capture and storage of the CO_2 produced by the fossil fuels.

6.3 Process Design

To date (*i.e.*, beginning of 2021), there have been two primary approaches to DAC, one based on solid adsorption and one based on liquid absorption. The former system relies on using a weak base as the sorbent, usually a type of amine, while the later system uses a strong base, usually a hydroxide. In these systems, CO_2, a weak acid, is removed from the air by chemically binding with the base.

A major challenge of DAC systems is the small driving force for adsorption or absorption due to the low CO_2 concentration in the air. A strong base has the advantage of more easily grabbing the CO_2 from the air. However, it forms a stronger bond with the CO_2 compared to a weak base, making regeneration more difficult. This was illustrated in Figure 6.5, where the minimum work for regenerating hydroxides is about five times that of regenerating amines.

It is important to point out that the use of amines for DAC is very different than the use of amines at power plants. At power plants, amines are in the form of a liquid solution (nominally 30% solution in water). Large absorption towers are used to contact the flue gas with the amine solution, with the flue gas rising up the tower and the amine flowing down. For DAC, adsorber beds are used. Adsorption typically takes place in a fixed bed filled with a substrate of porous dry solids impregnated with amines. The substrate creates the large surface areas required, while the amine attracts and binds with the CO_2. The use of an amine adsorber for DAC *versus* an amine absorber for flue gases is driven by the differences in the feed stream; most notably the 300 times more volumetric gas needed to be processed for DAC and the factor of 300 lower CO_2 concentration in air compared to flue gas.

While some of the DAC literature likes classifying DAC systems as either solid or liquid, the more fundamental decision is choosing the sorbent. The choice of sorbent, up to now either a weak base or a strong base, then dictates much of the process design choices. In this section, each of these two approaches will be illustrated with specific commercial designs. Climeworks illustrates a weak base sorbent using solid adsorption beds. Carbon Engineering illustrates a strong base system, using a liquid absorption system. It must be noted that these companies are continuing their process development, so the following sections characterise where their designs are as of this writing. These processes will continue to evolve over time.

6.3.1 Climeworks

Climeworks is a Swiss company that was founded in 2009. Data about the company and their process reported in this section was provided to the author directly by Climeworks or comes from Beuttler *et al.*[10] The company has taken a modular design approach, with a single module having a nominal capacity of 50 t CO_2 per year at a utilisation of 100%. The first demonstration prototype was built in 2011. The first commercial DAC plant was commissioned in Hinwil, Switzerland in 2017 (see Figure 6.6). This plant contains 18 modules and uses the captured CO_2 to fertilise a greenhouse. In 2017, Climeworks partnered with Reykjavik Energy and their CarbFix project in Iceland for their first negative emissions plant.[11] Termed CarbFix2, the CO_2 captured by a Climeworks DAC module is injected into underground basaltic rock formations where it eventually turns into carbonate rocks over

Figure 6.6 Climeworks 18 module direct air capture plant in Hinwil, Switzerland. Photo by Climeworks. Reproduced from ref. 13 with permission from Climeworks, Copyright 2015.

Table 6.2 Plants built by Climeworks based on their 50 t CO_2 per year module. Data provided directly to the author from Climeworks.

Application	Number of modules	Capacity (t CO_2 per year)	Start date	Location
Power to methane[a]	1	50	2016	Oerlikon, Switzerland
Greenhouse fertilisation	18	900	2017	Hinwil, Switzerland
Negative emissions	1	50	2017	Hellisheidi, Iceland
Power to methane	3	150	2018	Troia, Italy
Beverage carbonation	12	600	2018	Hinwil, Switzerland
Power to liquids	1	50	2020	Karlsruhe, Germany

[a]The CO_2 captured from the air can be converted to fuels, such as methane or liquid fuels. The conversion requires hydrogen, which can be produced from the electrolysis of water. The resulting fuel can be considered carbon neutral if the electrolysis is powered by carbon-free electricity.

a period of a couple of years.[12] As of September 2020, Climeworks has built 6 DAC plants based on the 50 t CO_2 per year module (see Table 6.2). In addition, they have built 5 demonstration modules of 1–3 t CO_2 per year.

Each module, referred to as a "CO_2 collector", works on a cycle. For the current version of the module, each cycle lasts around 3.5 hours. Most of this time, the collector operates in adsorption mode. The fan blows air through the bed, which contains "alkaline-functionalised adsorbents" that chemically bind with the CO_2, thus removing it from the air. Eventually, the sorbent gets saturated, requiring regeneration. This is accomplished by a temperature-vacuum-swing desorption process, where the sorbent bed is heated to 80–120 °C and the pressure is lowered to 0.2–0.3 atmospheres, releasing the CO_2 from the sorbent and allowing it to be collected. The desorption cycle, including cooling down of the sorbent bed after desorption is complete, takes about 30–40 minutes. Now the bed is ready to return to adsorption mode.

The linear velocity of the air flowing to the sorbent bed is in the range of 2–5 m s^{-1}. The capture fraction is highest at the beginning of an absorption cycle, but decreases as the cycle continues and the sorbent bed becomes more and more saturated. The overall capture fraction is typically 0.6–0.7. The pressure drop through the bed is in the range of 150–200 Pa.

The exact system performance will vary depending on the ambient meteorological conditions, with humidity being a major factor. The current sorbent used by Climeworks will adsorb water vapour from the air, as well as the CO_2. The more water vapour adsorbed by the sorbent, the more energy to regenerate the sorbent per unit of CO_2 recovered. Therefore, dry conditions are preferred to humid conditions. During regeneration, the water can be recovered as a by-product. Under moderate climate conditions, about one tonne of water will be produced for each tonne of CO_2. This means that more than two molecules of water vapour are adsorbed for every molecule of CO_2.

The current energy requirements for Climework's DAC process are 500 kW h per t CO_2 electric energy and 2000 kW h per t CO_2 thermal energy. The primary purpose of the electric energy is to operate the fans and the vacuum pumps, while the thermal energy is used primarily to heat up the sorbent beds during regeneration.

Based on these numbers, the following observations can be made:

- The thermal requirement for this DAC plant, 2000 kW h per t CO_2, is about 2.6 times higher than the thermal energy requirement for a power plant CCS amine process, which is about 770 kW h per t CO_2.
- From Table 6.1, the minimum work for a DAC process is about 130 kW h per t CO_2. To be able to calculate a second law efficiency, the thermal energy must be converted to work. This can be done by calculating the work required by a heat pump with a coefficient of performance (COP) of 3.5.[†] The total work requirement is then 1071 kW h per t CO_2,[‡] about 8.2 times the minimum work for DAC. This yields a second law efficiency of 0.12, which is consistent with the analysis in Section 6.2.2

Another important and well-known issue is amine degradation, which occurs for several reasons, including exposure to oxygen (especially at elevated temperatures) and poisoning by pollutants such as SO_2 and NO_x. Even though the levels of these pollutants are small in ambient air, the large amount of air processed can make that exposure significant. Currently, the sorbent needs to be replaced after about 24 months and represents a significant maintenance cost.

In my discussions with Climeworks, they stressed that the next generation DAC system will look different compared to the current design. One item they are looking at is developing better sorbents. Goals include reducing the amount of sorbent degradation and better operation performance in humid conditions. In terms of energy use, they have a longer-term goal of reducing both the thermal and electric energy requirements by 20%. They are also working on new ways to manufacture the modules and system to reduce capital costs.

In late 2021, Climeworks started up a 4000 t CO_2 per year negative emission system in Iceland, incorporating a new module design.[§] They are also looking to the future and developing a design for a million t CO_2 per year DAC plant. Their modular approach has allowed them to get small DAC plants out into the market quickly and have benefited from learning by doing so. This approach is very different to that of Carbon Engineering, described in the next section.

[†]This is the COP being used by Climeworks in their plans to run the entire process on solar photovoltaics.

[‡]$500 + 2000/3.5 = 1071$.

[§]https://climeworks.com/news/climeworks-has-signed-groundbreaking-agreements-with

6.3.2 Carbon Engineering

Carbon Engineering is a Canadian-based company founded in 2009. Data about the company and their process reported in this section were either provided to the author directly by Carbon Engineering or comes from Keith *et al.*[6] Since 2015, the company has operated a 1 t CO_2 per day pilot facility in Squamish, British Columbia, Canada. A pilot plant that captures 5 t CO_2 per day is currently under construction at the same site.

The basic Carbon Engineering process is shown schematically in Figure 6.7. The process is designed to operate continuously, differing from Climeworks which operates in a cyclic manner. Note that Carbon Engineering has several variations of this process, with the base case design from Keith *et al.*[6] being described here.

The absorption step uses potassium hydroxide in solution to pull the CO_2 out of the air *via* the following reaction:

$$CO_2(g) + 2KOH(aq) = H_2O(l) + K_2CO_3(aq)$$

The contactor is not a traditional absorber, but a new design based on cooling tower technology. The absorber bed is filled with structured plastic packing and the sorbent flows down through the packing, wetting the surface. This is a cross-flow design, so the air flows horizontally through the bed, which is about 7 m deep. The estimated pressure drop through the bed is 68 Pa. The basic contactor unit is a cell that has a cross-sectional area of about 175 m^2. The linear velocity of the incoming air is 1.4 m s^{-1} with a capture fraction of 0.745. The air flow through the absorber will be driven by a 30-foot diameter fan. One cell, operating 90% of the time, can capture

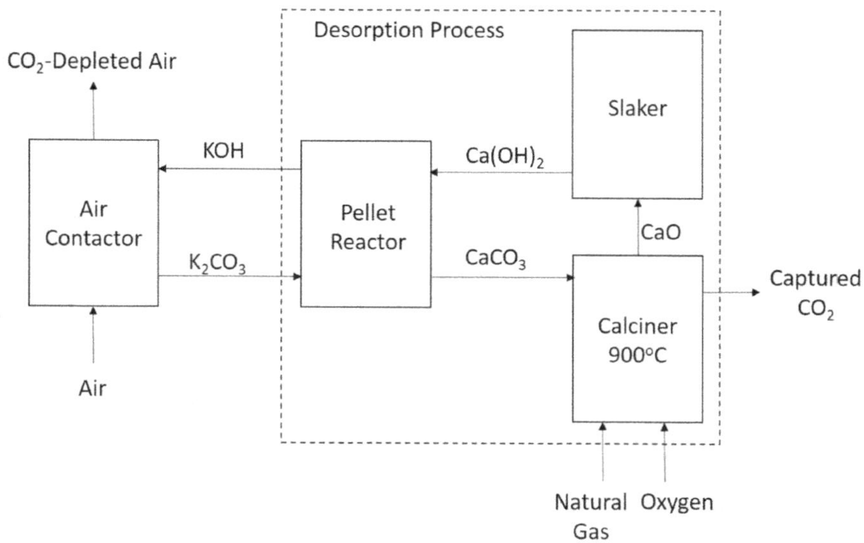

Figure 6.7 Schematic of Carbon Engineering's absorption and regeneration process. Based on material from ref. 6.

about 3800 t CO_2 per year. For commercial units, ten cells will be integrated into a module. Plants will consist of multiple modules properly spaced, all connected to a single central regeneration unit. For example, a million t CO_2 per year plant would require 26 modules.

Sorbent regeneration is more complex than the Climeworks process, requiring chemical processing at high temperatures. This is necessary because of the strong bond between the sorbent and CO_2 as discussed in Section 6.2.2. The regeneration process has 3 major steps, as shown in Figure 6.7.

The CO_2-rich sorbent from the absorber is sent to a crystalliser, termed a pellet reactor by Carbon Engineering. The CO_2 in the sorbent is in the form of K_2CO_3, which is transformed into KOH in the pellet reactor by the following reaction:

$$K_2CO_3(aq) + Ca(OH)_2(s) = 2KOH(aq) + CaCO_3(s)$$

The CO_2-lean sorbent containing the KOH is sent back to the air contactor. The $Ca(OH)_2$ comes from the slaker. The CO_2 leaves the reactor as $CaCO_3$, which is in the form of a solid pellet, and is sent to the calciner after drying and pre-heating.

The calciner is an oxy-fired circulating fluidised bed reactor (CFB) operating at 900 °C. Its job is to decompose the $CaCO_3$ into CaO (a solid which is sent to the slaker) and the product CO_2. The reaction requires a significant amount of energy, which is supplied by natural gas fired directly in the calciner. Because combusting the gas with air would dilute the product CO_2 with nitrogen, the gas is combusted in high-purity oxygen coming from a cryogenic air separation plant. The natural gas combustion produces additional CO_2, which leaves the calciner with the CO_2 that was removed from the air.

The slaker is a "refractory lined bubbling/turbulent fluid bed" that takes the CaO from the calciner and reacts it with steam to form $Ca(OH)_2$, which is sent to the pellet reactor. The slaking reaction is highly exothermic. The energy released is used to dry and preheat the $CaCO_3(s)$ pellets on their way from the pellet reactor to the calciner. In addition, the generated steam is sent to a steam turbine to produce electricity needed by the process.

The process is a net consumer of water, primarily from evaporation in the CO_2 contactor. The exact amount of water needed depends on the ambient meteorological conditions. At 20 °C and 64% relative humidity, the plant requires 4.7 tH_2O per t CO_2 captured from the atmosphere. The drier and hotter the conditions, the more the water loss will be. Low temperatures approaching freezing also impact operations because the viscosity of the liquid solvent increases.

The base case version of the process is powered entirely by natural gas. The natural gas is used to directly fire the calciner, as well as being fed to a gas turbine combined cycle facility to produce electricity required by the process. Most of the CO_2 produced by the combustion of the natural gas is captured and combined with the CO_2 captured from the air. The CO_2 from the natural gas accounts for about 35% of the final CO_2 product.

Other versions of the process eliminate the electricity production by importing the electricity from the grid or from a dedicated renewable energy facility.

For the base case design, the amount of natural gas required is 8.81 GJ per t CO_2 captured from the air. To calculate a second law efficiency, this quantity must be converted to work. Using the heat rate from Keith *et al.*[6] of 6.8 MJ $kW^{-1} h^{-1}$ (52.7% conversion efficiency) results in a work input of 1287 kW h per t CO_2, which includes the energy for compression. To be on the same basis as earlier analyses in this chapter of capture without compression, we subtract the compression energy of 120 kW h per t CO_2 to yield an energy requirement of 1167 kW h per t CO_2. This is slightly larger than the Climeworks energy requirement, yielding a slightly lower second law efficiency of 0.11. However, as shown in Figure 6.5, a more appropriate value of minimum work for hydroxide sorbents would be 123 kJ mol^{-1} (774 kW h per t CO_2); using that value, the second law efficiency is 0.66.

The above second law analysis results raise significant questions about the proposed Carbon Engineering process as presented in Keith *et al.*[6] A second law efficiency of 0.66 is well outside the experience with separation processes[7] and is probably unattainable without excessive costs. The fact that the energy requirement is about the same as Climeworks also raises questions. One would expect a higher energy requirement for the Carbon Engineering process because of the strong bond formed between the CO_2 and the sorbent and the multi-step high temperature regeneration process. One can say that the uncertainty in the energy numbers from Carbon Engineering is very much greater than those from Climeworks. This is because Climeworks can measure their energy use directly from complete DAC systems in commercial operation, whereas Carbon Engineering generates their numbers based on data from small-scale pilot units that are not fully integrated and uses simulations to extrapolate to commercial scale. Even small errors in estimates of unit performance can lead to large variations in energy requirements. Thus, in order to provide greater confidence, data from a fully integrated plant operating successfully for a prolonged period will be required.

Carbon Engineering is actively scaling up their process. One advantage their approach has over Climeworks is that their process should show significantly better economies of scale. As mentioned earlier, a 5 t CO_2 per day pilot plant is under construction at their Squamish site. Also, in partnership with Occidental Petroleum (Oxy), they are designing a million t CO_2 per year demonstration plant for the Permian Basin in Texas. The current design version of the Oxy plant will not generate its own electricity, but instead contract to purchase renewable energy from the grid, which is inexpensive in Texas. The captured CO_2 will be used for Enhanced Oil Recovery (EOR), producing oil that qualifies for California's low carbon fuel standard credits in addition to revenue from CO_2 sales and the 45Q tax credits from the US government, resulting in total revenues estimated to be \$250 per t CO_2.

6.4 Costs

As the previous sections have shown, there is no question about the technical feasibility of removing CO_2 from the atmosphere. The real question is whether it is economically feasible to do so at the gigatonne scale, the scale required to have a mitigating impact on climate change. This section will address that question by analysing DAC costs and related issues.

There is only one "real" cost number today for negative emissions from DAC. That number is \$1200 per t CO_2,[¶] the price at which Climeworks sells negative emissions generated by their DAC facility in Iceland (see Section 6.3.1). While informative, this is a price associated with a single facility, in a single location, at a specific point in time. As discussed in more detail below, care must be taken in extrapolating any cost number to different locations and times and scales of operations.

The Intergovernmental Panel on Climate Change Special Report *Global Warming of* 1.5 °C (IPCC SR15)[14] sums up what the literature says about DAC costs as follows: *However, the literature shows low agreement and is fragmented. This fragmentation is reflected in a large range of cost estimates: from \$*20–1000 *per t CO_2. There is lower agreement and a smaller evidence base at the lower end of the cost range. Fuss et al.* (2018)[15] *narrow this range to \$*100–300 *per t CO_2.* Here is exactly what *Fuss et al.* (2018)[15] have said: *Based upon our literature review, it appears that a first-of-a-kind plant may be on the order of \$*600–1000 *per t CO_2 initially, but that as advances are recognized through the building of more plants, this cost may decrease to \$*100–300 *per t CO_2.* Notice the use of "may" twice in the above statement. There is no guarantee that costs will come down that low. Also, the price of the first plant is now known; that price of \$1200 per t CO_2 is above the \$600–1000 per t CO_2 range suggested by Fuss *et al.* (2018).[15]

It is important to acknowledge that there is no single cost for DAC. Identical plants built in different locations will have different costs and these cost differences can be very significant. Rubin *et al.*[16] document many variables that should be provided with a cost number so that cost can be put into proper context. The list literally contains dozens of variables, including fuel costs, cost of capital, plant location, plant size, and whether it is for a plant built today or sometime in the future. To illustrate further, three items important to understanding DAC costs are:

- What is included in the estimate? If we are using DAC for negative emissions, then compression and storage costs need to be included. In the literature, some cost studies include these, but others do not.
- The cost in terms of \$ per t CO_2 must be on a net CO_2 removed basis. In Section 6.2.2, the implications of carbon content of the fuel were discussed. To get an accurate understanding of the "net CO_2 removed" quantity, a life-cycle carbon balance of the process must be conducted.

¶From https://www.climeworks.com/subscriptions, visited on May 11, 2022.

While some studies attempt this, almost all report the costs as \$/t of gross CO_2 removed from the air. A simple formula that relates the two quantities is:

\$ per t net CO_2 removed = \$ per t gross CO_2 removed/$(1 - x)$

where x = lifecycle CO_2 emitted by DAC/gross CO_2 captured by DAC. The formula is valid only if $x \leq 1$. If $x > 1$, then the DAC process produces no negative emissions, but is a CO_2 emitter.

- As discussed previously, ambient meteorological conditions can greatly impact DAC process performance. Costs for favourable sites may not be indicative for costs at the multitude of sites required for getting to the gigatonne scale.

In summary, it is critical to understand the full context of any cost number claimed for DAC processes. Without that understanding, it is impossible to make a credible assessment of DAC costs.

Climeworks has reported the cost of their DAC unit in Switzerland (see Figure 6.6) to be \$600 per t CO_2.[1] However, information to put this number in context is limited. We know compression and storage costs are not included, as the captured CO_2 is piped to an adjacent greenhouse at atmospheric pressure. Climeworks has stated that its carbon footprint on a lifecycle basis is about 90%, so the cost on a "net CO_2 removed" basis is 11% higher than their quoted \$600 per t CO_2 cost (see the formula in the previous paragraph). While no energy prices have been reported, one can infer that the project had favourable energy costs because (1) it is reported that the project uses "waste heat" for the thermal energy, which implies low cost and (2) Switzerland has inexpensive low-carbon electricity available due to hydroelectric power in the country. On the other hand, one can infer that the project was burdened with high capital costs. The DAC unit was "hand made in Switzerland", so significant capital cost reductions can be expected in the future. There is an additional Climeworks data point; the projected cost for their 4000 t CO_2 per year negative emissions DAC plant under construction in Iceland is reported to be \$700 per t CO_2.[‖]

In Keith *et al.*,[6] Carbon Engineering has estimated cost numbers of \$94–232 per t CO_2 and has provided the most supporting information of any DAC cost estimate presented to date. The numbers are based on detailed simulations with a very transparent listing of the input parameters. However, though their methodology is very commendable, the results are only as good as the inputs. Carbon Engineering is estimating the cost of a million t CO_2 per year plant based on performance data from a pilot plant that is not fully integrated and that is at a scale four orders of magnitude less than the commercial plant.

‖ https://www.cleanenergyministerial.org/sites/default/files/2020-04/CCUS%20-%20Direct%20Air%20Capture%20webinar.pdf page 35.

Some of the input data used in their costing model that may be too optimistic include:

- The energy requirement of their process, which seems excessively low based on the minimum work analysis presented in Section 6.3.2.
- The estimated pressure drop of 68 Pa.
- The contingency factor of 20%, where processes at this stage of development should use a contingency factor of about 100%.[16]

Rubin states[17] "Experience teaches us that cost estimates for early-stage technologies tend to be optimistic and poorly predict the actual cost of those technologies that reach commercialisation" (see Figure 6.8). There is no reason to think that DAC plants will not follow that pattern. A very public and relatively recent cost overrun is that of Southern Company and their Kemper County coal gasification power plant (see Table 6.3).[18] Note that the cost for

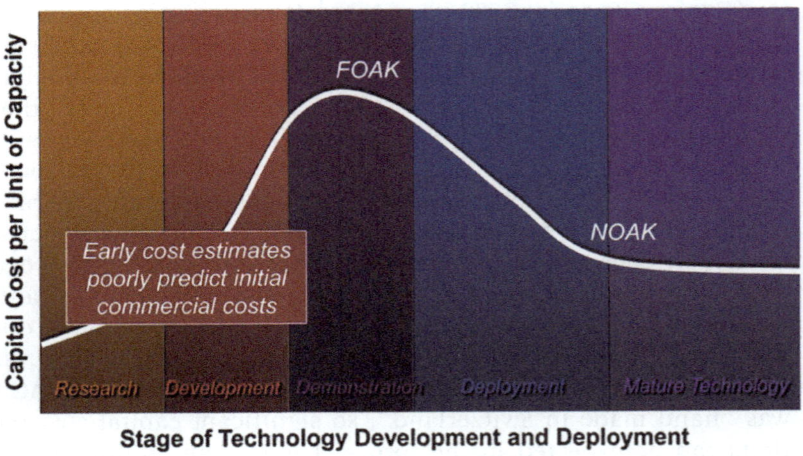

Figure 6.8 Typical cost trajectory of a new energy technology (adapted from ref. 19). FOAK is first-of-a-kind; NOAK is *n*th-of-a-kind. Reproduced from ref. 17 with permission from Elsevier, Copyright 2019.

Table 6.3 Comparing the scale-up of Southern Company's Kemper County gasification project to the scale-up proposed by Carbon Engineering for their DAC process.

	Southern/Kemper	Carbon Engineering
Pilot plant size	50 tons coal per day	0.6 tonne CO_2 per day
Full plant size	4 million tons coal per year	112 tonne CO_2 per hour
Scale-up ratio	~220	~4500
Pilot operations	10 + years	0 years integrated (some units ~1 year)
Cost estimate	\$2.2 billion	94–232 \$ per t CO_2
Actual cost	> \$7 billion	To be determined
Final status	Abandoned	To be determined

Kemper turned out to be over three times the estimate that was made at the beginning of their project. Compared to Carbon Engineering, Kemper had the advantage of a detailed FEED study, far more operating experience with the process at a pilot level, and a much smaller scale-up ratio.

Moving from pilot scale and small demonstration units to commercial-scale units will bring new challenges for DAC developers that will impact costs. For example, for a one million t CO_2 per year plant, the placing of the air contactors may require several square kilometres of land. Besides the land costs and permitting issues, the multiple contactors all need to be connected by piping, wiring, *etc.* In addition, DAC plants will need to be hardened for the elements, such as high winds, cold weather, dust storms, or whatever else mother nature may have in store.

As shown in Section 6.2.2, it is imperative for DAC processes to use energy that produces very low to zero CO_2 emissions. As a result, some DAC developers are considering future DAC installations to include a dedicated renewable energy source, such as a wind farm or solar array. While this sounds good in theory, intermittency of solar and wind will present challenges. Because DAC plants are capital-intensive, they need to run at a high utilisation rate; otherwise capital charges will become excessive. One solution to the problem is to include energy storage, but this significantly raises the cost of the renewable energy. Providing significant amounts of low-cost, low-carbon energy to DAC systems will be a major challenge as the size and number of DAC installations grow.

Finally, over the past decade several reasons have been proposed as to why DAC may have some cost advantages. I consider these "cost myths", as explained below.

- *Since DAC uses air as a feed stream and it is available anywhere, siting issues are less acute for air capture facilities.* Siting plants is a complicated process with many dimensions. Items like land availability, access to low-carbon energy and other utilities like water, permitting issues, acceptable meteorological conditions, accessibility of CO_2 storage options, *etc.* still put many constraints on the siting of DAC plants. It has specifically been suggested that DAC plants be sited close to CO_2 storage reservoirs. However, CO_2 transport costs tend to be small relative to the cost of capture, so even this advantage for DAC projects will produce only small savings compared to the overall project costs.
- *Since there is an unlimited feedstock of air, you are free to choose any capture fraction you want.* This is a major red herring. As discussed earlier, the optimal capture fraction is a result of a trade-off of factors. It is one of several operating parameters whose values are determined from the design process in order to come up with an optimal (*i.e.*, low-cost, high-reliability, *etc.*) design. This is standard operating procedure considered during process design and there is nothing special about the capture fraction of DAC processes.

- *The thermal energy for regeneration can be supplied by waste heat.* There may be some opportunities for a plant to take advantage of excess thermal energy, but if one is talking gigatonnes per year, those opportunities will be a very small fraction. Calling thermal energy at 80–120 °C "waste heat" is a misnomer. "Stranded heat" is a better description because infrastructure to distribute thermal energy is expensive and the distance it can be economically transported is fairly limited. A significant amount of energy each year is used for heating buildings at 25 °C; that is a huge market that could use waste heat that is an even lower quality than that required by DAC.

So, returning to the question this section started with: What is the cost for DAC to provide negative emissions on a gigatonne scale? The answer is no one really knows at this time. Here is what we do know:

- The only real cost number today is \$1200 per t CO_2.
- Climeworks claim cost numbers of \$600–700 per t CO_2 for their plants, but there is not enough detail to extrapolate sensibly. We do know the \$600–700 per t CO_2 are on a "gross CO_2 removed" basis, but the "net CO_2 removed" cost, which will be larger, is the real cost of negative emissions. Also, the \$600 per t CO_2 number does not include compression and storage costs.
- There are plenty of estimates out there, some better than others.

I feel comfortable in concluding the following:

- The cost for DAC is not a single number, but a range. There will be some outliers on the low end of the range due to favourable circumstances, but those opportunities will be limited. The cost range of interest needs to represent the costs of the bulk of the DAC plants required to get to a gigatonne negative emissions level.
- I see no basis to believe that the \$100–300 per t CO_2 range from the IPCC[14] report can be reached anytime soon, if at all. It is based on estimates from early-stage development, which from past experiences turn out to be unreliable.
- Costs today for negative emissions from DAC are on the order of \$1000 per net t CO_2 removed.
- There are opportunities to have significant cost reductions from today's costs, especially capital cost reductions. However, as plants scale up in size, there will be additional costs arising that the early small plants did not have to contend with.
- There is a floor for costs because there are two fundamental issues that DAC cannot avoid: (1) the large amount of air that must be processed that requires large air contactors and (2) the large amount of low- to zero-carbon energy required to capture CO_2 from the air.

6.5 Assessment and Takeaway Messages

Technologies to remove CO_2 from the air have been in commercial use for over 80 years, but this is different from DAC, which not only needs to remove the CO_2 from the air, but also capture and store it so it does not return to the atmosphere. Research into DAC started less than 25 years ago and companies formed to commercialise DAC are barely more than a decade old.

There is no question that DAC is technically feasible, with commercial units in operation today. The real question is one of cost. Today, the price to buy negative emissions from DAC is $1200 per t CO_2.

DAC technology will be continually evolving as it moves from small pilot projects, currently on the scale of a thousand t CO_2 per year, to plants on the million t CO_2 per year scale. This will provide developers with both opportunities as well as challenges, both technical and economic.

In order to significantly impact climate change, DAC will eventually need to deliver negative emissions on a billion t CO_2 per year scale. While technically feasible, it is unclear whether it will be economical.

Nature presents DAC with two major, non-negotiable challenges. First, large amounts of air must be processed, requiring large arrays of air contactors. Second, there is a minimum amount of work required to separate CO_2 from the air, resulting in significant energy requirements for DAC processes.

The correct units for DAC costs if you are interested in negative emissions is dollars (or other currency) per *net* amount of CO_2 removed from the atmosphere. Most DAC costs today are presented as dollars per *gross* amount of CO_2 removed from the atmosphere. These studies yield cost estimates that are too low. A life-cycle carbon balance is required to relate the two metrics.

To keep costs for negative emissions (*i.e.*, the net removed basis) from becoming excessive, DAC processes will need to use very low or zero carbon energy sources. Commercial companies are investigating different options including dedicated renewable energy as part of their projects. The future trajectory of carbon-free energy costs will have a significant impact on DAC costs.

DAC will face the same issues all large engineering facilities face. These include, but are not limited to:

- Project siting: The project must be sited with adequate land, in a place that meets technical requirements, satisfies regulatory and permitting rules, and has public acceptance.
- Water: For DAC processes that have significant water consumption, access to water can be a serious concern.
- Environment: As DAC deployment grows, it will come under more environmental scrutiny. A cautionary tale is provided by climate friendly technologies like wind and solar, which have run into environmental issues as they have grown.

There are many cost estimates for DAC in the literature. This chapter presented information to help better understand the literature and make informed assessments of the claims. A major message in this chapter is that a cost number for DAC is not very useful without details about what went into the cost estimate. Also, there is no single number for DAC costs, but there will be a range.

At this early stage of technology development, it is very difficult to estimate what costs will be for large-scale deployment sometime in the future. To repeat a message from Rubin,[17] "Experience teaches us that cost estimates for early-stage technologies tend to be optimistic and poorly predict the actual cost of those technologies that reach commercialisation".

My assessment of where we are on DAC costs looking forward: I do not find the range of \$100–300 per net t CO_2 removed to be credible for the multitude of reasons presented in this chapter. If I had to give a range of DAC costs in 2030, my educated guess would be \$600–1000 per net t CO_2 removed. There may be plants with favourable conditions that cost less, but they will be exceptions and not the rule.

Acknowledgements

I have been following and analysing developments in DAC from its inception over 20 years ago. In those early years, I am indebted to my colleague and mentor Prof. Ken Smith of MIT (Chemical Engineering Department) for his advice and the many conversations we had about DAC. More recently, I want to thank Carbon Engineering for hosting me at their pilot plant in Squamish, British Columbia in December, 2019 and Christoph Beuttler of Climeworks for meeting me on Zoom. Both companies were very open in sharing details of their technology with me and I came away very impressed with the quality of their engineering. Finally, I want to thank my colleagues at the MIT Energy Initiative and the wider CCS community for both their encouragement and advice regarding my analysis of DAC over these many years.

References

1. *Negative Emissions Technologies and Reliable Sequestration: A Research Agenda*, The National Academies Press, Washington, DC, 2019. https://nap.nationalacademies.org/catalog/25259/negative-emissions-technologies-and-reliable-sequestration-a-research-agenda.
2. H. Herzog, *Carbon Capture*, MIT Press, Cambridge, MA, 2018.
3. M. Ranjan and H. Herzog, *Energy Procedia*, 2011, **4**, 2869.
4. K. Lackner, P. Grimes, and H.-J. Ziock, presented in part at The *24*th Annual Conference on Coal Utilization and Fuel Systems, Clearwater, FL, 1999.
5. *Direct Air Capture of CO_2 with Chemicals: A Technology Assessment for the APS Panel on Public Affairs*, American Physical Society, 2011. https://www.aps.org/policy/reports/assessments/upload/dac2011.pdf.
6. D. Keith, G. Holmes, D. St Angelo and K. Heidel, *Joule*, 2018, **2**, 1.

7. K. House, A. Baclig, M. Ranjan, E. van Nierop, J. Wilcox and H. Herzog, *Proc. Natl. Acad. Sci.*, 2011, **108**, 20428.
8. H. Caram, R. Gupta, H. Thomann, F. Ni, S. Weston and M. Afeworki, *Int. J. Greenhouse Gas Control*, 2020, **97**, 102986.
9. *The Future of Coal: Options for a Carbon-Constrained World*, Massachusetts Institute of Technology, Cambridge, MA, 2007. http://web.mit.edu/coal/.
10. C. Beuttler, L. Charles and J. Wurzbacher, *Front. clim.*, 2019, **1**, 1.
11. V. Gutknecht, S. Snaebjörnsdottir, B. Sigfusson, E. Aradottir and L. Charles, *Energy Procedia*, 2018, **146**, 129.
12. D. Clark, E. Oelkers, I. Gunnarsson, B. Sigfússon, S. Snæbjörnsdóttir, E. Aradóttir and S. Gíslason, *Geochim. Cosmochim. Acta*, 2020, **279**, 45.
13. Climeworks, *News: Climeworks builds first commercial-scale direct air capture plant*, 2015. https://climeworks.com/news/climeworks-ag-builds-first-commercial-scale-co2-capture.
14. Summary for Policymakers. In: Global Warming of 1.5 °C. An IPCC Special Report on the impacts of global warming of 1.5 °C above pre-industrial levels and related global greenhouse gas emission pathways, in the context of strengthening the global response to the threat of climate change, sustainable development, and efforts to eradicate poverty, 2018.
15. S. Fuss, W. Lamb, M. Callaghan, J. Hilaire, F. Creutzig, T. Amann, T. Beringer, W. de Oliveira Garcia, J. Hartmann, T. Khanna, G. Luderer, G. Nemet, J. Rogelj, P. Smith, J. Vicente Vicente, J. Wilcox, M. del Mar Zamora Dominguez and J. Minx, *Environ. Res. Lett.*, 2018, **13**, 063002.
16. E. Rubin, C. Short, G. Booras, J. Davison, C. Ekstrom, M. Matuszewski and S. McCoy, *Int. J. Greenhouse Gas Control*, 2013, **17**, 488.
17. E. Rubin, *Int. J. Greenhouse Gas Control*, 2019, **88**, 1.
18. *Kemper IGCC Plant with CCS, Georgia Tech Energy Policy and Innovation Center*, Atlanta, GA, 2018. https://cpn-us-w2.wpmucdn.com/sites.gatech.edu/dist/6/1908/files/2021/03/kemper-technicalreview_aug2017-mar2018.pdf.
19. Electric Power Research Institute (EPRI) and TAGTM, *Technical Assessment Guide, Volume 1: Electricity Supply – 1993. TR-102276-V1R1*, Electric Power Research Institute, Palo Alto, CA, 1993.

CHAPTER 7

Geochemical Negative Emission Technologies

OLIVIA HAWROT,[a] JAMES CAMPBELL,[a] FRANCES BUCKINGHAM[b] AND PHIL RENFORTH*[a]

[a] Research Centre for Carbon Solutions, School of Engineering and Physical Sciences, Heriot-Watt University, Edinburgh EH14 4AS, United Kingdom; [b] Department of Earth Sciences, University of Oxford, Oxford OX1 3AN, United Kingdom
*Emails: p.renforth@hw.ac.uk; okh2@hw.ac.uk

Weathering of alkaline rocks is a geochemical process which is fundamental to the Earth's natural carbon cycle. It is responsible for removing CO_2 from the atmosphere over geological time periods by transforming it into bicarbonate ions or solid carbonate minerals. Geochemical negative emission technologies (geochemical NETs) seek to harness and accelerate this process. The alkaline materials involved in geochemical NETs (*e.g.*, basic or ultrabasic rocks or alkaline wastes/by-products from some industrial activities) are generally abundant, and, whilst the kinetics of the reaction are relatively slow, it is thermodynamically favourable. Furthermore, storage of CO_2 as solid carbonate minerals or dissolved bicarbonate is permanent in human relevant timescales. The challenge is implementing technically feasible processes or technologies that sufficiently accelerate rock weathering in a way that is environmentally responsible and socially acceptable. In this chapter, we broadly define the concept and evolution of geochemical negative emission technologies, and the associated environmental benefits and risks.

Energy and Environment Series No. 31
Greenhouse Gas Removal Technologies
Edited by Mai Bui and Niall Mac Dowell

Published by the Royal Society of Chemistry, www.rsc.org

7.1 Historical Context

7.1.1 Introduction

Urey (1952)[1] is often cited as being the first to link the weathering of silicate rocks with the carbon cycle. However, the concept can be traced back a century prior to the work of Ebelmen (1845)[2] who made the link between geological processes and atmospheric composition.[3] Expanded from Urey's equations, eqn (7.1) shows the process by which carbon dioxide (CO_2) hydrates to form carbonic acid (H_2CO_3), which disassociates to form bicarbonate (HCO_3^-) and carbonate (CO_3^{2-}) ions. The dissolution (or chemical weathering) of the mineral olivine (Mg_2SiO_4) is shown in eqn (7.2), and the precipitation of the mineral magnesite ($MgCO_3$) is shown in eqn (7.3).

$$CO_2 + H_2O \rightleftharpoons H_2CO_3 \rightleftharpoons HCO_3^- + H^+ \rightleftharpoons CO_3^{2-} + 2H^+ \tag{7.1}$$

$$Mg_2SiO_4 + 4H_2CO_3 \rightleftharpoons 2Mg^{2+} + 4HCO_3^- + SiO_2 + 2H_2O \tag{7.2}$$

$$2Mg^{2+} + 4HCO_3^- \rightleftharpoons 2MgCO_3 + 2CO_2 + 2H_2O \tag{7.3}$$

Throughout the 20th century the influence of global geochemical cycles in controlling long-term climate change has gained increasing attention. Early Earth system models that examined the interaction of weathering and climate[4] suggest that atmospheric CO_2 elevated by tectonic activity is removed by weathering in a negative feedback cycle (Figure 7.1). Subsequent models simulate and couple additional geochemical cycles.[5] In these models, the carbon within the Earth system is ultimately removed as magnesium (Mg) and calcium (Ca) carbonate minerals (eqn (7.3)).

The interaction between rock weathering and climate has prompted work exploring the geochemical weathering response to anthropogenic climate

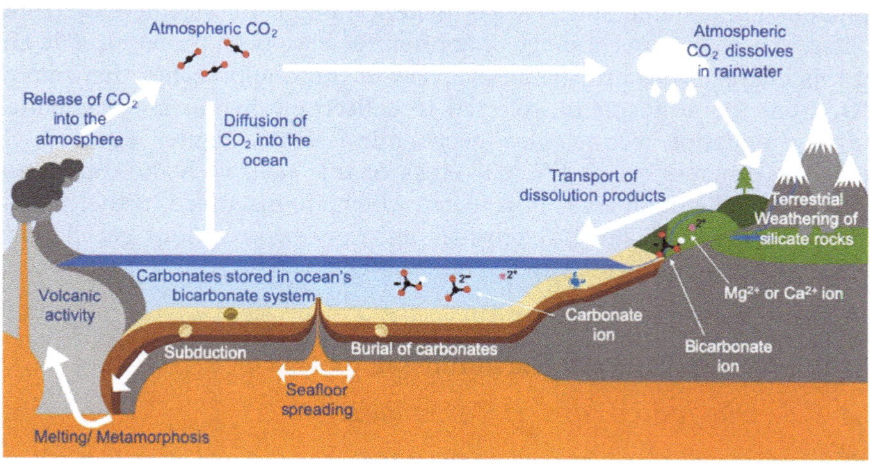

Figure 7.1 The negative feedback relationship between weathering and climate.

change showing that weathering accelerates the removal of fossil fuel CO_2 from the atmosphere. Unfortunately, this could take >100 000 years to reduce CO_2 to preindustrial levels. For instance, Lenton and Britton[6] used an Earth system box model to assess emissions of ~1100–15 000 billion tonnes of carbon (Gt C) running for 1 million years. They estimated that for excess atmospheric carbon emissions to halve, it would take around 200 thousand years. Similarly, Lord *et al.*[7] used a 3D Ocean-based Earth system model testing CO_2 pulse emissions of 1000–20 000 Gt C and found it would take 245–278 thousand years to reach the half-life. While chemical weathering is one of the largest natural regulators of atmospheric CO_2,[6] it is too slow to sequester anthropogenic CO_2 necessary to mitigate climate change over decades.

Weathering of silicate rocks (eqn (7.1)–(7.2)) is responsible for consuming atmospheric and soil CO_2 at a rate of 250–300 million tonnes of C per year (916–1098 Mt CO_2 per year).[8] While this is relatively small compared to the natural carbon cycle (100s Gt C per year), it is responsible for sequestering CO_2 released by volcanic degassing over thousands of years. Methods to increase the rate of this reaction in the context of climate change mitigation was first discussed by Seifritz in a short letter to Nature.[9] This publication proposed the addition of crushed silicate minerals to water that could then react with pressurised CO_2 to form hard water and silica, which could subsequently be released into the deep ocean. This idea was taken further by Dunsmore[10] emphasising the need to "...return the CO_2 to the lithosphere" to mitigate climate change. The idea was taken further again, by Lackner *et al.*[11,12] who offered an initial insight of the resources, reaction thermodynamics, and techno-economics.

Mineral weathering and carbonation are considered interesting methods of CO_2 storage for the following reasons: (i) the overall reaction for silicate minerals is exothermic indicating that it occurs spontaneously at ambient conditions, (ii) the minerals are abundant and accessible, and (iii) carbonate minerals are a stable and near-permanent method for storing CO_2. To facilitate these reactions, a range of approaches have been proposed. The aim of this chapter is to provide an overview of those approaches that remove CO_2 from the atmosphere, referred to collectively herein as geochemical negative emission technologies (geochemical NETs). Figures 7.2–7.4 highlight the progression of different areas of this field with the foundation studies and the main areas of research which are discussed further in subsequent sections. Figure 7.5 provides an overview of various geochemical NETs and how they are related to each other.

7.1.2 *Ex-situ* Mineral Carbonation

7.1.2.1 Single-step (direct) Carbonation

Direct carbonation involves minerals reacting in a single step with CO_2 to form carbonates. It is the simplest reaction pathway and consists of two

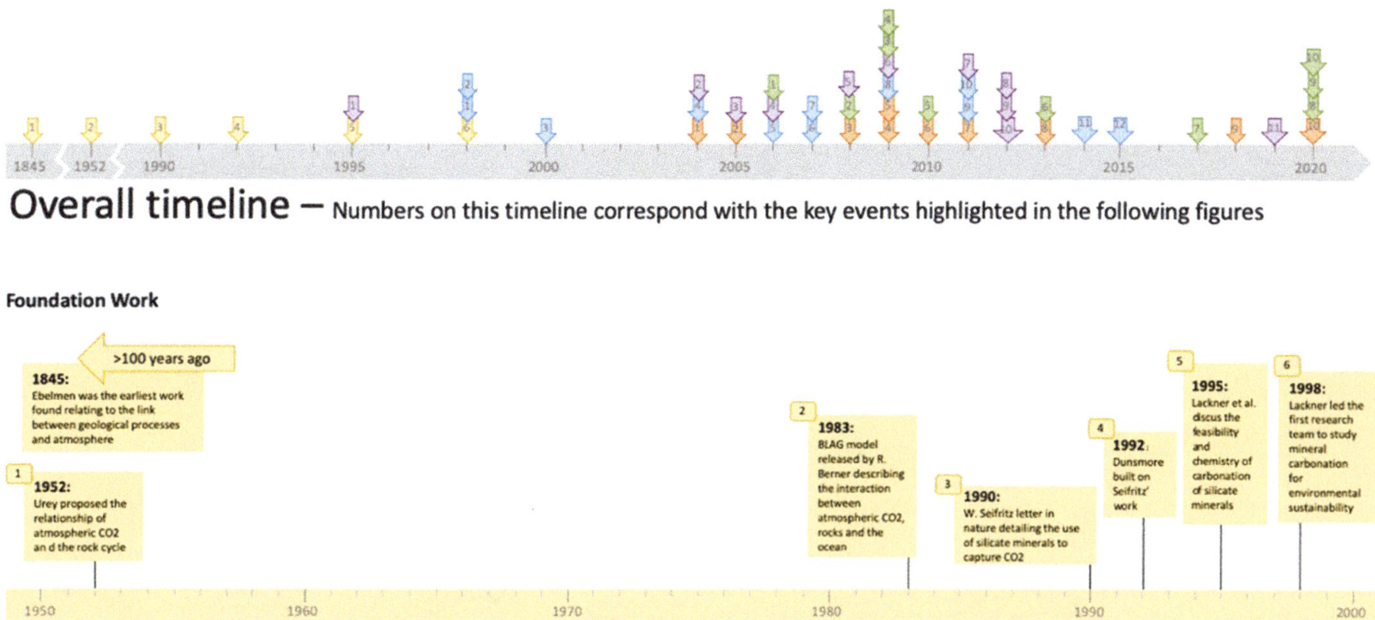

Figure 7.2 Timelines of field progression for various aspects of geochemical NETs. These consist of main findings judged by the frequency a study has been cited per year since publication and based on studies that first discussed a novel idea/process.

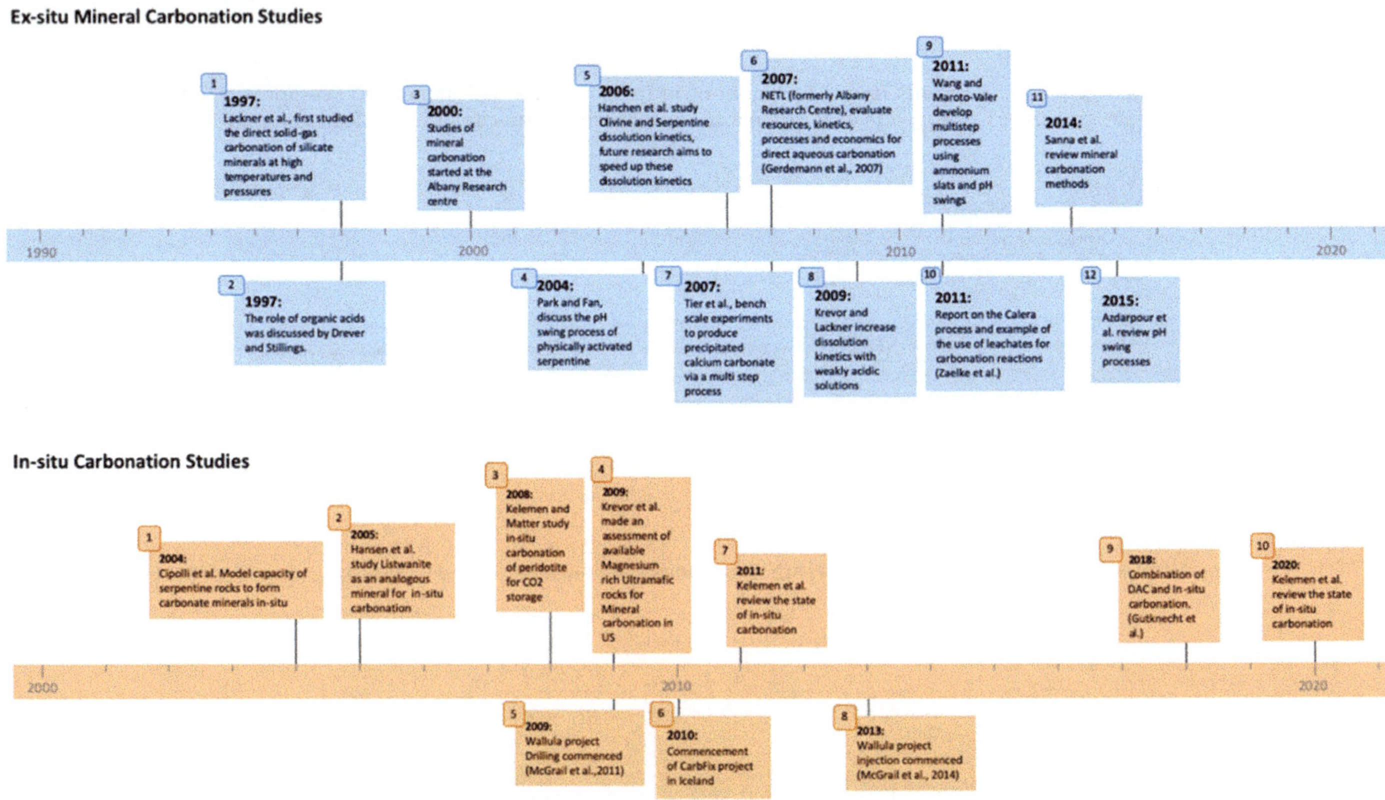

Figure 7.3 Timelines of field progression for various aspects of geochemical NETs. These consist of main findings judged by the frequency a study has been cited per year since publication and based on studies that first discussed a novel idea/process.

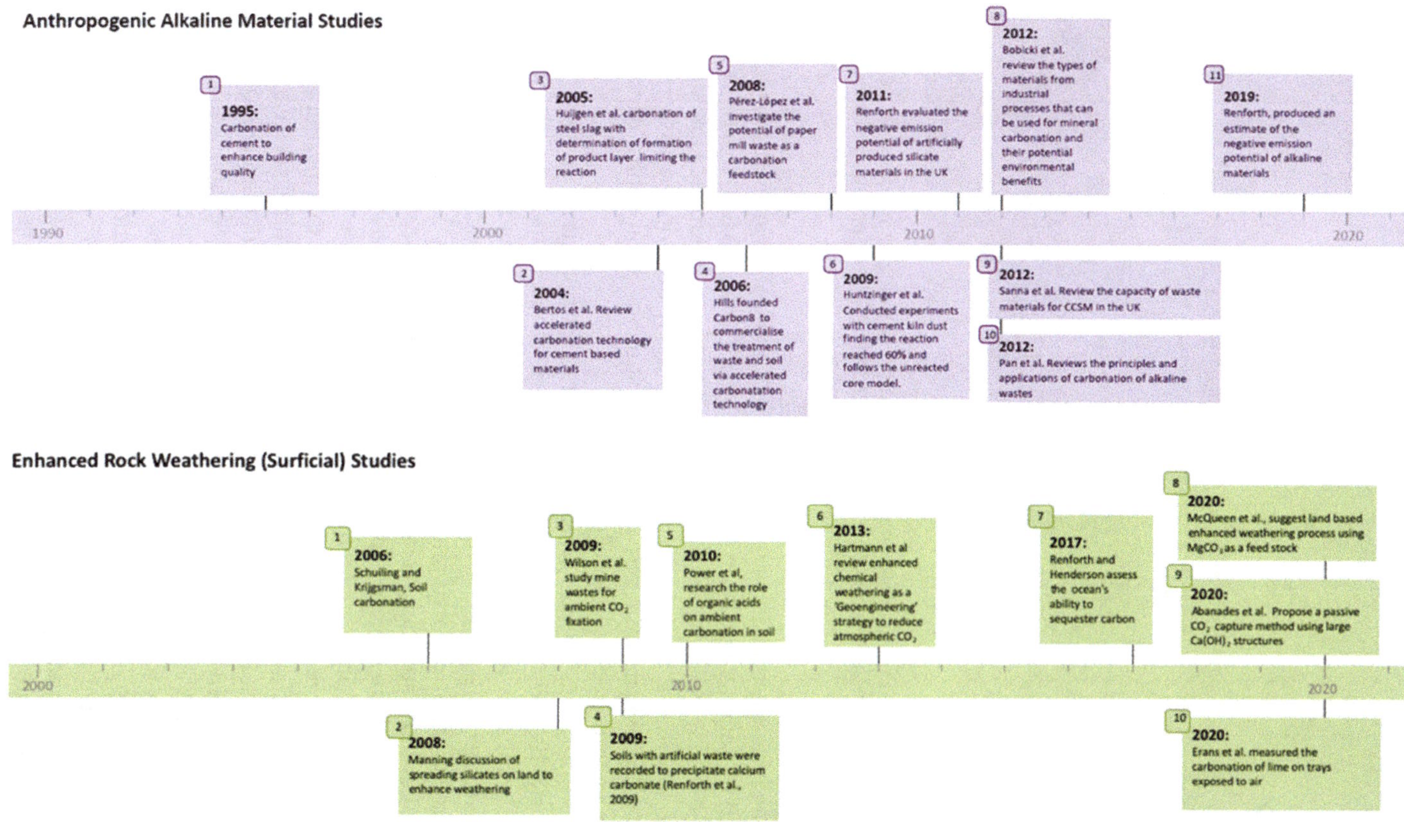

Figure 7.4 Timelines of field progression for various aspects of geochemical NETs. These consist of main findings judged by the frequency a study has been cited per year since publication and based on studies that first discussed a novel idea/process.

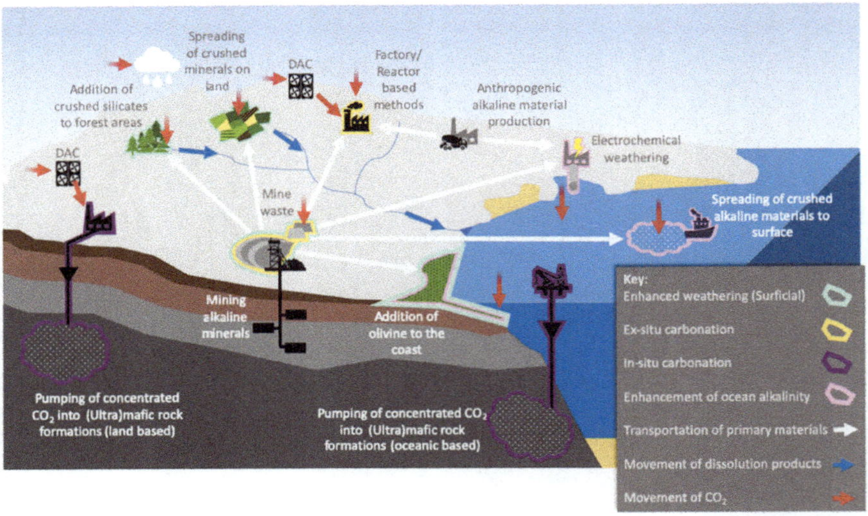

Figure 7.5 Overall picture of the processes and technologies involved in geochemical negative emission technologies (NETs). Reactor/factory based methods include Ca and Mg looping technologies.

approaches: gas-solid and aqueous.[13] Direct gas–solid carbonation was first studied by Lackner *et al.*[12] in high temperature and pressure reactions using minerals such as olivine and serpentine. The reaction kinetics are slow and conversion extent poor, such that it was largely abandoned as a research pathway for silicate minerals in favour of aqueous approaches, which have been reviewed in detail elsewhere.[13]

To summarise, direct aqueous mineral carbonation involves simultaneously dissolving the mineral and CO_2 in solution and precipitating carbonate minerals. Research has focused on the carbonation of magnesium silicate minerals such as olivine and serpentine into magnesium carbonates inside reactors at elevated temperatures and pressures.[14] Dissolution of the mineral into the aqueous medium is considered to be the rate-limiting step.[15] This may be limited by the creation of a passivating layer that forms on the mineral surface, and a reduction in CO_2 partial pressure in the reactor.[13,16] However, for all naturally occurring minerals, elevated CO_2 partial pressure (~30–50 bar) and temperature (>100 °C) is required to drive the reaction, as such, the applicability to geochemical NETs is potentially limited, although possible if combined with direct air capture (DAC), see Section 7.4.

Pre-treatments have been explored as a means of increasing conversion rates, which aim to either increase reactive surface area ('comminution'),[17] or the reactivity of the mineral (*e.g.*, through heating). Thermal pre-treatment of hydrated magnesium silicates drives off chemically bound water, which is thought to increase porosity and thus surface area leading to increased reaction kinetics.[18,19] The energy penalties incurred by pre-treatments[15,20] are discussed further in Section 7.4.

7.1.2.2 Multi-step (indirect) Carbonation

A limitation of single-step carbonation is that the primary weathering agent (CO_2) is, by intention, consumed in the process. Multi-step mineral carbonation overcomes this by extracting the Ca and Mg from the mineral before reaction with CO_2.[13] In addition, it is possible to form relatively pure products (carbonates, iron oxides, silica) and with reaction rates that are generally faster than direct carbonation. However, these processes are more complex and involve reagents that require financial, material, and energy resources to regenerate.[21]

Strong acids and bases such as sulphuric acid, nitric acid, and hydrochloric acid have been used as weathering agents,[22–25] as well as weaker organic acids.[26,27] A pH swing (*i.e.*, a rapid change from low to high pH) is required to facilitate CO_2 uptake and regenerate the acid.[28] Tier *et al.*[23] conducted bench-scale laboratory experiments to precipitate calcium carbonate from steel slag using acetic acid. This study was performed at atmospheric pressure, highlighting that extraction of Ca from iron and steel slags is feasible at temperatures below 156 °C and carbonates precipitated above 45 °C. These experiments showed that steel and iron slags dissolved rapidly (within minutes) in acetic acid with the production of heat. The production of calcium carbonate is seen as a way to partially offset the cost of the technology, with higher purities being more valuable.[29] It is reported that the volume of the global calcium carbonate markets was ~116 Mt in 2020, this is predicted to rise to ~145 Mt in 2027.[30] Exploiting significant CO_2 capture while using the sale of calcium carbonates to reduce the cost of the technology would rapidly saturate this market.

Krevor and Lackner[31] showed that dissolution of serpentine was greatly enhanced by citric, oxalic, or ethylenediaminetetraacetic acid, leading to a large extent of dissolution. Wang and Maroto-Valer[32] proposed a 5-step process using ammonium salts at varying temperatures and pH to capture CO_2 from flue gas. However, an aqueous pH swing using ammonium salts uses large amounts of water (50–56 t H_2O per t CO_2) during the salt regeneration steps. Fagerlund *et al.*[33,34] developed a 2-step process using ammonium salts to address this issue by recovering the heat of reaction in the carbonation step, but this approach was limited by the Mg recovery rate. Some of these processes produce alkaline minerals initially intended for reaction with flue gases,[35] but similar to alkaline wastes (see below) may be able to react directly with atmospheric CO_2 or in a hybrid direct air capture system (see Section 7.4).

7.1.3 *In-situ* Carbonation

In-situ carbonation requires the injection of CO_2 into geological formations where carbonation occurs; this is notably different from more conventional carbon capture storage methods, which predominantly physically trap CO_2 in rock formations. *In-situ* carbonation seeks to take advantage of natural deposits of minerals, which are rich in divalent metals such as Mg and Ca, without the need to mine or extract them. Fluid laden with CO_2 is injected into fractures below ground into mafic or ultramafic rocks where the CO_2 reacts to form stable carbonates. In nature, these deposits take the form of silicates such as olivine ($(Mg,Fe)_2SiO_4$), clinopyroxene ($CaMgSi_2O_6$–$CaFeSi_2O_6$) and serpentine ($(Mg,Fe)_3Si_2O_5(OH)_4$).[13] Carbonation occurs naturally around these minerals forming alkaline aquifers that are rich in divalent ions and dissolved HCO_3^-. These dissolved ions subsequently emerge at the surface as springs and react with atmospheric CO_2. This further enrichment of CO_2 can cause the carbonates to precipitate from solution. Additional mixing with deeper aquifers can enhance the precipitation of more carbonates.[13]

Paukert *et al.*[36] measured the aqueous solute concentrations in water of a natural peridotite system in Oman. Their findings suggest that the supply of inorganic carbon was the limiting factor of carbonation and injecting CO_2 in these systems could increase the carbonation rate by 16 000 times. Unlike *ex-situ* mineral carbonation, which involves the movement of large quantities of solids, *in-situ* methods require relatively straightforward pumping and injection of liquid. In addition, *in-situ* carbonation has, potentially, greater CO_2 sequestration capacity than *ex-situ* methods.[13] Kelemen and Matter[37] suggest that increasing the weathering rates of natural peridotites in Oman has the potential to capture Gt CO_2 per year. The Columbia river basalt group has an estimated storage potential of over 100 Gt CO_2,[38,39] but theoretically the global storage potential of oceanic basaltic rocks is estimated to be $>10^5$ Gt C.[40,41] This method could potentially have low financial costs similar to surficial carbonation with 'giant' storage capacities.[42]

There have been 2 projects to test *in-situ* carbonation. First, the Wallula project which injected supercritical CO_2 saturated with water into Columbia River basalt at a depth of 828–886 meters.[38] The conversion of the injected supercritical CO_2 is unknown, however side cores from the main borehole show newly formed carbonates. Secondly, the CarbFix experiment based in Iceland co-injected ~200 tonnes of CO_2 with water into fractured basalt at 400–800 metres depth.[43] An EU-funded project saw Climeworks (a DAC company) partner with CarbFix to combine DAC and *in-situ* mineralisation in an attempt to optimise these processes.[44] This could make DAC more viable as a 'hybrid approach' (see Section 7.4.5) with *in-situ* carbonation.[42,44] This method seeks to enrich CO_2 for *in-situ* carbonation, which would in turn reduce the financial cost of each technology involved. Fundamental research on injectivity, permeability, geomechanics and microstructure are needed to progress this technology[42] (this is discussed more in Section 7.4.2).

7.1.4 Enhanced Rock Weathering (Surficial)

Enhanced rock weathering (ERW) is the process by which powdered silicate rocks and/or minerals are spread over the land, coastal areas or the ocean.[45–54] Although the application of minerals to agricultural soils to adjust their pH or provide nutrients has been performed for millennia,[45,55] the addition of silicate minerals would need to be an order of magnitude larger, whereby the addition of 10–30t basalt per ha per year is equivalent to a ~1 mm layer of basalt powder that is predicted to remove 0.5–4 Gt CO_2 per year by 2100 when spread over 2/3 of the global cropland.[56] The distinction between this and *ex-situ* mineral carbonation (Section 7.1.2) is that the latter uses contained engineered systems to precipitate carbonate minerals, whereas the former describes mineral dissolution within terrestrial and marine environments. While the dissolution rate of minerals is slower at ambient temperatures (*e.g.*, olivine dissolves 4 orders of magnitude slower at 25 °C than it does at 180 °C), the reaction time for ERW is not constrained by the requirements of chemical reactants and as such is 3–4 orders of magnitude larger.[42] The intention of ERW is to promote mineral dissolution (eqn (7.2)) whereby dissolution products are transported to the ocean, increase alkalinity and, in turn, draw down atmospheric CO_2.[45] Alternatively, ERW may promote the precipitation of carbonate minerals (eqn (7.3)) in soils or drainage waters depending on soil conditions, *i.e.*, "passive carbonation".[49,57] The contribution to ocean alkalinity may also help to ameliorate ocean acidification (see Section 7.3).

The application of olivine to coastal environments seeks to further enhance olivine weathering by utilising the flow regimes of tides and wave energy, whereby particle attrition accelerates dissolution.[58] This has been studied in batch reaction experiments where agitation in seawater enhanced olivine dissolution, increasing alkalinity and dissolved inorganic carbon.[59] It is also thought that benthic organisms will further increase dissolution when ingesting the sediment during feeding.[60] Similar biological enhancements of silicate weathering have been documented in terrestrial soils, for example organic acids produced by plant roots and microbes can increase the dissolution rate of minerals by up to an order of magnitude.[61] Application of crushed silicate rocks directly to the ocean surface has received little attention,[62] however early work suggests olivine would need to be crushed to ~1 μm to ensure that its sinking velocity is sufficiently slow that it dissolves in the surface of the ocean. Increasing ocean alkalinity also raises concerns and challenges around monitoring and verification, governance,[63] social acceptability[64] and justice.[65]

7.1.5 Anthropogenic Alkaline Materials

Many existing industries, such as steel and cement manufacturing, produce alkaline materials with the potential to be used as geochemical NETs. These materials include blast furnace and steel slag, fly ash/residues, cement kiln dust, demolition waste, cement products, red mud, waste ultramafic rocks, alkaline paper mill wastes and mine tailings. These are typically rich in Ca and Mg, are often conveniently located next to industrial sources of CO_2. As such, most of the previous work has focused on the direct reaction of these materials with simulated flue gases or pure CO_2.[13,66] However, these materials have been shown to also react with atmospheric CO_2.[67–69]

The benefits of using alkaline materials in carbonation processes include:

1) The low cost and ready availability both from historical stockpiles (~95 Gt)[70] and future production, which is estimated to be tens of thousands of Gt by 2100.[71]
2) The produced alkaline materials are generally more reactive than primary minerals.[25]
3) This reactivity may mean that the engineering requirements (pretreatment, temperature, pressure) may be reduced.[72,73]
4) No additional mining of virgin rock is required.[13]

Anthropogenic alkaline materials have the potential to sequester 2.9–8.5 Gt CO_2 per year by 2100.[71] These artificial or industrial materials behave similarly to natural minerals used for mineral carbonation in that their reaction rates were found to increase with greater surface area, pressure, temperature, lower pH and smaller particle size.[18,74–76] However, mass transfer of CO_2 is often initially rate-limiting followed by occlusion with the increasing thick layer of reaction products.[15,73,77,78]

Cement is created *via* heating limestone with a source of silicon, and the resulting product (clinker) its then hydrated for use in construction as mortar and concrete. This energy intensive process results in the cement industry accounting for 5% of CO_2 emissions globally.[79] Cement kiln dust is a by-product from Portland cement production and 0.15–0.2 tonnes is produced per tonne of cement. This waste is high in CaO (38–48%) and MgO (1.5–2%).[25,80,81] Ambient temperature but elevated CO_2 experiments suggest a carbonation degree of 0.08–0.25 t CO_2 per tonne of kiln dust.[77] In the US, most of the cement kiln dust is recycled into the kiln, but a large proportion ($45 \pm 10\%$) is disposed of presenting opportunities for geochemical NETs to utilise this material as a potential feedstock.[82]

Construction and demolition (C&D) waste in soils have been studied in this context as carbonation has been observed around these wastes in brown field sites,[61,67,83,84] which supports research into developing urban soils into carbon sinks.[68,85] Renforth *et al.*[68] explored the use of artificial substances such as waste materials spread on the land, however, issues arise with waste material addition to soil when unwanted chemicals within the waste are released into the natural environment. Furthermore, the increase in pH could mobilise potentially toxic substances such as heavy metals. Similarly to cement, lime (CaO) is created through the process of calcining limestone. Hanak *et al.*[86] proposed a system which produces electricity whilst simultaneously capturing CO_2 directly from the air using lime as a solid sorbent. The techno-economics of this process are comparable to other direct air capture technologies.

Slag is produced in steel making, which is the industry responsible for 6–7% of global CO_2 emissions.[87] Slag is composed of minerals and semicrystalline or amorphous materials, mainly Ca, Mg and Al-silicates and oxides,[88] however, the variable composition results in a wide range of carbonation rates and extents. Slurry experiments show that carbonation of slag can occur at ambient temperature and pressure,[89–91] however, most work has focused on reaction at elevated temperature (~80 °C).[15,73,92–95] Some studies have enhanced the rates of these reactions using ultrasound or additives (*e.g.*, $MgCl_2$), which improve mass transport and diffusion[96,97] but this increased the cost of the process significantly (4000 euros per t slag).[13] Other studies have investigated the potential of increasing the commercial value of the slag carbonation products, *via* indirect carbonation, producing high purity precipitated calcium carbonate.[92,98]

The mining industry produces vast quantities of wastes during mineral extraction, some of which are favourable for passive CO_2 sequestration. For example, 1 tonne of asbestos mined produces 20 tonnes of fine processing waste called 'tailings',[25] which have large CO_2 sequestration potential due to their high (~40%) MgO content.[18] Mine tailings that are left to react naturally often form carbonate crusts on the surface of the tailing pile, which can impede diffusion of CO_2.[99] Wilson *et al.*[69,100] found that 164 000 tonnes of atmospheric CO_2 had been sequestered in mine tailings at Clinton Creek in Canada over the past 26 years and suggests that with proper management the tailings could be modified to increase this uptake. It has been proposed that the addition of microorganisms could enhance carbonation rates *via* bioleaching.[101] The addition of certain bacteria and fungi can increase the porosity and change the mineralogy of mine wastes, which helps provide a stream of Ca and Mg in an aqueous solution.[102] Ultramafic mine tailings (~40% MgO) are often produced during nickel mining.[25] Experiments have shown that indirect carbonation of these mine tailings using a variety of acids can give a high degree of carbonation (94%).[23] Although this comes at a considerable cost of ~\$600–\$1600 per tonne of CO_2.[103]

Red mud, or bauxite tailings, is an alkaline by-product that is produced by the alumina industry, with large stockpiles of material distributed near existing and former production sites. Production of red mud is predicted to increase from 150 Mt per year to 500–1100 Mt per year by 2100.[71] Red mud is a caustic (sodium hydroxide) liquor primarily composed of suspended iron and aluminium oxides.[104] Carbonating red mud may reduce its toxicity, and some suggest that the carbonated red mud is appropriate for agricultural soil amendment or, as a construction material (providing the chemistry of the material is suitable).[90] Direct carbonation experiments at ambient temperature have shown that red mud can be utilised for CO_2 sequestration but the carbonation extent to which this occurs varies (5–60%) based on the mineral phases present in the red mud.[104–106] The maximum sequestration potential is 128 kg CO_2 per tonne of unsintered/causticised red mud and may offer a suitable remediation strategy for this alkaline waste. Renforth *et al.*[107] explored the effects of gypsum dosing on a bauxite spill as a form of emergency remediation, noting that 36% of precipitated carbonates were formed by reaction with atmospheric CO_2.

Ash, a product of combustion, is rich in alkaline minerals and amorphous materials (*e.g.*, glasses),[108,109] which is usually produced at fine particle sizes (10–15 μm). There are different types of ash some non-hazardous and some hazardous. The hazardous types require treatment before their use in other products, for example bottom ash can be 'naturally weathered' (reacted with CO_2 which stabilises the ash) then used as a secondary building material.[110] Air Pollution Control Residue (APCR) has also been carbonated, potentially reducing its environmental risk.[111] Other studies have investigated the carbonation of ash from municipal solid waste incineration[108,112–114] and coal combustion.[115] Although we currently only carbonate hazardous varieties of ash to reduce their toxicity, all varieties could potentially be carbonated to sequester CO_2.

Kraft pulping is the process used to create chemical pulp for the paper industry. This process ultimately produces several waste streams which are highly alkaline, they are collectively referred to as alkaline paper mill wastes (APMW).[116] Their alkalinity and high volume of CaO (45–82%[117]) makes them good candidates for mineralisation.[13] The production of the APMW releases CO_2, and these emissions could be reacted with the waste to reduce the overall emissions of the process and produce $CaCO_3$. Moreover, this reaction is fast with a high sequestration capacity suggesting these materials are promising for mineralisation of CO_2.[116]

7.2 Resources

7.2.1 Chemistry

Geochemical NETs are driven by chemical weathering processes which sequester atmospheric CO_2 into solution. During chemical weathering, H^+ ions are adsorbed onto the mineral surface causing silicate minerals to dissolve and release base cations into solution (eqn (7.1)–(7.2)). This increases total alkalinity[118] and causes the equilibria in the carbonate system to shift towards the production of bicarbonate ions ($HCO_3{}^-$) and carbonate ions ($CO_3{}^{2-}$).[119]

The maximum CO_2 capture potential of a mineral or rock is strongly influenced by its base cation content. Renforth[71] related the maximum carbon sequestration potential through either carbonate mineral formation (C_{pot}) or enhanced weathering† (E_{pot}) to the moles of charge released from liberated cations (eqn (7.2)). These correlations are shown in eqn (7.4) and (7.5), expressed in tonnes of CO_2 per tonne of material, whereby M_x is the molecular mass and η is molar ratio of CO_2 to base cations consumed during weathering (typically 1.5–1.8, eqn (7.6)).[119]

$$C_{pot} = \frac{M_{CO_2}}{100} \cdot \left(\alpha \frac{CaO}{M_{CaO}} + \beta \frac{MgO}{M_{MgO}} + \gamma \frac{SO_3}{M_{SO_3}} + \delta \frac{P_2O_5}{M_{P_2O_5}}\right) \cdot 10^3 \quad (7.4)$$

†Where mineral dissolution leads to an increase in alkalinity in the ocean.

$$E_{\text{pot}} = \frac{M_{CO_2}}{100} \cdot \left(\alpha \frac{CaO}{M_{CaO}} + \beta \frac{MgO}{M_{MgO}} + \varepsilon \frac{Na_2O}{M_{Na_2O}} + \theta \frac{K_2O}{M_{K_2O}} + \gamma \frac{SO_3}{M_{SO_3}} + \delta \frac{P_2O_5}{M_{P_2O_5}} \right) \cdot 10^3 \cdot \eta \tag{7.5}$$

$$\frac{\Delta C_T}{\Delta A_T} = \eta = \left(S \cdot 10^{-3.009} + 10^{-1.519}\right)\ln\left(pCO_2\right) - \left(S \cdot 10^{-2.100}\right) - \left(T \cdot pCO_2\right)\left(S \cdot 10^{-7.501} - 10^{-5.598}\right) - \left(T \cdot 10^{-2.337}\right) + 10^{-0.102} \tag{7.6}$$

Thus, the higher the total Ca, Mg, K, or Na content, the higher the carbon capture potential of the rock. It is for this reason that research has focused on the use of cation-rich, basic (*e.g.*, basalt) and ultrabasic rocks (*e.g.*, olivine-rich dunite), rather than their felsic counterparts.[13,51,120–124] Table 7.1 shows a range of materials and their dissolution reactions.

7.2.2 Dissolution Minerals

For geochemical NETs, alkaline earth metal cations, namely Mg^{2+} and Ca^{2+}, must be supplied. Igneous and metamorphic rocks (*e.g.*, ultramafic and mafic rocks) are rich in alkaline earth metals, making them ideal sources of divalent metal cations for carbon mineralisation.[125] Given their higher concentration of basic minerals, *e.g.*, wollastonite, olivine, brucite and serpentine group minerals, aluminosilicates, and natural volcanic glasses, they have been investigated as potential feedstocks for carbon mineralisation (see Table 7.1). In addition, many industrial processes produce Ca- and Mg-rich wastes such as steel slag, fly ash, bottom ash, cement wastes, mining processing wastes, sludge ash or pulp and paper mill wastes that could provide readily available material for carbon mineralisation (see Table 7.1). Although alkaline solid wastes are less abundant than natural minerals, primary extraction is not required. In addition, alkaline solid wastes dissolve 4–5 orders of magnitude faster than natural minerals and therefore, notionally, require less pre-treatment.[42] In addition to silicate and oxide minerals, carbonates, *e.g.*, $CaCO_3$, might be used for geochemical NETs such as ocean alkalinity enhancement.[126–128] Addition of limestone or dolomite has a threefold benefit: reversing ocean acidification, reversing the decrease in calcite supersaturation (beneficial to calcifying organisms) and resulting in CO_2 drawdown.[129]

7.2.2.1 *Natural Silicate Resources*

As discussed above, mafic and ultramafic rocks are commonly associated with geochemical NETs due to their high total alkaline cation content (see Box 7.1) and are comprised from a range of mafic minerals: dark coloured minerals enriched in magnesium and iron, such as olivine and pyroxene (see Table 7.1). Ultramafic and mafic minerals are globally

Table 7.1 Dissolution reactions of natural occurring and anthropogenically produced minerals.

Mineral				
Type	Name	Occurrence	Dissolution reaction	Eq.
Naturally occurring minerals				
Carbonates	Calcite	Sedimentary & metamorphic rocks	$CaCO_3 + CO_2 + H_2O \rightarrow Ca^{2+} + 2HCO_3^-$	(7.7)
	Dolomite	Metamorphic rocks	$CaMg(CO_3)_2 + 2CO_2 + 2H_2O \rightarrow Ca^{2+} + Mg^{2+} + 4HCO_3^-$	(7.8)
	Magnesite	Metamorphic rocks	$MgCO_3 + CO_2 + H_2O \rightarrow Mg^{2+} + 2HCO_3^-$	(7.9)
Felsic silicates	Anorthite	Basic & felsic rocks	$CaAl_2Si_2O_8 + 2CO_2 + 3H_2O \rightarrow Ca^{2+} + 2HCO_3^- + Al_2Si_2O_5(OH)4$	(7.10)
	Muscovite	Metamorphic rocks	$2KAl_2(Si_3Al)O_{10}(OH)_2 + 2CO_2 + 5H_2O \rightarrow 2K^+ + 2HCO_3^- + 3Al_2Si_2O_5(OH)_4$	(7.11)
	Orthoclase	Felsic rocks	$2KAlSi_3O_8 + 2CO_2 + 3H_2O \rightarrow 2K^+ + 2HCO_3^- + 4SiO_2 + Al_2Si_2O_5\ (OH)_4$	(7.12)
	Wollastonite	Metamorphic rocks	$CaSiO_3 + 2CO_2 + H_2O \rightarrow Ca^{2+} + 2HCO_3^- + SiO_2$	(7.13)
Mafic silicates	Chrysotile	Metamorphic rocks	$Mg_3Si_2O_5(OH)_4 + 6CO_2 + H_2O \rightarrow 3Mg^{2+} + 6HCO_3^- + 2SiO_2$	(7.14)
	Diopside	(Ultra) basic rocks	$MgCaSi_2O_6 + 4CO_2 + 2H_2O \rightarrow Mg^{2+} + Ca^{2+} + 4HCO_3^-$	(7.15)
	Enstatite	(Ultra) basic rocks	$+ MgSiO_3 + 2CO_2 + H_2O \rightarrow Mg^{2+} + 2HCO_3^- + SiO_2$	(7.16)
	Forsterite	(Ultra) basic rocks	$MgSiO_4 + 4CO_2 + 4H_2O \rightarrow 2Mg^{2+} + 4HCO_3^- + H_4SiO_4$	(7.17)
	Phlogopite	(Ultra) basic rocks	$2KMg_3(AlSi_3)O_{10}(OH)_2 + 14CO_2 + 7H_2O \rightarrow 2K^+ + 6Mg^{2+} + 14HCO_3^- + 4SiO_2 + Al_2Si_2O_5\ (OH)_4$	(7.18)
	Tremolite	Metamorphic rocks	$Ca_2Mg_5Si_8O_{22}(OH)_2 + 14CO_2 + 6H_2O \rightarrow 2Ca^{2+} + 5Mg^{2+} + 14HCO_3^- + 8SiO_2$	(7.19)
Anthropogenically produced minerals				
Akermanite		Slag	$Ca_2MgSi_2O_7 + 6CO_2 + 3H_2O \rightarrow 2Ca^{2+} + Mg^{2+} + 6HCO_3^- + 2SiO_2$	(7.20)
Brucite[a]		Refractory bricks	$Mg(OH)_2 + 2CO_2 + H_2O \rightarrow Mg^{2+} + 2HCO_3^-$	(7.21)
Larnite		Cement clinker	$Ca_2SiO_4 + 2H_2O + 4CO_2 \rightarrow 2Ca^{2+} + 4HCO_3^- + SiO_2$	(7.22)
Lime		Lime	$CaO + 2CO_2 + H_2O \rightarrow Ca^{2+} + 2HCO_3^-$	(7.23)
Portlandite		Portland cement	$Ca(OH)_2 + 2CO_2 + H_2O \rightarrow Ca^{2+} + 2HCO_3^-$	(7.24)
Periclase		Lime slaking	$MgO + 2CO_2 + H_2O \rightarrow Mg^{2+} + 2HCO_3^-$	(7.25)
Rankinite		Cement clinker/slag	$Ca_2Si_2O_7 + 6CO_2 + 3H_2O \rightarrow 3Ca^{2+} + 6HCO_3^-$	(7.26)

[a]Brucite ($Mg(OH)_2$) occurs both naturally and as a product of anthropogenic processes.

Box 7.1 Rocks and minerals for geochemical NETs

A "mineral" is a solid inorganic substance with a definite chemical composition and crystal structure, whereas a "rock" is a material made up of many minerals. Deposits of pure minerals are relatively rare, therefore for a geochemical NET to be applicable at a global scale, rocks rather than specific minerals will be used as feedstocks for processes. Further definitions are included within the glossary in Chapter 7.7.

Felsic rocks: Comprised predominantly from quartz (>63% silica), K-feldspar, and Na-rich plagioclase feldspar. Felsic rocks include granite and rhyolite. Given the low concentration of cations, they are unlikely to be suitable for geochemical NETs.

Basic rocks: Comprised from 45–52% silica, 20–85% Ca-rich plagioclase feldspar, 15–50% pyroxene, 0–35% olivine. Basic rocks include:

Basalt – an extrusive igneous rock comprised primarily of anorthite, augite and possibly olivine.

Gabbro – an intrusive igneous rock comprised primarily of Ca-rich plagioclase feldspar (labradorite/ bytownite), augite and to a lesser extent olivine and orthopyroxene.

Ultrabasic rocks: Igneous rocks that contain <45% silica,>18% MgO (Mg content expressed as its oxide), high iron, and low potassium. Comprised from >90% mafic minerals. Ultrabasic rocks include:

Dunite – a peridotite that is composed predominantly of olivine and may contain significant amounts of chromite, pyroxene and spinel.

Harzburgite – a peridotite composed primarily of olivine and orthopyroxene with small amounts of spinel and garnet.

Kimberlite – a peridotite composed of at least 35% olivine with significant amounts of other minerals that might include: phlogopite, pyroxenes, carbonates, serpentine, diopside, monticellite, garnet and diamonds.

Lherzolite – a peridotite composed primarily of olivine with significant amounts of orthopyroxene and clinopyroxene.

Wehrlite – a peridotite composed mainly of orthopyroxene and clinopyroxene with olivine and hornblende.

abundant, hosted within large ophiolite complexes and flood basalt provinces, respectively.[14] Ultramafic rock units are estimated to occur over 736 000 km^2,[50] whereas flood basalt complexes cover 100 000–1 000 000 km^2, globally. Substantial volumes of mafic minerals occur within each flood basalt, for example, the volume of mafic rocks in the Siberian Traps (Russia) and Columbia River basalt (USA) exceeds 1 000 000 km^3 and 200 000 km^3, respectively.[130] However, a proportion of natural ultrabasic and basic minerals are located in protected areas or mountainous regions, which may be unsuitable for extraction for geochemical NETs.[45,131] In these regions, *in-situ* mineral carbonation could be more favourable. Specifically, the United States hosts three flood basalt provinces suitable for *in-situ* mineral carbonation,[14] including the Columbia River flood basalt with an estimated storage capacity of 100 Gt CO_2.[38] The US also contains 200 km^3 of ultramafic rock,[45] principally found in coastal regions proximal to large point sources of CO_2 emissions.[14] In the United Kingdom, natural silicate rock formations suitable for geochemical NETs (1669 Gt) are dominated by basic rock units and could react with hundreds of Gt CO_2.[131]

7.2.2.2 *Anthropogenic Alkaline Material Resources*

Cation-rich, fine-grained, artificial materials produced as products, by-products or waste from industrial processes could provide an effective alternative resource for geochemical NETs as they are abundant, do not entail extensive mining and comminution and require less pre-treatment than their natural counterparts (see Section 7.1.5 for further details).[73,132,133] Calcium- or magnesium-rich alkaline materials include cement wastes, biomass and paper sludge ash, sewage sludge ash, steel slag, concrete waste, and municipal solid waste.[13,80,134] Around 7 Gt of Ca and Mg bearing fine silicate-rich waste are produced annually in industrial processes, with the potential to sequester 2.9–8.5 Gt CO_2 per annum through enhanced weathering by 2100.[71] Furthermore, global stockpiles of unused silicate waste may be in the order of 100 Gt,[70] however information on the location and availability of these legacy materials is globally limited. The potential of legacy steel waste in the UK may have a CO_2 capture potential of 60–140 Mt CO_2.[135]

7.2.3 Secondary Minerals

Weathering of primary silicate minerals results in the drawdown and storage of atmospheric CO_2 by the formation of secondary minerals, *e.g.*, $CaCO_3$. A carbonate mineral is any mineral that contains a substantial concentration of carbonate anions (CO_3^{2-}). Carbonate minerals can exist at the Earth's surface and subsurface for millions of years and are thus considered a stable storage repository for CO_2. A variety of mineral carbonates exist but not all are suitable for CO_2 sequestration. Although transition metal carbonates such as siderite ($FeCO_3$) are conceivable candidates for mineral storage[136] decomposition by redox processes makes the permanence of this storage pathway uncertain. The carbonates and bicarbonates of the alkali metals, *e.g.*, $NaHCO_3$ and K_2CO_3, are highly soluble and may not be suitable for long-term secondary mineral storage, but may contribute to alkalinity (see Section 7.2.4). Although $BaCO_3$, and $SrCO_3$ are both insoluble and more thermally stable than $CaCO_3$, these elements are not abundant enough to contribute significantly towards large-scale CO_2 storage. Thus, geochemical

NETs are limited mainly to sequestration *via* magnesium and calcium carbonates.

The stability of a mineral is determined by the temperature and pressure, as well as the concentration of any proximal reactive phases, *e.g.*, water, carbon dioxide, sulphur, oxygen, or other minerals. At surface conditions, the main limitation to stability of a carbonate will be the proximity of acidic phases. Specifically, compounds that may replace the carbonate base, for instance organic anions and inorganic acids (*e.g.*, HCl, HNO_3, H_2SO_4) (eqn (7.27)).

$$CaCO_3 + 2HNO_3 \rightarrow Ca(NO_3)_{2(aq)} + CO_2 + H_2O \tag{7.27}$$

Given that the total natural and anthropogenic sulphur and nitrogen cycles are ~100s of Mt per year[137] and that 1.2–1.3 g of N or S are needed to replace 1 g of C, it would not be geophysically possible to generate sufficient alternative base to displace large Gt of carbon in carbonate. However, some reversal may be possible in niche environments.

Secondary minerals are not only limited to carbonates. Weathering of primary silicate minerals may produce a wide variety of aluminosilicate minerals referred to as clays. For example, the clay kaolinite is produced by weathering of a plagioclase feldspar.

$$2KAlSi_3O_{8(s)} + 9H_2O_{(l)} + 2H^+{}_{(aq)} \rightarrow Al_2Si_2O_5(OH)_{4(s)} + 4H_4SiO_{4(aq)} + 2K^+{}_{(aq)} \tag{7.28}$$

Once formed, these clays play an important role in the control of the weathering rate of primary minerals and the flux of basic cations to the oceans. Clay formation is also known to occur in so-called reverse weathering reactions. For example, formation of sepiolite:

$$4Mg^{2+}{}_{(aq)} + 6H_4SiO_{4(aq)} + 8HCO_{3(aq)}^- \rightarrow Mg_4Si_6O_{15}(OH)_{2(s)} + 8CO_2 + 15H_2O_{(l)} \tag{7.29}$$

The role of reverse weathering in the control of ocean acidity and long-term carbon storage is still a topic of controversy.[138,139] Nevertheless, in addition to quantification of the formation of carbonate minerals, it is important to understand and quantify any secondary mineral formation, as these may counteract the sequestration efficiency of geochemical NETs.

7.2.4 Ocean Alkalinity

7.2.4.1 Function of Natural Alkalinity in the Ocean

The oceans can be considered a natural resource for geochemical NETs, as they sequester CO_2 in similar ways to mineral carbonation and enhanced weathering, *via* the transformation of atmospheric CO_2 into carbonates and dissolved inorganic carbon. The ocean naturally stores a large quantity of carbon (~38 000 Gt C) as dissolved carbonate and bicarbonate ions

(CO_3^{2-} and HCO_3^-).[140] Carbon can be added to this reservoir when CO_2 in the atmosphere diffuses across the ocean's surface where it reacts with water to form carbonic acid (eqn (7.1)), which subsequently dissociates into bicarbonate and carbonate ions in two deprotonation steps (eqn (7.1)).

The extent or rate of forward reaction (eqn (7.1)) is governed by the solution's total alkalinity (TA) (eqn (7.30)). The greater the TA of the solution, the greater its ability to neutralise an acid. TA is a measurable conservative parameter, its value does not vary with changes in temperature and pressure but through changes in chemical equilibria (a review of ocean alkalinity can be found in Zeebe and Wolf-Gladrow[118]). The TA of a natural water sample can be defined as a measure of the proton deficit when compared with an arbitrarily defined value of zero protons. This level of zero is defined such that bases comprised of weak acids (*e.g.*, carbonic acid), which have a disassociation constant of $K < 10^{-4.5}$ (25 °C and zero ionic strength) are classed as proton acceptors, whereas strong acids (*e.g.*, hydrochloric acid), which have a disassociation constant of $K > 10^{-4.5}$ are classed as proton donors.[141] After defining the constituent of the solution *via* its K value as either a proton donor or a proton acceptor, the TA is then a simple equation where the sum of the concentration of the proton donors is subtracted from the sum of the concentration of the proton acceptors, see eqn (7.30), reproduced from Dickson.[141]

$$\begin{aligned} TA = {} & [HCO_3^-] + 2[CO_3^{2-}] + [B(OH)_4^-] + [OH^-] + [HPO_4^{2-}] + 2[PO_4^{3-}] \\ & + [H_3SiO_4^-] + [NH_3] + [HS^-] - [H^+] - [HSO_4^-] - [HF] - [H_3PO_4] \end{aligned} \tag{7.30}$$

Weathering of silicate minerals increases ocean alkalinity as the dissolution of silicate minerals (eqn (7.2)) consumes H^+ ions and releases divalent cations.[121] To achieve electroneutrality, positively charged cations are balanced with negative charges (*i.e.*, bicarbonate and carbonate ions) by a forward shift in the equilibria in eqn (7.1). This process consumes $CO_{2(aq)}$,[119] which creates a greater pressure gradient between atmospheric and aqueous CO_2 and, in turn, draws down atmospheric CO_2 into the solution.

7.2.4.2 *Weathering and Alkalinity*

Cations produced from terrestrial weathering may react with CO_2 to precipitate pedogenic carbonate minerals (discussed previously). However, dissolved cations transported to the ocean *via* run-off in streams and rivers will increase ocean alkalinity *via* the mechanisms described above.[119] Most of the carbon naturally stored in the ocean is in the form of the monovalent anion HCO_3^- (~85%), therefore when silicate weathering adds divalent cations to the ocean, their charges are balanced with the production of two HCO_3^- ions. Thus, when weathering products reach the ocean, they have the potential to draw down twice as much CO_2 than in the formation of solid carbonate minerals. In reality, the presence of CO_3^{2-} and other variations in temperature, pH and salinity mean that the addition of divalent ions to the ocean result in a draw-down of 1.4–1.7 CO_2 molecules per ion.[119] The possibility of carbon storage as bicarbonate ions offers the possibility of carbonate minerals as a feedstock (see Section 7.4).

7.2.4.3 Stability of Increased Alkalinity

Dissolved inorganic carbon (DIC) has a residence time of 100 000 years in the ocean and is therefore considered to be a permanent store in relation to anthropogenic climate change.[119] Alkalinity is naturally reduced *via* the precipitation of carbonate minerals. However, abiotic carbonate precipitation is largely inhibited in seawater despite seawater already being supersaturated with respect to Mg and Ca carbonate.[142] The majority of carbonate precipitation in the ocean is thought to be controlled by biological activity. However, the mechanisms by which organisms biomineralise varies,[143] meaning whole ecosystem responses are hard to predict. Fluxes of $CaCO_3$ burial are promoted by coral growth in continental seas and calcifying phytoplankton in open oceans, thus these organisms have a profound influence on total alkalinity.[144] Ocean acidification caused by increased atmospheric CO_2 has put considerable stress on some calcifying organisms.[145] This biological response has led to extensive research around ocean acidification,[146–148] however, studies on ocean alkalinisation are far fewer, see Renforth & Henderson[119] for review.

The "carbonate compensation depth" is the depth in the ocean in which calcite dissolves quicker than it is supplied. Increasing alkalinity will increase this depth, which may increase calcification, thus countering the effectiveness of alkalinisation.[149] Nonetheless, this is considered to be a long-term negative feedback, which is not thought to greatly diminish the effectiveness of ocean alkalinity enhancement as a geochemical NET on policy relevant timescales.[119]

7.3 The Kinetics of Geochemical NETs

The barrier to widespread adoption of mineral carbonation and enhanced weathering systems as carbon management tools is primarily an issue of overcoming slow reaction rates rather than material or thermodynamic limitations. Therefore, this section is devoted to explication of the factors of relevance to the kinetics of geochemical NETs, and how these barriers can be overcome.

7.3.1 Gas–Solid Processes

In "dry" carbonation, CO_2 reacts directly with magnesium or calcium silicate minerals:

$$(Ca/Mg)_2SiO_{4(s)} + 2CO_{2(g)} \rightleftharpoons 2(Ca/Mg)CO_{3(s)} + SiO_{2(s)} \qquad (7.31)$$

The negative free energy for eqn (7.31) indicates the reaction is thermodynamically favourable. However, kinetic limitations prevent rapid conversion.[11] Despite improvements in conversion rates achieved by increasing temperature, pressure, and mineral surface area, the approach is considered unviable for capture of CO_2 from industrial flue gas streams. Calcium and magnesium oxides (and hydroxides) have more favourable carbonation kinetics than silicates. However, they are not as abundant in nature. Thus, oxides or hydroxides would first need to be produced by separating the basic cations from the mineral matrix. For example, via staged solid/solid reaction with ammonium sulphate (AS) salt followed by precipitation in aqueous ammonia solution.[35] The extracted alkali (hydr)oxides can then be reacted with CO_2 according to:

$$(Ca/Mg)O_{(s)} + CO_{2(g)} \rightleftharpoons (Ca/Mg)CO_{3(s)} \tag{7.32}$$

$$(Ca/Mg)(OH)_{2(s)} + CO_{2(g)} \rightleftharpoons (Ca/Mg)CO_{3(s)} + H_2O \tag{7.33}$$

Generally, the carbonation of hydroxides proceeds more rapidly than carbonation of oxides.[150] As with silicate carbonation, the carbonation rate can be improved by increasing temperature, pressure and surface area. In addition, the presence of humidity has been shown to significantly accelerate the carbonation of MgO (and CaO),[151,152] although the effect for $Mg(OH)_2$ and $Ca(OH)_2$ is more complex.[33] The initial stages of carbonation of oxides and hydroxides is considered to be under chemical control.[153] However, as the reaction progresses, the rate becomes limited by intraparticle CO_2 diffusion due to obstruction by a growing product layer.[154] Processes of attrition, which occur in fluidised beds for example, could help break down these product layers, opening new reactive sites, improving carbonation rates, although fluidisation performance may be diminished.[155,156]

Gas–solid mineralisation pathways might be more viable as "passive", large-scale, slow-reacting systems rather than as centralised industrial processes.[157] For example, Myers and Nakagaki[158] propose directly mineralising finely-ground calcium and magnesium silicate minerals *via* reaction with atmospheric CO_2 in a tiered greenhouse. Conversion to carbonate products occurs over months to years, in ambient conditions, thus avoiding the practical, kinetic and regulatory limitations of industrial gas–solid carbon mineralisation.

7.3.2 Aqueous Processes

Most geochemical NETs involve reactions in the aqueous phase. Aqueous carbon mineralisation reactions can be further divided into in three main processes: (1) mineral dissolution; (2) CO_2 dissolution, and (3) carbonate mineral precipitation. Here, the factors affecting the rates of these three steps will be discussed.

7.3.2.1 Mineral Dissolution

In aqueous conditions, the mineral surfaces are attacked by H_2O and/or H^+, which causes basic cations (*e.g.*, Mg^{2+}/Ca^{2+}) and silicic acid (H_4SiO_4) to enter the solution. For example, dissolution of forsterite (Mg_2SiO_4) or larnite (Ca_2SiO_4) occurs according to:

$$(Ca/Mg)_2SiO_{4(s)} + 4H_2O_{(l)} \rightleftharpoons 2(Ca^{2+}/Mg^{2+}) + H_4SiO_4 + 4OH^- \quad (7.34)$$

$$(Ca/Mg)_2SiO_{4(s)} + 4H^+{}_{(aq)} \rightleftharpoons 2(Ca^{2+}/Mg^{2+}) + H_4SiO_4 \quad (7.35)$$

Mineral composition and structure are also fundamental to dissolution rates and mechanisms. The dissolution rate is controlled by breaking of the shortest and strongest bonds in the minerals, which is usually the Si–O bond.[159] Thus, as a rule of thumb, silicate minerals with a lower degree of silica polymerisation (*e.g.*, olivine) weather at faster rates than minerals with a higher degree of silica polymerisation (*e.g.*, quartz).[160] Generally, calcium silicates dissolve at faster rates than magnesium silicates owing to the comparatively weaker Ca–O bond.[159,161] As a result of their industrial pre-treatments, artificial materials such as steel slags and mine tailings are physically and chemically less stable than their comparable natural silicates, giving greater dissolution rates.[73] The variety of bond types and thus strengths in silicate minerals mean that these tend to dissolve incongruently, with their more soluble components released preferentially.[159]

Dissolution in acidic conditions may lead to formation of a silica-rich passivating layer *via* preferential dissolution of basic cations, whereas dissolution at high pH may lead to Mg- or Ca-rich surface layers due to enhanced release of silicate.[15,162,163] In either case, these passivating layers block reactive sites and inhibit diffusion of reactants and products.[164] The formation of carbonate products also acts to inhibit further dissolution. For instance, precipitation of $CaCO_3$, with its greater crystal volume than the reactant minerals, creates a product solid phase with lower overall porosity.[108] The formation of other non-carbonate secondary minerals can also inhibit further mineral dissolution. For example, in carbonation of steel slag, oxidation of the Fe(II) content can form passivating iron(III) oxide phases,[165] whereas in enhanced weathering of silicate minerals in soils, the formation of clay minerals retard further silicate dissolution.[166–168] On the other hand, oxidation of Fe(II) produces acidity which contributes to mineral dissolution.[169] Various forms of agitation, such as milling, mixing, and sonication, have been investigated to prevent formation of these occluding layers.[170] Decreasing the initial particle size has been shown to increase dissolution rates by orders of magnitude owing to the greater reactive surface area of contact between the mineral and aqueous phases, as well as the reduced thickness of the passivating layer.[171]

The composition of the aqueous phase also has a strong bearing on mineral dissolution rate. In most cases, a decrease in pH of an aqueous solution has been found to increase dissolution rates of minerals[159] although some minerals, *e.g.*, aluminosilicates demonstrate a minimum dissolution rate at neutral pH.[159,161] The presence of organic ligands such as citrate and oxalate may increase the dissolution rates significantly,[172] whereas inorganic ligands such as sulphate, carbonate and phosphate have more complex effects.[173]

Finally, temperature also has a significant impact on dissolution kinetics. For instance, a 10 °C increase in temperature can bring about a three-fold increase in the olivine dissolution rate[174] due to greater collision frequencies between mineral surfaces and reactive species in the aqueous phase. Thus, hotter tropical coasts and soils are desirable locations for spreading and enhancing the weathering of silicate minerals.[59]

7.3.2.2 CO_2 Dissolution

In aqueous geochemical NETs, CO_2 must first enter solution, becoming dissolved gas (7.36). The extent to which gaseous CO_2 enters the solution phase depends on the CO_2 partial pressure and temperature. Essentially, doubling the CO_2 partial pressure doubles its solubility in accordance with Henry's Law.[175] Henry's coefficient is a function of temperature and can be described by different empirical correlations.[176,177] For CO_2 in water, the maximum value of Henry's coefficient corresponds to the minimum solubility, and thus, increasing temperature will reduce the amount of dissolved CO_2. Salinity also has a noteworthy effect on the solubility of gases. Generally, solubility decreases with increasing salinity due to the "salting out effect".[178] Indeed, this effect is observed for CO_2 in brine.[179,180] Thus, increased salinity may have a negative impact on the rate of carbonate formation in aqueous processes.[181] Although, in some applications, increased salinity has been found to enhance carbonation rates.[182]

$$CO_{2(g)} \rightleftharpoons CO_{2(aq)} \tag{7.36}$$

Dissolution of CO_2 into aqueous solution also involves chemical reaction. At low and neutral values of pH, CO_2 reacts with water forming carbonic acid, H_2CO_3, (eqn (7.1)) with deprotonation to form bicarbonate, HCO_3^-, and carbonate, CO_3^{2-} (eqn (7.1)).[177,183] The greater the pH of the aqueous phase, the greater the rightward shift of dissolved CO_2 towards HCO_3^- and CO_3^{2-}, which are the essential product species in aqueous geochemical NETs. Similarly, the greater the alkalinity, *i.e.*, the capacity of the solution to neutralise acid, the greater the amount of CO_2 that can enter solution and form dissolved inorganic carbon.[118]

In addition to thermodynamic equilibrium considerations, the kinetics of CO_2 dissolution play an important role. At low to moderate pH, the forward rate of eqn (7.1) is thought to be the slowest step.[118] At higher pH >8 the mechanism of CO_2 reaction changes,[177] with bicarbonate forming directly *via* the much faster reaction (eqn (7.37)):

$$CO_{2(aq)} + OH^-_{(aq)} \rightleftharpoons HCO^-_{3(aq)} \tag{7.37}$$

In a similar fashion, the zinc-containing metalloenzyme carbonic anhydrase can catalyse the formation of carbonic acid by creating locally a high OH^- concentration and subsequently increasing the concentration of dissolved bicarbonate. The rate constant of this reaction is 8 orders of magnitude higher than that of the CO_2 hydrolysis reaction in neutral water without the catalyst.[184] Several authors have investigated the application of carbonic anhydrase for enhancing precipitation of carbonate minerals.[185,186]

The equilibrium and kinetic considerations thus far have implicitly assumed closed, well-mixed systems. In geochemical NETs, this is often not the case. Instead, significant CO_2 concentration gradients between the carrier gas (usually air), and the reactive phases may exist. For example, two bodies of water may be capable of absorbing the same amount of CO_2. However, if one of those bodies is vigorously mixed, the CO_2 will reach equilibrium faster than the unperturbed body. The rate of CO_2 transfer across the gas–solution boundary can be determined by the reaction-diffusion equations involving the dimensionless parameter, the specific gas enhancement factor.[118] The enhancement factor is the ratio between the true flux (diffusion plus reaction) and the flux if there were no reaction. In geochemical NETs, the increased alkalinity of the aqueous phase created by mineral dissolution provides this chemical driving force, drawing CO_2 into solution. Where diffusion is limiting, this can be overcome by increasing the gas-to-solution contact area, *e.g.*, *via* bubbling systems, or, by creating thin liquid films.[187] It has been found that water bubbled with micrometre-sized CO_2 bubbles resulted in 1.4 times more CO_3^- content than water bubbled with milli-sized bubbles causing an increase in the rate of calcite precipitation.[188]

7.3.2.3 Carbonate Precipitation

In geochemical NETs where solid carbonates are the desired product, better understanding of the factors affecting precipitation is essential. Even in applications where dissolved bicarbonate is the desired product, it is vital to quantify the rate of carbonate mineral precipitation since it reduces the overall CO_2 sequestration efficiency of the technology (see eqn (7.38)). Thus, preventing precipitation may actually be beneficial in some geochemical NETs.[157] The solubility of the carbonate is perhaps the most important factor determining the likelihood of precipitation. For example, for calcium carbonate precipitation:

$$Ca^{2+}_{(aq)} + CO^{2-}_{3(aq)} \rightleftharpoons CaCO_{3(s)} \tag{7.38}$$

the stoichiometric solubility product, K^*_{SP}, is defined as:

$$K^*_{SP} = [Ca^{2+}]_{sat}[CO_3{}^{2-}]_{sat} \tag{7.39}$$

where $[Ca^{2+}]_{sat}$ and $[CO_3{}^{2-}]_{sat}$ are the equilibrium concentrations of each species in a solution saturated with $CaCO_3$ (at a specific temperature, pressure and salinity). The saturation state, Ω, is then defined as:

$$\Omega = \frac{[Ca^{2+}][CO_3^{2-}]}{K_{SP}}$$

where $[Ca^{2+}]$ and $[CO_3{}^{2-}]$ are the concentrations of each species in solution. If $\Omega = 1$, the solution is at equilibrium with the mineral phase. If $\Omega < 1$, the aqueous phase is undersaturated, and calcium carbonate is expected to dissolve, whereas if $\Omega > 1$ then aqueous phase is oversaturated, and calcium carbonate is expected to precipitate.[118] The precipitation rate increases with increasing supersaturation, due to the increasing rate of nucleation.[189] However, kinetic factors also play a role. For example, a solution that is supersaturated with respect to calcite may not result in precipitation due to inhibition by the presence of dissolved Mg.[190] Similarly, ligands present in the aqueous phase can inhibit calcite formation *via* complexation and adsorption,[190] although some ligands enhance the calcite precipitation rates by accelerating the dissolution kinetics.[159]

The precipitation kinetics may be influenced by a variety of other factors. Most importantly, the rate of calcium or magnesium carbonate precipitation increases with increasing pH owing to the shift of aqueous equilibrium towards $CO_3{}^{2-}$.[191–193] Precipitation also has a strong temperature dependence.[194] Calcium carbonate is an unusual salt in that higher temperatures result in greater precipitation.[118] Similarly, for magnesium carbonates, increasing temperature accelerates precipitation.[37] It was found that the precipitation rate of calcite depended strongly on the aqueous volume to mineral surface area ratio,[195] however these findings are disputed and instead, may be caused by the formation of nuclei on other active surface sites.[196] The flow regime also has a strong bearing upon the precipitation rate.[197,198] In turbulent conditions, eddy formation increases the diffusion rates of species leading to large increases in precipitation rates.[195] The rate of carbonate precipitation is proportional to the rate of exchange of water molecules in the cation coordination sphere. Since Ca^{2+} is much larger and "softer" than Mg^{2+}, the water molecules in its coordination sphere are held more loosely enabling faster exchange.[159] Thus, precipitation of $CaCO_3$ is approximately 4 orders of magnitude faster than precipitation of $MgCO_3$.[159] Instead, under ambient conditions, magnesium precipitates from solution as various hydrated magnesium carbonates, *e.g.*, nesquehonite, hydromagnesite and dypingite.[199] Kinetic factors may also influence the crystal structure of the product carbonates. Although calcite is the most thermodynamically stable polymorph of calcium carbonate, kinetic factors can lead to the formation of less stable aragonite or vaterite.[29]

7.3.3 Summary

Understanding geochemical NETs requires understanding of equilibrium thermodynamics, kinetics, and mass transfer processes. In many geochemical NETs there are three stages (1) mineral dissolution; (2) CO_2 dissolution; and (3) carbonate precipitation. Where these processes occur simultaneously, an attempt to enhance one of these processes may hamper the rate of another. For example, maintaining a low pH can significantly increase the dissolution rate of silicate minerals but will slow the formation of carbonate minerals. Thus, a trade-off will be required. Separating dissolution and precipitation allows each process to be optimised independently. However, this adds significantly to the complexity of the method. On the other hand, geochemical NETs which make use of the inherent production of alkalinity during mineral dissolution may enable the system to operate without pH manipulation, provided that the rates of mineral dissolution, CO_2 invasion and precipitation reactions are balanced appropriately.

7.4 Geochemical NET systems

Numerous proposals have been put forward to facilitate the reaction for rocks and CO_2. Much of the previous work has focused on approaches that help to reduce the emissions from point sources by reacting with purified CO_2 gas or directly with flue gas. This section examines how some of these systems can be adapted for CO_2 removal from the atmosphere, and the technical challenges associated with that. Table 7.2 provides an overview of previous work.

7.4.1 *Ex-situ* Mineral Carbonation

Reacting silicate minerals to form new carbonate minerals in chemically engineered processes has been extensively researched, and the life-cycle costs have been reviewed elsewhere.[13] The following presents a brief summary of this work, with particular focus on their pathway within geochemical NETs.

Ex-situ mineral carbonation requires the extracting, transporting, processing, and disposal/reuse of minerals. Kirchofer *et al.*[214] examine the life-cycle impact of a direct (single-step) mineral carbonation approach. Direct reaction of a silicate requires the input of energy, water, and, most importantly, a concentrated supply of CO_2. A long-standing constraint on the development of single-step mineral carbonation is that in most situations, it would be cheaper to compress and inject the CO_2 into underground geological formations than to react directly with minerals.

However, given that direct single-step mineral carbonation also produces energy (as low-grade heat) and secondary minerals, the integration of these products with existing markets could provide a means by which this process is economically viable.[215] It is especially pertinent in locations that have

Table 7.2 An overview of geochemical NET studies and systems along with the predicted sequestration potentials and cost.

Type of technology		Study description	Study type			Total CO_2 sequestration potential		Per tonne of CO_2					Ref.
			Labs and theory	Concept testing	Well developed	Cumulative	Per year	Land demand	Resource requirements	Cost ($)	Electricity (GJ)	Thermal (GJ)	
Mineral Carbonation (MC)	**Full potential estimates**	Global current production estimates of various artificial silicate materials and their associated sequestration potential	X				690–1217 Mt	Processing factories	7–17 Gt of aggregate and mine waste, cement kiln dust, construction and demolition waste, iron and steel slag and fuel ash				70
		Global predicted potential of alkaline materials (in 2100) to sequester carbon	X				2.9–8.5Gt	Processing factories		**'low'**			71
	***Ex-situ* - Direct**	Steel slag suspension in water was carbonated at high temperatures and pressures to assess the sequestration potential	X					Processing factories	4 t of steel slag (depending upon the Ca content)	**low**			15
		A cost evaluation of CO_2 sequestration by aqueous carbonation using wollastonite and steel slag as the feed stock	X					Processing factories		**90**			200
	***Ex-situ* - Indirect**	Multistep process where asbestos mine tailings are reacted with CO_2 rich flue gas, with thermal pre-treatment and		X				Processing factories	4.3 t of rock	**142–205**	7.8		201

		precipitation reactions											
	In-situ	CarbFix Project – *In-situ* carbonation of basalt rocks *via* the injection of CO_2 to the depth of >800m		X			2200 t	Under ground	CO_2 dissolved in water (approx. 2.86 t of water required)				202
		CarbFix2 – deeper and hotter injection site and propose to use a gas mix instead of pure CO_2		X				Under ground	30–100 t of water	**25–42**			203
Enhanced weathering (EW)	**Full potential estimates**	Carbon dioxide from the atmosphere is captured when silicates on the surface of the land dissolve.	X			430 Gt		1.32–8.25 ha per year (application rate dependent)	3.3 t of Basic rocks Comminution and transport and application, Extensive mining	**24–578**	0.1–8.4	0.8–4.2	131
		Predicted potential of alkaline wastes to sequester carbon through their dissolution storing CO_2 as aqueous bicarbonate ions	X				2.6–3.8 Gt	Potentially large areas	Waste materials	**Presumed low**	Low	Low	71
		Potential for large-scale CO_2 removal *via* enhanced rock weathering with croplands	X				>2 Gt	Potentially large areas		**60–220**			54
		Maximum potential for CO_2 sequestration by 'simulated' weathering globally	X				65 Mt C	Potentially large areas					46
	Mine tailings	Weathering of chrysotile mine tailings in Clinton creek			X	164 000 t (26 y after mining)	6300 t						69
				X			1 Pg C						47

Table 7.2 (*Continued*)

Type of technology		Study description	Study type: Labs and theory	Concept testing	Well developed	Total CO_2 sequestration potential: Cumulative	Per year	Per tonne of CO_2: Land demand	Resource requirements	Cost ($)	Electricity (GJ)	Thermal (GJ)	Ref.
	Crushed minerals on croplands	Olivine as a fine powder spread over land areas of humid tropics						Potentially large areas					
		Ultramafic silicate rock 'flour' on to land		X				Potentially large areas	t of rock		1.6–9.9		204
		Weathering of olivine in soil, specifically looking at plant growth in that soil (perennial rye grass)		X				0.37–3.44 ha	833 kg of pure olivine				51
Enhanced ocean alkalinity	**coastal EW**	Olivine weathering on coastal environments		X		2.5–9 Mt		Coastal land and mines					59
		Olivine weathering on coastal environments		X				Coastal land and mines	~1 t of ground olivine (100–300μm)				60
	Full potential estimates	Increasing the alkalinity *via* different methods: accelerated weathering of limestone (AWL), enhanced weathering, electrochemical weathering and ocean liming	X			T		Potentially large areas	Upscaling production of silicate minerals at varying amounts	**20–100/ 50–150**	1–10		119
		Simulated ocean response to increased ocean alkalinisation (from 2010–2100)	X			644–662.4 Gt		Minimal		Ref to Renforth and Henderson (2017)			205
			X								3–18		206

Electro-chemical weathering	Direct electrolysis of sea water to increase aqueous NaOH, known as the Chlor-alkali method. The high pH solution produced at the cathode can be added to ocean to increase alkalinity			Processing factories	A method of disposal of Cl_2 gas				
	Use $CaCO_3$ solution as the electrolyte to produce a high pH solution rich in $Ca(OH)_2$	X		Processing factories		**14–190**	5		207
	Use silicate mineral in the anode, dissolved in sea water for electrolysis. H_2 is also produced	X		Processing factories	1.2 t of crushed silicate mineral 0.4 t of waterH_2 priced and sold at $1500 per tonne	**<100**	5		208
	$MgCl_2$- rich brine is thermally decomposed to make MgO and HCl	X	1 Mt	Processing factories			0.8	13	209
AWL	Carbonate mineral mixed with water and CO_2 rich gas stream to store CO_2 as bicarbonate in solution, the process also produces Ca^{2+} ions	X	Framed as CO_2 reduction rather than NET	Processing factories	2.3 t of limestone High water demand	**18–128**			126
	Carbonate mineral mixed with water and CO_2 rich gas stream to store CO_2 as bicarbonate in solution, the process also	X		Processing factories	2.5 t of low grade or waste limestone	**3–66**			128

Table 7.2 (*Continued*)

Type of technology	Study description	Study type: Labs and theory	Study type: Concept testing	Study type: Well developed	Total CO_2 sequestration potential: Cumulative	Total CO_2 sequestration potential: Per year	Per tonne of CO_2: Land demand	Per tonne of CO_2: Resource requirements	Per tonne of CO_2: Cost ($)	Per tonne of CO_2: Electricity (GJ)	Per tonne of CO_2: Thermal (GJ)	Ref.
	produces Ca^{2+} ions											
Ocean liming (OL)	Increasing alkalinity to the ocean *via* addition of lime (CaO) to ocean. This lime is produced *via* calcination of limestone using a range of materials and calcination methods	X				1 Gt	Processing factories	1.4–1.7 t of limestone to be crushed and distributed Method of storing produced CO_2	**72–159**	0.1–1.2	0.6–5.6	210
	Oxy-fuel flash calciner: limestone								**126**	1.3	4.8	
	Endex CFC: Limestone								**100**	− 0.1	5.5	
	Oxy-fuel flash calciner: dolomite								**95**	0.7	3.2	
	Endex CFC: dolomite								72	− 0.1	4.2	
	Solar calciner: limestone								**159**	0.4	0.6	
	Calciner + amine flue gas absorption: limestone								**261**	2	17.2	
	Increase of alkalinity *via* the addition of $Ca(OH)_2$ to chemically increase ocean CO_2 uptake, the lime is	X					Processing factories	10 Pg year of $Ca(OH)_2$				

		produced from limestone								
Integrated approaches	**OL + MC**	Mg silicate minerals are reacted with CO_2 (mineral carbonation) then calcined to produce a pure stream of CO_2 which can be buried, the MgO produced can be used as feed stock for ocean liming	X		Processing factories	1.9 t of olivine rich rocks, ships to spread minerals	**98–143**	2.2	4.9	211
	Ambient looping of CaO	A fuel cell powered with natural gas produces electricity and heat which is used to calcine lime or magnesite and then captures CO_2 from the air. The calcination process produces pure CO_2 which is captured and stored.	X	436.5 gCO_2/kW h_{el}	Processing factories		**52.26–99.97 (C tax)171.98–491.67(LCOA)**			86
		1–2 kg of CaO was left in trays in ambient conditions to carbonate in air	X		Possibly large	Possibly large amounts of lime (CaO)				212

Table 7.2 (*Continued*)

Type of technology	Study description	Study type: Labs and theory	Study type: Concept testing	Study type: Well developed	Total CO_2 sequestration potential: Cumulative	Total CO_2 sequestration potential: Per year	Per tonne of CO_2: Land demand	Per tonne of CO_2: Resource requirements	Per tonne of CO_2: Cost ($)	Per tonne of CO_2: Electricity (GJ)	Per tonne of CO_2: Thermal (GJ)	Ref.
DAC + MC	DAC system that stores the collected CO_2 with *in-situ* carbonation *via* injection into basalts	X			> 950 Gt	10 000 t	Underground				low grade waste heat	44
Ambient looping of MgO	Utilising magnesite ($MgCO_3$) to repeatedly capture CO_2 from the atmosphere *via* spreading on land, collecting and calcining		X			2–3 Gt	0.61–0.75 Mha per Gt CO_2		**46–159**	0.3		213

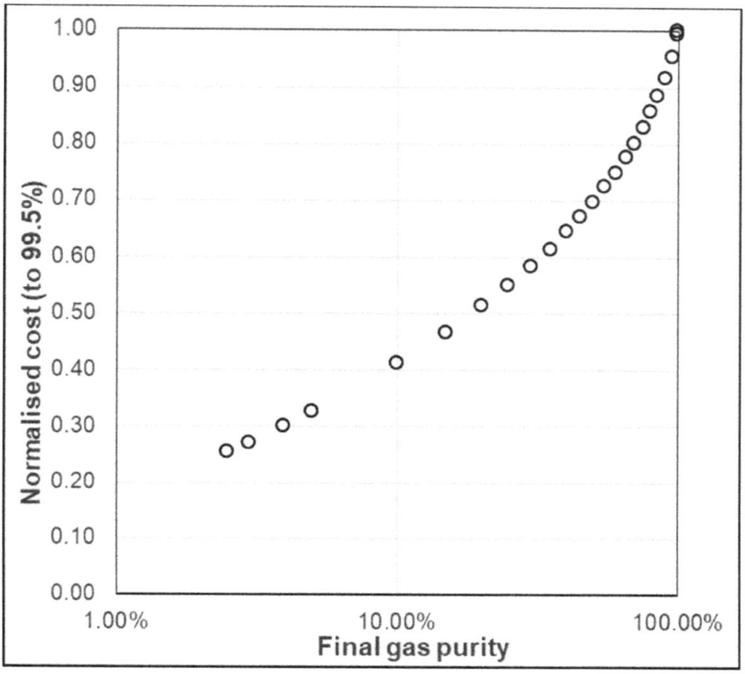

Figure 7.6 The predicted cost of CO_2 removal from a hypothetical DAC system against CO_2 purity in the effluent gas stream (adapted from Kelemen *et al.*[42]). This assumes that 60% of the CO_2 in the air passing through the DAC system is captured.

abundant mineral resources but limited access to geological storage. The role of direct mineral carbonation in geochemical NETs may be minimal given that direct air capture (DAC) would be needed to produce high purity CO_2, and DAC's relative geospatial freedom to operate.

Some multi-step mineral carbonation approaches were proposed as a means of reacting lower concentration CO_2 gas streams (*e.g.*, directly with flue gases), without the need for gas purification, and as such offer some opportunities for integration within geochemical NET pathways.

For instance, approaches that produce magnesium hydroxide from magnesium silicate[35] were initially proposed to react the product directly with flue gas.[216] However, this material may also be suitable for reaction with atmospheric CO_2, similar to the anthropogenic alkaline materials discussed below (Section 7.4.3), or the magnesium hydroxide may form the feedstock of an MgO–$MgCO_3$ looping system.[213] A major cost driver for DAC is that high purity CO_2 is required for geological storage, but some alkaline minerals can react with lower concentration CO_2 (see Section 7.1.4). By combining low purity CO_2 DAC with mineralisation, it may be possible to substantially cut overall costs. For instance, mineralisation with 10% pure CO_2 gas could reduce the cost of DAC by >60%[42] (Figure 7.6).

Box 7.2 Case-Study – CarbfFix & Climeworks

CarbFix began development in 2007 to explore the feasibility of underground injection of CO_2. The project initially made use of a CO_2 rich gas produced by the local Hellisheidi geothermal power plant in SW-Iceland. The results of the initial experimentation suggest that the majority of CO_2 injected into the site was mineralised.[43]

Gunnarsson *et al.*[203] and Snæbjörnsdóttir *et al.*[41] have outlined the feasibility of this approach. They suggest a system in which water is used to scrub a dilute (20–63% by volume) CO_2 gas stream, at a ratio of ~30–100 tonnes of water per tonne CO_2 removed, at a capture efficiency of around ~36–56%. The resulting mixture is pressurised and injected into the rock formation. Preliminary cost estimates by Gunnarsson *et al.*[203] suggest up to \$42 for capture, \$1 for water/CO_2 transport, \$4 for pumping and injection, and \$1 for monitoring per tonne of CO_2 injected (excluding water makeup). A comprehensive technoeconomic assessment remains to be published.

In 2017, CarbfFix partnered with Climeworks to sequester atmospheric CO_2, and are currently offering offsetting rates at approximately \$1200 per t CO_2.

7.4.2 *In-situ* Mineral Carbonation

Reacting atmospheric CO_2 with basic or ultrabasic rocks underground removes the need for mineral extraction, processing and transport. The key engineering challenges associated with *in-situ* mineral carbonation are (i) CO_2 gas separation (similar to *ex-situ* mineral carbonation discussed previously) and transport to injection site, (ii) water make-up and transport to the injection site if a co-injection method is used as it is by CarbFix (Box 7.2), (iii) injection and inspection well drilling, (iv) monitoring during the site operational phase, and (v) well closure and long-term monitoring.

Kelemen *et al.*[42] explored the feasibility of atmospheric CO_2 injection into peridotite and suggest a minimum cost that is highly sensitive to reservoir permeability and CO_2 injection rate. The estimate costs $<$\$500 per t CO_2 are possible for the co-deployment of direct air capture and *in-situ* mineralisation, with minimum values $<$\$200 per t CO_2. Upscaling to Gt CO_2 per yr would require the creation of 100 000 to 1 million new wells, a scale similar to boreholes in the oil and gas sector.

7.4.3 Anthropogenic Alkaline Material Carbonation

Alkaline materials, but specifically wastes, have been considered as feedstocks for direct aqueous mineral carbonation, and costs ~\$90 per t CO_2.[200] Numerous commercial ventures exploit this or similar processes to carbonate waste materials, reduce their environmental liability, and create secondary construction products. These materials can react with lower purity CO_2 gas, and possibly as low as atmospheric concentrations (see Section 7.1.5). Pasquier *et al.*[201] estimated that contacting asbestos rich mine tailings with flue gas concentrations of CO_2 in a series of reactors (including thermal pre-treatment and precipitations/drying) to cost \$142–205 per t CO_2. Alternatively, reacting alkaline materials with atmospheric CO_2 in tailings facilities or heaps of processed material may reduce the cost. For instance, Kelemen *et al.*[42] estimate that over a multi-decade timescale, the cost of reacting distributed ultramafic mine tailings with atmospheric CO_2 may be <\$100 per t CO_2. The feasibility assessment of reacting other alkaline materials (*e.g.*, slag, cement, red mud) in this way has yet to be published.

7.4.4 Proposals for Increasing Ocean Alkalinity

It is estimated that the ocean has already consumed 40% of anthropogenic carbon emissions[140] and could reduce atmospheric CO_2 to preindustrial levels within 100 000–200 000 years.[7] However, human intervention is needed to enhance the rate of CO_2 uptake in the oceans if it is to sequester additional anthropogenic carbon on a timescale relevant to solving climate change.

Kheshgi[217] first proposed calcining abundant $CaCO_3$ and adding the resulting lime (CaO) to the ocean to increase total alkalinity. The technoeconomic viability of ocean liming was then revised to suggest that the economic and energy costs of this process were comparable to other air capture technologies,[210] but noted that calcination of limestone may not be the most efficient option. To overcome the inefficiencies seen in calcination, other work[211] suggest the creation of MgO from Mg-silicates through *ex-situ* mineral carbonation coupled with calcination.

Alternatively, accelerated weathering of limestone (AWL) was suggested by Rau and Caldeira,[126] in which carbonate minerals (*e.g.*, $CaCO_3$) are reacted with elevated concentrations of CO_2 (from flue gas or possibly direct air capture) and sea water, which is then discharged back to the ocean. This ultimately stores the CO_2 in the bicarbonate system of the ocean and increases ocean alkalinity. Thousands of tonnes of sea water are required per tonne of CO_2 sequestered. However, this could be reduced to around 400 tonnes of seawater per tonne of CO_2 by utilising the cooling water that is already being used in some power stations.[126] Kirchner *et al.*[218] modelled the effects of AWL on the ocean chemistry and discussed the possible side effects on the ocean ecosystem. They conclude that there is a potential for calcium carbonate to precipitate out of solution at discharge points. Therefore, further consideration of local discharge sites and the size of flue gas scrubbing is needed.

Increasing ocean alkalinity may also be achieved through electrolysis and the creation of $Cl_{2(g)}$ and $H_{2(g)}$ in a fuel cell to produce electricity.[206] The acidity produced can then be neutralised by silicate mineral dissolution. This reaction is rapid but requires vast amounts of energy. A different electrochemical approach was suggested in which $CaCO_3$[207] or magnesium silicate[208] are dissolved at an acidic anode creating Ca–OH and Mg–OH solutions at the cathode. CO_2 is then captured at the high pH cathode and the resulting solution is used for ocean alkalinity enhancement.

Several studies have investigated the process of adding naturally occurring silicate minerals to the land surface,[45–47,49,51–53,57,131,204] coastal systems[59,60] and the open oceans.[62,129] Crushing rocks and minerals to a fine powder increases their surface area and, in turn, their dissolution rate, however, this can come at a large energetic cost. For instance, a comminution circuit for reducing rock to ~mm, 100s μm, or 10s μm sized particles may require on the order of 0.1–1, 1–10, and 10–100s kW h per t, respectively.[131] A life cycle assessment[123] for enhanced weathering leading to carbonate formation in soils in Brazil concluded that the most carbon intensive steps were rock grinding and transport. It was calculated that, after transport distances of 540 ± 65 km (for carbonation) and 990 ± 116 km (for enhanced weathering), more CO_2 would be released than sequestered. Although others[122] have noted that 80% of temperate and 95% of tropical agricultural land is located within 300 km of a suitable rock outcrop. By integrating a soil weathering model to nationally resolved enhanced weathering supply chains, Beerling *et al.*[54] were able to estimate a global enhanced weathering operation at >2 Gt CO_2 per year scale to cost between $60 and $220 per t CO_2.

7.4.5 Hybrid Approaches

Hybrid methods of geochemical NETs integrate the reaction of CO_2 with a mineral, or have minerals present in the flowsheet, but the CO_2 is stored elsewhere. Alternatively, proposals that use direct air capture or biomass energy with carbon capture to concentrate atmospheric CO_2 for accelerated reaction with minerals could also be considered a hybrid approach.

McQueen *et al.*[213] explored a method in which magnesium oxide or hydroxide are specifically created for reaction with atmospheric CO_2 but the produced carbonates are recycled back into a kiln to regenerate the oxide/hydroxide and produce pure CO_2 for geological storage. Similarly, Origen Power have proposed the gas–solid carbonation of CaO under ambient conditions.[86] This method reduces the amount of land required for this technology. This is conceptually similar to the use of 'large $Ca(OH)_2$ structures' proposed by Abanades *et al.*[219]

7.5 Risks and Co-benefits

7.5.1 Valuable Products

Re-purposing the end-products of mineral carbonation into marketable goods is expected to improve the economic viability of the lifecycle.[13,80] Specific post-processing steps create high purity products with a high market value; for example, carbonate products can be re-processed into high purity calcite and hydromagnesite,[103,220] micro-silica can be used as a partial cement replacement or as a filler,[80] and iron oxide can be used in iron/steel manufacture.[80] However, reprocessing carbonated products into high value products with rigorous market specifications reduces net profit and could oversaturate the market.[13,80] Instead, low value products could be utilised on a larger scale; for example, silica and carbonate products are suitable for construction and land reclamation, the latter of which could consume 100–200 Gt CO_2.[80,81] Additionally, carbonate products could replace conventional agricultural lime and/or be applied in ocean liming.[80,211] However, prior to the successful deployment of geochemical NETs and concomitant production of marketable, carbon neutral goods, the quality of high-value products and large-scale production of low-value products remains to be proven.

7.5.2 Valuable Services

In addition to reducing atmospheric CO_2, geochemical NETs have the potential to provide a plethora of valuable services to the soil, plant and marine environment. The addition of cation-rich silicate powder to agricultural soils during enhanced weathering is thought to improve an array of soil health parameters; for example, the consumption of H^+ ions during dissolution increases soil pH[221] which, in turn, reduces heavy metal toxicity and improves cation exchange capacity.[222,223] Furthermore, silicate addition replenishes eroded soil and increases soil organic carbon[56] whilst the liberation of bioavailable plant nutrients (K, Ca, Mg, P, Si) enhances plant growth, increases the nutrient density of crops and enhances plant resistance to abiotic and biotic stress.[45,56,224–226] Improving crop yields is not only essential to meet rising food demand,[227] but also plays a crucial role in reducing the application of finite, emission intensive and often inaccessible conventional fertilisers.

Although the impact of enhanced weathering on the marine ecosystem remains uncertain, the transfer of dissolution products into coastal waters could increase the total alkalinity, pH and $CaCO_3$ saturation state.[45,56,228] Furthermore, enhancing the flux of dissolved Si into the ocean is expected to stimulate diatom growth, which plays a critical role in the biological pump transporting organic carbon to the deep ocean.[45]

7.5.3 Risks

Geochemical NETs have relatively low land and water requirements in comparison to other NETS, needing 2–10 Mha and 0.3–1.5 km^3 per year for 0.2–1 Gt C per year removal, respectively. However, processing stages (mining, comminution, transport) make ERW & MC (not including anthropogenic alkaline wastes) both carbon and energy intensive processes (requiring 46 EJ per year).[221] It is expected that a substantial increase in mining and transport infrastructure will be necessary if geochemical NETs

Table 7.3 Benefits and side effects of geochemical NETs, where ERW refers to enhanced rock weathering; MC, mineral carbonation; SSC, single step carbonation; MSC, multi-step carbonation; SC, surficial carbonation; InC, *in-situ* carbonation.

Impact			Geochemical NET
Benefits	Economic	Use of end products	MSC
	Environmental	Long term carbon storage	ERW, SSC, MSC, SC, MC
		Low water requirements	ERW, SSC, MSC, SC, MC
		Low land requirements	ERW, SSC, MSC, SC, MC
		Reduced stockpiling of waste materials	ERW, MSC, SC, MC
		Negligible interference with soil and water environment	SSC, MSC, SC
		Improved soil health • Reduced heavy metal toxicity • Increased soil pH • Increased soil cation exchange capacity • Increases soil organic carbon • Reduced soil erosion • Release of nutrients	ERW
		Improved crop yield & crop resistance to abiotic & biotic stress	ERW
		Increased ocean alkalinity	ERW
Adverse Impacts	Economic	Energy & CO_2 emissions associated with mineral extraction & comminution	ERW, SSC, MSC, MC
		Additional heating required	SSC, MSC
		Additional reagents required	MSC
	Environmental	Ecological destruction associated with extraction of source minerals	ERW, SSC, MSC
		Release of heavy metals into soil & water	ERW
		Increased water pH (terrestrial, coastal, marine/deep aquifer)	ERW, InC
		Changes in soil physical and chemical properties	ERW
	Social	Maintenance requirements for future generations	ERW, InC
		Moral hazard	ERW, SSC, MSC, SC, MC
		Release of respirable particles	ERW

are to significantly reduce atmospheric greenhouse gases[131] which, in turn, will reduce the net CO_2 drawdown associated with each technology.

Geochemical NETs could adversely impact the environment in a number of ways: ecological destruction associated with extraction,[229] changes in water pH,[47,230] release of heavy metals, *e.g.*, Ni and Cr,[120,124] changes in ground, river and coastal water as well as soil physical and chemical properties.[231] The severity of these environmental impacts will be determined by the scale of deployment, silicate source, regional climate and soil properties.

Finally, understanding the ethical and social impact of geochemical NETs is imperative to their successful deployment and is highlighted by a number of authors.[231–233] NETs present a moral hazard whereby political interest in greenhouses gas removal technologies could impair efforts to reduce direct greenhouse gas emissions and therefore risk the successful curtailment of global temperature rise.[233,234] Furthermore, the unquantified impact of geochemical NETs on Earth's terrestrial, freshwater and marine ecosystems displaces risk both spatially and temporally, whereby undesirable consequences could impact far field sites and future generations.[233] Considering the scale of deployment and concomitant side effects, international legislation, governance and monitoring will be essential to minimise risk. Table 7.3 above summarises the potential benefits and impacts of different geochemical NETs.

7.6 Conclusions

The interaction of CO_2 and weathering of rocks has been discussed for over 150 years, and the purposeful acceleration of these reactions to help prevent climate change has been considered for over 30 years. Initially, this work focused on the reaction of minerals with pure CO_2 at elevated pressures and temperatures. This chapter explores the potential of 'geochemical negative emission technologies' that propose to use reaction with minerals to remove CO_2 from the atmosphere.

Geochemical NETs are based on the fundamental principles of chemical weathering in which the production of divalent cations sequesters atmospheric CO_2 as a stable bicarbonate ion in the ocean or a solid carbonate mineral. Secondary carbonate formation reduces the CO_2 sequestration potential by half. A variety of naturally occurring minerals or rocks and anthropogenically produced materials are appropriate for geochemical NETS based on their cation composition. Anthropogenic alkaline materials show promise due to their availability and reactivity, however caution must be taken if their weathering products are released into the natural environment. On the other hand, carbonation of environmental hazards can offer a solution, which simultaneously reduces the hazard and captures CO_2 at the same time.

To be considered a geochemical NET, *ex-situ* mineral carbonation, in which CO_2 is directly reacted with a naturally occurring silicate mineral, would require coupling to direct air capture. This may be feasible within niche applications but may not be competitive at scale. Alternatively, multistep processes that produce alkaline minerals may offer greater potential as geochemical NETs. *In-situ* carbonation has the potential to store vast quantities of CO_2 without the need to extract and process minerals, however, only two field tests have been conducted to date in basalt. Therefore, much more work is required to quantify the long-term evolution of the storage reservoir. Similarly, enhanced weathering, the process by which minerals are added to soils, sediments or waters, have received comparatively little attention. Proposals to increase ocean alkalinity are at an early stage of development, but numerous studies highlight a large storage potential together with relatively low-cost methods, and the additional benefit of helping to reduce the impact of ocean acidification. However, the potential impact on wider marine ecosystems is not well understood.

Recycling carbonated products into marketable goods can increase the economic viability of geochemical NETs, which have the potential to safely store carbon on a timescale relevant for anthropogenic climate change whilst consuming minimal land and water resources. In addition, some geochemical NETs have the potential to contribute to soil health, enhance crop yield and increase ocean alkalinity. However, their impact on the terrestrial and marine environment is under-researched, placing a potential burden on future generations to monitor geochemical fluxes.

Glossary

Alkalinity is the excess of proton acceptors (bases) in a solution and influences the ability of a solution to buffer changes in pH.

Alkaline minerals are deficient in silica and alumina in comparison to alkali minerals.

Amorphous describes a mineral with no definite crystalline structure.

Aquifer is an underground layer of water-bearing permeable rock, rock fractures or unconsolidated materials (gravel, sand, or silt).

Basalt is a dark, fine-grained, extrusive volcanic rock that sometimes displays a columnar structure, typically composed of 42–53% silica and dominated by the minerals plagioclase, pyroxene and olivine.

Basic rock has a relatively high concentration of iron, magnesium, and calcium, and with 45–53 per cent of silica by weight.

Biomineralisation is the process by which living organisms produce minerals, often to harden or stiffen existing tissues.

Benthic organisms are the community of organisms that live on, in, or near the seabed, river, lake, or stream bottom, also known as the benthic zone.

Brine is water with a salt concentration greater than 35 parts per thousand (3.5%). Sea water has a similar average concentration. Brine in underground formations can contain 20% or more salt.

Calcination in cement manufacturing calcination is carried out in furnaces or reactors, which heat limestone leading to its decomposition into calcium oxide (lime) and CO_2 as a by-product.

Calcite is one polymorph of the mineral calcium carbonate ($CaCO_3$).

Calcifying hardens by deposition of, or conversion into, calcium carbonate or another insoluble calcium compound.

Carbonates rocks are composed of an anion attached to a CO_3^{2-} cation (*e.g.*, $CaCO_3$). For example, limestone, chalk, marble, dolomite, calcite.

Carbon credit is a convertible and transferable metric that allows an organisation to benefit financially from an emission reduction.

Carbon Dioxide Removal (CDR), also known as greenhouse gas removal, is a process in which carbon dioxide gas (CO_2) is removed from the atmosphere and sequestered for long periods of time.

Caustic is a strong corrosive alkali (such as sodium hydroxide).

Calcine is to reduce, oxidise, decompose or desiccate by roasting or exposing to strong heat.

Chemical weathering is the erosion or disintegration of rocks, building materials, *etc.*, caused by chemical reactions (chiefly with water and substances dissolved in it) rather than by mechanical processes.

Clay soil particles with a diameter <0.005 mm; also, a rock that is composed essentially of clay particles.

Cement is a powdery substance made by calcining lime and clay, mixed with water to form mortar or mixed with sand, gravel, and water to make concrete.

Concrete is a building material made from a mixture of broken stone or gravel, sand, cement, and water, which can be spread or poured into moulds and forms a mass resembling stone on hardening.

Comminution is the size reduction of a material, especially of a mineral ore, to minute particles or fragments.

Direct carbonation for example of $Ca(OH)_2$ refers to the reaction of $Ca(OH)_2$ with CO_2 to produce $CaCO_3$ and H_2O.

Direct air capture is a technological method that uses chemical reactions to capture carbon dioxide (CO_2) from the atmosphere.

Dissolved inorganic carbon is the sum of the aqueous species of inorganic carbon in a solution. Inorganic carbon is found primarily in simple compounds such as carbon dioxide, carbonic acid, bicarbonate, and carbonate (CO_2, H_2CO_3, HCO_3^-, CO_3^{2-}, respectively).

Dolomite is an anhydrous carbonate mineral (*e.g.*, $CaMgCO_3$). It also refers to a sedimentary rock comprised largely of the mineral dolomite.

Enhanced weathering (EW) refers to a method of CDR in which fine minerals/rocks are spread in terrestrial or marine environments to remove atmospheric CO_2 in the form of dissolved inorganic carbon or secondary carbonates. EW often involves the use of natural or artificially created silicates or carbonates.

Ex situ in *ex situ* mineralisation, the carbonation reaction occurs above ground, within a separate reactor or industrial process.

Felsic minerals are silicates rich in Si, O, Al, Na, and K, *e.g.*, quartz, orthoclase, plagioclase, and muscovite.

Felsic rocks are igneous rocks relatively rich (>75%) in felsic minerals which tend to be light in colour, *e.g.*, granite and rhyolite. They contrast with darker, mafic rocks which are richer in magnesium and iron.

Flood basalt is the result of a giant volcanic eruption or series of eruptions that covers large stretches of land or the ocean floor with basalt lava.

Flue gas a mixture of gases produced by the burning of fuel or other materials in power stations and industrial plants and extracted *via* ducts.

Fly ash is produced in small dark flecks by the burning of powdered coal or other materials and carried into the air.

Geological formation is a formally named rock stratum or geological unit. It is a rock unit that is distinctive enough in appearance that a geologist can tell it apart from the surrounding rock layers and is a fundamental unit of lithostratigraphy (the scientific study and categorisation of rock strata based on their lithology – colour, texture, and composition). The concept of formally defined layers or strata is central to the geologic discipline of stratigraphy. Some geological formations are suitable for storing CO_2. Non-geologists may refer informally to outcroppings of rock or interesting geological features as geological formations, even though this is not technically correct.

Geological storage is the injection of CO_2 into suitable deep rock formations where it can remain stored.

Host rock in geology, is a rock formation that contains a foreign material (such as metal mineralisation, for example). In CCS it is a rock that receives stored CO_2.

Igneous rock is rock formed from the cooling of molten magma, either from a volcano, or deep intrusion. As the magma cools it crystallises and solidifies. Granite and basalt are examples.

Indirect carbonation for example of $Ca(OH)_2$ refers to a process of dehydration of $Ca(OH)_2$ to form CaO and H_2O, with the resulting CaO reacting with CO_2 to form $CaCO_3$.

***In-situ* mineralisation** is the process of injecting carbon dioxide into a rock formation where it reacts with minerals to form silica and carbonates. It is also known as mineral trapping (qv).

Kinetics is the branch of chemistry concerned with measuring and studying the rates of reactions.

Limestone is a carbonate sedimentary rock, composed from different forms of $CaCO_3$, *e.g.*, calcite and aragonite, formed when these minerals precipitate out of water.

Lime is a mineral composed mainly of calcium oxide (CaO).

Mafic mineral is a silicate mineral rich in magnesium and iron, often dark in colour, *e.g.*, olivine, pyroxene, amphibole, and biotite.

Mafic rock is an igneous rock rich in mafic minerals. Common mafic rocks include basalt, diabase and gabbro. Mafic rocks often also contain calcium-rich varieties of plagioclase feldspar.

Metamorphic is denoting or relating to rock that has undergone transformation by heat, pressure, or other natural agencies.

Mineral is an inorganic solid with a distinct chemical and crystal structure.

Mineral dissolution is the process in which a mineral dissolves in a solvent as a result of the liberation of its ions (*i.e.*, transformation of the compound into free ions). The most important solvent in nature is water, and the minerals which dissolve most readily or easily are halides, nitrates, carbonates and sulphates.

Mineral trap is a geological formation that retains fluids through the reaction of mineral components with a fluid, forming a stable mineral.

Mineral Carbonation is a process in which CO_2 reacts with magnesium or calcium oxide to form mineral carbonates. The mineral carbonates are unreactive solids and provide permanent carbon storage. Challenges include slow reaction rates and the large volume of mineral-rich earth that must be mined for each unit of CO_2 sequestered.

Negative CO_2 emissions is a process that results in permanent removal of CO_2 from the Earth's atmosphere through removal technologies which might include bioenergy with CCS, biochar, 'blue carbon' habitat restoration and enhanced weathering.

Negative Emissions Technology (NET) is an engineered system which either directly removes atmospheric CO_2 or enhances natural carbon sinks.

Passive carbonation refers to the carbonation reactions which require no additional energy inputs for example ambient processes that occur spontaneously at standard temperatures and pressures.

Passivating product layers that inhibit further conversion of reactants.

Pedogenic is relating to or denoting processes occurring in soil or leading to the formation of soil.

Peridotite a dense, coarse-grained igneous rock consisting mostly of olivine and pyroxene. Peridotite is ultramafic, as the rock contains less than 45% silica. It is high in magnesium, reflecting the high proportions of magnesium-rich olivine, with appreciable iron.

Primary minerals are minerals formed during the original solidification process.

Red mud also known as bauxite residue, is an industrial waste generated during the processing of bauxite into alumina using the Bayer process.

Reservoir a subsurface body of rock with the porosity and permeability to store and transmit fluids or gases.

Rock a a naturally occurring, lithified solid comprised of one or more mineral.

Saline formation is a geological strata where brackish water or brine occupies the tiny spaces between the grains of rock.

Saline groundwater is groundwater in which a high quantity of salts is dissolved.

Saturation state is a measure of the thermodynamic potential of a mineral to dissolve or precipitate.

Sedimentary rock is a type of rock that is formed by the accumulation or deposition of mineral or organic particles at the Earth's surface, followed by cementation.

Slag is alkaline waste matter separated from metals during the smelting or refining of ore.

Supersaturated is when the concentration of solute exceeds the concentration necessary for precipitation, *i.e.*, the saturation point.

Tailings are the materials left over after the process of separating the valuable fraction from the uneconomic fraction (gangue) of an ore.

Total alkalinity is the charge difference between conservative cations and anions in a solution, which is equivalent to the charge difference between non-conservative anions and non-conservative cations.

Ultrabasic is igneous rock with a low silica content (<45%) but may not be rich in Fe and Mg.

Ultramafic rock is igneous rocks with a low silica content (<45%) and >90% mafic minerals. Generally, >18% MgO, high FeO, low potassium. The Earth's mantle is composed of ultramafic rocks.

Secondary minerals are minerals formed as a by-product of weathering primary minerals.

Sequestration is to store something so that it is no longer available. Carbon sequestration involves the removal or storage of carbon dioxide so that it cannot be released into the atmosphere. The term is often regarded as synonymous with carbon storage but has other confusing meanings and should be avoided in favour of storage. Terrestrial sequestration is the absorption and storage of CO_2 by vegetation and soils in terrestrial ecosystems. In biology sequestration includes both capture and storage – a further complication to be avoided.

Soil carbon sequestration occurs through direct and indirect fixation of atmospheric CO_2. Direct soil carbon sequestration occurs by inorganic chemical reactions that convert CO_2 into soil inorganic carbon compounds such as calcium and magnesium carbonates. Direct plant carbon sequestration occurs as plants photosynthesise atmospheric CO_2 into plant biomass. Subsequently, some of this plant biomass is indirectly sequestered as soil organic carbon (SOC) during decomposition processes. Worldwide, SOC in the top 1 meter of soil comprises about 3/4 of the earth's terrestrial carbon.

Solubility trapping see "Trapping, Solubility".

Stable geological formation is a formation that has not recently been disturbed by tectonic movement.

Surficial is when CO_2-bearing fluid or gas is reacted with mine tailings, alkaline industrial wastes, or sedimentary formations rich in reactive rock fragments, all with a high proportion of reactive surface area.

Trapping, Geochemical is where CO_2 reacts with *in-situ* fluids and the host rock. First, CO_2 dissolves in water over 100s to 1000s of years and becomes denser and sinks down into the formation. Next, chemical reactions between the dissolved CO_2 and rocks +/minerals form ionic species, so that a fraction of the injected CO_2 is converted into solid carbonate minerals (millions of years).

Trapping, Mineral is where CO_2-rich water sinks to the bottom of a reservoir and reacts to form minerals.

Trapping, Physical a cap rock which prevents the upwards migration of a gas or fluid above the storage location. Normally clay or shale.

Trapping, Residual is when CO_2 is trapped in rock pores and cannot move.

Trapping, Solubility is the process in which CO_2 dissolves into formation waters. The extent of dissolution generally decreases with increasing temperature and salinity and increases with increasing pressure. The process removes CO_2 as a separate buoyant phase.

Weathering is the physical, biological or chemical break down of rocks, soils, and minerals as well as wood and artificial materials through contact with the Earth's atmosphere, water, and biological organisms.

Acknowledgements

The authors would like to acknowledge funding from the the UK Greenhouse Gas Research Programme (NE/P019730/2, NE/P019943/2), EPSRC Heriot-Watt Doctoral Training Partnership (EP/N509474/1), NERC Oxford Doctoral Training Partnership (NE/L002612/1), and philanthropic support from Patrick Collison, Veronica Furey, Prof. Niall Mac Dowell, Dr Mai Bui, Kirsty Harrington, and Pol Knops are thanked for their helpful comments during drafting.

References

1. H. C. Urey, *Geophysics*, 1952, **38**, 351–363.
2. J.-J. Ebelmen, in *Annales des Mines*, 1845, 7, 66.
3. R. A. Berner and K. A. Maasch, *Geochim. Cosmochim. Acta*, 1996, **60**, 1633–1637.
4. R. A. Berner, A. C. Lasaga and R. M. Garrels, *Am. J. Sci.*, 1983, **283**, 641–683.
5. R. S. Arvidson, F. T. Mackenzie and M. Guidry, *Am. J. Sci.*, 2006, **306**, 135–190.
6. T. M. Lenton and C. Britton, *Global Biogeochem. Cycles*, 2006, **20**, 1–12.
7. N. S. Lord, A. Ridgwell, M. C. Thorne and D. J. Lunt, *Global Biogeochem. Cycles*, 2016, **30**, 2–17.
8. J. Hartmann, N. Jansen, H. H. Dürr, S. Kempe and P. Köhler, *Glob. Planet. Change*, 2009, **69**, 185–194.
9. W. Seifritz, *Nature*, 1990, **345**, 486.
10. H. E. Dunsmore, *Energy Convers. Manag.*, 1992, **33**, 565–572.
11. K. S. Lackner, C. H. Wendt, D. P. Butt, E. L. Joyce and D. H. Sharp, *Energy*, 1995, **20**, 1153–1170.
12. K. S. Lackner, D. P. Butt and C. H. Wendt, ~ *Pergamon Energy Com'ers. Mgmt*, 1997, **38**, 259–264.
13. A. Sanna, M. Uibu, G. Caramanna, R. Kuusik and M. M. Maroto-Valer, *Chem. Soc. Rev.*, 2014, **43**, 8049–8080.
14. W. K. O'Connor, D. C. Dahlin, G. E. Rush, S. J. Gerdemann, L. R. Penner and D. N. Nilsen, *Aqueous Mineral Carbonation*, 2005.
15. W. J. J. Huijgen and R. N. J. Comans, *Environ. Sci. Technol.*, 2005, **39**, 9676–9682.
16. D. Daval, I. Martinez, J. Corvisier, N. Findling, B. Goffé and F. Guyot, *Chem. Geol.*, 2009, **265**, 63–78.
17. T. A. Haug, R. A. Kleiv and I. A. Munz, *Appl. Geochem.*, 2010, **25**, 1547–1563.

18. S. J. Gerdemann, W. K. O'Connor, D. C. Dahlin, L. R. Penner and H. Rush, *Environ. Sci. Technol.*, 2007, **41**, 2587–2593.
19. L. Penner, W. K. O'Connor, D. C. Dahlin, S. Gerdemann and G. E. Rush, *3rd Annual Conference on Carbon Capture and Sequestration*, 2004, pp. 1–18.
20. J. Sipilä, S. Teir and R. Zevenhoven, *Carbon dioxide sequestration by mineral carbonation, Literature Review Update 2005–2007*, Åbo Akademi University, 2018. http://users.abo.fi/rzevenho/MineralCarbonation LiteratureReview05-07.pd.
21. A. Azdarpour, M. Asadullah, E. Mohammadian, H. Hamidi, R. Junin and M. A. Karaei, *Chem. Eng. J.*, 2015, **279**, 615–630.
22. M. M. Maroto-Valer, D. J. Fauth, M. E. Kuchta, Y. Zhang and J. M. Andrésen, *Fuel Process. Technol.*, 2005, **86**, 1627–1645.
23. S. Teir, H. Revitzer, S. Eloneva, C. J. Fogelholm and R. Zevenhoven, *Int. J. Miner. Process.*, 2007, **83**, 36–46.
24. P. C. Lin, C. W. Huang, C. T. Hsiao and H. Teng, *Environ. Sci. Technol.*, 2008, **42**, 2748–2752.
25. E. R. Bobicki, Q. Liu, Z. Xu and H. Zeng, *Prog. Energy Combust. Sci.*, 2012, **38**, 302–320.
26. A.-H. H. Park, R. Jadhav and L.-S. Fan, *Can. J. Chem. Eng.*, 2003, **81**, 885–890.
27. J. Bałdyga, M. Henczka and K. Sokolnicka, *Mater. Lett.*, 2010, **64**, 702–704.
28. A. Azdarpour, M. Asadullah, R. Junin, M. Manan, H. Hamidi and E. Mohammadian, *Fuel Process. Technol.*, 2014, **126**, 429–434.
29. R. Chang, S. Kim, S. Lee, S. Choi, M. Kim and Y. Park, *Front. Energy Res.*, 2017, 5.
30. Grand. View. Research, Calcium Carbonate Market Size, Share & Trends Analysis Report By Application (Paper, Plastics, Adhesives & Sealants), By Region (APAC, Europe, North America, Central & South America, MEA), And Segment Forecasts, 2020 – 2027, 2020, 1–118.
31. S. C. Krevor and K. S. Lackner, in *Energy Procedia*, 2009, vol. 1, pp. 4867–4871.
32. X. Wang and M. M. Maroto-Valer, *Fuel*, 2011, **90**, 1229–1237.
33. J. Fagerlund, E. Nduagu, I. Romão and R. Zevenhoven, *Energy*, 2012, **41**, 184–191.
34. J. Fagerlund, E. Nduagu and R. Zevenhoven, in *Energy Procedia*, Elsevier Ltd, 2011, vol. 4, pp. 4993–5000.
35. E. Nduagu, I. Romão and R. Zevenhoven, *25th Int. Conf. Effic. Cost, Optim. Simul. Environ. Impact Energy Syst.*, 2012, 233–250.
36. A. N. Paukert, J. M. Matter, P. B. Kelemen, E. L. Shock and J. R. Havig, *Chem. Geol.*, 2012, **330–331**, 86–100.
37. P. B. Kelemen and J. Matter, *Proc. Natl. Acad. Sci.*, 2008, **105**, 17295–17300.
38. B. P. McGrail, H. T. Schaef, A. M. Ho, Y. J. Chien, J. J. Dooley and C. L. Davidson, *J. Geophys. Res. Solid Earth*, 2006, **111**, 1–13.
39. R. S. Jayne, H. Wu and R. M. Pollyea, A probabilistic assessment of geomechanical reservoir integrity during CO_2 sequestration in flood basalt formations, *Greenhouse Gases: Sci. Technol.*, 2019, **9**(5), 979–998.
40. S. R. Gislason and E. H. Oelkers, *Science*, 2014, **344**, 373 LP–374.

41. S. Ó. Snæbjörnsdóttir, B. Sigfússon, C. Marieni, D. Goldberg, S. R. Gislason and E. H. Oelkers, *Nat. Rev. Earth Environ.*, 2020, **1**, 90–102.
42. P. B. Kelemen, N. McQueen, J. Wilcox, P. Renforth, G. Dipple and A. P. Vankeuren, *Chem. Geol.*, 2020, 119628.
43. J. M. Matter, M. Stute, S. O. Snaebjornsdottir, E. H. Oelkers, S. R. Gislason, E. S. Aradottir, B. Sigfusson, I. Gunnarsson, H. Sigurdardottir, E. Gunnlaugsson, G. Axelsson, H. A. Alfredsson, D. Wolff-Boenisch, K. Mesfin, D. F. de la, R. Taya, J. Hall, K. Dideriksen and W. S. Broecker, *Science*, 2016, **352**, 1312–1314.
44. V. Gutknecht, S. Ó. Snæbjörnsdóttir, B. Sigfússon, E. S. Aradóttir and L. Charles, in *Energy Procedia*, Elsevier Ltd, 2018, vol. 146, pp. 129–134.
45. J. Hartmann, A. J. West, P. Renforth, P. Köhler, C. L. D. La Rocha, D. A. Wolf-Gladrow, H. H. Dürr and J. Scheffran, *Rev. Geophys.*, 2013, **51**, 113–149.
46. J. Hartmann and S. Kempe, *Naturwissenschaften*, 2008, **95**, 1159–1164.
47. P. Köhler, J. Hartmann and D. A. Wolf-Gladrow, *Proc. Natl. Acad. Sci. U. S. A.*, 2010, **107**, 20228–20233.
48. D. A. C. Manning, *Elements*, 2008, **4**, 105–108.
49. D. A. C. Manning, P. Renforth, E. Lopez-Capel, S. Robertson and N. Ghazireh, *Int. J. Greenhouse Gas Control*, 2013, **17**, 309–317.
50. N. Moosdorf, P. Renforth and J. Hartmann, *Environ. Sci. Technol.*, 2014, **48**, 4809–4816.
51. H. F. M. ten Berge, H. G. van der Meer, J. W. Steenhuizen, P. W. Goedhart, P. Knops and J. Verhagen, *PLoS One*, 2012, **7**, e42098.
52. R. D. Schuiling and P. Krijgsman, *Clim. Change*, 2006, **74**, 349–354.
53. L. L. Taylor, J. Quirk, R. M. S. Thorley, P. A. Kharecha, J. Hansen, A. Ridgwell, M. R. Lomas, S. A. Banwart and D. J. Beerling, *Nat. Clim. Change*, 2016, **6**, 402–406.
54. D. J. Beerling, E. P. Kantzas, M. R. Lomas, P. Wade, R. M. Eufrasio, P. Renforth, B. Sarkar, M. G. Andrews, R. H. James, C. R. Pearce, J. F. Mercure, H. Pollitt, P. B. Holden, N. R. Edwards, M. Khanna, L. Koh, S. Quegan, N. F. Pidgeon, I. A. Janssens, J. Hansen and S. A. Banwart, *Nature*, 2020, **583**, 242–248.
55. A. F. White and S. L. Brantley, *Rev. Mineral. Geochem.*, 1995, 1–22.
56. D. J. Beerling, J. R. Leake, S. P. Long, J. D. Scholes, J. Ton, P. N. Nelson, M. Bird, E. Kantzas, L. L. Taylor, B. Sarkar, M. Kelland, E. DeLucia, I. Kantola, C. Müller, G. H. Rau and J. Hansen, *Nat. Plants*, 2018 **4**, 392.
57. D. A. C. Manning, *Mineral. Mag.*, 2008, **72**, 639–649.
58. S. J. T. Hangx and C. J. Spiers, *Int. J. Greenhouse Gas Control*, 2009, **3**, 757–767.
59. F. Montserrat, P. Renforth, J. Hartmann, M. Leermakers, P. Knops and F. J. R. Meysman, *Environ. Sci. Technol.*, 2017, **51**, 3960–3972.
60. F. J. R. Meysman and F. Montserrat, *Biol. Lett.*, 2017, **13**, 20160905.

61. D. A. C. Manning and P. Renforth, *Environ. Sci. Technol.*, 2013, **47**, 135–141.
62. P. Köhler, J. F. Abrams, C. Völker, J. Hauck and D. A. Wolf-Gladrow, *Environ. Res. Lett.*, 2013, **8**, 14009.
63. W. Burns and C. R. Corbett, *One Earth*, 2020, **3**, 154–156.
64. E. Cox, E. Spence and N. Pidgeon, *Nat. Clim. Change*, 2020, **10**, 744–749.
65. C. Baatz, C. Heyward and H. Stelzer, *Environ. Values*, 2016, **25**, 1–5.
66. J. C. Abanades, R. Allam, K. S. Lackner, F. Meunier, E. Rubin, J. C. Sanchez, K. Yogo and R. Zevenhoven, *IPCC Special Report on Carbon dioxide Capture and Storage*, 2008, pp. 1–20. https://www.ipcc.ch/site/assets/uploads/2018/03/srccs_chapter7-1.pdf.
67. C.-L. Washbourne, P. Renforth and D. A. C. Manning, *Sci. Total Environ.*, 2012, **431**, 166–175.
68. P. Renforth, D. A. C. Manning and E. Lopez-Capel, *Appl. Geochem.*, 2009, **24**, 1757–1764.
69. S. A. Wilson, G. M. Dipple, I. M. Power, J. M. Thom, R. G. Anderson, M. Raudsepp, J. E. Gabites and G. Southam, *Econ. Geol.*, 2009, **104**, 95–112.
70. P. Renforth, C.-L. Washbourne, J. Taylder and D. A. C. Manning, *Environ. Sci. Technol.*, 2011, **45**, 2035–2041.
71. P. Renforth, *Nat. Commun.*, 2019, **10**, 1–8.
72. W. J. J. Huijgen and R. Comans, *Carbon dioxide sequestration by mineral carbonation, literature review*, Energy research Centre of the Netherlands (ECN), Petten, The Netherlands, Report ECN-C--03-016. 2003, pp. 1–52.
73. W. J. J. Huijgen and R. N. J. Comans, *Environ. Sci. Technol.*, 2006, **40**, 2790–2796.
74. S. Y. Pan, E. E. Chang and P. C. Chiang, *Aerosol Air Qual. Res.*, 2012, **12**, 770–791.
75. G. Costa, R. Baciocchi, A. Polettini, R. Pomi, C. D. Hills and P. J. Carey, *Environ. Monit. Assess.*, 2007, **135**, 55–75.
76. A. Iizuka, M. Fujii, A. Yamasaki and Y. Yanagisawa, *Ind. Eng. Chem. Res.*, 2004, **43**, 7880–7887.
77. D. N. Huntzinger, J. S. Gierke, L. L. Sutter, S. K. Kawatra and T. C. Eisele, *J. Hazard. Mater.*, 2009, **168**, 31–37.
78. M. Uibu and R. Kuusik, *Oil Shale*, 2009, **26**, 40–58.
79. D. N. Huntzinger and T. D. Eatmon, *J. Cleaner Prod.*, 2009, **17**, 668–675.
80. A. Sanna, M. R. Hall and M. Maroto-Valer, *Energy Environ. Sci.*, 2012, **5**, 7781–7796.
81. D. N. Huntzinger, J. S. Gierke, S. K. Kawatra, T. C. Eisele and L. L. Sutter, *Environ. Sci. Technol.*, 2009, **43**, 1986–1992.
82. US EPA, CKD GENERATION AND CHARACTERISTICS 3.0.
83. M. E. Jorat, M. A. Goddard, P. Manning, H. K. Lau, S. Ngeow, S. P. Sohi and D. A. C. Manning, *Sci. Total Environ.*, 2020, **703**, 135573.
84. C.-L. Washbourne, E. Lopez-Capel, P. Renforth, P. L. Ascough and D. A. C. Manning, *Environ. Sci. Technol.*, 2015, **49**, 5434–5440.

85. P. Renforth, J. Edmondson, J. R. Leake, K. J. Gaston and D. A. C. Manning, *Proc. Inst. Civ. Eng. - Urban Des. Plan.*, 2011, **164**, 121–128.
86. D. P. Hanak, B. G. Jenkins, T. Kruger and V. Manovic, *Appl. Energy*, 2017, **205**, 1189–1201.
87. F. J. Doucet, *Miner. Eng.*, 2010, **23**, 262–269.
88. S. Teir, S. Eloneva, C. J. Fogelholm and R. Zevenhoven, *Energy*, 2007, **32**, 528–539.
89. E. E. Chang, A. C. Chiu, S. Y. Pan, Y. H. Chen, C. S. Tan and P. C. Chiang, *Int. J. Greenhouse Gas Control*, 2013, **12**, 382–389.
90. D. Bonenfant, L. Kharoune, S. Sauvé, R. Hausler, P. Niquette, M. Mimeault and M. Kharoune, *Ind. Eng. Chem. Res.*, 2008, **47**, 7617–7622.
91. M. Uibua, R. Kuusik, L. Andreas and K. Kirsimäe, in *Energy Procedia*, Elsevier Ltd, 2011, vol. 4, pp. 925–932.
92. S. Kodama, T. Nishimoto, N. Yamamoto, K. Yogo and K. Yamada, *Energy*, 2008, **33**, 776–784.
93. E. E. Chang, S. Y. Pan, Y. H. Chen, C. S. Tan and P. C. Chiang, *J. Hazard. Mater.*, 2012, **227–228**, 97–106.
94. S. N. Lekakh, C. H. Rawlins, D. G. C. Robertson, V. L. Richards and K. D. Peaslee, *Metall. Mater. Trans. B Process Metall. Mater. Process. Sci.*, 2008, **39**, 125–134.
95. J. K. Stolaroff, G. V Lowry and D. W. Keith, *Energy Convers. Manag.*, 2005, **46**, 687–699.
96. R. M. Santos, J. Van Bouwel, E. Vandevelde, G. Mertens, J. Elsen and T. Van Gerven, *Int. J. Greenhouse Gas Control*, 2013, **17**, 32–45.
97. R. M. Santos, M. Bodor, P. N. Dragomir, A. G. Vraciu, M. Vlad and T. Van Gerven, *Miner. Eng.*, 2014, **59**, 71–81.
98. S. Eloneva, S. Teir, J. Salminen, C. J. Fogelholm and R. Zevenhoven, *Energy*, 2008, **33**, 1461–1467.
99. H. C. Oskierski, B. Z. Dlugogorski and G. Jacobsen, *Chem. Geol.*, 2013, **358**, 156–169.
100. S. A. Wilson, M. Raudsepp and G. Dipple, *Am. Mineral.*, 2006, **91**, 1331–1341.
101. I. M. Power, G. M. Dipple and G. Southam, *Environ. Sci. Technol.*, 2010, **44**, 456–462.
102. Y. W. Chiang, R. M. Santos, A. Monballiu, K. Ghyselbrecht, J. A. Martens, M. L. T. Mattos, T. Van Gerven and B. Meesschaert, *Miner. Eng.*, 2013, **48**, 116–125.
103. S. Teir, S. Eloneva, C. J. Fogelholm and R. Zevenhoven, *Appl. Energy*, 2009, **86**, 214–218.
104. R. C. Sahu, R. K. Patel and B. C. Ray, *J. Hazard. Mater.*, 2010, **179**, 28–34.
105. V. S. Yadav, M. Prasad, J. Khan, S. S. Amritphale, M. Singh and C. B. Raju, *J. Hazard. Mater.*, 2010, **176**, 1044–1050.
106. C. Si, Y. Ma and C. Lin, *J. Hazard. Mater.*, 2013, **244–245**, 54–59.
107. P. Renforth, W. M. Mayes, A. P. Jarvis, I. T. Burke, D. A. C. Manning and K. Gruiz, *Sci. Total Environ.*, 2012, **421–422**, 253–259.

108. M. Fernández Bertos, S. J. R. Simons, C. D. Hills and P. J. Carey, *J. Hazard. Mater.*, 2004, **112**, 193–205.
109. R. Baciocchi, G. Costa, E. Lategano, C. Marini, A. Polettini, R. Pomi, P. Postorino and S. Rocca, *Waste Manag.*, 2010, **30**, 1310–1317.
110. E. Rendek, G. Ducom and P. Germain, *J. Hazard. Mater.*, 2006, **128**, 73–79.
111. P. J. Gunning, C. D. Hills, A. Antemir and P. J. Carey, *Proc. Inst. Civ. Eng. - Constr. Mater.*, 2011, **164**, 231–239.
112. T. Van Gerven, E. Van Keer, S. Arickx, M. Jaspers, G. Wauters and C. Vandecasteele, *Waste Manag.*, 2005, **25**, 291–300.
113. J. A. Meima, R. D. Van Der Weijden, T. T. Eighmy and R. N. J. Comans, *Appl. Geochem.*, 2002, **17**, 1503–1513.
114. A. Polettini and R. Pomi, *J. Hazard. Mater.*, 2004, **113**, 209–215.
115. G. Montes-Hernandez, R. Pérez-López, F. Renard, J. M. Nieto and L. Charlet, *J. Hazard. Mater.*, 2009, **161**, 1347–1354.
116. R. Pérez-López, G. Montes-Hernandez, J. M. Nieto, F. Renard and L. Charlet, *Appl. Geochem.*, 2008, **23**, 2292–2300.
117. P. J. Gunning, C. D. Hills and P. J. Carey, *Waste Manag.*, 2009, **29**, 2722–2728.
118. R. E. Zeebe and D. Wolf-Gladrow, *CO_2 in Seawater: Equilibrium, Kinetics, Isotopes*, Elsevier Science, 1st edn., 2001, vol. 65.
119. P. Renforth and G. Henderson, *Rev. Geophys.*, 2017, **55**, 636–674.
120. P. Renforth, P. A. E. Pogge von Strandmann and G. M. Henderson, *Appl. Geochem.*, 2015, **61**, 109–118.
121. E. H. Oelkers, J. Declercq, G. D. Saldi, S. R. Gislason and J. Schott, *Chem. Geol.*, 2018, **500**, 1–19.
122. J. Strefler, T. Amann, N. Bauer, E. Kriegler and J. Hartmann, *Environ. Res. Lett.*, 2018, **13**, 034010.
123. D. Lefebvre, P. Goglio, A. Williams, D. A. C. Manning, A. C. de Azevedo, M. Bergmann, J. Meersmans and P. Smith, *J. Cleaner Prod.*, 2019, **233**, 468–481.
124. T. Amann, J. Hartmann, E. Struyf, W. D. O. Garcia, E. K. Fischer, I. Janssens, P. Meire and J. Schoelynck, *Biogeosciences*, 2020, **17**, 103–119.
125. K. S. Lackner, *Ssrn*, 2011, 193–232.
126. G. H. Rau and K. Caldeira, *Energy Convers. Manag.*, 1999, **40**, 1803–1813.
127. K. Caldeira and G. H. Rau, *Geophys. Res. Lett.*, 2000, **27**, 225–228.
128. G. H. Rau, K. G. Knauss, W. H. Langer and K. Caldeira, *Energy*, 2007, **32**, 1471–1477.
129. L. D. D. Harvey, *J. Geophys. Res. Ocean*, 2008, **113**, 1–21.
130. E. H. Oelkers, S. R. Gislason and J. Matter, *Elements*, 2008, **4**, 333–337.
131. P. Renforth, *Int. J. Greenhouse Gas Control*, 2012, **10**, 229–243.
132. M. Dri, A. Sanna and M. M. Maroto-Valer, *Fuel Process. Technol.*, 2013, **113**, 114–122.
133. M. Dri, A. Sanna and M. M. Maroto-Valer, *Appl. Energy*, 2014, **113**, 515–523.
134. W. M. Mayes, A. L. Riley, H. I. Gomes, P. Brabham, J. Hamlyn, H. Pullin and P. Renforth, *Environ. Sci. Technol.*, 2018, **52**, 7892–7900.

135. A. L. Riley, J. M. MacDonald, I. T. Burke, P. Renforth, A. P. Jarvis, K. A. Hudson-Edwards, J. McKie and W. M. Mayes, *J. Geochemical Explor.*, 2020, **219**, 106630.
136. J. L. Palandri and Y. K. Kharaka, *Chem. Geol.*, 2005, **217**, 351–364.
137. D. E. Canfield, A. N. Glazer and P. G. Falkowski, *Science*, 2010, **330**, 192 LP–196.
138. T. T. Isson and N. J. Planavsky, *Nature*, 2018, **560**, 471–475.
139. P. Renforth and J. C. Campbell, *Philos. Trans. R. Soc. B Biol. Sci.*, 2021, **376**, 20200174.
140. C. L. Sabine and T. Tanhua, *Ann. Rev. Mar. Sci.*, 2010, **2**, 175–198.
141. A. G. Dickson, *Deep Sea Res. Part A. Oceanogr. Res. Pap.*, 1981, **28**, 609–623.
142. J. V Strauss and N. J. Tosca, *Geology*, 2020, **48**, 599–603.
143. M. Seifan, Z. Sarabadani and A. Berenjian, *Appl. Microbiol. Biotechnol.*, 2020, **104**, 2029–2037.
144. J. Kerr, R. Rickaby, J. Yu, H. Elderfield and A. Y. Sadekov, *Earth Planet. Sci. Lett.*, 2017, **471**, 42–53.
145. T. Cyronak, A. J. Andersson, C. Langdon, R. Albright, N. R. Bates, K. Caldeira, R. Carlton, J. E. Corredor, R. B. Dunbar, I. Enochs, J. Erez, B. D. Eyre, J. P. Gattuso, D. Gledhill, H. Kayanne, D. I. Kline, D. A. Koweek, C. Lantz, B. Lazar, D. Manzello, A. McMahon, M. Meléndez, H. N. Page, I. R. Santos, K. G. Schulz, E. Shaw, J. Silverman, A. Suzuki, L. Teneva, A. Watanabe and S. Yamamoto, *PLoS One*, 2018, **13**(1), e0190872.
146. S. C. Doney, D. S. Busch, S. R. Cooley and K. J. Kroeker, *Annu. Rev. Environ. Resour.*, 2020, **45**(1), 83–112.
147. L. Zhao, K. Shirai, K. Tanaka, S. Milano, T. Higuchi, N. Murakami-Sugihara, E. O. Walliser, F. Yang, Y. Deng and B. R. Schöne, *Estuar. Coast. Shelf Sci.*, 2020, **235**, 106620.
148. K. Tan and H. Zheng, *Sci. Total Environ.*, 2020, **701**, 134794.
149. C. Basak and E. E. Martin, *Nat. Geosci.*, 2013, **6**, 121–124.
150. R. Zevenhoven, J. Fagerlund and J. K. Songok, *Greenh. Gases Sci. Technol.*, 2011, **1**, 48–57.
151. V. Manovic and E. J. Anthony, *Ind. Eng. Chem. Res.*, 2010, **49**, 9105–9110.
152. R. Zevenhoven, T. Björklöf, J. Fagerlund, I. Romão, J. Highfield and J. Bu, *ACEME10 Conference Proceedings*, 2010, 41–49.
153. R. H. Borgwardt, *Ind. Eng. Chem. Res.*, 1989, **28**, 493–500.
154. S. K. Bhatia and D. D. Perlmutter, *AIChE J.*, 1983, **29**, 79–86.
155. F. Scala, P. Salatino, R. Boerefijn and M. Ghadiri, *Powder Technol.*, 2000, **107**, 153–167.
156. Z. Chen, C. J. Lim and J. R. Grace, *Elsevier*, 2007, **62**, 867–877.
157. H. Pullin, A. W. Bray, I. T. Burke, D. D. Muir, D. J. Sapsford, W. M. Mayes and P. Renforth, *Environ. Sci. Technol.*, 2019, **53**, 9502–9511.
158. C. Myers and T. Nakagaki, *Environ. Res. Lett.*, 2020, **15**, 124018.
159. J. Schott, O. S. Pokrovsky and E. H. Oelkers, *Rev. Mineral. Geochem.*, 2009, **70**, 207–258.
160. S. S. Goldich, *J. Geol.*, 1938, **46**, 17–58.

161. S. L. Brantley, in *Treatise on Geochemistry*, Elsevier, 2003, pp. 73–117.
162. W. K. O'Connor, D. C. Dahlin, G. E. Rush, C. L. Dahlin and W. K. Collins, *Miner. Metall. Process.*, 2002, **19**, 95–101.
163. O. S. Pokrovsky and J. Schott, *Geochim. Cosmochim. Acta*, 2000, **64**, 3299–3312.
164. H. Béarat, M. J. McKelvy, A. V. G. Chizmeshya, D. Gormley, R. Nunez, R. W. Carpenter, K. Squires and G. H. Wolf, *Environ. Sci. Technol.*, 2006, **40**, 4802–4808.
165. G. P. Assima, F. Larachi, J. Molson and G. Beaudoin, *Can. J. Chem. Eng.*, 2014, **92**, 2029–2038.
166. P. Michalopoulos and R. C. Aller, *Science*, 1995, **270**, 614–617.
167. A. Stefánsson, S. R. Gıslason and S. Arnórsson, *Chem. Geol.*, 2001, **172**, 251–276.
168. K. Maher, C. I. Steefel, F. A. White and D. A. Stonestrom, *Geochim. Cosmochim. Acta*, 2009, **73**, 2804–2831.
169. G. J. Churchman and D. J. Lowe, *Alteration, Formation, and Occurrence of Minerals in Soils*, 2012.
170. R. Santos, P. Ceulemans, D. François and T. Van Gerven, *Symp. Ser.*, 2011, **2011**, 1–9.
171. R. Baciocchi, G. Costa, A. Polettini and R. Pomi, *Energy Procedia*, 2009, **1**, 4859–4866.
172. V. Prigiobbe and M. Mazzotti, *Elsevier*, 2011, **66**, 6544–6554.
173. O. S. Pokrovsky, J. Schott and A. Castillo, *Geochim. Cosmochim. Acta*, 2005, **69**, 905–918.
174. E. H. Oelkers, *Chem. Geol.*, 2001, **175**, 485–494.
175. W. Henry, *Philos. Trans. R. Soc. London*, 1803, **93**, 29–42.
176. J. J. Carroll, J. D. Slupsky and A. E. Mather, *J. Phys. Chem. Ref. Data*, 1991, **20**, 1–10.
177. F. M. M. Morel and J. G. Hering, *Principles and Applications of Aquatic Chemistry*, 1993.
178. J. Setschenow, *Zeitschrift für Physikalische Chemie*, 1889, **4**(1), 117–125.
179. A. Yasunishi and F. Yoshida, *J. Chem. Eng. Data*, 1979, **24**, 11–14.
180. D. Koschel, J. Y. Coxam, L. Rodier and V. Majer, *Fluid Phase Equilib.*, 2006, **247**, 107–120.
181. S. A. Wilson, S. L. L. Barker, G. M. Dipple and V. Atudorei, *Environ. Sci. Technol.*, 2010, **44**, 9522–9529.
182. S. Mignardi, C. De Vito, V. Ferrini and R. F. Martin, *J. Hazard. Mater.*, 2011, **191**, 49–55.
183. W. Knoche, in *Biophysics and Physiology of Carbon Dioxide*, ed. Bauer C., Gros G., and Bartels H., Proceedings in Life Sciences. Springer, Berlin, Heidelberg, 1980, 1, 3–11.
184. L. Bao and M. C. Trachtenberg, *J. Memb. Sci.*, 2006, **280**, 330–334.
185. N. Favre, M. L. Christ and A. C. Pierre, *J. Mol. Catal. B: Enzym.*, 2009, **60**, 163–170.
186. I. M. Power, A. L. Harrison, G. M. Dipple and G. Southam, *Int. J. Greenhouse Gas Control*, 2013, **16**, 145–155.

187. R. Socolow, M. Desmond, R. Aines and J. Blackstock, *APS Panel on Public Affairs*, 2011, 1–91.
188. K. Suzuki, C. Li and S. Zhang, *Chinese J. Chem. Eng.*, 2015, **23**, 1871–1874.
189. K. Zeppenfeld, *J. Water Supply Res. Technol.*, 2010, **59**, 482–491.
190. J. W. Morse, R. S. Arvidson and A. Lüttge, *Chem. Rev.*, 2007, **107**, 342–381.
191. E. Ruiz-Agudo, C. V Putnis and C. Rodriguez-Navarro, *Geochim. Cosmochim. Acta*, 2011, **75**, 284–296.
192. G. D. Saldi, J. Schott, O. S. Pokrovsky and E. H. Oelkers, *Geochim. Cosmochim. Acta*, 2010, **74**, 6344–6356.
193. G. D. Saldi, J. Schott, O. S. Pokrovsky, Q. Gautier and E. H. Oelkers, *Geochim. Cosmochim. Acta*, 2012, **83**, 93–109.
194. M. Hänchen, V. Prigiobbe, R. Baciocchi and M. Mazzotti, *Chem. Eng. Sci.*, 2008, **63**, 1012–1028.
195. W. Dreybrodt, L. Eisenlohr, B. Madry and S. Ringer, *Geochim. Cosmochim. Acta*, 1997, **61**, 3897–3904.
196. Y. Zhang and C. A. Grattoni, *Geochim. Cosmochim. Acta*, 1998, **62**, 3789–3790.
197. Z. Liu and W. Dreybrod, *Geochim. Cosmochim. Acta*, 1997, **61**, 2879–2889.
198. R. P. Cosmo, F. A. R. Pereira, D. C. Ribeiro, W. Q. Barros and A. L. Martins, *J. Pet. Sci. Eng.*, 2019, **181**, 1–8.
199. I. M. Power, A. L. Harrison, G. M. Dipple, S. A. Wilson, P. B. Kelemen, M. Hitch and G. Southam, *Rev. Mineral. Geochem.*, 2013, **77**, 305–360.
200. W. J. J. Huijgen, R. N. J. Comans and G. J. Witkamp, *Energy Convers. Manag.*, 2007, **48**, 1923–1935.
201. L. C. Pasquier, G. Mercier, J. F. Blais, E. Cecchi and S. Kentish, *Int. J. Greenhouse Gas Control*, 2016, **50**, 147–157.
202. J. M. Matter, W. S. Broecker, S. R. Gislason, E. Gunnlaugsson, E. H. Oelkers, M. Stute, H. Sigurdardóttir, A. Stefansson, H. A. Alfreðsson, E. S. Aradóttir, G. Axelsson, B. Sigfússon and D. Wolff-Boenisch, *Energy Procedia*, 2011, **4**, 5579–5585.
203. I. Gunnarsson, E. S. Aradóttir, E. H. Oelkers, D. E. Clark, M. Ô. Arnarson, B. Sigfússon, S. Snæbjörnsdóttir, J. M. Matter, M. Stute, B. M. Júlíusson and S. R. Gíslason, *Int. J. Greenhouse Gas Control*, 2018, **79**, 117–126.
204. N. Moosdorf, P. Renforth and J. Hartmann, *Environ. Sci. Technol.*, 2014, **48**, 4809–4816.
205. A. Lenton, R. J. Matear, D. P. Keller, V. Scott and N. E. Vaughan, *Earth Syst. Dyn.*, 2018, **9**, 339–357.
206. K. Z. House, C. H. House, D. P. Schrag and M. J. Aziz, *Environ. Sci. Technol.*, 2007, **41**, 8464–8470.
207. G. H. Rau, *Environ. Sci. Technol.*, 2008, **42**, 8935–8940.
208. G. H. Rau, S. A. Carroll, W. L. Bourcier, M. J. Singleton, M. M. Smith and R. D. Aines, *Proc. Natl. Acad. Sci.*, 2013, **110**, 10095–10100.
209. P. A. Davies, *Environ. Sci.: Water Res. Technol.*, 2015, **1**, 131–137.

210. P. Renforth, B. G. Jenkins and T. Kruger, *Energy*, 2013, **60**, 442–452.
211. P. Renforth and T. Kruger, *Energy Fuels*, 2013, **27**, 4199–4207.
212. M. Erans, S. A. Nabavi and V. Manović, *J. Cleaner Prod.*, 2020, **242**, 118330.
213. N. McQueen, P. Kelemen, G. Dipple, P. Renforth and J. Wilcox, *Nat. Commun.*, 2020, **11**, 3299.
214. A. Kirchofer, A. Brandt, S. Krevor, V. Prigiobbe and J. Wilcox, *Energy Environ. Sci.*, 2012, **5**, 8631–8641.
215. G. F. Brent, *US Pat.*, 20090305378, 12-10-2009, 1–20.
216. L. Zhao, L. Sang, J. Chen, J. Ji and H. H. Teng, *Environ. Sci. Technol.*, 2010, **44**, 406–411.
217. H. S. Kheshgi, *Energy*, 1995, **20**, 915–922.
218. J. S. Kirchner, K. A. Lettmann, B. Schnetger, J. O. Wolff and H. J. Brumsack, *Int. J. Greenhouse Gas Control*, 2020, **92**, 102855.
219. J. C. Abanades, Y. A. Criado and J. R. Fernández, *Sustain. Energy Fuels*, 2020, **4**, 3409–3417.
220. X. Wang and M. M. Maroto-Valer, *Energy*, 2013, **51**, 431–438.
221. P. Smith, S. J. Davis, F. Creutzig, S. Fuss, J. Minx, B. Gabrielle, E. Kato, R. B. Jackson, A. Cowie, E. Kriegler, D. P. Van Vuuren, J. Rogelj, P. Ciais, J. Milne, J. G. Canadell, D. McCollum, G. Peters, R. Andrew, V. Krey, G. Shrestha, P. Friedlingstein, T. Gasser, A. Grübler, W. K. Heidug, M. Jonas, C. D. Jones, F. Kraxner, E. Littleton, J. Lowe, J. R. Moreira, N. Nakicenovic, M. Obersteiner, A. Patwardhan, M. Rogner, E. Rubin, A. Sharifi, A. Torvanger, Y. Yamagata, J. Edmonds and C. Yongsung, *Nat. Clim. Change*, 2015, **6**, 42–50.
222. M. Anda, J. Shamshuddin and C. I. Fauziah, *Soil Tillage Res.*, 2013, **132**, 1–11.
223. M. Anda, J. Shamshuddin and C. I. Fauziah, *Catena*, 2015, **124**, 147–161.
224. G. P. Gillman, D. C. Burkett and R. J. Coventry, *Soil Res.*, 2001, **39**, 799.
225. G. P. Gillman, D. C. Burkett and R. J. Coventry, in *Applied Geochemistry*, Pergamon, 2002, 17, 987–1001.
226. M. E. P. De Souza, I. M. Cardoso, A. M. X. De Carvalho, A. P. Lopes, I. Jucksch and A. Janssen, *Commun. Soil Sci. Plant Anal.*, 2018, **49**, 1–12.
227. J. C. Minx, W. F. Lamb, M. W. Callaghan, S. Fuss, J. Hilaire, F. Creutzig, T. Amann, T. Beringer, W. De Oliveira Garcia, J. Hartmann, T. Khanna, D. Lenzi, G. Luderer, G. F. Nemet, J. Rogelj, P. Smith, J. L. Vicente Vicente, J. Wilcox and M. Del Mar Zamora Dominguez, *Environ. Res. Lett.*, 2018, **13**(6), 063001.
228. I. B. Kantola, M. D. Masters, D. J. Beerling, S. P. Long and E. H. DeLucia, *Biol. Lett.*, 2017, 13.
229. J. C. Minx, W. F. Lamb, M. W. Callaghan, S. Fuss, J. Hilaire, F. Creutzig, T. Amann, T. Beringer, W. D. O. Garcia, J. Hartmann, T. Khanna, D. Lenzi, G. Luderer, G. F. Nemet, J. Rogelj, P. Smith, J. L. V. Vicente, J. Wilcox and M. D. M. Z. Dominguez, *Environ. Res. Lett.*, 2018, **13**, 63001.

230. L. L. Taylor, J. Quirk, R. M. S. Thorley, P. A. Kharecha, J. Hansen, A. Ridgwell, M. R. Lomas, S. A. Banwart and D. J. Beerling, *Nat. Clim. Change*, 2016, **6**, 402–406.
231. S. Fuss, W. F. Lamb, M. W. Callaghan, J. Hilaire, F. Creutzig, T. Amann, T. Beringer, W. De Oliveira Garcia, J. Hartmann, T. Khanna, G. Luderer, G. F. Nemet, J. Rogelj, P. Smith, J. V Vicente, J. Wilcox, M. Del Mar Zamora Dominguez and J. C. Minx, *Environ. Res. Lett.*, 2018, **13**(6), 063002.
232. G. Lomax, M. Workman, T. Lenton and N. Shah, *Energy Policy*, 2015, **78**, 125–136.
233. H. Lawford-Smith and A. Currie, *Biol. Lett.*, 2017, **13**, 20160859.
234. C. Holz, L. S. Siegel, E. Johnston, A. P. Jones and J. Sterman, *Environ. Res. Lett.*, 2018, **13**, 64028.

CHAPTER 8

Soil Carbon Sequestration and Biochar

SYLVIA H. VETTER,* MOHAMED ABDALLA, MATTHIAS KUHNERT AND PETE SMITH

Institute of Biological and Environmental Sciences, University of Aberdeen, Aberdeen AB24 3UU, UK
*Email: sylvia.vetter@abdn.ac.uk

The greenhouse gas removal (GGR) potential of soil carbon sequestration (SCS) and biochar is achieved *via* a deliberate increase in the soil organic carbon (SOC) stocks. This can be realised by improving management practices in agriculture and grassland systems by increasing the organic carbon (C) input or reducing soil disturbances. Soil C sequestration potential differs depending on climate, soil properties and management, with the highest potential for poor soils (SOC stock farthest from saturation). The main downside of using SCS for CO_2 removal is the limited time C can be sequestrated until a new C equilibrium is reached. SCS management needs to continue even after C saturation in the soil occurs, otherwise, the newly gained SOC stock might lose the C again, reversing the SCS impact. However, agricultural and grassland soils have the feasible technical potential of 2–5 Gt CO_2eq per year globally and an economic potential of 1.5–2.6 Gt CO_2eq per year (at prices up to 100 $ per t CO_2eq). They can act as a C sink over the next 20 to 30 years following SCS management implementation. Biochar is more recalcitrant, and a new C equilibrium is expected to take much longer, so, biochar will probably be a long-term option for CO_2 removal. Biochar's feasible technical potential is 2.6 Gt CO_2eq per year. Both SCS and biochar

Energy and Environment Series No. 31
Greenhouse Gas Removal Technologies
Edited by Mai Bui and Niall Mac Dowell

Published by the Royal Society of Chemistry, www.rsc.org

can be implemented on existing land and bring additional benefits for soil quality and health, which provides further advantages like an increase in productivity. Thus, SCS and biochar can contribute greatly to sustainable productivity and food security.

8.1 Introduction

This chapter focuses on greenhouse gas removal (GGR) technologies that increase soil organic carbon (SOC) by changing agricultural practices to increase soil carbon sequestration (SCS) and the use of biochar as soil amendment.

Globally, soil holds 2480 Gt CO_2eq† of SOC in the top 30 cm and up to 5500 Gt CO_2eq in the depth of 1 m.[1,2] Natural SCS is "in balance" with 220 Gt CO_2eq flowing into soils and 220 Gt CO_2eq flowing back to the atmosphere from heterotrophic respiration.[3]

Soils are depleted of their SOC stocks through agricultural production and lose typically ~30–50% of their topsoil (0–30 cm) carbon (C) stock until reaching a new C equilibrium when converted from native ecosystems to agricultural land.[4] Simultaneously, agricultural soils release greenhouse gas (GHG) emissions through production, including the most relevant gases carbon dioxide (CO_2), nitrous oxide (N_2O) and methane (CH_4).[5] The processes in conventional agriculture emitting these GHG emissions include the preparation of the soil through tillage processes, the use of synthetic fertilisers, pesticides and herbicides, heavy irrigation and concentrated monoculture production. Improved soil management can be seen as a greenhouse gas removal (GGR) option, as this can significantly reduce GHG emissions and sequester CO_2 from the atmosphere as C in soil organic matter (SOM). At the same time, improved management helps to keep carbon stored as SOC by reducing losses to the atmosphere due to decomposition. In addition to the CO_2 removal effects, improved soil management can help enhance soil fertility and productivity, increase soil biodiversity, reduce erosion, runoff and water pollution and can increase the resilience of crop and pasture systems against climate change and provide a way to increase food security.[6,7]

Soil C stocks can be increased by adding organic matter to the soil, reducing losses by decomposition or both[8] (Figure 8.1). Practices that increase C inputs include cultivation with deeper rooting plants,[9] crop rotations,[10] residue management[11] and cover cropping during fallow periods,[10,12] improved irrigation in water-limited systems[5] as well as improved nutrient management and adding organic amendments, such as biochar application.[5,7] Reducing the decomposition rate can be achieved by practices that reduce soil disturbance, like reduced tillage, and practices that reduce erosion like restoration of degraded land.[2,5]

†Giga tonnes of carbon dioxide equivalent.

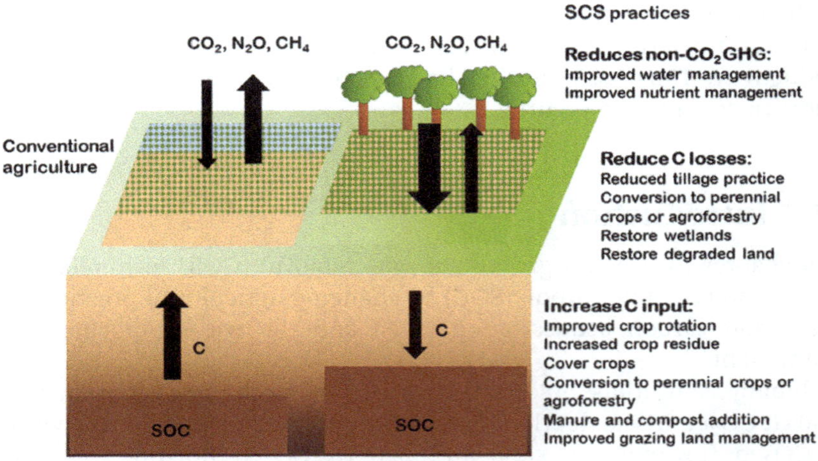

Figure 8.1 Comparison of carbon sequestration impacts between conventional agriculture (left) and SCS practices (right). Arrows show the fluxes from and to the soil and the thickness of the arrows indicates the relative amount of the fluxes.

Addition of biochar, a C rich material formed by pyrolysis of organic material, as an organic amendment can increase SOC and influence soil quality.[13] Biochar has been used for centuries to improve soil quality and increase crop productivity.[14,15] In recent years, it has become more prominent in its role as a carbon dioxide removal (CDR) technology; and more research show the impacts of biochar on SOC, agricultural production and other environmental aspects.[16,17]

Increasing the SOC stocks, with the aim of reducing the overall GHG emissions, is one of the strategies that can be applied on a large scale.[18] One policy initiative to encourage soil carbon sequestration is the "4 per 1000" initiative proposed under the UN Framework Convention on Climate Change (UNFCCC) with the aim to globally increase the soil C concentration in agricultural soils by 0.4% each year[19] (see Section 8.4.1). It is claimed that this could potentially produce a C sink of 10.2 Gt CO_2eq per year in the top (30 cm) soil depth.[20]

In the following section, we will give an overview of SCS practices that increase SOC and decrease GHG emissions, including the current scientific state of the GGR potential and limitations of SCS. In Section 8.3, biochar will be introduced as a CO_2 removal technology, *i.e.*, negative emissions technology (NET), and an amendment to improve soil health and soil quality, both traditionally as well as in modern times. The co-benefits and trade-offs with ecosystem services, seen as Nature's Contribution to People (NCP) and the United Nation's Sustainable Development Goals (SDG), will be described.[21]

Management practices of afforestation and reforestation are not considered in this chapter as these will be discussed in Chapter 9. In addition, fire management, which affects GHG emissions and soil properties in conflicting ways, is not discussed here.[5]

8.2 Soil Carbon Sequestration (SCS)

This section focuses on improved management practices to increase SOC in crop cultivation and grasslands, including perennials and mixed cropping systems like agroforestry. Management changes increase SOC and decrease GHG emissions, as one aspect cannot be viewed separately in the context of global warming and the potential to mitigate climate change. First, the management delivering SCS (see Section 8.2.1) will be presented. Then, the potential and impacts of SCS as a whole and as a GGR technology will be discussed, including limitations like C sink saturation (see Sections 8.2.2 and 8.2.3).

8.2.1 SCS Management

There are several management practices that can achieve SCS in agricultural and grassland systems[5,7] (Figure 8.1, Table 8.1). This section briefly outlines the broad range of practices and management options as well as their impact on the soil, C sequestration and GHG emissions in general. For all management options, the extent to which the soil C sequesters depends strongly on soil properties, climate and specific local conditions.[5,6,22,23] Thus, practices and management decisions should be adapted to the local requirements to achieve the best possible SCS.[24]

Previous studies on SOC changes following management change mainly consider the topsoil (0–30 cm), mostly for reasons of research feasibility. Although agricultural management mainly affects topsoil, it can also impact deeper layers and change SOC stocks. There is certainly a demand for more research including deeper soil layers over longer time periods and standardised methodology.

8.2.1.1 Tillage Management

Tillage is the process of preparing the soil for growing crops by mechanically loosening and overturning the soil (*e.g.*, digging, stirring, overturning *etc.*). Less intensive tillage management and no tillage reduces disturbances and disruption on the aggregated structure of the soil, which supports C sequestration and increases SOC stocks.[4] The shift from conventional tillage to no tillage practice is effectively protecting soils under cultivation, improving their quality, reducing the decline rate of soil organic matter and enhancing the resilience of cropping systems.[33]

There is a gradient between the intensity or reduction of tillage (*e.g.*, reduced tillage, no tillage) and the increase in SOC. A review of 351 studies of tillage practices in a temperate climate found higher SOC stocks under no tillage compared to conventional tillage in the upper soil (0–30 cm) by about

Table 8.1 Overview of SCS practices with mitigation estimates based on recent literature, the main impact on soil (direct impact) and other advantages.

	Practice	Description	Mitigation potential	Reference	Direct impact	Other advantages
Agronomy	**Reduced till**	No till for at least 10 years, 0–30 cm, temperate region	1.69 t CO_2eq per ha (0.286-3.091 t CO_2eq per ha)	Haddaway *et al.* 2017[25]	Reduced soil disturbance	Less energy requirements
	Residue retention/ incorporation	Based on reviews, average for topsoil	0.2 to 2.2 t CO_2eq per ha per year	Bolinder *et al.* 2020[26]	Increase organic matter input	Improved soil quality and soil health, soil moisture retention, enhanced nutrient cycling, decreased soil loss
	Improved crop rotations and cover crops	Based on reviews, average for topsoil	1 to 2.1 t CO_2eq per ha per year	Abdalla *et al.* 2019; Bolinder *et al.* 2020; Kaye and Quemada, 2017[26–28]	Increase organic matter input	Possible decrease of albedo; compensation of SOC loss through tillage practice; decreasing N leaching, reduce erosion, fix atmospheric nitrogen, and improve soil health
	Improved nutrient management	Average for different climate zones	0.26 to 0.33 t CO_2eq per ha per year	Smith *et al.* 2008[5]	Reduces non-CO_2 GHG emissions	Yield increase (depending on initial management)
	Organic amendment	Manure application	0.81 to 5.2 t CO_2eq per ha per year (5%–95% confidence interval)	Ren *et al.* 2018[29]	Increase organic matter input	Rebuilt microbial biomass and improve soil functions
		Change from NPK to manure	0.77 to 1.9 t CO_2eq per ha per year (5%–95% confidence interval)			

Grasslands	**Improved grassland systems**	Rangeland	0.26 to 1.1 t CO_2eq per ha per year	Morgan *et al.* 2010[30]	Increase organic matter input	Vegetation cover
		Pastures	1.1 to 5.2 t CO_2eq per ha per year			
Growing Trees	**Perennials**[a]	Arable to perennial (0–30 cm)	1.1 t CO_2eq per ha per year	Ledeo *et al.* 2020[31]	Increased root biomass	Less till, year around vegetation cover, higher residue input, reduces erosion
		Grassland to perennial (0–30 cm)	− 0.46 t CO_2eq per ha per year			
		Forest to perennials (0–30 cm)	3.1 t CO_2eq per ha per year			
	Agroforestry	Change from different land uses to agroforestry	− 2.9 to 16.1 t CO_2eq per ha per year	Feliciano *et al.* 2018[32]	Increased root biomass	Reverting land degradation and ensuring food security; improving soil fertility, control erosion, and conserve biodiversity
Restoration	**Restoration of degraded land**	Salt-affected and chemically degraded soils	1.1 to 2.6 t CO_2eq per ha per year	Lal 2018[2]	Increased root biomass, increase organic matter input	Reduces the release of C and N
		From water erosion	0.73 to 1.83 t CO_2eq per ha per year			
		From wind erosion	0.18 to 0.73 t CO_2eq per ha per year			

[a]Perennials include the group of woody plants and perennial grasses.

16.9 t CO_2eq per ha (2.86–30.91 t CO_2eq per ha, 95% confidence interval) over a period of 10 years.[25] The differences between no tillage practice to intermediate and high tillage were significantly in the upper 15 cm of the soil. Positive effects could be found for no tillage practices under rice-based cropping systems.[34]

Powlson *et al.*[35] suggested that the smaller increase in C content in the surface horizon could be offset by C losses in deeper soil layers. Other studies showed an impact on N_2O emissions when a change in tillage practice in combination with water management alters the microenvironment in the soil.[36,37] A global study by Abdalla *et al.*[38] reported that conservation tillage (*i.e.*, non-inversion tillage, eco-tillage, minimum tillage, mulch-tillage, reduced tillage, zone tillage or no-tillage) reduced CO_2 emissions, increased SOC and improved soil structure but enhanced N_2O emissions. They suggested that conservation tillage should be applied within the context of specific climates and soils. Generally, there is a lack of studies on the effect of tillage in deeper soil horizons and still a lot to learn regarding how to refine the no tillage practice to maximise benefits for the C balance.[33,39]

Considering that the soil reaches a new C equilibrium after some time under the new practice, there is little known about the impacts of additional management changes or reverse to former management practice on the soil and SOC stock. A study by Blanco-Canqui and Wortmann[40] concluded that occasional tillage practice (once in 5–10 years) may not undo the benefits achieved through no tillage as long as erosion is controlled.

The benefits of SCS depend on climate zone and soil properties. Sun *et al.*[23] showed the importance of regional climate on SCS and crop yields and found that arid regions can benefit most from conservation agriculture with an enhanced C sequestration and increased crop yield. Humid regions are more likely to increase SOC only, whereas some colder regions can have yield losses and soil C loss instead of soil C gains.[23]

8.2.1.2 Residue Management

Residue management includes residue retained on the field and residue incorporation, both increasing the organic matter input to the soil. Residue removal lowers the carbon input to the soil, leads to a decrease in SOC and enhances the soil erosion risk.[26] Retaining crop residues on the field or incorporating them increases SOC by amounts from 194 to 2160 kg CO_2eq per ha per year.[26] The higher value is an assumption that refers to a study about bare fallow combined with large amounts of straw application.

As the most residue-producing crop in Asia, rice residue management is important for SOC and the GHG balance in South Asia. Traditionally, residue is removed from the fields, but recent studies showed the benefit of on-field residue retention on SOC, productivity and the environment, although there are several challenges of implementing the management on a large scale.[34,41]

In addition to higher SOC, residue retention provides several benefits to soil quality and soil health, such as higher soil moisture retention, enhanced nutrient cycling, and decreased soil loss.[42] However, there are differences in the effect of surface retention *vs.* incorporation on various soil properties; and the importance of local region-specific factors need to be taken into account (climate, soil texture, study duration, sampling methods, and agronomic practices).[42,43] The impact that residue incorporation has on SOC may strongly differ with the climatic zone.[44]

8.2.1.3 Improved Crop Rotations and Cover Crops

Adopting crop rotations that provide greater C inputs to the soil can increase SOC stocks.[5,10,45] Specifically, rotations with legume crops or using perennials (see Section 8.2.1.7) as cover crops can help allocate more C belowground. For example, intensifying and diversifying crop rotations in dry climate crop systems can increase average annual C inputs, leading to higher soil C stocks compared to high fallow frequency systems.[46–49]

Cover crops make a temporary vegetative cover during fallow periods and provide year-round C inputs.[10,12] They also have other benefits like decreasing N leaching, reducing erosion, fixing atmospheric nitrogen (N), and improving soil health.[5,27,28,50] Additionally, cover cropping might compensate SOC losses due to residue removal in the short term. However, more data and long-term studies are needed to quantify the effect. Cover crops could negatively affect grain yield; however, this can be avoided by selecting mixed cover crops with a range of legumes and non-legumes.[27]

The mitigation potential of cover crops was recently reported in several studies with an overall range from 1000 to 2060 kg CO_2eq per ha per year (based on 100 to 150 g CO_2eq per m^2 per year,[28] 270 to 430 kg C per ha per year,[26] 2.06 ± 2.10 t CO_2eq per ha per year[27]). Estimates from Kaye and Quemada[28] also consider a change in albedo[‡] due to cover cropping what could potentially increase the mitigation potential by ~120 to 460 kg CO_2eq per ha per year.

Poeplau and Don[12] calculated a global SOC sequestration potential for cover crops of 0.44 Gt CO_2eq per year, which would compensate for 8% of the direct annual GHG emissions from agriculture (not considering effects from albedo change due to cover crop cultivation). The same study gave a longer time span until the soil reaches a new C equilibrium compared to previous studies and concluded it could take up to 155 years.[12]

Combined practices of tillage and cover cropping could offset the benefits of the cover crops to SOC.[51] However, cover crops can compensate the loss in SOC due to tillage practice. Abdalla *et al.*[27] noted that the impacts of cover crops are soil and management specific and should be adapted to regional climatic conditions.

[‡]Albedo gives the proportion of incoming solar radiation that is reflected by the surface back into the atmosphere.

8.2.1.4 Nutrient Management

Adding nutrients in the form of nitrogen, phosphorus and potassium (NPK) fertilisers to increase productivity of low-yielding, nutrient-deficient systems can also stimulate SOC.[5,52] However, N fertiliser input may increase N_2O emissions and there are CO_2 emissions associated with the fertiliser manufacturing process, which in turn, limits the benefits of adding nutrients if applied inefficiently. Management options for optimising fertilisation include improving fertiliser application timing, adjusting application rates to the precise needs of crops and using fertiliser inhibitors to slow the microbial processes leading to N_2O formation.[5,7] Nitrification inhibitors can reduce N_2O fluxes by up to 40% in some soils[7] but the effect is variable and more research on this topic is needed.[53] Other impacts on N_2O emissions include tillage and water management, which alter the soil microenvironment[36,37] and uncertainty related to climate change, *e.g.*, changing temperature and precipitation patterns.[54]

Fertiliser application over longer periods can increase SOC and result in sustainable productivity.[55] Waqas *et al.*[55] found that SOC sequestration is highly time-dependent. Irrespective of fertilisation mode, SOC gradually increased and attained the peak of sequestration rate in the initial two decades rather than at later stages of fertiliser addition.[55]

8.2.1.5 Organic Amendment

Organic amendments like manure, compost or biochar can increase soil C when added to the soil. Compost decomposes more slowly compared to fresh plant residues; its mean residence time is several times greater than un-composted organic matter.[56] This means that a large fraction of added C is retained in the soil over several decades or longer.

Manure application has the potential to rebuild microbial biomass and improve soil functions. A meta-analysis for China indicated that manure application is an effective way to restore the loss of microbial biomass due to intensive application of NPK fertilisers. Nevertheless, the impact varies with manure type, application rate and region as well as region-specific climate and soil conditions.[57]

An analysis of 80 European long-term field experiments, including 107 papers, showed a clear increase in SOC with farmyard manure (32.9%) and slurry (17.4%) application.[58] However, the organic fertiliser was slightly less effective on the crop yield compared to synthetic fertiliser.[58] Zhou *et al.*[59] concluded that application of manure as an organic fertiliser not only increases SOC and sustain agricultural production but also affects the soil N cycle and N_2O emissions. The study showed a 30% increase in N_2O emissions due to manure application compared to synthetic fertiliser. Thus, the benefit of increasing SOC through manure application may be offset by stimulating soil N_2O emissions (and CH_4 emissions, particularly in paddy rice cultivation[59,60]).

However, because the organic matter originates from outside the ecosystem boundary, the whole life-cycle, including transport, processing and interactions with other GHG-producing processes, will need to be assessed to estimate the impact on the environment and GHG balance.[7] Only then can the real benefit of adding organic amendments be evaluated, which will vary for each farming system.

Organic amendments in the form of biochar are discussed in Section 8.3.

8.2.1.6 Grasslands

Worldwide, grasslands cover about 40% of the land surface[61] and account for 70% of agricultural area.[62] They produce about 10% of terrestrial plant biomass and represent about 20–30% of SOC.[63,64] These amounts could further increase when considering the additional C inputs derived from animal manure. Grasslands contribute to mitigation of climate change[65] by potentially absorbing atmospheric CO_2 during the grass-plant's growth and ultimately sequestering it in the soil.[66] The rates of C sequestration vary with climate, soil, and management with an overall annual sequestration rate of 1.8 GtCO_2eq per year.[67,68] Grass biomass eaten by animals or grass remaining in the field with their roots will eventually decompose and store C in the soil. Previous studies reported that grassland soils could potentially act as C sinks.[5,69] When grasslands burn, the carbon fixed underground tends to stay in the roots and soils, making them more adaptive to climate change. However, Smith[70] found that simply having grassland does not result in a carbon sink, but the judicious management of grasslands is responsible for increasing carbon sink capacity.[70] Soil carbon accumulates in grasslands when there is a change (i) from suboptimal to optimal field management, (ii) from cropland to grassland, or (iii) due to stopping of human interventions like ploughing and reseeding.[70,71] Improved grassland management (including biomass removal from grazing or increasing forage production through improved species, irrigation and fertilisation) has the potential to increase SOC stocks by 10%.[72] Other management options, such as adjusting animal stocking rates and managing plant species, could increase SOC stock by 0.26–1.1 t CO_2eq per ha per year on rangelands and 1.1–5 t CO_2eq per ha per year on pastures.[30]

8.2.1.7 Perennials

There is a large variety of perennial crops, which can be divided in the two main groups of woody plants (*e.g.*, fruit and nut crops, beverage crops and oil crops) and perennial grasses (*e.g.*, sugarcane, switchgrass and miscanthus).[31] Perennial crops provide year-round vegetation cover, ensuring C input to the soil. They increase SOC stocks through a higher root biomass compared to annual crops.[73] Perennials also have stable roots (longer residence times of root carbon in the soil), which influence SOC to a greater extent compared to shoot contributions.[74] In addition, there is a

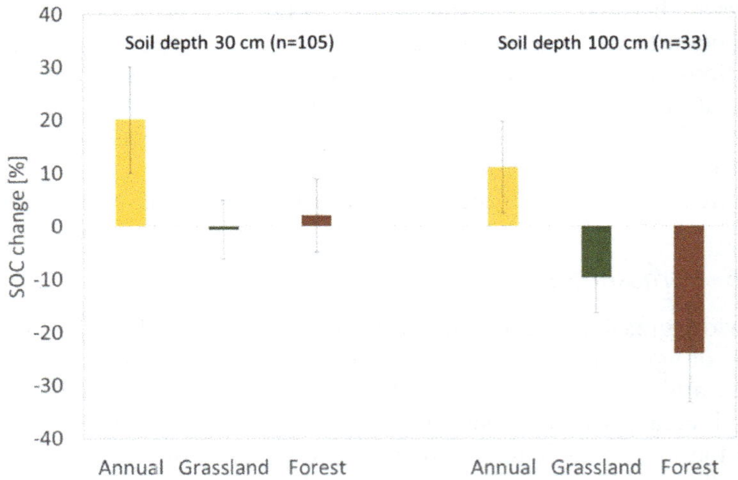

Figure 8.2 Percentage changes of SOC after 20 years by changing from annual crops, grassland or forest to perennial crops (data from Ledo *et al.*[31]). The data is differentiated into studies that investigated the top 30 cm of the soil ($n = 105$ studies) and the top 100 cm of the soil ($n = 33$ studies). Positive values indicate a carbon gain from changing to perennial crops and negative values correspond to carbon loss.

higher C input to the soil compared with annual crops, as perennial crops provide more residues[75] and might add materials provided by pruning.[76] Soil disturbances through tillage are reduced as the crops stay longer than one year on the field, decreasing GHG emissions.[31,77] Perennial crops often establish better plant coverage on the soil surface, which reduces soil erosion.[78]

There is potential for high SOC content using perennial crops, however, studies showing the impacts are scarce. A meta-analysis by Ledo *et al.*[31] analysed the land use change from annual crops, grasslands and forests to perennial crops.[31] The analysis showed an increase of SOC by 11 to 20% (Figure 8.2) for a change from annual to perennial crops. In contrast, a change from grassland to perennials decreased SOC by 0.6 to 9.6% (Figure 8.2). The analysis considered studies of both the top 30 cm ($n = 111$) and 100 cm soil depth ($n = 10$). For a change from grassland or arable land, the trend was the same for both soil groups, but the impact varies (Figure 8.2). Furthermore, the change from forest to perennials showed an increase in SOC of 2% for studies that considered the topsoil only (0–30 cm) and a loss of 24% for studies considering deeper layers (*i.e.*, 0–100 cm). Although depths of 1 m are preferable for analysis of carbon stocks, the higher number of studies considering the top 30 cm makes the analysis more robust. However, single studies also show strong losses in SOC in the top 30 cm for land use change from forest to perennials. For instance, Don *et al.*[79] found a decrease of 30% in SOC for a change from tropical forest to perennial crops.[79] For that reason, more studies that investigate the SOC change in soil depths of 100 cm are needed.

The change in SOC after a land use change is not linear. After an initial carbon gain, a loss of SOC can be observed over a few years, this is followed by a strong increase of SOC, but the annual increase gets smaller over time until the SOC reaches a new C equilibrium.[31] The SOC loss after the initial gain can be explained by the disturbance due to the land use change and possible priming effects that enhance the respiration, because of fresh biomass being added to the soil.

Sequestration rates should be provided as long-term averages, rather than for single year or short periods. Ledo *et al.*[31] found that 20 years is a robust period to quantify long-term C sequestration rates for land use change to perennials. The C sequestration rates for a change from arable, grassland and forest to perennials are quantified to be 22.7 ($\pm$17.2) t CO_2eq per ha, $-$9.2 ($\pm$15) t CO_2eq per ha and 62 ($\pm$202) t CO_2eq per ha, respectively (average over 20 years for studies on the top 30 cm of soil together with the standard deviation). The main driving factor of C sequestration rate is the temperature[31] and climate in general.[80] Other important factors are crop age, bulk density and clay content of the soil.[31]

In summary, perennials are an option to increase SOC by maintaining food and fibre production. The impacts on SOC depend mainly on weather and soil conditions. Land use change from natural ecosystems may have a negative impact on SOC. More research analysing crop-specific impacts is required to get a better understanding of the dynamics that affect the soil carbon sequestration.

8.2.1.8 Agroforestry

Agroforestry systems are defined as the presence of trees on external and internal boundaries, cropland, or on any other available niche of farmland.[81] These systems have the potential to increase soil C storage while having positive impacts on agricultural productivity and people's livelihoods. Benefits include reversing land degradation, ensuring food security, improving soil fertility, controlling erosion, and conserving biodiversity.[82] In agroforestry systems, woody perennials or trees are grown with crop cultivation and/or animal farming.[32,81] This adds to the C stocks above ground and increases the SCS.[32,83–85]

The change in SOC depends on the specific land use type before the transition to an agroforestry system. A meta-analysis on a transition from agriculture to agroforestry[86] of 53 studies showed a significant increase in SOC stocks by 26% at depths of 0–15 cm, 40% at 0–30 cm depths and 34% at 0–100 cm. A change from grassland and pasture systems to agroforestry increased the SOC by 9% and 10% at depths of 0–15 cm and 0–30 cm, respectively. A transition from uncultivated land showed an SOC increase of 25% in soil depths of 0–30 cm, but SOC decreases by 23% in deeper soil layers up to 60 cm. The study[86] showed a significant increase of SOC stocks for a change from agriculture to agrisilvicultural[§] and silvopastoral[¶]. Similar results were found by Feliciano *et al.*,[32] where the highest SCS potential was found for silvopastoral with an annual average of about 16.06 t CO_2eq per ha. Soil carbon sequestration *via* agroforestry is largely dictated by the climate region (*e.g.*, temperate, arid, tropical) has a large impact on and the previous land use.

In a global analysis of agroforestry, Zomer *et al.*[87] estimated that C storage contributed between 2000 to 2010 through trees was around 7.59 Gt CO_2eq globally, assuming a 3.7% increase in tree cover over the same time period. This corresponds to an annual rate of 0.73 Gt CO_2eq storage in biomass above and below ground.

8.2.1.9 Restoration of Degraded Land

Agricultural land becomes degraded due to erosion, excessive disturbances, organic matter loss, salinisation, acidification or other processes that curtail productivity.[5] The restoration process of this degraded land could potentially provide sequestration of carbon. Applicable practices include revegetation, improving fertility by nutrient amendments (see Section 8.2.1.4), applying organic amendments like manure (see Section 8.2.1.5) or biochar (see Section 8.3), reducing tillage (see Section 8.2.1.1) and retaining crop residues (see Section 8.2.1.2). Restoring degraded land to perennial forest (see Sections 8.2.1.7 and 8.2.1.8) or grassland (see Section 8.2.1.6) increases soil C storage, although usually at a smaller rate than the original conversion losses.[64,88] Restoring wetlands and peatlands or avoiding drainage of these lands, does not only increase SOC, but also reduces the release of C and N.[89] Restoring wetlands can result in rapid accumulation of soil carbon, although this conversion may stimulate CH_4 emissions.[5]

Degraded soils can be hotspots for SOC sequestration as they are furthest from C saturation and therefore, have the potential to store large amounts of C before the soils reach a new C saturated equilibrium.[2] There is around 2000 Mha of degraded land that could be restored and used for SCS at a rate of 1.1–2.6 t CO_2eq per ha per year in salt-affected and chemically degraded soils, 0.7–1.8 t CO_2eq per ha per year in soils physically degraded and prone to water erosion, and 0.2–0.7 t CO_2eq per ha per year in soil susceptible to wind erosion.[2]

8.2.1.10 Water Management

Currently, 20% of world's cropland is irrigated.[90] Improving the effectiveness of irrigation systems can enhance C storage in soils.[5] An increase of irrigation in water-limited systems in addition to fertiliser input to increase

§This system involves simultaneously growing crops and trees on the same piece of land.[32]
¶Agroforestry practice that integrates livestock, forage production, and forestry on the same land-management unit.[32]

productivity in nutrient-deficit systems can increase C sequestration.[5] However, the additional nutrients and water may increase non-CO_2 emissions from the field or increase GHG emissions due to the energy required to pump the water.

Paddy rice systems emit significant amounts of CH_4 to the atmosphere.[91] These emissions can be reduced through a change in irrigation system. For example, transitioning from flooded fields to a system that alternates wetting and drying, however, the benefits may be offset by increasing N_2O emissions.[60,91] Other methods of reducing CH_4 emissions include keeping the soil dry in the off-rice season and avoiding waterlogging, adjusting the timing of organic residue additions by incorporating the organic material during the dry period and composting the residues.[5,92–94]

8.2.2 Impacts of SCS Managements

The previous section reviewed the individual practices for delivering SCS, summarising the benefits and limitations. SCS strongly depends on soil type and climate region, and consequently, it differs with region and management.[6] Overall, SCS management impacts on the environment and other ecosystem services.[17,21] This section will review the main effects of SCS management, which will include impacts on delivered soil functions, land use and GHG balances, water, nutrients, albedo, energy and costs. In addition, the effect of SCS on Nature's Contributions to People (NCPs) will be discussed, as well as the impacts on ecosystem services[95] and on the United Nation's Sustainable Development Goals (SDGs).[96]

8.2.2.1 Impact of SCS on Soil Functions

SCS management provide positive impacts to the soil, which can be summarised as the seven soil functions as defined by the European Commission.[97] (1) Biomass production (*e.g.*, agriculture and forestry) can be enhanced through SCS,[20] which improves soil health and reduces erosion. This in turn has a positive impact on (2) storing, filtering and transforming nutrients, substances and water. (3) The biodiversity pool, such as habitats, species and genes benefit from SCS as soil biota is improved through enhanced habitats. The soil health improvement from SCS is the basis for a sustainable landscape, which can be used for (4) physical and cultural environment for humans and human activities. (5) Soils are a source of raw material for products such as pharmaceuticals, clay for bricks and ceramics, silicon from sand used in electronics, and other minerals.[98] Through improved soil health, SCS can have a positive impact on these materials. SCS increases soil organic matter, which therefore has a significant impact on soil (6) acting as a carbon pool.[17] Lastly, SCS increases soil protection against erosion, thus enhancing the ability of the soil to (7) archive geological and archaeological heritage.[99,100]

8.2.2.2 Impact of SCS on Land Use and GHG Emissions

Of the different SCS managements, the lowest per-area SCS potential was observed for cropland management (0.03 t C per ha per year); restoration of degraded land had the highest carbon sequestration potential with an average of 1 t C per ha per year.[5] Based on these numbers, the required land ranges from 1 to 33 ha per t C stored.[17] SCS can be applied without changing the current land use. An increase in soil organic matter through SCS can have an additional impact on other GHGs. The impact on CH_4 emissions is small to negligible.[5] However, N_2O emissions could rise because of an increase in organic N with higher soil C storage.

The requirement of land depends on the implementation level and the SCS practices. For example, considering a C price of up to ~100 $ per t CO_2eq, a SCS mitigation potential of 2.5 Gt CO_2eq per year would require a minimum of 100 Mha land area and could reach to 23 100 Mha.[17] For implementation globally and assuming a maximum technical potential, the required land area would be 1300 to 42 900 Mha.

8.2.2.3 Impact of SCS on Water

The impact of SCS on soil water is difficult to quantify. There is a general acceptance that increased soil organic matter storage improves the soil water holding capacity, which improves the retention of water in the soil/plant system.[17,101] Therefore, SCS is likely to have a net beneficial impact on the water footprint, however, the global impact on water is assumed to be neutral.[17] On a small scale, increased SOC can affect the soil properties and enhance water repellent soil conditions.[102] This can increase surface run-off and erosion in a minor way.

8.2.2.4 Impact of SCS on Nutrients

Nutrients are provided to the soil through soil organic matter from SCS. Adjusted numbers from Lal[101] provide a general overview of the additional nutrients: for every t C per ha of soil organic matter added, nutrients in the form of nitrogen (N), phosphorus (P) and potassium (K) increase by 80 kg N per ha, 20 kg P per ha and 15 kg K per ha, respectively. The nutrients need to be added for a stable soil organic matter. It is assumed that this is possible through adding a combination of organic matter and nitrogen fixation.

Additionally, SOC increases the cation exchange capacity.[103] As a result, the soil increases the potential to store nutrients, which can have positive impacts on soil productivity and soil health. On a global scale, if SCS achieves a CO_2 removal potential of 2.5 Gt CO_2eq per year, the soil would in addition hold 56 Gt N per year, 14 Gt P per year and 10.5 Gt K per year.[17]

8.2.2.5 Impact of SCS on Albedo

Some SCS management can have an impact on albedo due to the change of the soil surface conditions. For example, residue management can affect albedo when the residue is left on the field[104] or when cover crops are planted.[28] To some extent, remote sensing can detect if residue is applied to fields.[105,106] Light coloured residue (*e.g.*, cereal straw) could then increase albedo if left on bare soil and have an additional climate benefit to soil C storage.[17] A study by Kaye and Quemada[28] considered change in albedo effects due to cover cropping (see Section 8.2.1.3). De Simon *et al.*[107] changed the albedo on a maize field to analyse the impacts of changed temperature on respiration rates. They found only a small impact in their experiments for the annual carbon budget. However, the overall impact on albedo cannot be quantified with certainty as many SCS management practices can be applied together. More field research is needed to understand the impact of each practice individually and in combination. Therefore, the overall impact on albedo from SCS is assumed to be zero.

8.2.2.6 Impact of SCS on Energy

Different practices that promote SCS are either already practiced (*e.g.*, improved rotation, residue management) or can reduce energy input in farm operations (*e.g.*, reduced or no tillage). However, some practices like additional pumping for irrigation could increase energy use. Considering the management practices and the energy use that would change from baseline practice, there is only a small change (positive or negative), hence, SCS can be assumed to be energy neutral.[17]

8.2.2.7 Cost of SCS

Based on calculations by Smith *et al.*,[17] costs of implementing SCS managements range from a cost saving – 45 \$ per t CO_2eq for tillage and residue management up to 11 \$ per t CO_2eq for cropland and grazing land. About 20% of mitigation from SCS can be realised at negative costs (−45 to 0 \$ per t CO_2eq) whilst the remaining 80% can be realised for a price ranging from 0 \$ to 11 \$ per t CO_2eq.[17]

Globally, the costs of SCS to mitigate 2.5 Gt CO_2eq per year (at a C price of 100 \$ per t CO_2eq) could save 7.7 B\$, comprising 16.9 B\$ of savings and 9.2 B\$ of positive costs.[17]

8.2.2.8 The Impact of SCS on Nature's Contributions to People

SCS provides several ecosystem services, recently classified as Nature's Contributions to People (NCPs).[95] A summary of the intensity of the impacts is given in Table 8.2.

Table 8.2 Impact of SCS and Biochar on Nature's Contributions to People (NCP).[95]

NCP	Nature's contributions to people	SCS	Biochar
1	Habitat creation and maintenance	[a]	[b]
2	Pollination and dispersal of seeds and other propagules	No impact	No impact
3	Regulation of air quality	[b]	[b]
4	Regulation of climate	[a]	[b]
5	Regulation of ocean acidification	No impact	No impact
6	Regulation of freshwater quantity, location and timing	[c]	[c]
7	Regulation of freshwater and coastal water quality	[c]	[c]
8	Formation, protection and decontamination of soils and sediments	[a]	[b]
9	Regulation of hazards and extreme events	[b]	[b]
10	Regulation of organisms detrimental to humans	[a]	No impact
11	Energy	[c]	[a]
12	Food and feed	[a]	[a]
13	Materials and assistance	[c]	[c]
14	Medicinal, biochemical and genetic resources	[c]	[c]
15	Learning and inspiration	Indirect	[c]
16	Physical and psychological experiences	Indirect	No impact
17	Supporting identities	Indirect	[c]
18	Maintenance of options	No impact	No impact

[a]Direct strong impact on NCP.
[b]Direct impact on NCP.
[c]Small or indirect impact on NCP.

NCPs 1 and 10 are greatly supported by SCS through improved soil life,[108] *e.g.*, positive impacts that SCS has on soil health[99] include suppressing soil-borne crop diseases[109] and a potential positive impact on sediment particles.[110–113] SCS influences NCPs 6 and 9 through: (i) regulation of the timing and magnitude of peak water flow, particularly after heavy precipitation;[114] (ii) improved soil structure which makes soil less prone to water and wind erosion;[100] and (iii) an increase in ground cover through SCS practices, which reduces the vulnerability of soils to degradation and landslides.[115] NCPs 7 and 8 benefit through the positive impact of SCS on the filtering capacity and storing function of soils. The positive effects of SCS on soil health[99] can reduce soil erosion[116] and buffer various pollutants, decreasing the determination of pollutants downstream in freshwater and coastal waters. The impact of SCS implementation on decontamination and sediments is indirect as it determines soil carbon and consequently soil formation.[117] In addition, SCS protects soils from global change pressure.[‖][118] SCS has a positive impact on NCP 11, 12 and 13. SCS leads to

[‖]Soils are subject to varying degrees of direct or indirect human disturbances (*e.g.*, land-use change, land management and land degradation like erosion, compaction, sealing and salinization) constituting a major global change driver.[118]

improved productivity of food,[100,119] energy and fibre crops, and can provide a stable yield improvement over years.[120] Looking at vulnerable agroecosystems, SCS can provide stability and support production as it helps to prevent degradation and desertification.[121] SCS supports NCP 14 through increased soil health,[99] also maintaining resources for medicines isolated from soil organisms.[108] The impact on NCP 3 can be large as SCS reduces wind erosion[122] and therefore, reduces dust in the air. However, a potential negative impact is an increase in ammonia emissions.[123] SCS also contributes to NCP 4 as SCS practices can provide significant GGR[7,17,124] and can be used to remove CO_2 from the atmosphere, *i.e.*, as a NET (see Section 8.2.3). SCS managements increase the SOC sink, and this positive impact may reverse if other GHG emissions increase at the same time.[125] Also, SCS improves the resilience to climate change.[99,118] Some of the 18 NCPs are not impacted by soil and SCS, for instance, NCP 2, NCP 5 and NCP 18 (see Table 8.2). Although NCP 15, NCP 16 and NCP 17 are not directly influenced by SCS, the landscapes providing these NCPs are shaped by their underlying soils (*e.g.*, peatlands, deserts, forests).[21] Thus, indirect impacts could be provided to these NCPs when SCS managements are applied.

Overall, the ecosystem services provided by soil can be seen as natural capital, which gives the opportunity to value and track the changes in services provided by soils for a given human use.[126] The natural capital of soils can be characterised by soil properties like C stocks as one important component. There is some framework done to quantify and map soil natural capital and their potential to provide ecosystem services.[126,127] However, more techniques and research are needed to provide the value of soil natural capital estimates as well as more effective engagement of the soil science decision-makers.[127] An alternative way to quantify the value is *via* the carbon price (see Section 8.2.2.7).

8.2.2.9 *Impact of SCS on the United Nation's Sustainable Development Goals*

Several functions provided by SCS can have an impact or supportive role on the SDGs, as summarised in Table 8.3. Specifically, the impact that SCS has on improving soil health[99] supports agricultural productivity and hence specifically supports SDG 2 and SDG 1 as it helps to secure income in the world's poorest regions. SDG 15 also benefits from improved soil health as it helps to enhance the potential for biodiversity and healthy ecosystems.

SCS provides a soil C sink for atmospheric CO_2 and improves resilience to climate change[99,100] and therefore, contributes largely to SDG 13. SCS contributes to SDG 6 through reducing soil erosion.[122] The buffer function of healthy soils towards various pollutants increases air quality. SCS has a positive impact on SDG 3 through the combined effects of healthier soils, which can enhance agricultural production, improve water and air quality

Table 8.3 Impact of SCS and biochar on the United Nation's sustainable development goals (SDGs).[96]

SDG	Sustainable development goals	SCS	Biochar
Goal 1	No poverty	*b*	*b*
Goal 2	Zero hunger	*a*	*a*
Goal 3	Good health and well-being	*b*	*b*
Goal 4	Quality education	No impact	No impact
Goal 5	Gender equality	No impact	*c*
Goal 6	Clean water and sanitation	*c*	No impact
Goal 7	Affordable and clean energy	*c*	*a*
Goal 8	Decent work and economic growth	*b*	*b*
Goal 9	Industry, innovation and infrastructure	*c*	*c*
Goal 10	Reduced inequality	No impact	*c*
Goal 11	Sustainable cities and communities	*c*	*c*
Goal 12	Responsible consumption and production	*b*	*b*
Goal 13	Climate action	*a*	*b*
Goal 14	Life below water	*c*	No impact
Goal 15	Life on land	*a*	*a*
Goal 16	Peace and justice strong institutions	No impact	No impact
Goal 17	Partnerships to achieve the goal	No impact	No impact

[a]Direct strong impact on SDG.
[b]Direct impact on SDG.
[c]Small or indirect impact on SDG.

and maintain resources (*e.g.*, raw materials for medicine). At the same time, it supports SDG 8 and SDG 9, which in turn can have a positive impact on SDG 11. Improved production of energy crops is a positive factor for SDG 7. SDG 14 is supported by SCS through reducing erosion and pollutant substances from reaching water bodies. The SDGs 4, 5, 10, 16 and 17 are not directly impacted through SCS (Table 8.3).

8.2.3 SCS as a Greenhouse Gas Removal Technology

8.2.3.1 Potential and Implementation

When SCS is considered as a GGR technology, the sum of all SCS management approaches (Section 8.2.1) is reflected and the joint impact on the environment is accounted for. SCS managements not only have the potential to increase SOC and capture atmospheric CO_2, but many managements also help decrease GHG emissions from agricultural production.

Presenting a global mitigation potential of SCS has the challenge to include the different managements and assumptions about considered areas. So, a range of possible mitigation estimates can be calculated, considering specific aspects, managements or land uses.

The overall technical mitigation potential of SCS was recently estimated to be ~7 Gt CO_2eq per year based on data from 22 studies ranging from 0.5 to 11 Gt CO_2eq per year.[124] Estimates for the technical mitigation potential of SCS from different studies are shown in Figure 8.3; some studies focus on

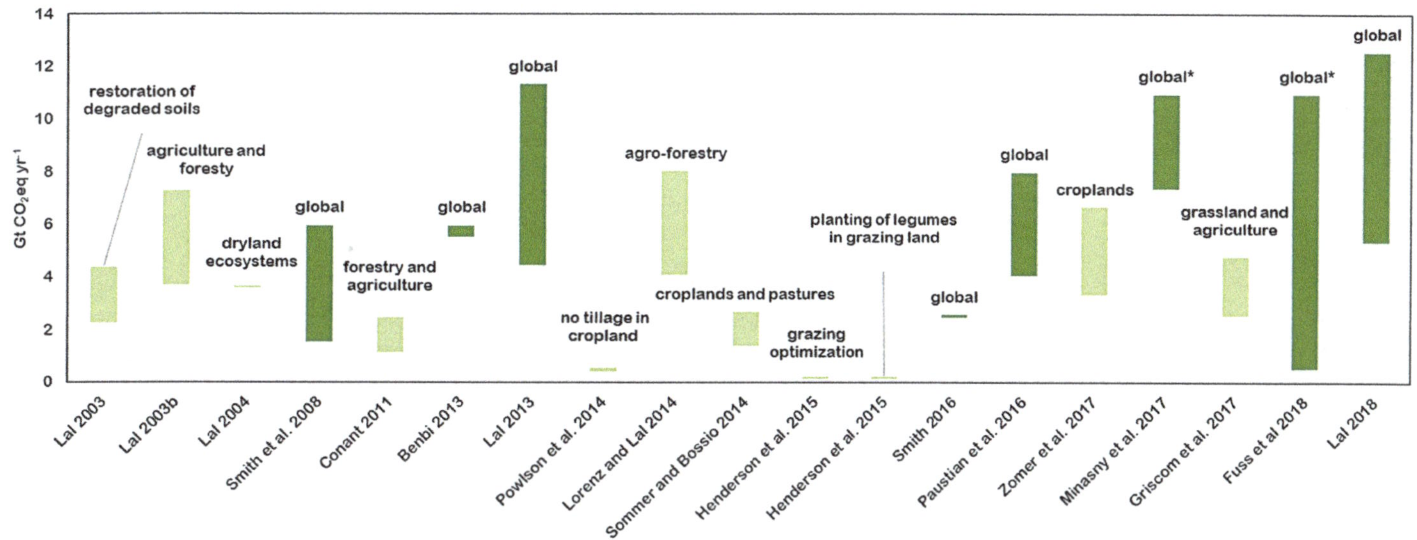

Figure 8.3 Estimated CO_2 removal potential of SCS as provided by previous studies, globally and for specific land uses or practices. Dark green bars correspond to global estimates for agriculture and grasslands; * = estimate based on review of the other studies shown in the graph.

specific managements or regions; others estimate an overall global mitigation potential. A feasible technical potential is summarized as 2–5 Gt CO_2eq per year,[124] and the economic potential (at prices up to 100 $ per t CO_2eq) with 1.5–2.6 Gt CO_2eq per year.[5,17]

Currently, the global land area where SCS management is feasible includes 4900 Mha of agricultural land where 332 Mha is equipped for irrigation, 400 Mha of urban lands, and ~2000 Mha of degraded lands.[2] Based on these land area values, Lal[2] calculated the global technical potential of SCS to be between 5.3–12.6 Gt CO_2eq per year.[2] Paustian *et al.*[6] gave a similar range, assuming the lowest potential of ~1.5 Gt CO_2eq per year at the lowest specific C price to 5.5 Gt CO_2eq per year for no economic constraints. The overall highest technical mitigation potential could be as high as ~8 Gt CO_2eq per year considering all existing (and potential future) soil management practices.[7] Generally, the majority of managements have a modest C sequestration rate of around 1.1 t CO_2eq per ha per year. When SCS is implemented globally on agricultural used soils and grasslands, the full GGR potential can be exploited.

Whilst summarising different scopes and assumptions in the studies shown in Figure 8.3, the global technical potential is estimated to be 1.8 to 12.5 Gt CO_2eq per year. The main factors considered for calculations include the land area and the C sequestration rate of certain practices. Initiatives such as "4 per 1000"[19] promote and support the implementation of SCS managements globally to increase SOC stocks (see Section 8.4.1). The estimates and studies support an adaption of SCS management, which suggest a mitigation potential at the upper end of the estimates for the next 20 to 30 years until the soil reaches a new C equilibrium.[46]

Reaching the full potential of SCS as a GGR technology depends on the effectiveness of implementation strategies, as well as socioeconomic and policy constraints.[7] Challenges of implementing SCS practices include engaging the people from all sectors. For instance, the farmer who works on the field and will implement the SCS managements, as well as the local and national politicians, who can interact through taxations and subsidies. There are possibilities to engage with farmers and land managers *via* decision support tools about mitigation scenarios. These tools can inform stakeholders about GHG emissions of current practices and production as well as showing the impact of SCS management (*e.g.*, CoolFarmTool;[128,129] COMET-Farm tool[130]).

There are local and regional differences, which can only be assessed locally to implement the best strategy. These specific conditions are hard to predict and need local expert knowledge which goes beyond the capability of the models. A successfully implementation of SCS managements requires an understanding of the context, for instance, the cultural, political and socioeconomic aspects for the region and the livelihood of the farmers to deploy the full potential of climate change mitigation.[7,131]

8.2.3.2 Limitations and Uncertainties

The C sink saturation is the limit of organic matter that can be stabilised in the soil under certain conditions. A key element in soil C dynamics is the relationship between soil structure and the ability to stabilise SOM.[132] This relationship will determine whether SOC stocks increase and a possible saturation.[132]

The saturation of SOC is the finite capacity of a soil to stabilise organic matter,[132–134] although this cannot be estimated with absolute confidence yet.[135] Most estimates and models suggest a linear increase of SOC stocks, but it is more likely that soil will reach its SOC saturation level asymptotically.[136] The final phase of soil C sequestration is the slowest and not fully understood.[137] Therefore, setting a limit for SCS after a certain time may not mean that the C sequestration potential is zero, instead, it may just be very small and thus negligible in terms of GGR.

Other studies showed that the soil C saturation varies among soils.[2,138,139] When C input increases, the mineral fraction of a soil saturates and consequently, additional C input accumulates only in labile soil C pools which have a faster turnover.[138] The time until a soil reaches its potential C saturation differs for practices, soil types and climate. For example, an increase in C following a change in rotation intensity could take a slightly longer period (26 years) than a change to no tillage (21 years) and sequestration strategies for grasslands have on average a longer duration (33 years) than that for croplands.[139]

The soils with the highest C sequestration potential are those farthest from C saturation, for instance, eroded,** degraded†† and depleted‡‡ soils.[2,137] The saturation C sink capacity might also vary when atmospheric concentrations of CO_2 change, which can enhance SOC turnover and potentially limit long-term C sequestration.[2]

Research into C saturation of soils and the role of mineral surface in the context of an upper limit or "saturation level" has only begun recently,[7,46,140,141] with more studies emerging to address the matter. A study of tropical soils showed that the C saturation deficit in the fine fraction was the largest for croplands and a change in management did not fulfil the C deficit.[142] Bai *et al.*[22] found that coarse-textured soils are not carbon-saturated and as a result, have great potential for C uptake. Feng *et al.*[143] even reported an underestimation of SCS in current models, as the process of organic matter that can be stabilised in the fine soil fraction is not adequately represented. More research is needed to quantify the effect with certainty.

However, the time soil can effectively sequestrate C under implementation of a certain SCS management is limited.[2,138,139] Soil carbon change is

---

**Soil erosion is caused through the impact of water and wind by removing soil particles gradually, causing deterioration of the soil.

††Soil degradation is the decline in soil condition (physical, chemical and biological) caused by its improper use or poor management.

‡‡Soils which are depleted of their SOC stocks.

assumed to occur over 20 years by the IPCC.[144] This is likely to be longer in colder climates and shorter in warmer climates, and there is also a large variation with soil type and climate zone. This means that, once the soil reaches (or nearly reaches) this new C equilibrium, the chosen SCS loses its CO_2 removal potential. Furthermore, the accumulated C stock is reversible,[6] hence the practices need to be maintained even after saturation occurs.

Once all soils suitable for SCS management reach their C sink saturation under current conditions, SCS will lose its GGR potential. This is of interest for the future, as GGR will be important in the second half of the century,[117,145] however, SCS may become less effective by then if implemented soon.

One barrier of large-scale implementation of SCS is the monitoring of SOC changes. Measurements are the best option, but they are costly and labour intensive. Due to the high heterogeneity in space and in depth there is a demand for a high number of samples, which is unrealistic for most monitoring approaches. There is the need for credible and reliable measurement/monitoring, reporting and verification (MRV) platforms, both for national reporting and for emissions trading.[146]

8.3 Biochar

Over the last decade, interest and research in biochar have increased immensely.[16] Biochar became more popular not only to improve soil functions and soil fertility for an increase in yield production, but also as a GGR technology to sequestrate C in the soil and reduce the overall GHG emissions. However, adding biochar to soil and in potting mixes goes back centuries and is part of traditional management concepts in many regions worldwide.[16]

This section will first introduce biochar as an organic soil amendment, both the traditional use and as modern technology. Following this, the impact of biochar as a GGR technology will be explored and its consequences on different aspects of the environment will be discussed.

8.3.1 What is Biochar?

Biochar is produced through pyrolysis, a process where organic biomass is thermally degraded under zero to low oxygen conditions. It is distinguished from charcoal, which is created under the same procedure, through the use as soil amendment or for the intention to use in environmental management.[13]

During pyrolysis, a certain level of organic C is created, which ensures a high organic C content in biochar.[16] Factors influencing the biochar outcome include the type of biomass that is used (the so-called feedstock) and the technology, mainly the reaction retention time and reaction temperature.[147] Consequently, the technological process has a fundamental impact on the physical and chemical properties of biochar and on the content of carbonised organic matter.[148]

The process of pyrolysis is a technology of providing energy.[149,150] In most cases, it will be a trade-off to biochar production, and it might be a way to contribute to a future energy solution.[151] Generally, the operation of pyrolysis can occur at different scales, from stoves to cook meals or heating individual homes, to large bioenergy installations.[150]

Biochar summarises a range of materials produced from a variety of feedstocks, each providing different benefits and constraints.[16] The chemical components differ between the feedstock sources from woody biomass, agricultural biomass, animal and human waste, industrial waste biomass and aquatic biomass, with the highest C content provided by woody and agricultural feedstocks.[147,152] The choice of material may be limited by the requirements of being safe for the soil and the appropriateness to address certain soil constraints.[149,150] Until now, there is no decision tool available to identify biochar types suitable to meet certain soil constraints.[16] Waste management can be a starting opportunity to produce biochar locally, although economics could dictate the materials of disposal.[153,154]

Currently, the largest charcoal producers in the world as of 2015[150] include Brazil (9.9 Mt per year), Thailand (3.9 Mt per year), Ethiopia (3.2 Mt per year), Tanzania (2.5 Mt per year), India (1.7 Mt per year) and Democratic Republic of Congo (1.7 Mt per year).

8.3.1.1 Traditional Use of Biochar

Biochar as a soil amendment has been commonly used in many parts of the world for centuries or even millennia.[15] One prominent example is the *Terra Preta (do Indio)* soils of central Amazonia originated by pre-Columbian native populations by adding large amounts of charred residues, organic waste excrements and bones in the ground. The result is a carbon enriched soil (three times more organic matter) with high concentrations of nutrients such as N, P, K, and calcium (Ca) and 70 times more charcoal compared to adjacent infertile soils.[155]

Other examples in history can be found around the world and include the Plaggen[§§] soil cultivation in the Medieval Period in the Netherlands, Germany and Belgium with higher SOC content than reference soils.[156] The formiguer[¶¶] technique was used to produce fertiliser material for soil improvement in Spain up to the 1960s and still used in India and Bhutan.[157] There are ancient charred soils in China based on burned rice and straw[158,159] and dark-coloured soils as a result of slash-and-burn and slash-and-char practices in Borneo.[15,160] In Japan, biochar has been used in

[§§]Plaggen are pieces of heath and pasture grass mixed with manure (or similar) and biochar before it is applied to the fields.[156]

[¶¶]A *formiguer* is a stacked heap made from dried woody vegetation and covered by soil which leads to slow and incomplete combustion.[157]

agriculture for soil improvement for centuries.[161] In Australia, several soils were found similar to the Amazonian Dark Earths,[162] which could be connected to the earthen ovens that aborigines used for cooking. In Africa, farmers enriched the soils through a mixture of excrements from people, domestic animals and poultry, also residues from harvest, fishing and cooking and added ash from wood fires over long periods of time.[163]

In modern times, biochar has been discussed as a soil amendment in agricultural textbooks and scientific journals in the middle of the 19th century. In the second half of the 19th century and in the 20th century, the studies about biochar increased and proved why biochar improves nutrient availability.[164] The interest in biochar ceased in the middle of the 20th century, possibly with the development and marketing of inorganic fertiliser.[16] Only in the 1980s did notable research about biochar start in Japan.[161] More recent research is motivated by the Amazonian *Terra Preta* and focuses on investigating whether biochar can provide broader soil benefits in its own right.[165,166]

8.3.2 Biochar as a Greenhouse Gas Removal Technology

When biochar is used in agriculture as a mitigation option or for GGR, the management should be accounted as a system rather than a material. The key process in reducing CO_2 in the system is the lower mineralisation of biochars compared to the material it was produced from. However, not only does the C cycle depend on the composition of the biochar, but the life-cycle of all GHGs is generated through the soil amendment and can be crucial for N_2O and CH_4 emissions from soil. On one hand, it is not always entirely clear how different biochars affect soil biota.[167] What is clear, on the other hand, is the important role certain properties of biochar play for pore structure, surface area and mineral matter to determine such affects.

When viewing biochar management as a GGR system, the decomposition of biomass, transport, infrastructure and indirect land use change have to be considered to ensure that all relevant aspects of biochar are accounted for in the GHG balance. Hence, the mitigation potential depends not only on the addition of biochar to the soil, but also on the environmental sustainability, social acceptance, technological implementation and economic competitiveness.[16]

The global mitigation potential of biochar application was presented to be between 0.5 and 11.9 Gt CO_2eq per year (presented in Fuss *et al.* 2018[124] and Figure 8.4). The upper value is based on the assumption that 80% of current harvested biomass is converted to biochar.[168] Based on the availability of annual unused waste biomass, the mitigation potential was estimated to be 6 Gt CO_2eq per year in Lee and Day.[169] Roberts *et al.*[170] calculated an achievable global annual C sequestration of 0.7 Gt CO_2eq per year based on only late stover being used as a feedstock for biochar. Based on the higher mitigation potential generally projected for the future and the assumption of an increase in biochar application,[124] the upper estimates for technical

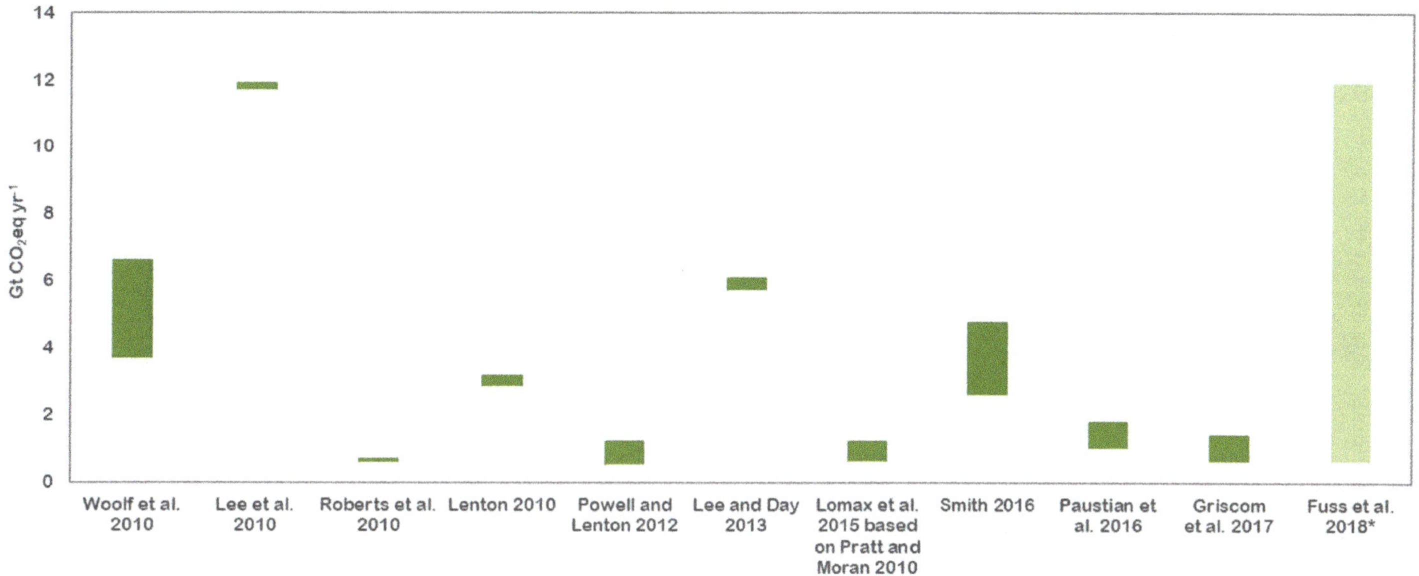

Figure 8.4 Estimated global CO_2 removal potential of biochar application as provided by previous studies (* = range based on review of other studies shown in the graph).

mitigation potential are 4.8 Gt CO_2eq per year[17] and 6.6 Gt CO_2eq per year.[171] Overall, the feasible technical potential for biochar is 2.6 Gt CO_2eq per year.[17] At present the economic data is too poor to estimate the economic potential for biochar.

8.3.2.1 Biochar Impact on GHG Emissions

Net CO_2 removal from the atmosphere could be achieved due to an increase in plant productivity. As the traditional use of biochar is for the increase of soil quality and plant productivity, the application of biochar will improve production in the long term.[15] However, if the additional yield is removed and consumed as food or feed, it would not contribute to net CO_2 removal from the atmosphere.[17]

Other non-CO_2 GHG emissions are influenced by the addition of biochar to the soil. Depending on the chemical composition of the biochar and the environmental conditions, it can have opposite impacts on emissions.[172] The influences of biochar on the soil N cycle depends on soil type and biochar properties. Most previous studies reported an overall reduction of N_2O emission[173,174] with an averaged reduction range between 32%[175] to 38%[176] but the effect seems to be negligible after one year.[176] Others did not find robust evidence for N_2O emission reductions by biochar under field conditions.[177,178]

There are only a few studies showing the impact on the N cycle in the long term: Hageman *et al.*[179] investigated long-term impacts on CO_2 and N_2O to estimate the mitigation potential and found effective reductions in these emissions over the first 3 years. The study highlights key research gaps and limitations, specifically in the understanding of how biochar altered soil functions and missing knowledge of the underlying mechanism for the suppression of N_2O emissions. Therefore, more research is needed for a full understanding.

Emissions in paddy soils and paddy rice production are strongly affected by biochar application with an overall reduction of GHG emissions over several years.[180,181] The reduction was mainly achieved through a reduction in N_2O emissions but at the same time CH_4 emissions can increase.[181,182] The effect can be reduced when fertiliser N application is adjusted. A fine tuning between biochar application and fertiliser rates is needed to get the optimum benefits for mitigation and yield production.[183] On one side, there is adjustment needed with the biochar composition, and the application of fertiliser, manure and fertiliser inhibitor on the other side; the combination of these determines the balance of GHG emissions in paddy rice fields.[183–186] Soil properties like soil pH also have an impact on GHG emissions related to biochar application.[187]

The soil type and the biochar production temperature will determine whether the process takes decades or centuries.[188,189] Lower residence times occur under tropical and sub-tropical conditions with higher temperatures[190] and acid soils.[191]

Other studies showed the importance of the agricultural system and the biochar feedstock on N_2O emissions.[192–197] The quantification of non-CO_2 GHG emissions through biochar amendments are difficult due to a lack of studies and high uncertainties. The variation of impacts due to the type of biochar adds further uncertainty.

8.3.2.2 Biochar Impact on Soil Carbon Sequestration

When added to soil, biochar retards the return of biomass C to the atmosphere as CO_2 and acts therefore as a GGR technology and a high-recalcitrance carbon pool.[17,166] An increase in soil organic carbon was reported in several studies.[185,198] Liu *et al.*[199] showed a significantly enhanced SOC content of 40% through biochar amendment but no significant effect on soil CO_2 fluxes. A meta-analysis by Bai *et al.*[22] reported a 39% increase in SOC through biochar application, but the data included field and pot experiments where SOC storage was enhanced by 28% and 57%, respectively. The study compared data for different climate zones and showed a stronger increase in SOC under biochar application for arid and cool climate regions. In general, biochar supports long-term carbon storage, and the biochar should remain in the soil for as long as possible for the best outcome.[198]

8.3.2.3 Land Required for Biochar Application

The application of biochar does not require a land use change and does not compete with other land uses. Biochar application to the soil can be as high as 30–60 t per ha,[182,200] which can result in high GGR. Considering its potential for CO_2 removal, the required land for biochar will be less than 0.27 ha per t CO_2eq.[17] Land use change is not a requirement for the use of biochar.

Globally, biochar implementation with a potential to emit 2.6 Gt CO_2 per year would require a land area of 14 Mha (assuming a biochar application of 50 t per ha).[200] For a theoretically possible implementation potential of 4.7 Gt CO_2 per year, the land footprint would be 26 Mha. However, as the application of biochar does not require a land use change, the land footprint that needs to be considered is the requirement of the feedstock to produce biochar. Around 43% of the 2.6 Gt CO_2 per year derives from dedicated biomass crops, which is proposed to occur on 50% of abandoned cropland that is not used for other purposes.[171] The area requirements to deliver the CO_2 removal potential of 1.1 Gt CO_2 per year ranges between 40–260 Mha, this is based on the assumptions that biomass yields of lignocellulosic energy crops (*e.g.*, miscanthus) are used, and that 20% through fast pyrolysis and 88% through torrefaction[‖] of the biomass C will be converted to

[‖]Torrefaction is a pyrolysis process at low temperature.

biochar C.[148] If the yields on these fields were lower, the area requirements will be higher to deliver the same CO_2 removal potential.

Considering the 1.5 °C goal through mitigation strategies, Werner *et al.*[201] calculated the area that would be needed for large-scale CO_2 removal with plantation-based pyrogenic carbon capture and storage (PyCCS) and permanent pyrogas. The study considered the conversion of natural vegetation to biomass plantations, which subsequently estimated that an area of 133–3280 Mha would be required globally for biomass plantations[201]. Other considerations in this study included biochar as soil amendment, bio-oil*** (pumped into geological storages) and permanent-pyrogas††† (with capture and storage of CO_2 from gas combustion).

8.3.2.4 *Application of Biochar in Grassland Systems*

Considering application of biochar to mitigate GHG emissions from grasslands, literature data are scarce and variable. Rondon *et al.*[202] found that application of biochar reduced N_2O emissions by 80% from low fertile grassland systems and from soil amendment with bovine urine.[203] In forage grass in Colombia, Renner[204] found that adding biochar completely suppressed CH_4 emissions. In contrast, Clough *et al.*[205] found that adding wood biochar to pasture soil had no effect on cumulative N_2O–N loss from chemical fertiliser or bovine urine. Moreover, for a managed grassland in Korea, Chen *et al.*[206] reported an overall increase in N_2O emissions due to the interactions between biochar and urea N fertiliser. They suggested that biochar may have stimulated microbial N mineralisation or provided soil with the ability to retain mineral N and thereby increase N_2O emissions. Furthermore, Angst *et al.*[207] reported that pine wood biochar had no significant impact on GHG emissions in systems with high manure application rates. Fidel *et al.*[208] noted that biochar could reduce soil N_2O emissions under specific, but not all, conditions. Nonetheless, biochar amendments that have high residence times are important approaches to increase C sequestration in soils and reduce greenhouse gas emissions.[209]

8.3.3 Impact of Biochar Beyond the Function as GGR

The use of biochar as a GGR technology has an impact on different resources at different scales (Figure 8.5, Table 8.4). This section discusses the impact that biochar has on different aspects of the environment and people including yield, waste, water, nutrients and soil fertility, biophysical climate (represented by surface albedo), toxic elements, energy produced, energy demand and costs. Like SCS, the impact of biochar goes beyond the environmental effects and the impact will also be viewed under the

***A pyrolytic liquid (bio-oil) pumped into depleted fossil oil repositories.[201]

†††Permanent-pyrogas (dominated by the combustible gases CO,H_2 and CH_4) that may be transferred as CO_2 to geological storages after combustion.[201]

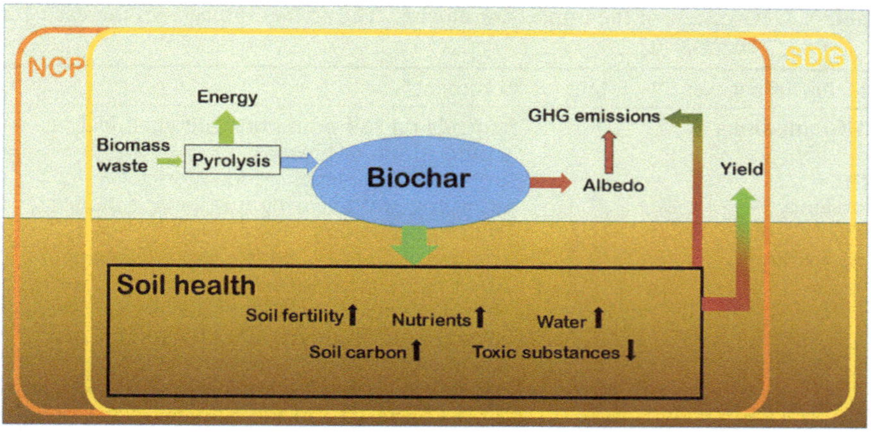

Figure 8.5 Interactions and impacts of biochar production (pyrolysis) and application on soil, greenhouse gas (GHG) emissions and production. The green arrows indicate a positive impact, the red arrows indicate a negative impact, and the black arrows indicate an increase ↑ or decrease ↓. The overarching and overlapping impacts on ecosystem services as Nature's Contributions to People (NCP), orange box, and on the United Nation's Sustainable Development Goals (SDG), yellow box, are indicated.

premise of Nature's Contribution to People (NCP) and the United Nation's Sustainable Development Goals (SDG) - previously analysed in Smith *et al.*[21] The contribution of biochar to NCPs and SDGs are summarised in Tables 8.2 and 8.3.

8.3.3.1 Impact of Biochar on Nutrients and Soil Fertility

There are several soil properties that biochar amendment has a positive impact on, which support the overall soil health. Biochar can improve soil fertility and productivity by absorbing and releasing nutrients and increasing soil pH and the cation exchange capacity in poor, acidic soils, decreases soil compaction, increases soil porosity, reduces bulk density and increases soil water content.[210–212] Further impacts on soil properties include the increase of soil microbial biomass,[199,213] root growth and root biomass and legume module formation[214] as well as symbiotic N_2 fixation.[175] Hagemann *et al.*[215] identified a complex, nutrient-rich organic coating on co-composted biochar that covers the outer and inner pore surfaces of biochar particles; this coating strengthened the biochar–water interaction and hence enhanced nutrient retention. Kammann *et al.*[216] demonstrated that co-composting considerably promoted biochars' positive effects, largely by nitrate (nutrient) capture and delivery. Biochar addition can increase the resistance of both the bacterial and fungal networks to drought.[217]

The wide variation of biochar also results in variation of nutrients that will be increased in the soil through the application. However, using a study of

Table 8.4 Overview of the impacts of biochar. The arrows indicate an increase ↑ or decrease ↓.

Biochar impact on	Impact	Comment
GHG emissions	↑↓	Depends on soil properties and chemical composition of biochar
SOC	↑	Supports a long-term carbon storage
Nutrients	↑	Absorbing and releasing nutrients, enhanced nutrient retention
Soil fertility	↑	Increase of soil pH and the cation exchange capacity, decrease of soil compaction, increase of soil porosity, reduced bulk density and increase of soil water content, root growth and root biomass and legume module formation, symbiotic N_2 fixation, increase of the resistance of the bacterial and fungal networks to drought
Yield	(↑)	Depending on climate zone, soil properties and crop; increase in crop more likely when biochar is applied to poor soils
Soil contamination and toxic substances	↓	Effective in the sorption of organic contaminants and toxic elements
Waste	(↓)	Recycling of biomass waste through the production of biochar through pyrolysis
Albedo	↓	The reduced albedo through darker soils can reduce the mitigation potential
Energy	↑	Energy is produced when biochar is generated through pyrolysis
Costs	↑↓	Depends on feedstock and technology

lignocellulosic biochar,[218] the average ranges for C, N, P and K can be given for biochar containing 1 t C per ha with ~30 kg N per ha, ~10 kg P per ha and ~70 kg K per ha. This would add up to 21 Mt N per year, 7 Mt P per year and 49 Mt K per year of additional nutrients to the soil for a total global GGR potential of 2.5 Gt CO_2 per year. Additional fertiliser that could be used to grow biochar feedstocks were not considered in the study.

8.3.3.2 Impact of Biochar on Productivity and Yield

The impacts of biochar on soil health and soil fertility can impact on agricultural productivity. There are several studies and reviews reporting on an increase of yield and yield intensity, specifically in the tropics and paddy rice.[184,219,220] Other studies could find no impact on yield[221] or a reduction in productivity.[172,222] The impact of biochar application depends not only on the soil properties and biochar composition, but also on additional fertiliser and the temperature that is used in the biochar production (*i.e.*, pyrolysis temperature).[214,223] In temperate soils, there are moderately negative to positive yield effects recorded.[216] Yield improved mainly through an increase in nutrient use efficiency and water-holding capacity. However, the improved yield is often recorded where biochar was added to poor soils; in fertile or healthy soils, yield did not improve as much.[224]

Long-term studies are needed to understand fully the underlying mechanisms of the biochar impact on yield and GHG emissions.

8.3.3.3 *Impact of Biochar on Soil Contamination and Toxic Substances*

Biochar has a considerable ability to mitigate the transfer of toxic elements in the soil.[225,226] Peng *et al.*[226] found a noticeable reduction of toxic elements in plants when biochar is applied. And so, biochar amendments reduce toxic elements in edible plant parts and food, thereby reducing the associated health risks. In regard to the uptake of contaminants from compost, biochar reduced the harmful effect of feedstock on living organisms.[227] The absorption of toxic elements and contaminants depend on pyrolysis temperature, residence time, heat transfer rate, and feedstock type.[228] Ahmad *et al.*[228] reported that high pyrolysis temperatures could produce biochar that is effective in the sorption of organic contaminants by increasing surface area, micro-porosity, and water repellence. In contrast, low temperatures produced biochar suitable for removing inorganic/polar organic contaminants by oxygen-containing functional groups, electrostatic attraction, and precipitation.[228]

8.3.3.4 *Impact of Biochar on Waste*

Biomass production uses agricultural and forestry biomass wastes. The waste material is recycled in the form of feedstocks for biomass, therefore, there is no competition for land. It also contributes to the development of self-sufficient communities as they do not need to set aside land for feedstock production.[21]

Traditional residue burning includes burning for recycling ash nutrients, and sanitising residues for pest and pathogen control for cooking. New technologies like Kon-Tiki flame curtain pyrolysis can substantially reduce emissions of CO, NO_x and soot emerging through pyrolysis as they combine the simplicity of a traditional kiln[‡‡‡] with combustion of pyrolysis gases in the flame curtain,[§§§229] The technology has the advantage of not only reducing emissions but also improving health by reducing indoor air pollution. The use of cooking stoves improves women's health and reduces the risks of injuries.[21]

Using food waste to create biochar, gives the opportunity to transform food waste into valuable commodities. However, the process is challenging as biochar production depends on the type of waste and the composition of processed mixed foods can vary.[230]

[‡‡‡]Traditional kiln technologies are used to produce charcoal slowly and without treatment of the pyrolysis gases, resulting in emissions of gases.[229]

[§§§]Similar to retort kilns pyrolysis, where gases are led back to a combustion chamber.[229]

8.3.3.5 *Impact of Biochar on Albedo*

Biochar tends to be dark and can be applied in large amounts. Consequently, biochar can darken the soil surface and have an impact on surface albedo. Verheijen *et al.*[231] showed the importance of considering the change in soil surface through the application of biochar.

Studies showed a decrease of surface albedo by 40% through biochar application of 30–60 t per ha over the crop season.[200] The control measurements of the experiment gave an albedo from 0.2 in the early growing season up to 0.3 in the late growing season. The 40% reduction in albedo represents the absolute decrease of around 0.08–0.12.[200] Other studies showed similar results for a field in Germany, where 30–32 t per ha biochar application resulted in a mean annual albedo reduction of 0.05.[232]

An analysis of MODIS data confirmed the decrease of albedo on a larger scale and assumes a 30% lower mitigation potential for biochar application if specific measures of albedo-increasing strategies to compensate this effect is not practiced.[233]

One study found differences for biochar application in grasslands and fallows with an albedo change after rain events.[234] For grassland, albedo increased after rain events, but in fallows a substantial decrease was recorded. Regardless of land use and biochar application rate, the albedo varied in the range of 0.1 to 0.23.

Globally, the mitigation potential of biochar could be reduced by 13–22% as a consequence of the reduction in albedo[232]. Based on that assumption, a mitigation potential of 2 Gt CO_2eq per year for biochar (see Section 8.3.2) could be reduced to 1.5 Gt CO_2eq per year.

8.3.3.6 *Impact of Biochar on Water*

The impact on water through biochar is generally positive. There is no water needed in the production of biochar (except by hydrothermal carbonisation[148]) which reduces the water footprint. The quantification of the impact of biochar on soil water is difficult. However, it is widely accepted that biochar increases soil water holding capacity[13,171] and thereby, increases the retention of water in the soil/plant system.[17] Additional water that might be used to irrigate biomass feedstock could have a water footprint but is not considered here.

8.3.3.7 *Impact of Biochar on Energy*

In the production of biochar lies the potential of generating large amounts of energy. Based on Roberts *et al.*,[218] for the production of lignocellulosic biochar of 16.4–35.3 MJ kg^{-1} including a 10% process energy cost and 20% energy loss in pyrolysis plants,[235] the generated energy potential is 5.5 to 13.6 GJ per t CO_2eq of biochar.

Considering its CO_2 removal potential of 2.5 Gt CO_2eq per year, biochar implementation could produce 14–35 EJ year^{-1}. A theoretical possible mitigation potential of 4.7 Gt CO_2eq per year could produce 65 EJ year^{-1}.

The co-product of permanent pyrolysis is the heat generation by combustion, which can be useful in locations where demand for heat exists and could be used as a replacement for fossil energy sources.

8.3.3.8 *Biochar Costs*

The cost of biochar production and application can have a large range depending on the process and the benefits in one region. Considering UK conditions, costs vary between −225 to 330 \$ per t CO_2eq with an average of 150\$ per t CO_2eq (converted from GB£ per t CO_2eq in Shackley *et al.*[235]). Based on numbers presented in Meyer *et al.*,[148] for 95% C content processed under slow pyrolysis from yard waste and retort charcoal, a range of 51–386 \$ per tbiochar results in 15–110 \$ per t CO_2eq with an average of 62 \$ per t CO_2eq. For a 51% C content of biochar from torrefaction of the same feedstock results in a cost range of 27–206 \$ per t CO_2eq with an average of 117 \$ per t CO_2eq.

Assuming a 2.5 Gt CO_2eq per year GGR removal potential for the implementation of biochar globally, costs range between −581 B\$ to 1560 B\$ depending on the processes and material. The resulting mean costs would be around 130 B\$ (for 50 \$ per t CO_2eq in Shackley *et al.*[235]).

8.3.3.9 *Impact of Biochar on Nature's Contributions to People*

A summary of the impact biochar has on NCPs is given in Table 8.2. Under the premise that biochar is produced with well-established clean technical standards and without negative impacts on planetary boundaries or food safety, biochar addition can have multiple regulating features. For instance, the improvement in soil fertility has a positive impact on habitats for soil organisms (NCP 1). NCP 8 benefits from the high reactivity of the surface of biochar which could have a positive effect on the immobilisation of soil and water; also, the improved soil structure and porosity can increase nutrient cycling and retention. The improved soil properties due to biochar addition will make them more resistant against erosion, degradation and contamination and hence helps NCP 9. Biochar can have a positive impact on NCPs 11 and 12 as energy is generated through pyrolysis and yields of food and bioenergy crops could increase (specifically in tropical regions). The effect that biochar has on soil properties improves soil health, which thereby increases crop productivity and stability. Biochar can have a positive impact on NCP 15 as there are multiple existing methodologies for pyrolysis from cooking stoves to farm- and industrial-scales, which can be explored to find the best needed outcome. Future generations could learn and be inspired by the different possibilities of pyrolysis to adapt to local needs and conditions for agroecosystems and recycling biomass residues. Given the tradition of using forms of biochar for millennia, further fostering of this production will prevent the loss of a traditional technique in agriculture and support NCP 17. NCP 4 can be impacted by biochar application through enhanced SCS (see Section 8.3.2.2), depending on the amount of land surface that is dedicated to this pathway and the C fraction of biochar that is used for storage.

8.3.3.10 Impact of Biochar on the United Nation's Sustainable Development Goals

Application of biochar can have several positive and supporting impacts on SDGs (Table 8.3). An increase in soil fertility and productivity would help farmers to be more self-sufficient and increase their income (SDG 1) and can improve farmers life in rural areas (SDG 8), while increasing food security (SDG 2) and contributing to people's nutritional health (SDG 3). Application of biochar can reduce air and water pollution by absorbing pollutants (SDG 6). Biochar contributes to SDG 7 as it is an affordable source of energy and is produced from accessible feedstock. Biochar production through different pyrolysis technology from smallholder farmers to large-scale industrial production contributes to SDG 9. Using and producing biochar will contribute to reduced inequality (SDG 10) through an increase in self-sufficient farmers. It also improves the health of women through clean cooking technology and improved yields in their garden. Biochar could contribute to SDG 11 when applied in macadam[¶¶¶]-based urban planting substrates for improving city tree growth. An increase in crop yields from biochar will contribute positively to SDG 12. The use of biochar in animal husbandry may permit reduced use of antibiotics and contribute to improvement in animal health and therefore, also human health[236] (SDG 3). The high recalcitrance of the organic C in biochar, together with the decrease in N_2O and CH_4 emission from soils amended with biochar, contributes to climate regulation (SDG 13). In addition, biochar application can impact future climate change as it improves the resilience of agroecosystem and the mitigation potential. The positive impact of biochar in soil includes improved soil fertility, decreased effects of soil and water pollutant, increased capacity of ecosystems to provide habitats for (soil) organism and improved soil health (SDG 15). Some SDGs are not directly impacted by biochar production and application (see Table 8.3).

8.4 Implementation Frameworks

8.4.1 The "4 per 1000" Initiative – Soils for Food Security and Climate

While the technology and knowledge for increasing SOC are available, the options are not yet applied in extension to reach the targets of a maximum

[¶¶¶]A type of road construction consisting of layers of crushed stones.

2 °C temperature increase by the end of the century. There is a requirement for a knowledge transfer from research to stakeholders. On 1st December 2015, the international initiative "4 per 1000"[19] was officially launched at COP21.[237] The initiative encourages stakeholders from the public and private sector to transition towards a productive and highly resilient agriculture by managing land and soils appropriately and ensure a sustainable development. The main focus of the "4 per 1000 initiative", also known as "4 per mille" or "4‰", is the increase of SOC stocks at a rate of 0.4% per year. The 0.4% is based on the annual anthropogenic emissions that need to be compensated to achieve the targets of the Paris Agreement.[238] The ambitious aspirational target is focused on actively managed soils for agriculture and managed forest.

Several studies analysed the 4 per 1000 (4‰) goal and its feasibility for different regions as well as globally and calculated the technical potential of C sequestration. Minasny *et al.*[239] presented a study of 20 regions worldwide, highlighting region specific efforts and scopes and showed the possibility of a 4‰ SOC increase per year, including even higher rates up to 10‰ (*i.e.*, 10 per 1000) for soils with low initial soil C in the first 20 years after implementation of best managements. Cobeels *et al.*[240] analysed the feasibility of the 4‰ goal under agroforestry and conservation agriculture systems in sub-Saharan. Francaviglia *et al.*[241] analysed data of Mediterranean agricultural soils and showed that application of organic amendments led to significantly higher SOC storage rates than conventional management. The main factors influencing the results were the initial SOC content, experiment duration, soil texture, and climate regime. Under the 4‰ goal, Soussana *et al.*[20] calculated the global technical potential of C sequestration over the top 30 cm and top 40 cm soil depth with 10 Gt CO_2eq per year and 12.5 Gt CO_2eq per year, respectively. However, the study acknowledges that achievable potential is likely to be substantially lower given socio-economic constraints.

There is some critique and discussion on the general statement of an increase of 4‰ in all soils, which questioned the uncertainties of C sequestration, limitations and feasibility of the initiative goal.[242–244]

However, the initiative has been widely welcomed and since its launch, more than 250 organisations have endorsed the initiative by signing the Paris Declaration.[245] It is seen as a contribution to climate change mitigation and at the same time, any increase in SOC is certain to improve the quality and functioning of many soils, which is especially relevant for land that is in danger of degradation. The initiative highlights the important role of soil organic matter (SOM) and presents a way to connect agriculture and climate policy.[246] 4‰ addresses the three-fold challenge of (i) food and nutritional security, (ii) adaptation to climate change, and (iii) mitigation of human-induced GHG emissions.[20]

8.5 Summary and Conclusions

The GGR potential of SCS and biochar may not be as large as that of other GGR technologies like bioenergy with carbon capture and storage (BECCS, see Chapter 5) and direct air capture (DAC, see Chapter 6), but it is comparable to afforestation and reforestation (AR, see Chapter 9) and greater than enhanced weathering (EW, see Chapter 7). The advantage of both SCS and biochar is the accompanying benefits to soil health and soil quality that has the potential to increase agricultural production. Therefore, it is a sustainable option not only as a GGR approach but also with regard to food security over longer periods. Specifically, the increase in nutrients in the soil and the ability to reduce toxic substances through biochar application are positive side effects.

SCS and biochar can be applied to current land use and thus, do not compete with other land uses, like AR and BECCS. Although specific to climate zones and soil properties, SCS managements can decrease GHG emissions and add therefore to the mitigation potential. The impact on GHG emissions through biochar is highly variable with region and climate as well as over time. More research is needed to quantify the impact on GHG emissions through biochar application in the field. The required feedstock for biochar production could be an environmental burden, particularly if the feedstock needs to be specifically produced. However, if the biochar production is based on waste, other advantages are the reduction of waste and generation of energy in the process. BECCS also delivers energy at around 5–12 times higher than for biochar production. Both processes could be combined with BECCS delivering energy and CO_2 capture and producing biochar in the process.[171] That way, the mitigation potential of biochar can be enhanced as it is produced as a co-product.[247]

The costs of biochar (15–120 \$ per t CO_2eq)[17] are in a wide range but are comparable to the costs for implementing BECCS (15–400 \$ per t CO_2eq).[124] In contrast, DAC (25–1000 \$ per t CO_2eq)[124] and EW (15–3460 \$ per t CO_2eq)[124] have much higher costs and AR (0–50 \$ per t CO_2eq)[124] is lower. Although practice specific, SCS (−45–11 \$ per t CO_2eq)[17] is one GGR option that is able to deliver CO_2 removal at negative costs.

The main downsides of using SCS for CO_2 removal is the limited time C can be sequestrated in the soil until a new equilibrium is reached, and the soil is saturated under these conditions. Also, there is the need to maintain the management as the C sink is reversible, *i.e.*, to ensure C is not rereleased. Consequently, SCS may not be an option for GGR in the second half of the century if SCS managements are prematurely implemented on a large scale. This is less a problem for biochar application as biochar is more recalcitrant and it is expected to take much longer for the soil to reach a new C equilibrium.

However, the soils which are furthest from C sink saturation, *e.g.*, degraded lands, also have the potential to store much more C and deliver the required mitigation potential while having other positive benefits for the soil and land as a habitat.

Overall, agricultural and grassland soils have large potential to act as a C sink over the next 20 to 30 years if SCS managements and biochar applications are implemented while considering specific climatic conditions and soil properties. Many SCS practices are already implemented and show potential for GGR as well as other benefits. This will help to increase knowledge and update the expectations for SCS and biochar application as a GGR option in the future.

References

1. N. H. Batjes, *Eur. J. Soil Sci.*, 2014, **65**, 10–21.
2. R. Lal, *Global Change Biol.*, 2018, **24**, 3285–3301.
3. R. Gonçalves-Araujo, *Tracing Environmental Variability in the Changing Arctic Ocean with Optical Measurements of Dissolved Organic Matter*, 2016.
4. S. M. Ogle, F. J. Breidt and K. Paustian, *Biogeochemistry*, 2005, **72**, 87–121.
5. P. Smith, D. Martino, Z. Cai, D. Gwary, H. Janzen, P. Kumar, B. McCarl, S. Ogle, F. O'Mara, C. Rice, B. Scholes, O. Sirotenko, M. Howden, T. McAllister, G. Pan, V. Romanenkov, U. Schneider, S. Towprayoon, M. Wattenbach and J. Smith, *Philos. Trans. R. Soc., B*, 2008, **363**, 789–813.
6. P. Smith, *Soils and Climate Change*, 2012.
7. K. Paustian, J. Lehmann, S. Ogle, D. Ready, G. P. Robertson and P. Smith, *Nature*, 2016, **532**, 49.
8. K. Paustian, O. Andrén, H. H. Janzen, R. Lal, P. Smith, G. Tian, H. Tiessen, M. Van, Noordwijk and P. L. Woomer, *Soil Use Manage*, 1997, **13**, 230–244.
9. D. B. Kell, *Philos. Trans. R. Soc., B*, 2012, **367**, 1589–1597.
10. J. A. Burney, S. J. Davis and D. B. Lobell, *Proc. Natl. Acad. Sci.*, 2010, **107**, 12052–12057.
11. W. W. Wilhelm, J. M. F. Johnson, J. L. Hatfield, W. B. Voorhees and D. R. Linden, *Agron. J*, 2004, **96**, 1–17.
12. C. Poeplau and A. Don, *Agric., Ecosyst. Environ.*, 2015, **200**, 33–41.
13. J. Lehmann and S. Joseph, *Biochar for Environmental Management: Science and Technology*, earthscan by Routledge, London; Sterling, VA, 2009.
14. B. Glaser, L. Haumaier, G. Guggenberger and W. Zech, *Naturwissenschaften*, 2001, **88**, 37–41.
15. K. Wiedner and B. Glaser, in *Biochar for Environmental Management—Science, Technology and Implementation*, ed. J. Lehmann and S. Joseph, earthscan by Routledge, London, New York, 2015, p. 15.
16. J. Lehmann and S. Joseph, in *Biochar for Environmental Management—Science, Technology and Implementation*, ed. J. Lehmann and S. Joseph, earthscan by Routledge, London, New York, 2015, p. 1.
17. P. Smith, *Glob Change Biol*, 2016, **22**, 1315–1324.
18. P. E. A. Ciais, in *Climate Change 2013: The Physical Science Basis. Contribution of Working Group I to the Fifth Assessment Report of the*

Intergovernmental Panel on Climate Change, ed. T. F. Stocker, *et al.*,, Cambridge Univ. Press, 2013, p. 465–570.
19. https://www.4p1000.org/, (accessed 01/09/ 2020).
20. J. Soussana, S. Lutfalla, F. Ehrhardt, T. Rosenstock, C. Lamanna, P. Havlík, M. Richards, E. Wollenberg, J. Chotte, E. Torquebiau, P. Ciais, P. Smith and R. Lal, *Soil Tillage Res.*, 2019, **188**, 3–15.
21. P. Smith, J. Adams, D. J. Beerling, T. Beringer, K. V. Calvin, S. Fuss, B. Griscom, N. Hagemann, C. Kammann, F. Kraxner, J. C. Minx, A. Popp, P. Renforth, J. L. Vicente Vicente and S. Keesstra, *Annu. Rev. Environ. Resour.*, 2019, **44**, 255–286.
22. X. Bai, Y. Huang, W. Ren, M. Coyne, P. Jacinthe, B. Tao, D. Hui, J. Yang and C. Matocha, *Global Change Biol.*, 2019, **25**, 2591–2606.
23. W. Sun, J. G. Canadell, L. Yu, L. Yu, W. Zhang, P. Smith, T. Fischer and Y. Huang, *Global Change Biol.*, 2020, **26**, 3325–3335.
24. F. Abbas, H. M. Hammad, W. Ishaq, A. A. Farooque, H. F. Bakhat, Z. Zia, S. Fahad, W. Farhad and A. Cerdà, *J. Environ. Manage.*, 2020, **268**, 110319.
25. N. R. Haddaway, K. Hedlund, L. E. Jackson, T. Kätterer, E. Lugato, I. K. Thomsen, H. B. Jørgensen and P. Isberg, *Environ. Evid.*, 2017, **6**, 30.
26. M. A. Bolinder, F. Crotty, A. Elsen, M. Frac, T. Kismányoky, J. Lipiec, M. Tits, Z. Tóth and T. Kätterer, *Mitigation Adapt. Strategies Global Change*, 2020, **25**, 929–952.
27. M. Abdalla, A. Hastings, K. Cheng, Q. Yue, D. Chadwick, M. Espenberg, J. Truu, R. M. Rees and P. Smith, *Global Change Biol.*, 2019, **25**, 2530–2543.
28. J. P. Kaye and M. Quemada, *Agron. Sustainable Dev.*, 2017, **37**, 4.
29. F. Ren, X. Zhang, J. Liu, N. Sun, Z. Sun, L. Wu and M. Xu, *Catena*, 2018, **171**, 1–10.
30. J. A. Morgan, R. F. Follett, L. H. Allen, S. Del Grosso, J. D. Derner, F. Dijkstra, A. Franzluebbers, R. Fry, K. Paustian and M. M. Schoeneberger, *J. Soil Water Conserv.*, 2010, **65**, 6A–13A.
31. A. Ledo, P. Smith, A. Zerihun, J. Whitaker, J. Vicente-Vicente, Z. Qin, N. P. McNamara, Y. L. Zinn, M. Llorente, M. Liebig, M. Kuhnert, M. Dondini, A. Don, E. Diaz-Pines, A. Datta, H. Bakka, E. Aguilera and J. Hillier, *Global Change Biol.*, 2020, **26**, 4158–4168.
32. D. Feliciano, A. Ledo, J. Hillier and D. R. Nayak, *Agric., Ecosyst. Environ.*, 2018, **254**, 117–129.
33. P. Mehra, J. Baker, R. E. Sojka, N. Bolan, J. Desbiolles, M. B. Kirkham, C. Ross and R. Gupta, *Adv. Agron*, 2018, **150**, 185–230.
34. R. Ghimire, S. Lamichhane, B. S. Acharya, P. Bista and U. M. Sainju, *J. Integr. Agric.*, 2017, **16**, 1–15.
35. D. S. Powlson, C. M. Stirling, M. L. Jat, B. G. Gerald, C. A. Palm, P. A. Sanchez and K. G. Cassmann, *Nat. Clim. Change*, 2014, **4**, 678.
36. E. Aguilera, L. Lassaletta, A. Sanz-Cobena, J. Garnier and A. Vallejo, *Agric., Ecosyst. Environ.*, 2013, **164**, 32–52.

37. C. van Kessel, R. Venterea, J. Six, M. Adviento-Borbe, B. Linquist and K. J. van Groenigen, *Global Change Biol.*, 2013, **19**, 33–44.
38. M. Abdalla, B. Osborne, G. Lanigan, D. Forristal, M. Williams, P. Smith and M. B. Jones, *Soil Use Manage.*, 2013, **29**, 199–209.
39. A. N. Kravchenko and G. P. Robertson, *Soil Sci. Soc. Am. J.*, 2011, **75**, 235–240.
40. H. Blanco-Canqui and C. S. Wortmann, *Soil Tillage Res.*, 2020, **198**, 104534.
41. S. B. Goswami, R. Mondal and S. K. Mandi, *null*, 2020, **66**, 1218–1234.
42. M. Turmel, A. Speratti, F. Baudron, N. Verhulst and B. Govaerts, *Agric. Syst.*, 2015, **134**, 6–16.
43. D. L. Karlen, M. R. Schmer, S. Kaffka, D. E. Clay, M. Q. Wang, W. R. Horwath, A. Kendall, A. Keller, B. J. Pieper, S. Unnasch, T. Darlington, F. Vocasek and A. G. Chute, *Agron. J.*, 2019, **111**, 93–98.
44. C. Poeplau, L. Reiter, A. Berti and T. Kätterer, *Soil Res.*, 2017, **55**, 1–9.
45. A. Freibauer, M. D. A. Rounsevell, P. Smith and J. Verhagen, *Geoderma*, 2004, **122**, 1–23.
46. K. Paustian, E. Larson, J. Kent, E. Marx and A. Swan, *Front. Clim.*, 2019, **1**, 8.
47. T. O. West and W. M. Post, *Soil Sci. Soc. Am. J.*, 2002, **66**, 1930–1946.
48. L. A. Sherrod, G. A. Peterson, D. G. Westfall and L. R. Ahuja, *Soil Sci. Soc. Am. J.*, 2005, **69**, 1600–1608.
49. J. K. O'Dea, C. A. Jones, C. A. Zabinski, P. R. Miller and I. N. Keren, *Nutr. Cycl. Agroecosyst.*, 2015, **102**, 179–194.
50. C. Tonitto, M. B. David and L. E. Drinkwater, *Agric., Ecosyst. Environ.*, 2006, **112**, 58–72.
51. S. Daryanto, B. Fu, L. Wang, P. Jacinthe and W. Zhao, *Earth-Sci. Rev.*, 2018, **185**, 357–373.
52. R. L. Lemke, A. J. VandenBygaart, C. A. Campbell, G. P. Lafond and B. Grant, *Agric., Ecosyst. Environ.*, 2010, **135**, 42–51.
53. H. Akiyama, X. Yan and K. Yagi, *Global Change Biol.*, 2010, **16**, 1837–1846.
54. D. S. Reay, E. A. Davidson, K. A. Smith, P. Smith, J. M. Melillo, F. Dentener and P. J. Crutzen, *Nat. Clim. Change*, 2012, **2**, 410–416.
55. M. A. Waqas, Y. Li, R. Lal, X. Wang, S. Shi, Y. Zhu, J. Li, M. Xu, Y. Wan, X. Qin, Q. Gao and S. Liu, *Land Degrad. Dev.*, 2020, **31**, 1926–1941.
56. R. Ryals, M. Kaiser, M. S. Torn, A. A. Berhe and W. L. Silver, *Impacts of Organic Matter Amendments on Carbon and Nitrogen Dynamics in Grassland Soils*, 2014.
57. F. Ren, N. Sun, M. Xu, X. Zhang, L. Wu and M. Xu, *Soil Tillage Res.*, 2019, **194**, 104291.
58. L. Zavattaro, L. Bechini, C. Grignani, F. K. van Evert, J. Mallast, H. Spiegel, T. Sandén, A. Pecio, J. V. Giráldez Cervera, G. Guzmán, K. Vanderlinden, T. D'Hose, G. Ruysschaert, ten Berge and F. M. Hein, *Eur. J. Agron.*, 2017, **90**, 127–138.

59. M. Zhou, B. Zhu, S. Wang, X. Zhu, H. Vereecken and N. Brüggemann, *Global Change Biol.*, 2017, **23**, 4068–4083.
60. D. Nayak, E. Saetnan, K. Cheng, W. Wang, F. Koslowski, Y. Cheng, W. Y. Zhu, J. Wang, J. Liu, D. Moran, X. Yan, L. Cardenas, J. Newbold, G. Pan, Y. Lu and P. Smith, *Agric., Ecosyst. Environ.*, 2015, **209**, 108–124.
61. W. Wang and J. Fang, *Global Planet. Change*, 2009, **67**, 20–28.
62. R. T. Conant, in *Recarbonization of the Biosphere: Ecosystems and the Global Carbon Cycle*, ed. R. Lal, K. Lorenz, R. F. Hüttl, B. U. Schneider and J. von Braun, Springer Netherlands, Dordrecht, 2012, p. 275–302.
63. J. M. O. Scurlock and D. O. Hall, *Global Change Biol.*, 1998, **4**, 229–233.
64. R. T. Conant, K. Paustian and E. T. Elliott, *Ecol. Appl.*, 2001, **11**, 343–355.
65. V. Allard, J. Soussana, R. Falcimagne, P. Berbigier, J. M. Bonnefond, E. Ceschia, P. D'hour, C. Hénault, P. Laville, C. Martin and C. Pinarès-Patino, *Agric., Ecosyst. Environ*, 2007, **121**, 47–58.
66. R. S. Reid, P. K. Thornton, G. J. McCrabb, R. L. Kruska, F. Atieno and P. G. Jones, in *Tropical Agriculture in Transition — Opportunities for Mitigating Greenhouse Gas Emissions?*, ed. R. Wassmann and P. L. G. Vlek, Springer, Dordrecht, 2004.
67. R. F. Follett and D. A. Reed, *Rangeland Ecol. Manage.*, 2010, **63**, 4–15.
68. M. Abdalla, A. Hastings, D. R. Chadwick, D. L. Jones, C. D. Evans, M. B. Jones, R. M. Rees and P. Smith, *Agric., Ecosyst. Environ.*, 2018, **253**, 62–81.
69. J. F. Soussana, V. Allard, K. Pilegaard, P. Ambus, C. Amman, C. Campbell, E. Ceschia, J. Clifton-Brown, S. Czobel, R. Domingues, C. Flechard, J. Fuhrer, A. Hensen, L. Horvath, M. Jones, G. Kasper, C. Martin, Z. Nagy, A. Neftel, A. Raschi, S. Baronti, R. M. Rees, U. Skiba, P. Stefani, G. Manca, M. Sutton, Z. Tuba and R. Valentini, *Agric., Ecosyst. Environ.*, 2007, **121**, 121–134.
70. P. Smith, *Global Change Biol.*, 2014, **20**, 2708–2711.
71. A. E. Johnston, P. R. Poulton and K. Coleman, *Adv. Agron*, 2009 **101**, 1–57.
72. R. T. Conant, C. E. P. Cerri, B. B. Osborne and K. Paustian, *Ecol. Appl.*, 2017, **27**, 662–668.
73. U. M. Sainju, B. L. Allen, A. W. Lenssen and R. P. Ghimire, *Field Crops Res.*, 2017, **210**, 183–191.
74. D. P. Rasse, C. Rumpel and M. Dignac, *Plant Soil*, 2005, **269**, 341–356.
75. F. Ferchaud, G. Vitte and B. Mary, *GCB Bioenergy*, 2016, **8**, 290–306.
76. A. Ledo, R. Heathcote, A. Hastings, P. Smith and J. Hillier, *Environ. Modell. Software*, 2018, **102**, 292–305.
77. E. Toensmeier, R. Ferguson and M. Mehra, *PLoS One*, 2020, 15.
78. S. L. Cosentino, V. Copani, G. Scalici, D. Scordia and G. Testa, *BioEnergy Res.*, 2015, **8**, 1538–1547.
79. A. Don, J. Schumacher and A. Freibauer, *Global Change Biol.*, 2011, **17**, 1658–1670.
80. J. L. Vicente-Vicente, R. García-Ruiz, R. Francaviglia, E. Aguilera and P. Smith, *Agric., Ecosyst. Environ.*, 2016, **235**, 204–214.

81. P. Nair, *An Introduction to Agroforestry*, Kluwer Academic Publishers, Dordrecht, The Netherlands, 1993.
82. N. H. Batjes, *Environ. Dev. Sustainability*, 2004, **6**, 133–143.
83. P. K. R. Nair, *Agrofor. Syst.*, 2012, **86**, 243–253.
84. K. Lorenz and R. Lal, *Agron. Sustainable Dev.*, 2014, **34**, 443–454.
85. P. K. Mutuo, G. Cadisch, A. Albrecht, C. A. Palm and L. Verchot, *Nutr. Cycl. Agroecosyst.*, 2005, **71**, 43–54.
86. A. De Stefano and M. G. Jacobson, *Agrofor. Syst*, 2018, **92**, 285–299.
87. R. J. Zomer, H. Neufeldt, J. Xu, A. Ahrends, D. Bossio, A. Trabucco, M. van Noordwijk and M. Wang, *Sci. Rep.*, 2016, **6**, 29987.
88. L. B. Guo and R. M. Gifford, *Global Change Biol.*, 2002, **8**, 345–360.
89. J. Leifeld and L. Menichetti, *Nat. Commun.*, 2018, **9**, 1071.
90. UNESCO and UN-Water, *United Nations World Water Development Report 2020: Water and Climate Change*, UNESCO, Paris, 2020.
91. X. Yan, K. Yagi, H. Akiyama and H. Akimoto, *Global Change Biol.*, 2005, **11**, 1131–1141.
92. Z. C. Cai, H. Tsuruta and K. Minami, *J. Geophys. Res.*, 2000, **105**, 17231–17242.
93. H. Xu, Z. C. Cai, Z. J. Jia and H. Tsuruta, *Nutr. Cycl. Agroecosyst.*, 2000, **58**, 327–332.
94. H. Xu, Z. C. Cai and H. Tsuruta, *Soil Sci. Soc. Am. J.*, 2003, **67**, 1147–1157.
95. S. Díaz, U. Pascual, M. Stenseke, B. Martín-López, R. T. Watson, Z. Molnár, R. Hill, K. M. A. Chan, I. A. Baste, K. A. Brauman, S. Polasky, A. Church, M. Lonsdale, A. Larigauderie, P. W. Leadley, van Oudenhoven, P. E. Alexander, F. van der Plaat, M. Schröter, S. Lavorel, Y. Aumeeruddy-Thomas, E. Bukvareva, K. Davies, S. Demissew, G. Erpul, P. Failler, C. A. Guerra, C. L. Hewitt, H. Keune, S. Lindley and Y. Shirayama, *Science*, 2018, **359**, 270–272.
96. United Nations, *The United Nations Sustainable Development Goals*, UN, New York, 2018.
97. European Commission, Communication from the Commission to the Council, the European Parliament, the European Economic and Social Committee and the Committee of the Regions, Thematic Strategy for Soil Protection, Eur. Comm., Brussels, BE, 2006.
98. Soil Science Society of America, (SSSA), *Soils and Products we use. Soils Overview*, SSSA, Madison, WI, 2015.
99. R. Lal, *Food Energy Secur.*, 2016, **5**, 212–222.
100. S. D. Keesstra, J. Bouma, J. Wallinga, P. Tittonell, P. Smith, A. Cerdà, L. Montanarella, J. N. Quinton, Y. Pachepsky, W. H. van der Putten, R. D. Bardgett, S. Moolenaar, G. Mol, B. Jansen and L. O. Fresco, *Soil*, 2016, **2**, 111–128.
101. R. Lal, *Science*, 2004, **304**, 1623–1627.
102. S. H. Doerr, R. A. Shakesby and R. P. D. Walsh, *Earth-Sci. Rev.*, 2000, **51**, 33–65.
103. N. H. Batjes and W. G. Sombroek, *Global Change Biol.*, 1997, **3**, 161–173.

104. S. Luyssaert, M. Jammet, P. C. Stoy, S. Estel, J. Pongratz, E. Ceschia, G. Churkina, A. Don, K. Erb, M. Ferlicoq, B. Gielen, T. Grünwald, R. A. Houghton, K. Klumpp, A. Knohl, T. Kolb, T. Kuemmerle, T. Laurila, A. Lohila, D. Loustau, M. J. McGrath, P. Meyfroidt, E. J. Moors, K. Naudts, K. Novick, J. Otto, K. Pilegaard, C. A. Pio, S. Rambal, C. Rebmann, J. Ryder, A. E. Suyker, A. Varlagin, M. Wattenbach and A. J. Dolman, *Nat. Clim. Change*, 2014, **4**, 389–393.
105. C. S. T. Daughtry, P. C. Doraiswamy, E. R. Hunt, A. J. Stern, J. E. McMurtrey and J. H. Prueger, *Soil Tillage Res.*, 2006, **91**, 101–108.
106. A. Pacheco and H. McNairn, *Remote Sens. Environ.*, 2010, **114**, 2219–2228.
107. G. De Simon, G. Alberti, G. Delle Vedove, A. Peressotti, A. Zaldei and F. Miglietta, *Biogeosciences*, 2013, **10**, 5545–5553.
108. United States Department of Agriculture, (USDA), *Soil Health Nuggets*, USDA, Washington, DC, 2018.
109. A. G. Stone, S. J. Scheurell and H. M. Darby, in *Soil Organic Matter in Sustainable Agriculture*, ed. F. Magdoff and R. R. Weil, CRC Press LLC, Boca Raton, FL, 2004, p. 131–77.
110. F. Hassard, C. L. Gwyther, K. Farkas, A. Andrews, V. Jones, B. Cox, H. Brett, D. L. Jones, J. E. McDonald and S. K. Malham, *Front. Microbiol.*, 2016, 7, 1692.
111. Y. A. Pachepsky, *Int. Agrophys.*, 2008, **22**(66), 61.
112. W. Zhao, X. Liu, Q. Huang and P. Cai, *Chemosphere*, 2015, **119**, 52–58.
113. M. T. Callahan, S. A. Micallef and R. L. Buchanan, *J. Food Prot*, 2016, **80**, 189–197.
114. R. Munang, J. Andrews, K. Alverson and D. Mebratu, *Environ.: Sci. Policy Sustainable Dev.*, 2014, **56**, 18–24.
115. Food and Agriculture Organization of the United Nations, *(FAO) and Intergovernmental Technical Panel on Soils, (ITPS), Status of the World's Soil Resources. Main Report*, FAO, ITPS, Rome, 2015.
116. E. C. Brevik, A. Cerdà, J. Mataix-Solera, L. Pereg, J. N. Quinton, J. Six and K. Van Oost, *Soil*, 2015, 117–129.
117. P. Smith, M. F. Cotrufo, C. Rumpel, K. Paustian, P. J. Kuikman, J. A. Elliott, R. McDowell, R. I. Griffiths, S. Asakawa, M. Bustamante, J. I. House, J. Sobocká, R. Harper, G. Pan, P. C. West, J. S. Gerber, J. M. Clark, T. Adhya, R. J. Scholes and M. C. Scholes, *Soil*, 2015, **1**, 665–685.
118. P. Smith, J. I. House, B. Mercedes, J. Sobocká, H. Richard, P. Genxing, P. C. West, J. M. Clark, A. Tapan, R. Cornelia, P. Keith, K. Peter, C. M. Francesca, J. A. Elliott, M. Richard, R. I. Griffiths, A. Susumu, B. Alberte, A. K. Jain, M. Jeroen and T. A. Pugh, *Global Change Biol.*, 2016, **22**, 1008–1028.
119. R. Lal, *Land Degrad. Dev.*, 2006, **17**, 197–209.
120. G. Pan, P. Smith and W. Pan, *Agric., Ecosyst. Environ.*, 2009, **129**, 344–348.
121. R. Lal, *Land Degrad. Dev.*, 2001, **12**, 519–539.

122. D. Pimentel, C. Harvey, P. Resosudarmo, K. Sinclair, D. Kurz, M. McNair, S. Crist, L. Shpritz, L. Fitton, R. Saffouri and R. Blair, *Science*, 1995, **267**, 1117–1123.
123. M. A. Sutton, E. Nemitz, J. W. Erisman, C. Beier, K. B. Bahl, P. Cellier, W. de Vries, F. Cotrufo, U. Skiba, C. Di Marco, S. Jones, P. Laville, J. F. Soussana, B. Loubet, M. Twigg, D. Famulari, J. Whitehead, M. W. Gallagher, A. Neftel, C. R. Flechard, B. Herrmann, P. L. Calanca, J. K. Schjoerring, U. Daemmgen, L. Horvath, Y. S. Tang, B. A. Emmett, A. Tietema, J. Peñuelas, M. Kesik, N. Brueggemann, K. Pilegaard, T. Vesala, C. L. Campbell, J. E. Olesen, U. Dragosits, M. R. Theobald, P. Levy, D. C. Mobbs, R. Milne, N. Viovy, N. Vuichard, J. U. Smith, P. Smith, P. Bergamaschi, D. Fowler and S. Reis, *Environ. Pollut.*, 2007, **150**, 125–139.
124. S. Fuss, W. F. Lamb, M. W. Callaghan, J. Hilaire, F. Creutzig, T. Amann, T. Beringer, W. de Oliveira Garcia, J. Hartmann, T. Khanna, G. Luderer, G. F. Nemet, J. Rogelj, P. Smith, J. L. V. Vicente, J. Wilcox, M. M. Z. Dominguez and J. C. Minx, *Environ. Rese. Lett.*, 2018, **13**, 063002.
125. Y. Liao, W. L. Wu, F. Q. Meng, P. Smith and R. Lal, *Biogeosciences*, 2015, **12**, 1403–1413.
126. E. Dominati, M. Patterson and A. Mackay, *Ecol. Econ.*, 2010, **69**, 1858–1868.
127. A. Hewitt, E. Dominati, T. Webb and T. Cuthill, *Geoderma*, 2015, **241–242**, 107–114.
128. https://coolfarmtool.org/, (accessed 01/09/ 2020).
129. J. Hillier, C. Walter, D. Malin, T. Garcia-Suarez, L. Mila-i-Canals and P. Smith, *Environ. Modell. Software*, 2011, **26**, 1070–1078.
130. http://cometfarm.nrel.colostate.edu, (accessed 01/09/ 2020).
131. P. Smith, D. Martino, Z. Cai, D. Gwary, H. Janzen, P. Kumar, B. McCarl, S. Ogle, F. O'Mara, C. Rice, B. Scholes, O. Sirotenko, M. Howden, T. McAllister, G. Pan, V. Romanenkov, U. Schneider and S. Towprayoon, *Agric., Ecosyst. Environ.*, 2007, **118**, 6–28.
132. J. Six, R. T. Conant, E. A. Paul and K. Paustian, *Plant Soil*, 2002, **241**, 155–176.
133. J. Hassink, *Plant Soil*, 1997, **191**, 77–87.
134. H. Chung, J. H. Grove and J. Six, *Soil Sci. Soc. Am. J.*, 2008, **72**, 1132–1139.
135. S. M. O'Rourke, D. A. Angers, N. M. Holden and A. B. McBratney, *Global Change Biol.*, 2015, **21**, 3561–3574.
136. C. E. Stewart, A. F. Plante, K. Paustian, R. T. Conant and J. Six, *Soil Sci. Soc. Am. J.*, 2008, **72**, 379–392.
137. C. E. Stewart, K. Paustian, R. T. Conant, A. F. Plante and J. Six, *Biogeochemistry*, 2007, **86**, 19–31.
138. S. Gulde, H. Chung, W. Amelung, C. Chang and J. Six, *Soil Sci. Soc. Am. J.*, 2008, **72**, 605–612.
139. T. O. West and J. Six, *Clim. Change*, 2007, **80**, 25–41.

140. M. F. Cotrufo, M. D. Wallenstein, C. M. Boot, K. Denef and E. Paul, *Glob Change Biol*, 2013, **19**, 988–995.
141. C. Rumpel and I. Kögel-Knabner, *Plant Soil*, 2011, **338**, 143–158.
142. K. Fujisaki, L. Chapuis-Lardy, A. Albrecht, T. Razafimbelo, J. Chotte and T. Chevallier, *Geoderma*, 2018, **313**, 41–51.
143. W. Feng, A. F. Plante and J. Six, *Biogeochemistry*, 2013, **112**, 81–93.
144. IPCC, *Intergovernmental Panel on Climate Change: Guidelines for national greenhouse gas inventories*, 2006.
145. S. Fuss, J. G. Canadell, G. P. Peters, M. Tavoni, R. M. Andrew, P. Ciais, R. B. Jackson, C. D. Jones, F. Kraxner, N. Nakicenovic, C. Le Quéré, M. R. Raupach, A. Sharifi, P. Smith and Y. Yamagata, *Nat. Clim. Change*, 2014, **4**, 850–853.
146. P. Smith, J. Soussana, D. Angers, L. Schipper, C. Chenu, D. P. Rasse, N. H. Batjes, F. van Egmond, S. McNeill, M. Kuhnert, C. Arias-Navarro, J. E. Olesen, N. Chirinda, D. Fornara, E. Wollenberg, J. Álvaro-Fuentes, A. Sanz-Cobena and K. Klumpp, *Global Change Biol.*, 2020, **26**, 219–241.
147. Z. Tan, C. S. K. Lin, X. Ji and T. J. Rainey, *Applied Soil Ecology*, 2017, **116**, 1–11.
148. S. Meyer, B. Glaser and P. Quicker, *Environ. Sci. Technol.*, 2011, **45**, 9473–9483.
149. R. Brown, B. del Campo, A. A. Boateng, M. Garcia-Perez and O. Mašek, in *Biochar for Environmental Management—Science, Technology and Implementation*, ed. J. Lehmann and S. Joseph, earthscan by Routledge, London, New York, 2015, p. 39.
150. A. A. Boateng, M. Garcia-Perez, O. Mašek, R. Brown and B. del Campo, in *Biochar for Environmental Management—Science, Technology and Implementation*, ed. J. Lehmann and S. Joseph, earthscan by Routledge, London, New York, 2015, p. 63.
151. V. E. A. Dornburg, *Energy Environ. Sci.*, 2010, **3**, 258.
152. M. Tripathi, J. N. Sahu and P. Ganesan, *Renewable Sustainable Energy Rev.*, 2016, **55**, 467–481.
153. S. Shackley, A. Clare, S. Joseph, B. A. McCarl and H. Schmidt, in *Biochar for Environmental Management—Science, Technology and Implementation*, ed. J. Lehmann and S. Joseph, earthscan by Routledge, London, New York, 2015, p. 813.
154. S. Joseph, M. L. Anh, A. Clare and S. Shackley, in *Biochar for Environmental Management—Science, Technology and Implementation*, ed. J. Lehmann and S. Joseph, earthscan by Routledge, London, New York, 2015, p. 853.
155. B. Glaser, M. Parr, C. Braun and G. Kopolo, *Nat. Geosci.*, 2009, **2**, 2.
156. V. T. Holliday, *Soils in Archaeological Research*, Oxford University Press, New York, 2004.
157. J. R. Olarieta, R. Padrò, G. Masip, R. Rodríguez-Ochoa and E. Tello, *Agric., Ecosyst. Environ.*, 2011, **140**, 27–33.
158. Z. H. Cao, J. L. Ding, Z. Y. Hu, H. Knicker, I. Kögel-Knabner, L. Z. Yang, R. Yin, X. G. Lin and Y. H. Dong, *Naturwissenschaften*, 2006, **93**, 232–236.

159. L. Hu, Z. Chao, M. Gu, F. Li, L. Chen, B. Liu, X. Li, Z. Huang, Y. Li, B. Xing and J. Dai, *J. Archaeol. Sci.*, 2013, **40**, 72–78.
160. D. Sheil, I. Basuki, L. German, T. W. Kuyper, G. Limberg, R. K. Puri, B. Sellato, M. Van Noordwijk and E. Wollenberg, *Forests*, 2012, **3**, 207–229.
161. M. Ogawa and Y. Okimori, *Aust. J. Soil Res.*, 2010, **48**, 489–500.
162. P. Coutts, *The Prehistory of Victoria: A Review*, Aboriginal Affairs Victoria, Melbourne, 1976.
163. J. Fairhead and M. Leach, in *Amazonian Dark Earths: Explorations in Space and Time*, ed. B. Glaser and W. I. Woods, Springer Verlag, Heidelberg, Germany, 2009, p. 265–278.
164. J. Liebig, *Liebig's Complete Works on Chemistry*, Philadelphia, 1852, Peterson.
165. B. Glaser, J. Lehmann and W. Zech, *Biol. Fertility Soils*, 2002, **35**, 219–230.
166. J. Lehmann and S. Joseph, *Biochar for Environmental Management—Science, Technology and Implementation*, earthscan by Routledge, London, New York, 2015.
167. J. Lehmann, M. C. Rillig, J. Thies, C. A. Masiello, W. C. Hockaday and D. Crowley, *Soil Biol. Biochem.*, 2011, **43**, 1812–1836.
168. J. W. Lee, B. Hawkins, D. M. Day and D. C. Reicosky, *Energy Environ. Sci.*, 2010, **3**, 1695–1705.
169. J. W. Lee and D. M. Day, in *Advanced Biofuels and Bioproducts*, ed. J. W. Lee, Springer New York, New York, NY, 2013, p. 23–34.
170. K. G. Roberts, B. A. Gloy, S. Joseph, N. R. Scott and J. Lehmann, *Environ. Sci. Technol.*, 2010, **44**, 827–833.
171. D. Woolf, J. E. Amonette, F. Street-Perrott, J. Lehmann and S. Joseph, *Nat. Commun.*, 2010, **1**, 56.
172. A. Mukherjee and R. Lal, *Soil Res.*, 2014, **52**, 217–230.
173. N. Ameloot, P. Maenhout, S. De Neve and S. Sleutel, *Geoderma*, 2016, **267**, 10–16.
174. Y. Yuan, H. Chen, W. Yuan, D. Williams, J. T. Walker and W. Shi, *Soil Biol. Biochem.*, 2017, **113**, 14–25.
175. Q. Liu, Y. Zhang, B. Liu, J. E. Amonette, Z. Lin, G. Liu, P. Ambus and Z. Xie, *Plant Soil*, 2018, **426**, 211–225.
176. N. Borchard, M. Schirrmann, M. L. Cayuela, C. Kammann, N. Wrage-Mönnig, J. M. Estavillo, T. Fuertes-Mendizábal, G. Sigua, K. Spokas, J. A. Ippolito and J. Novak, *Sci. Total Environ.*, 2019, **651**, 2354–2364.
177. E. Verhoeven, E. Pereira, C. Decock, E. Suddick, T. Angst and J. Six, *J. Environ. Qual.*, 2017, **46**, 237–246.
178. S. Castaldi, M. Riondino, S. Baronti, F. R. Esposito, R. Marzaioli, F. A. Rutigliano, F. P. Vaccari and F. Miglietta, *Chemosphere*, 2011, **85**, 1464–1471.
179. N. Hagemann, J. Harter, R. Kaldamukova, I. Guzman-Bustamante, R. Ruser, S. Graeff, A. Kappler and S. Behrens, *GCB Bioenergy*, 2017, **9**, 953–964.

180. X. Qin, Y. Li, H. Wang, C. Liu, J. Li, Y. Wan, Q. Gao, F. Fan and Y. Liao, *Sci. Total Environ.*, 2016, **569–570**, 1390–1401.
181. A. Zhang, R. Bian, Q. Hussain, L. Li, G. Pan, J. Zheng, X. Zhang and J. Zheng, *Agric., Ecosyst. Environ.*, 2013, **173**, 37–45.
182. A. Zhang, L. Cui, G. Pan, L. Li, Q. Hussain, X. Zhang, J. Zheng and D. Crowley, *Agric., Ecosyst. Environ.*, 2010, **139**, 469–475.
183. M. Shaukat, K. Samoy-Pascual, Maas, D. V. L. Ellen and A. Ahmad, *J. Environ. Manage.*, 2019, **248**, 109242.
184. S. Wang, J. Shan, Y. Xia, Q. Tang, L. Xia, J. Lin and X. Yan, *Sci. Total Environ.*, 2017, **593–594**, 347–356.
185. Y. Sui, J. Gao, C. Liu, W. Zhang, Y. Lan, S. Li, J. Meng, Z. Xu and L. Tang, *Sci. Total Environ.*, 2016, **544**, 203–210.
186. H. Chen, C. Yin, X. Fan, M. Ye, H. Peng, T. Li, Y. Zhao, S. A. Wakelin, G. Chu and Y. Liang, *Sci. Total Environ.*, 2019, **694**, 133658.
187. M. Senbayram, E. P. Saygan, R. Chen, S. Aydemir, C. Kaya, D. Wu and E. Bladogatskaya, *Sci. Total Environ.*, 2019, **660**, 69–79.
188. Y. Fang, B. Singh, B. P. Singh and E. Krull, *Eur. J. Soil Sci.*, 2014, **65**, 60.
189. Y. Fang, B. P. Singh and B. Singh, *Temperature Sensitivity of Biochar and Native Carbon Mineralisation in Biochar-amended Soils*, 2014.
190. M. Zimmermann, M. I. Bird, C. Wurster, G. Saiz, I. Goodrick, J. Barta, P. Capek, H. Santruckova and R. Smernik, *Global Change Biol.*, 2012, **18**, 3306–3316.
191. Y. Sheng, Y. Zhan and L. Zhu, *Sci. Total Environ.*, 2016, **572**, 129–137.
192. C. Scheer, P. R. Grace, D. W. Rowlings, S. Kimber and L. Van Zwieten, *Plant Soil*, 2011, **345**, 47–58.
193. G. Yoo, Y. O. Lee, T. J. Won, J. G. Hyun and W. Ding, *Sci. Total Environ.*, 2018, **626**, 603–611.
194. T. J. Rose, B. Keen, S. G. Morris, P. Quin, J. Rust, L. Kearney, S. Kimber and L. Van Zwieten, *Agric., Ecosyst. Environ.*, 2016, **228**, 1–8.
195. Z. M. Lan, C. R. Chen, M. Rezaei Rashti, H. Yang and D. K. Zhang, *Geoderma*, 2019, **337**, 880–892.
196. E. Verhoeven and J. Six, *Agric., Ecosyst. Environ.*, 2014, **191**, 27–38.
197. T. F. Abbruzzini, C. A. Davies, F. H. Toledo and C. E. P. Cerri, *J. Environ. Manage.*, 2019, **252**, 109638.
198. J. Wang, Z. Xiong and Y. Kuzyakov, *GCB Bioenergy*, 2016, **8**, 512–523.
199. S. Liu, Y. Zhang, Y. Zong, Z. Hu, S. Wu, J. Zhou, Y. Jin and J. Zou, *GCB Bioenergy*, 2016, **8**, 392–406.
200. L. Genesio, F. Miglietta, E. Lugato, S. Baronti, M. Pieri and F. P. Vaccari, *Environ. Res. Lett.*, 2012, 7, 014025.
201. C. Werner, H. Schmidt, D. Gerten, W. Lucht and C. Kammann, *Environ. Res. Lett.*, 2018, **13**, 044036.
202. M. Rondon, J. A. Ramirez and J. Lehmann, *Charcoal Additions Reduce Net Emissions of Greenhouse Gases to the Atmosphere*, Univ. of Delaware, Baltimore, 2005.
203. A. Taghizadeh-Toosi, T. J. Clough, L. M. Condron, R. R. Sherlock, C. R. Anderson and R. A. Craigie, *J. Environ. Qual.*, 2011, **40**, 468–476.

204. R. Renner, *Environ Sci Technol*, 2007, **41**, 5932–5933.
205. T. J. Clough, J. E. Bertram, J. L. Ray, L. M. Condron, M. O'Callaghan, R. R. Sherlock and N. S. Wells, *Soil Sci. Soc. Am. J.*, 2010, **74**, 852–860.
206. J. Chen, H. Kim and G. Yoo, *PLOS One*, 2015, **10**, e0126841.
207. T. E. Angst, J. Six, D. S. Reay and S. P. Sohi, *Agric., Ecosyst. Environ.*, 2014, **191**, 17–26.
208. R. B. Fidel, D. A. Laird and T. B. Parkin, *Soil Syst.*, 2019, **3**, 8.
209. S. Bruun, S. L. Harmer, G. Bekiaris, W. Christel, L. Zuin, Y. Hu, L. S. Jensen and E. Lombi, *Chemosphere*, 2017, **169**, 377–386.
210. X. Tan, S. Liu, Y. Liu, Y. Gu, G. Zeng, X. Hu, X. Wang, S. Liu and L. Jiang, *Bioresour. Technol.*, 2017, **227**, 359–372.
211. P. G. Oguntunde, B. J. Abiodun, A. E. Ajayi and N. van de Giesen, *Z. Pflanzenernähr. Bodenk.*, 2008, **171**, 591–596.
212. A. Ribas, S. Mattana, R. Llurba, H. Debouk, M. T. Sebastià and X. Domene, *Sci. Total Environ.*, 2019, **685**, 1075–1086.
213. H. Zhou, D. Zhang, P. Wang, X. Liu, K. Cheng, L. Li, J. Zheng, X. Zhang, J. Zheng, D. Crowley, L. van Zwieten and G. Pan, *Agric., Ecosyst. Environ.*, 2017, **239**, 80–89.
214. Y. Xiang, Q. Deng, H. Duan and Y. Guo, *GCB Bioenergy*, 2017, **9**, 1563–1572.
215. N. Hagemann, S. Joseph, H. Schmidt, C. I. Kammann, J. Harter, T. Borch, R. B. Young, K. Varga, S. Taherymoosavi, K. W. Elliott, A. McKenna, M. Albu, C. Mayrhofer, M. Obst, P. Conte, A. Dieguez-Alonso, S. Orsetti, E. Subdiaga, S. Behrens and A. Kappler, *Nat. Commun.*, 2017, **8**, 1089.
216. C. I. Kammann, H. Schmidt, N. Messerschmidt, S. Linsel, D. Steffens, C. Müller, H. Koyro, P. Conte and S. Joseph, *Sci. Rep.*, 2015, **5**, 11080.
217. C. Liang, X. Zhu, S. Fu, A. Méndez, G. Gascó and J. Paz-Ferreiro, *Environ. Res. Lett.*, 2014, **9**, 064013.
218. D. A. Roberts, N. A. Paul, S. A. Dworjanyn, M. I. Bird and R. de Nys, *Sci. Rep.*, 2015, **5**, 9665.
219. S. Jeffery, D. Abalos, M. Prodana, A. C. Bastos, J. W. van Groenigen, B. A. Hungate and F. Verheijen, *Environ. Res. Lett.*, 2017, **12**, 053001.
220. H. Schmidt, B. H. Pandit, G. Cornelissen and C. I. Kammann, *Land Degrad. Develop.*, 2017, **28**, 2324–2342.
221. X. Liu, J. Zhou, Z. Chi, J. Zheng, L. Li, X. Zhang, J. Zheng, K. Cheng, R. Bian and G. Pan, *Catena*, 2019, **179**, 20–28.
222. A. M. Bass, M. I. Bird, G. Kay and B. Muirhead, *Sci. Total Environ.*, 2016, **550**, 459–470.
223. X. Liu, P. Mao, L. Li and J. Ma, *Sci. Total Environ.*, 2019, **656**, 969–976.
224. M. Hussain, M. Farooq, A. Nawaz, A. Al-Sadi, Z. M. Solaiman, S. S. Alghamdi, U. Ammara, Y. S. Ok and K. H. M. Siddique, *J. Soils Sediments*, 2017, **17**, 685–716.

225. S. Mandal, B. Sarkar, N. Bolan, J. Novak, Y. S. Ok, L. Van Zwieten, B. P. Singh, M. B. Kirkham, G. Choppala, K. Spokas and R. Naidu, *Crit. Rev. Environ. Sci. Technol.*, 2016, **46**, 1367–1401.
226. X. Peng, Y. Deng, Y. Peng and K. Yue, *Sci. Total Environ.*, 2018, **616–617**, 970–977.
227. P. Godlewska, H. P. Schmidt, Y. S. Ok and P. Oleszczuk, *Bioresour. Technol*, 2017, **246**, 193–202.
228. M. Ahmad, A. U. Rajapaksha, J. E. Lim, M. Zhang, N. Bolan, D. Mohan, M. Vithanage, S. S. Lee and Y. S. Ok, *Chemosphere*, 2014, **99**, 19–33.
229. G. Cornelissen, N. R. Pandit, P. Taylor, B. H. Pandit, M. Sparrevik and H. P. Schmidt, *PLOS One*, 2016, **11**, e0154617.
230. S. Elkhalifa, T. Al-Ansari, H. R. Mackey and G. McKay, *Resour. Conserv. Recycling*, 2019, **144**, 310–320.
231. F. G. A. Verheijen, S. Jeffery, M. van der Velde, V. Penížek, M. Beland, A. C. Bastos and J. J. Keizer, *Environ. Res. Lett.*, 2013, **8**, 044008.
232. S. Meyer, R. M. Bright, D. Fischer, H. Schulz and B. Glaser, *Environ. Sci. Technol.*, 2012, **46**, 12726–12734.
233. E. Bozzi, L. Genesio, P. Toscano, M. Pieri and F. Miglietta, *Environ. Res. Lett.*, 2015, **10**, 084014.
234. B. Usowicz, J. Lipiec, M. Łukowski, W. Marczewski and J. Usowicz, *Soil Tillage Res.*, 2016, **164**, 45–51.
235. S. Shackley, J. Hammond, J. Gaunt and R. Ibarrola, *Carbon Manage.*, 2011, **2**, 335.
236. C. E. A. Kammann, *J. Environ. Eng. Landsc. Manage.*, 2017, **25**, 114.
237. http://www.cop21paris.org/about/cop21/, (accessed 31/08/ 2020).
238. https://unfccc.int/process-and-meetings/the-paris-agreement/the-paris-agreement, (accessed 01/09/ 2020).
239. B. Minasny, B. P. Malone, A. B. McBratney, D. A. Angers, D. Arrouays, A. Chambers, V. Chaplot, Z. Chen, K. Cheng, B. S. Das, D. J. Field, A. Gimona, C. B. Hedley, S. Y. Hong, B. Mandal, B. P. Marchant, M. Martin, B. G. McConkey, V. L. Mulder, S. O'Rourke, A. C. Richer-de-Forges, I. Odeh, J. Padarian, K. Paustian, G. Pan, L. Poggio, I. Savin, V. Stolbovoy, U. Stockmann, Y. Sulaeman, C. Tsui, T. Vågen, B. van Wesemael and L. Winowiecki, *Geoderma*, 2017, **292**, 59–86.
240. M. Corbeels, R. Cardinael, K. Naudin, H. Guibert and E. Torquebiau, *Soil Tillage Res.*, 2019, **188**, 16–26.
241. R. Francaviglia, C. Di Bene, R. Farina, L. Salvati and J. Vicente-Vicente, *Mitigation Adapt. Strategies Global Change*, 2019, **24**, 795–818.
242. J. W. van Groenigen, C. van Kessel, B. A. Hungate, O. Oenema, D. S. Powlson and K. J. van Groenigen, *Environ. Sci. Technol.*, 2017, **51**, 4738–4739.
243. P. Poulton, J. Johnston, A. Macdonald, R. White and D. Powlson, *Global Change Biol.*, 2018, **24**, 2563–2584.
244. P. C. Baveye, J. Berthelin, D. Tessier and G. Lemaire, *Geoderma*, 2018, **309**, 118–123.

245. https://www.4p1000.org/4-1000-initiative-few-words, (accessed 31/08/2020).
246. A. Chabbi, J. Lehmann, P. Ciais, H. W. Loescher, M. F. Cotrufo, A. Don, M. SanClements, L. Schipper, J. Six, P. Smith and C. Rumpel, *Nat. Clim. Change*, 2017, 7, 307–309.
247. D. Woolf, J. Lehmann, E. M. Fisher and L. T. Angenent, *Environ. Sci. Technol.*, 2014, **48**, 6492–6499.

CHAPTER 9

Afforestation/Reforestation (AR)

SOLENE CHIQUIER[a,b] AND NIALL MAC DOWELL*[a,b]

[a] Centre for Environmental Policy, Imperial College London, Exhibition Road, London SW7 1NA, UK; [b] Centre for Process Systems Engineering, Imperial College London, Exhibition Road, London SW7 2AZ, UK
*Email: niall@imperial.ac.uk

This chapter defines afforestation/reforestation (AR) and provides insights into its role and value for achieving negative emissions. It contains a synthesis of the current research knowledge and gaps in the most up-to-date literature on AR, and a complete analysis of AR including forest growth, biogenic carbon sequestration, risk-adjusted carbon sequestration and forestry operations. More specifically, GHG balance, cost and water footprint analyses are carried out at the sub-regional level in five archetypal regions (Brazil, China, European Union, India and United States) over a 200-year time-period.

9.1 Introduction

9.1.1 Definition and Scope of the Study

Afforestation/reforestation (AR) was one of the first carbon dioxide removal (CDR) options proposed and discussed, concretely, by the international climate community. It is also one of the most well-established CDR options so far. AR is defined by the IPCC as the "direct human-induced conversion of non-forested land to forested land through planting, seeding and/or the human-induced promotion of natural seed sources".[1] Specifically,

Energy and Environment Series No. 31
Greenhouse Gas Removal Technologies
Edited by Mai Bui and Niall Mac Dowell

Published by the Royal Society of Chemistry, www.rsc.org

afforestation refers to the conversion of land that has not been afforested in recent history, commonly for a period of at least 50 years, whereas reforestation refers to the conversion of land that was previously deforested.[1]

Due to the simplicity and similarity of afforestation and reforestation definitions – distinguished only by the period during which the land is not forested – AR is often jointly categorised and dealt with. This has led to considerable discussion and confusion on their individual detailed definitions, implementations and carbon accounting methodologies. Key issues relate particularly to the lack of clear definitions around land use, land cover, and forestry activities.[1]

This chapter aims at clarifying the concept of AR by providing the following extended definition: In complement to the IPCC definition (*i.e.*, land cover-oriented definition), AR is defined explicitly as a land-based CDR option, which aims at maximising biogenic carbon sequestration potential, while creating and/or enhancing local and natural forested ecosystems (*i.e.*, land use-oriented definition), and this with reduced or minimum human intervention (*i.e.*, forestry activities-oriented definition).

9.1.2 Accounting for CO_2 Sequestration in AR Projects

The estimation of forests carbon (C) (and CO_2) stock is a key element in the evaluation of, and the discussion on, AR CO_2 sequestration potential. Forests C stock is usually broken down into three main components: tree/biomass (*i.e.*, above-ground and below-ground biomass), dead organic matter (*i.e.*, litter and dead wood) and soil C stocks. As soil CO_2 sequestration has been considered as a CDR option itself, and is discussed as such in Chapter 8, it will not be investigated further in this chapter.

A fourth component, the harvested wood product C stock, can also be considered, but the way it should be accounted for and the scope within which it should be allocated to, is more uncertain. There is currently a large range of methods (*i.e.*, bottom-up *versus* top-down) to evaluate each of these C stocks, and their appropriateness depends on the scale of the study, whether at the single-stand scale, national scale (*e.g.*, for national inventories) or global scale.[2]

Bottom-up Methods

At the single-stand scale, the bottom-up estimation of biomass C stock in existing forests typically requires the measurement of standard tree mensuration parameters (*e.g.*, diameter at breast height (DBH), tree or merchantable stem height) and tree density, using appropriate surveying and sampling protocols. Stem volume V_m (*i.e.*, the part of the tree that accounts for the merchantable biomass) is then evaluated with the use of standard tree mensuration tables or formulae. Appropriate empirical relationships are then used to estimate other biomass components volumes V_{nm} – including branches, foliage, stump and roots (*i.e.*, non-merchantable biomass) – in order to give a whole-tree biomass volume estimate. Finally, total tree C stock C_{bio} is evaluated using wood density D and carbon content C_c as follows:

$$C_{bio} = (V_m + V_{nm}) * D * C_c$$

At the national scale, forest growth models, based on standard tree mensuration tables, formulae and empirical relationships, can be directly applied to stand characteristics rather than tree characteristics, in order to estimate the biomass carbon stock in existing forests or in forests in creation (*i.e.*, afforestation or reforestation). Such models, like the Forest Research CSORT or BSORT ones in the UK, the USDA FVS one in the USA or the TIPSY one in British Columbia, Canada require tree species compositions, age classes, tree growth rates (*i.e.*, yield classes) and stand management regimes to evaluate the total biomass volume and its associated carbon stock.[2–5]

Although the estimation of stem volume and carbon stock is well-established, the estimation of the total biomass volume and its associated carbon stock is more uncertain. This is because the relationship between stem volume and other biomass components relies on limited datasets, owing to practical difficulties, specifically related to field measurements. Alternatively, the use of biomass expansion factors *BEFs* (*i.e.*, the ratio of above-ground biomass to merchantable timber biomass) and root-to-shoot ratios R (*i.e.*, the ratio of the below-ground biomass to the above-ground biomass) can be applied to both individual stands and national inventories. This approach is included in the IPCC Good Practice Guidance for LULUCF[6] and evaluates the total biomass carbon stock C_{bio} from the merchantable stem volume V_m as follows:

$$C_{bio} = V_m * \mathrm{BEF} * (1 + R) * D * C_c$$

The dead organic matter carbon stock is also particularly complicated to evaluate due to similar practical–field measurement-related–limitations. Estimates can be obtained using sampling protocols or surveying at the single-stand scale, or appropriate models, based on complex forest C dynamics, rates of decay and associated emissions factors. For large-scale estimations of the carbon stock in existing forests, the IPCC Good Practice Guidance for LULUCF usually recommends, for simplicity, that the dead organic matter carbon stock is at equilibrium, so that changes are assumed to be zero. In new forests (*i.e.*, afforestation/reforestation), the IPCC Good Practice Guidance for LULUCF provides default values for litter carbon stock, but recognises that these values are broad-scale estimates with considerable uncertainty when applied at the country level or higher.

Top-down Methods

At the global scale, another top-down approach relies on the Net Primary Production (NPP). The NPP is defined as the global CO_2 uptake from the atmosphere during photosynthesis (*i.e.*, Gross Primary Production (GPP)), minus the loss of CO_2 by plants (above and below ground) to provide the energy for biomass growth (*i.e.*, autotrophic respiration (R_A)). This approach is commonly used in IAMs as NPP can be obtained at high spatial and temporal resolutions (*i.e.*, globally and annually/monthly) from remote-sensing datasets.[7] The NPP can then be used with the heterotropic soil respiration (R_H) (*i.e.*, the loss of CO_2 by animals and microbes, mostly in soil) to determine the Net Ecosystem Productivity (NEP) (*i.e.*, the rate of carbon accumulation in the system). This approach is used, typically, in the terrestrial carbon cycle model of the IAM IMAGE 2.[8] Alternatively, the NPP can also be complemented by appropriate decay rates for litter, dead wood and soil. This second approach is used in the Global Forest Model (G4M) model.[9]

However, there are currently limitations in the widespread use of remote-sensing, particularly with passive sensors, as it cannot penetrate dense cloud cover. Frequent in tropic areas for example, this can distort the data acquisition and therefore alter the data analysis.

This chapter presents a detailed bottom-up approach to estimate AR CO_2 sequestration potential at the national and sub-national scales, following the IPCC good practice guidance for LULUCF national GHG Inventories Changes.

9.1.2.1 *AR Multi-dimensions Assessment: GHG, Energy, Cost and Water Balances*

A comprehensive analysis of the whole AR system needs to be undertaken in order to assess AR CO_2 sequestration potential, cost and side-effects (*e.g.*, water use). Such analysis should account for AR high spatially and temporally variability and include forestry operations occurring during the establishment of AR and throughout the forest lifetime, required to maintain forest carbon sinks (*e.g.*, forest management). Yet, it is often unclear, or unmentioned, if such forestry operations are included in IAMs carbon cycle/forest CO_2 sequestration models.

Several energy and GHG emissions studies on timber production and forestry residues extraction from forest systems have been performed in the context of commercial forest plantations and bioenergy/fuel production,[10–12] but they are complicated to compare due to differences in system boundaries, functional units and geographical regions.

Very few cost studies have been reported, essentially just on harvesting operations[13–15] or wood chips production,[16] but none on whole forest and forest residues supply-chain systems (*i.e.*, from the forest establishment to harvesting operations, and including forest residues biomass processing and transport).

Finally, some studies have reviewed and addressed globally the effects of afforestation on water yield and associated potential water stress and scarcities.[17,18] However, most water footprint analysis has been published for agricultural production (*i.e.*, crops for the production of food and bioenergy)[19,20] and scarcely for natural vegetation due to the lack of available field-data.[21,22] The development of remote-sensing over the last decade has given access to global datasets of high resolution (*e.g.*, 1×1 km, 500×500 m, to 250×250 m) on naturally vegetated areas and their geo-physical characteristics, and allowed for an increasing number of studies to be published recently. For instance, Corbari *et al.*[23] proposed the use of remote-sensing datasets to assess the evapotranspiration of natural vegetation in the USA.[23] However, it highlighted that significant differences could be observed between remote-sensing data-based and modelling-based analysis.

This chapter aims to provide insights from a global and region-specific assessment on AR GHG emissions, cost and water footprint analysis, subject to an appropriate forest management regime, involving conventional forestry operations, and based on the collection of representative data from the available literature.

9.1.3 Chapter Contributions

This chapter provides an explicit spatio-temporal model of AR's whole-system, in which 5 sub-models are integrated: 1) a forest growth model, 2) a forest management cycle model, 3) a biogenic carbon (C) (and CO_2) sequestration model and 4) its associated "fire-penalty" model, and 5) a forestry operations model.

Specifically, CO_2 (and N_2O) and cost balances are carried out for each step of the forestry operations model.

Spatial resolution of AR's whole-system model is at the climato-ecological level – ecological zones[†] – within each State/Province in Brazil, China, India and the USA and within each country in the EU (EU-27 & UK).

Temporal resolution of AR's whole-system model is double: 1) 10 years (decadal), ranging between 2020–2100, and 2) 1 year (annual), over a default 100-year time-period.

The decadal timescale is used for determining the establishment of newly afforested stands, then the annual timescale is used for computing and evaluating AR's whole-system model.

The different interactions between the AR sub-models are outlined in Figure 9.1.

9.2 Forest Growth Model

9.2.1 Model Scope & Characterisation

This chapter provides a detailed description of the forest growth model and its associated forest management cycle model.

Within the forest growth model, forest growth curves are characterised by ecological zone and forest type (broadleaves/conifers), to account for geographic, climatic and ecological variations.[24,25] Both the above-ground biomass – the vegetation above the soil, such as stems, branches, foliage or

---

[†]An ecological zone is defined as "a zone or area with broad yet relatively homogeneous natural vegetation formations, similar (not necessarily identical) in physiognomy".[24]

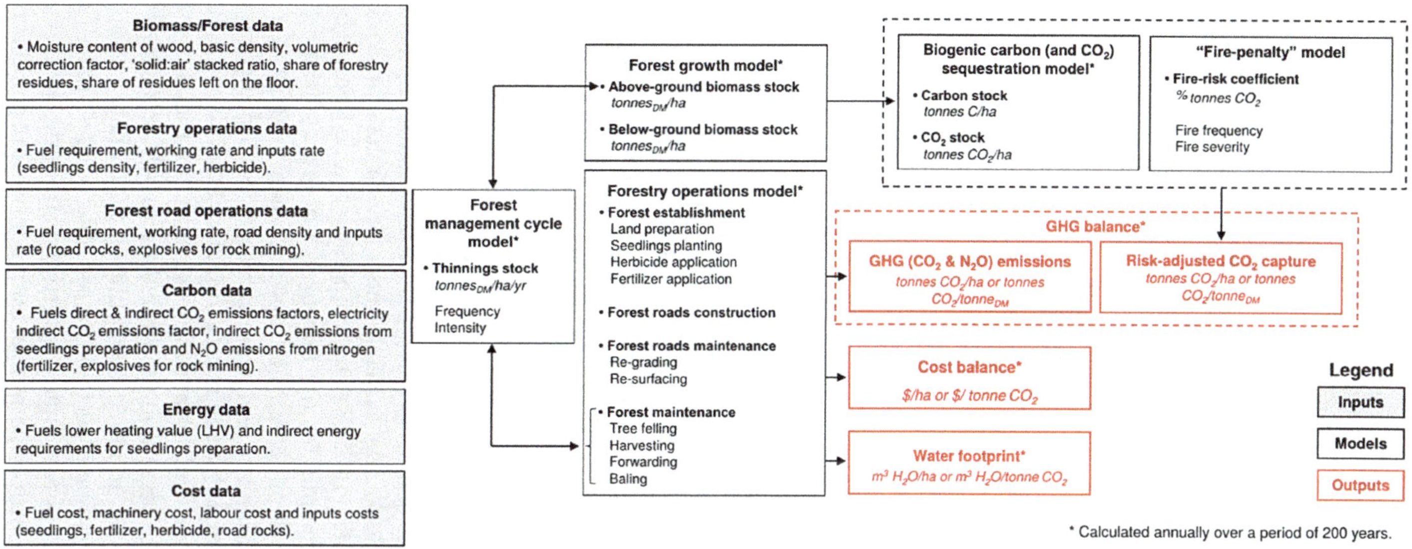

Figure 9.1 Schematic of the AR's whole-system model, outlining the interactions between 5 integrated sub-models: 1) a forest growth model, 2) a forest management cycle model, 3) a biogenic carbon (C) (and CO_2 sequestration model, and 4) its associated "fire-penalty" model, and 5) a forestry operations model.

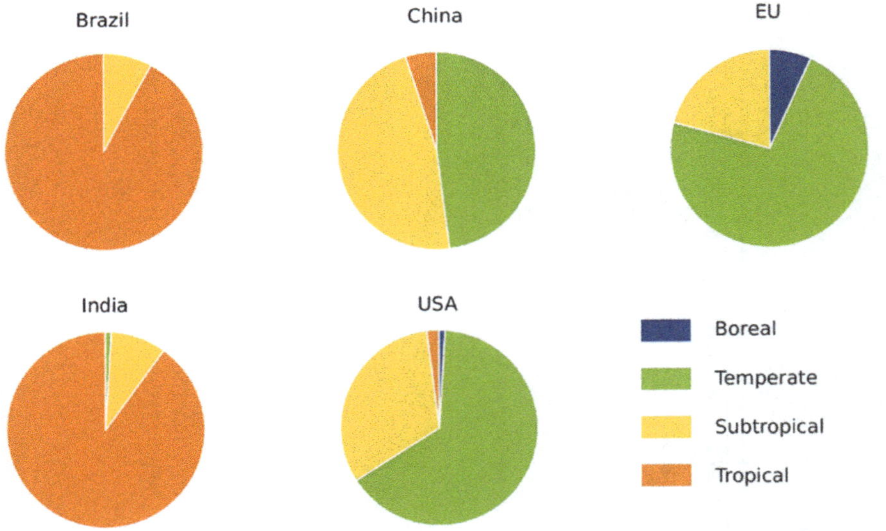

Figure 9.2 Distribution (per area) of climatic zones in Brazil, China, the European Union, India and the United States.

bark – without (of reference) and with forest management (managed), and the below-ground biomass – the roots – are included in the forest growth model.

Specifically, Brazil is dominated by a tropical climate (92%), and deciduous forests (>99%) (see Figure 9.2). The state of Para, with a tropical rainforest ecological zone, and broadleaves accounts for Brazil's archetype. In China, both temperate (48%) and subtropical (47%) climates are predominant, followed by tropical climate (5%). Tree species distribution is more equilibrated, with 58% of broadleaves and 42% of conifers. The province of Gansu, with a temperate mountain system ecological zone and conifers accounts for China's archetype. The European Union is dominated by a temperate climate (73%), with a subtropical climate around the Mediterranean Sea (21%), and boreal climate in Northern countries (7%). Broadleaves and conifers are equally present (48% and 52% respectively), and more or less predominant, with regards to the climate. France, with a temperate oceanic forest ecological zone and broadleaves accounts for the European Union's archetype. Similarly to Brazil, India is also dominated by a tropical climate (90%) and a subtropical climate (9%), but also has a temperate climate (1%). Broadleaves account for 90% and conifers for 10%. The state of Madhya Pradesh, with a tropical dry forest ecological zone and broadleaves accounts for Indian's archetype. Finally, all climates are present in the United States, where a temperate climate is predominant (65%), followed by a subtropical climate (32%), and where tropical and boreal climates account for the remaining climates (2% and 1%, respectively) at both of the country's vertical extremities (South and North). The state of Florida, with a subtropical humid forest and conifers accounts for the United States' archetype.

Finally, within the forest management cycle model, forest stands are subject to a non-intensive forest management – with reduced or minimum human intervention. The purpose of this forest management is to maximise and maintain the C (and CO_2) sequestration potential of the forest by clearing the forest of old and/or sick trees in order to let younger trees grow, more vigorously and with more space. Although the forest management cycle model consists of directly determining the proportion of above-ground biomass that needs to be thinned for the aforementioned reasons, it also impacts indirectly the proportion of below-ground biomass that remains after.

9.2.2 Above-ground Biomass

Forest growth depends on many parameters, such as climate, species, and forestry practices, themselves observing important geographic variations.[26]

The above-ground biomass (*i.e.*, trees) stock of reference can be defined as a sigmoid curve, which is typical in even-aged stands in the absence of forest management (without human intervention).

The above-ground biomass stock of reference $\mathrm{B}_{\mathrm{Ref}}^{\mathrm{AG}}$ (tonnes$_{\mathrm{DM}}$ per ha) is characterised by ecological zones and parametrised with the IPCC default biomass stock $\mathrm{B}_{\mathrm{Ref}}$ (tonnes$_{\mathrm{DM}}$ per ha) and net biomass growth rate G_{Ref} (tonnes$_{\mathrm{DM}}$ per ha year) of natural forests (see Table 9.1).[26]

$\mathrm{B}_{\mathrm{Ref}}^{\mathrm{AG}}$ is calculated for each ecological zone, *gez*, and each year, *yr*, over the forest growth period:

$$\forall \mathrm{gez}, \mathrm{yr} \leq T_{\mathrm{Ref}}(\mathrm{gez}), \mathrm{B}_{\mathrm{Ref}}^{\mathrm{AG}}(\mathrm{yr}, \mathrm{gez}) = \frac{L_{\mathrm{Ref}}(\mathrm{gez})}{1 + \exp(-k_{\mathrm{Ref}}(\mathrm{gez})(\mathrm{yr} - x_{0,\mathrm{Ref}}(\mathrm{gez})))}$$

with:

$$T_{\mathrm{Ref}}(\mathrm{gez}) = \frac{\mathrm{B}_{\mathrm{Ref}}(\mathrm{gez})}{G_{\mathrm{Ref}}(\mathrm{gez})}$$

$$L_{\mathrm{Ref}}(\mathrm{gez}) = \mathrm{B}_{\mathrm{Ref}}(\mathrm{gez})$$

$$x_{0,\mathrm{Ref}}(\mathrm{gez}) = \frac{T_{\mathrm{Ref}}(\mathrm{gez})}{2}$$

$$k_{\mathrm{Ref}}(\mathrm{gez}) = \frac{\ln\left(\frac{\mathrm{a}_{\mathrm{Ref}}}{100 - \mathrm{a}_{\mathrm{Ref}}}\right)}{x_{0,\mathrm{Ref}}(\mathrm{gez})}$$

where:

- T_{Ref} is the growing period of reference (years),
- L_{Ref} is the maximum biomass stock of reference (tonnes$_{\mathrm{DM}}$ per ha),
- $x_{0,\mathrm{Ref}}$ is the mid-point of the reference sigmoid curve (years),
- a_{Ref} is the asymptotic coefficient of the reference sigmoid curve (–), whose default value is set to 99,

Table 9.1 IPCC default biomass stock, net biomass growth and growing period of natural forests, characterised by ecological zone.[26]

Climate domain	Ecological zone	Biomass stock (tonnes$_{DM}$ per ha)	Biomass growth (tonnes$_{DM}$ per ha per year)	Growing period[a] (years)
Tropical	Tropical rainforest	300	7.0	43
	Tropical moist deciduous forest	180	5.0	36
	Tropical dry forest	130	2.4	54
	Tropical shrubland	70	1.0	70
	Tropical mountain system	140	1.0	140
Subtropical	Subtropical humid forest	220	5.0	44
	Subtropical dry forest	130	2.4	54
	Subtropical steppe	70	1.0	70
	Subtropical mountain system	140	1.0	140
Temperate	Temperate oceanic forest	180	4.4	41
	Temperate continental forest	120	4.0	30
	Temperate mountain system	100	3.0	33
Boreal	Boreal coniferous forest	50	1.0	50
	Boreal tundra woodland	15	0.4	38
	Boreal mountain system	30	1.0	30

[a]Calculated with the IPCC default values for biomass stock and biomass growth.

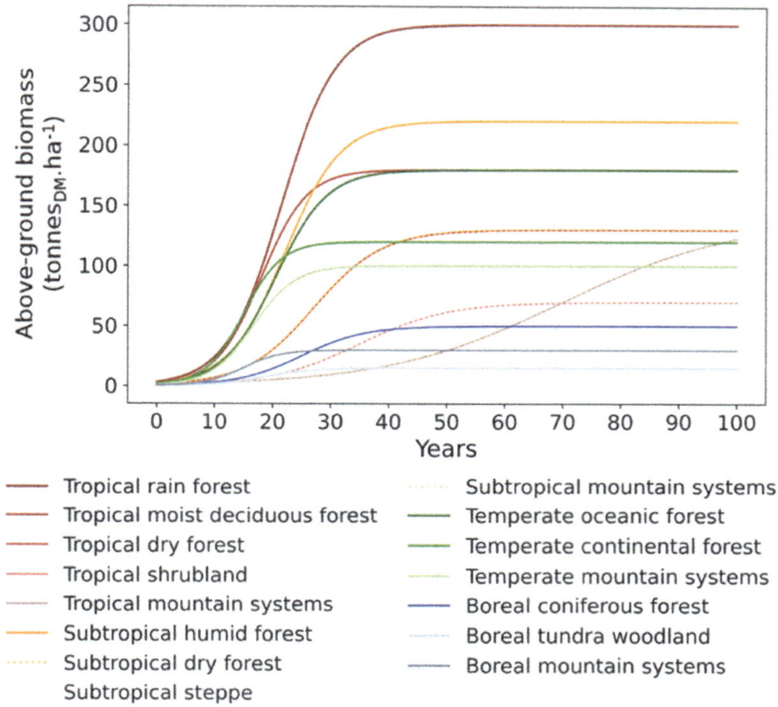

Figure 9.3 Growth of above-ground biomass stock of reference over 100 years. The above-ground biomass stock of reference saturates after 44 (30–140) years [median (min–max)].

- and k_{Ref} is the slope coefficient of the reference sigmoid curve (tonnes$_{\text{DM}}$ per ha per year), such as:

$$\frac{\mathrm{d}B_{\text{Ref}}^{\text{AG}}}{\mathrm{d}t}\left(\text{yr} = x_{0,\text{Ref}}(\text{gez}), \text{gez}\right) = \frac{k_{\text{Ref}}(\text{gez})}{4}$$

At the end of the growing period of reference, $B_{\text{Ref}}^{\text{AG}}$ is assumed to have reached its maximal value (*i.e.*, L_{Ref}).

$B_{\text{Ref}}^{\text{AG}}$ for all ecological zones are illustrated in Figure 9.3. $B_{\text{Ref}}^{\text{AG}}$ are observed to reach their maximum values and saturate after about 44 years (median). However, this implies that the forest must be maintained in perpetuity in order to preserve its climate benefit as well.

9.2.3 Afforested Above-ground Biomass

The managed above-ground biomass stock derives from the above-ground biomass stock of reference, subject to a forest management cycle.

This chapter introduces the concept of forest management cycle (FMC) and its associated phases, as developed within the forest management cycle model, and describes how it impacts the forest growth model.

9.2.3.1 *Forest Cycle Growth Phases*

In the context of timber production, a forest growth is usually broken down into 5 phases – the establishment, initial, full-vigour, mature, and old-growth phases. These 5 phases constitute the FMC, in which harvesting (and thinning) operations' characteristics – the frequency and the intensity – are set, and specific to each phase.

Because timber production is evaluated in terms of merchantable biomass volume, forest growth phases are usually determined based on the Mean Annual Increment (MAI) – the average rate of merchantable volume of biomass growth – and its Maximum Mean Annual Increment (MMAI). In the context of (biogenic) CO_2 removal, however, both merchantable and non-merchantable biomass stocks are considered, including both above-ground biomass and below-ground biomass stocks, and evaluated in terms of total biomass dry-mass.

The FMC modelled here is developed according to the aforementioned adaptations. This chapter introduces and defines the Mean Annual Growth (MAG) – the average rate of dry-mass of above-ground biomass growth – and the Maximum Mean Annual Growth (MMAG), which replace, respectively, the MAI and the MMAI in the determination of the forest growth phases. These forest growth phases are described in Table 9.2.

The annual (above-ground) growth (AG) AG^{AG}_{Ref}, the MAG MAG^{AG}_{Ref} and the MMAG $MMAG^{AG}_{Ref}$ are derived from the above-ground biomass stock of reference B^{AG}_{Ref}. They are calculated for each ecological zone gez and each year yr over a default period of 200 years:

$$AG^{AG}_{Ref}(yr, gez) = \begin{cases} B^{AG}_{Ref}(yr = 1, gez), \mid \text{if } yr = 1 \\ B^{AG}_{Ref}(yr, gez) - B^{AG}_{Ref}(yr - 1, gez), \mid \text{if } yr > 1 \end{cases}$$

$$MAG^{AG}_{Ref}(yr, gez) = \frac{B^{AG}_{Ref}(yr, gez)}{yr}$$

$$MMAG^{AG}_{Ref}(gez) = \underset{yr}{Max}\, MAG^{AG}_{Ref}(yr, gez)$$

Figure 9.4 illustrates a forest AG curve, its associated MAG curve and MMAG point, and its resulting forest growth phases.

9.2.3.2 *Forest Management Cycle (FMC)*

In the context of climate mitigation, AR is deployed while prioritising C (and CO_2) sequestration potential over timber production. Therefore, the FMC

Table 9.2 Forest growth cycle phases.

Phases	Description
Establishment	Seedlings are planted to create a new stand of trees. In this chapter, this phase is defined as lasting for the first five years after planting.
Initial	Once established, young trees grow from seedlings and their annual growth (AG) increases. This phase is defined here as lasting from age 5 up until (and including) the age of the first thinning, as specified here at 10 years after planting (age $5+10=15$ years). This phase has zero thinning.
Full vigour	During this period, trees grow with the greatest rate of AG. This phase is defined as lasting from the age of the first thinning (excluding) until (and including) the time at which the MMAG occurs. Thinning operations occur every 5 years, with an intensity of 10% of the above-ground biomass stock.
Mature	During this phase, the rate of AG declines progressively from its maximum value. This phase lasts from the time at which the MMAG is reached (excluding) up until (and including) the time at which the MAG has dropped to 50% of its maximum value. Thinning operations occur every 10 years with an intensity of 10% of the above-ground biomass stock.
Old growth	During this last phase, the biomass stock reaches its peak and stabilises, and the AG slowly levels off to zero. This phase is defined as lasting indefinitely from the time at which the MAG declines from half of the maximum (excluding). Stands in this phase also shift from an even-aged composition to a diverse structure of ages and sizes. Thinning operations occur every 15 years with an intensity of 10% of the above-ground biomass stock.

model introduced here is only comprised of thinning operations (no harvesting operations), in order to maximise and maintain the forest C (and CO_2) stock.

It is assumed that: 1) the frequency of thinning operations decreases with time – from every 5 years during the full-vigour phase to every 15 years during the old-growth phase, and 2) the intensity is set as 10% of the above-ground biomass stock at any time.

A schematic of the FMC workflow is described in Figure 9.5. The FMC workflow is defined for each ecological zone gez and for each year yr over a default period of 200 years, where:

- $yr_{E,END}$ is the last year of the establishment phase (years),
- $yr_{I,END}$ is the last year of the initial phase (years),
- MAG_{BT}^{AG} is the mean annual growth (MAG), before thinning (tonnes$_{DM}$ per ha per year),
- $MMAG_{Ref}^{AG}$ is the maximum mean annual growth (tonnes$_{DM}$ per ha per year), AT is the annual thinning stock (tonnes$_{DM}$ per ha),
- $\%_T$ is the thinning share of the above-ground biomass stock (%),
- B_{BT}^{AG} is the above-ground biomass stock, before thinning (tonnes$_{DM}$ per ha),

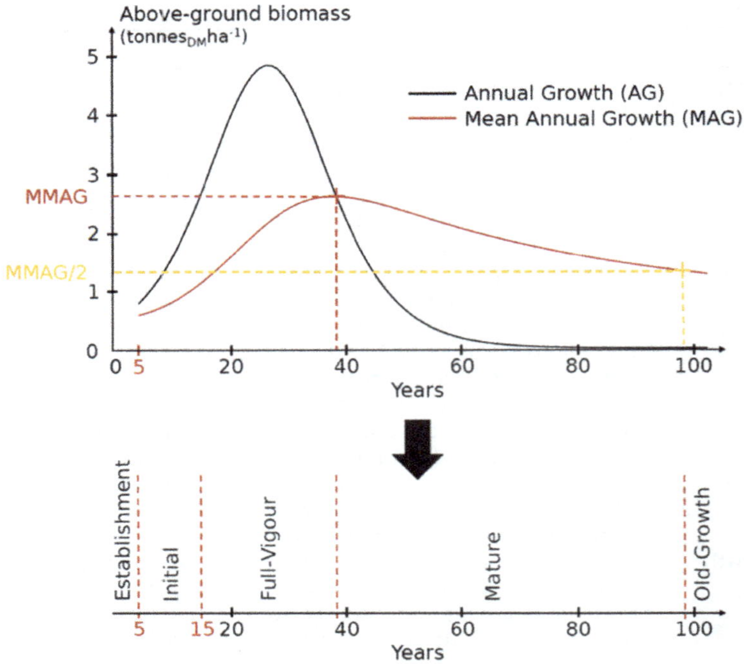

Figure 9.4 Illustration of a forest annual growth curve (AG), its associated mean annual growth curve (MAG) and maximum mean annual growth point (MMAG), and its resulting forest growth phases.

- $yr_{last\cdot FVT}$ is the year during which the last thinning of the full-vigour phase occurred (years),
- and $yr_{last\cdot MT}$ is the year during which the last thinning of the mature phase occurred (years).

$yr_{E,END}$ and $yr_{I,END}$ default values are respectively 5 and 15 years.

9.2.3.3 *Afforested Above-ground Biomass*

The managed above-ground biomass stock derives from the above-ground biomass stock of reference B^{AG}_{Ref} and the annual thinning stock AT.

Here, the managed above-ground biomass stock is evaluated in two steps– before thinning B^{AG}_{BT} and after thinning B^{AG}_{AT}, as illustrated in Figure 9.6.

B^{AG}_{BT} and B^{AG}_{AT} are calculated for each ecological zone gez and for each year yr over a default period of 200 years as follows.

Initialisation: $yr = yr_0$:

The initialisation year yr_0 is defined as the year of the first thinning.

In step (a), the above-ground biomass stock, before thinning, B^{AG}_{BT} is equal to the above-ground biomass stock of reference B^{AG}_{Ref}:

$$B^{AG}_{BT}(yr_0, gez) = B^{AG}_{Ref}(yr_0, gez)$$

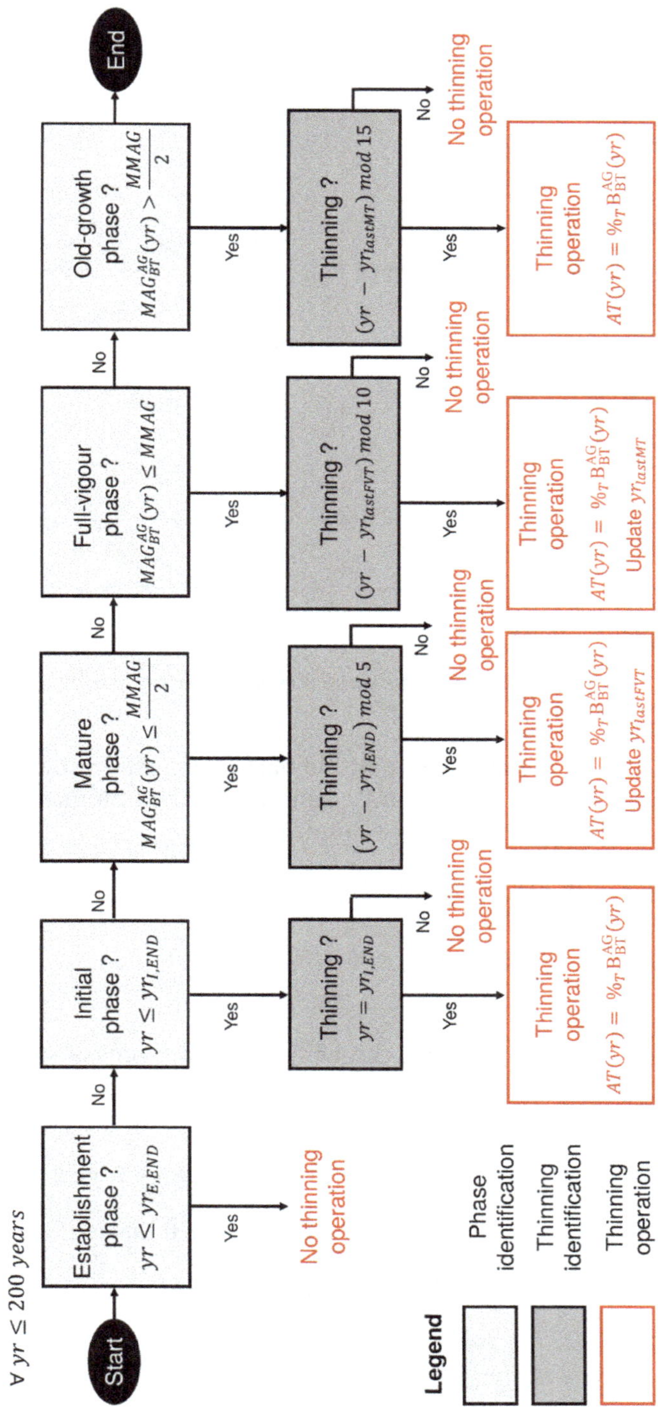

Figure 9.5 Schematic of the forest management cycle (FMC) workflow used for determining thinning operations' frequency and intensity.

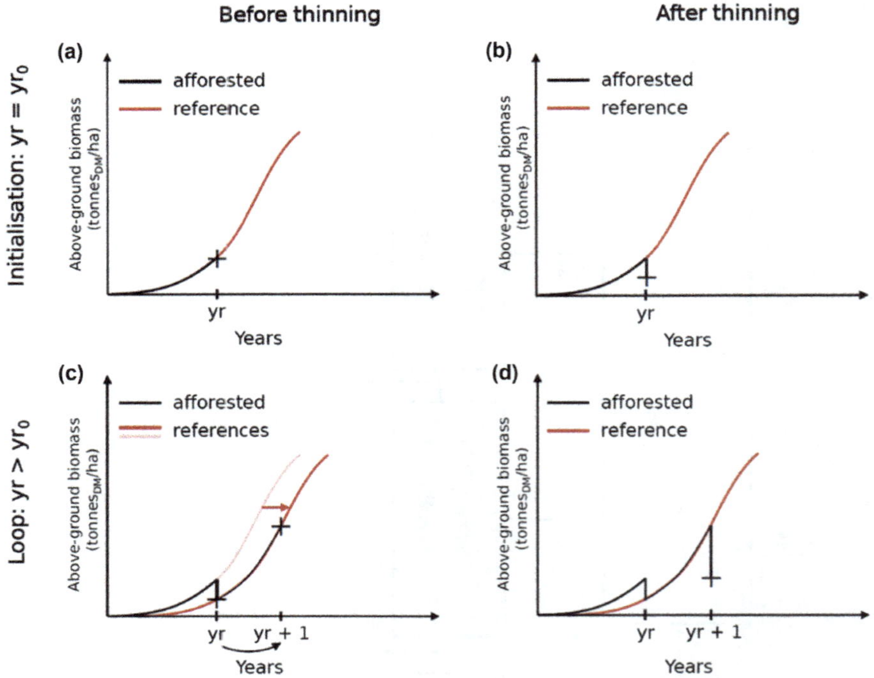

Figure 9.6 Illustration of the two-step workflow used for evaluating the managed above-ground biomass stock.

Then, in step (b), the above-ground biomass stock, after thinning, B_{AT}^{AG} is obtained by subtracting the amount of annual thinned biomass AT to the above-ground biomass stock before thinning:

$$B_{AT}^{AG}(yr_0, gez) = B_{BT}^{AG}(yr_0, gez) - AT(yr_0, gez)$$

$$AT(yr_0, gez) = \%_T \times B_{BT}^{AG}(yr_0, gez)$$

Loop: yr > yr$_0$:

In step (c), the following year, the above-ground biomass stock, before thinning, $B_{BT}^{AG}(yr_0 + 1, gez)$ has increased by $\Delta B^{AG}(yr_0 + 1, gez)$, following the above-ground biomass stock of reference:

$$B_{BT}^{AG}(yr_0 + 1, gez) = B_{AT}^{AG}(yr_0, gez) + \Delta B^{AG}(yr_0 + 1, gez)$$

where $\Delta B^{AG}(yr_0 + 1, gez)$ can be evaluated using the following equations:

$$yr_{Ref,1} = B_{Ref}^{AG-1}(B_{BT}^{AG}(yr_0, gez), gez)$$

$$yr_{Ref,2} = yr_{Ref,1} + 1$$

$$B_{AT}^{AG}(yr_0 + 1, gez) = B_{Ref}^{AG}(yr_{Ref,2}, gez)$$

$$\Delta B^{AG}(yr_0 + 1, gez) = B_{Ref}^{AG}(yr_{Ref\ 2}, gez) - B_{Ref}^{AG}(yr_{Ref\ 1}, gez)$$

Lastly, in step (d), the above-ground biomass stock, after thinning, B_{AT}^{AG} (yr_0+1, gez) is then obtained as previously in step (b):

Steps (c) and (d) are then repeated for each $yr > yr_0$ until the end of the default 200-year time-period. During the years of thinning operations, $\%_T = 10\%$, $\%_T = 0\%$ otherwise.

9.2.4 Below-ground Biomass

The below-ground biomass stock can be estimated from the above-ground biomass stock with the use of a "root-to-shoot" ratio.

A "root-to-shoot" ratio usually depends on climate, tree species, soil type and declines with stand age and/or productivity.

Extreme range values of 0.09–1.16 tonnes $root_{DM}$ per (tonnes $shoot_{DM}$) have been reported in the literature, although average range values of 0.20–0.56 tonnes $root_{DM}$ per (tonnes $shoot_{DM}$) might be more likely.[26]

In this chapter, the "root-to-shoot" ratio R evolves with the amount of above-ground biomass stock (before thinning) B_{BT}^{AG}. Specifically, R is interpolated from the IPCC default values[26] provided in Table 9.3.

R is characterised by ecological zones gez, tree species sp and by the amount of above-ground biomass stock a B_{BT}^{AG} and calculated as follows:

$$R(\text{gez}, \text{sp}, B_{BT}^{AG}(\text{yr}, \text{gez})) = \begin{cases} R_1(\text{gez}, \text{sp}) \times \ln\left(R_2(\text{gez}, \text{sp}) \times B_{BT}^{AG}(\text{yr}, \text{gez}) + 1\right) \mid R_2(\text{gez}, \text{sp}) \neq 0 \\ \hline R_1(\text{gez}, \text{sp}) \mid R_2(\text{gez}, \text{sp}) = 0 \end{cases}$$

where R_1 and R_2 coefficients are the coefficients interpolated[‡] from the IPCC, and provided in Table 9.4.

The below-ground biomass stock B^{BG} derives from the managed above-ground biomass stock (after thinning) B_{AT}^{AG} and the "root-to-shoot" ratio R, as follows:

$$B^{BG}(\text{yr}, \text{gez}) = B_{AT}^{AG}(\text{yr}, \text{gez}) \times R(\text{gez}, \text{sp}, B_{BT}^{AG}(\text{yr}, \text{gez}))$$

9.2.5 Total Biomass

Finally, the total biomass growth curve B^{Total} is defined for each ecological zone gez, each tree species sp and each year yr over a default period of 200 years and is calculated as follows:

$$B^{Total}(\text{yr}, \text{gez}, \text{sp}) = B_{AT}^{AG}(\text{yr}, \text{gez}) + B_{AT}^{BG}(\text{yr}, \text{gez}, \text{sp})$$

9.2.6 Overview

Overall, the forest growth, subject to the FMC, has a typical oscillating sigmoid pattern of growth and is characterised by ecological zones and tree species.

[‡] $R_1$and R_1 were obtained by solving a non-linear curve-fitting (data-fitting) problem in least-squares sense in Python 3.7 (function scipy.optimize.leastsq).

Table 9.3 IPCC default "root-to-shoot" ratio, characterised by ecological zone, forest type and above-ground biomass stock.[26]

Climate domain	Ecological zone	Above-ground biomass (tonnes$_{DM}$ ha^{-1})	Root-to-shoot R (tonne root$_{DM}$ per (tonne shoot$_{DM}$))	
Tropical	Tropical rainforest		0.37	
	Tropical moist deciduous forest	<125	0.20	
		>125	0.24	
	Tropical dry forest	<20	0.56	
		>20	0.28	
	Tropical shrubland		0.40	
	Tropical mountain system		0.27	
Subtropical	Subtropical humid forest	<125	0.20	
		>125	0.24	
	Subtropical dry forest	<20	0.56	
		>20	0.28	
	Subtropical steppe		0.32	
	Subtropical mountain system		No estimates[a]	
Temperate	Temperate oceanic forest, Temperate continental forest, Temperate mountain system	Conifers	<50	0.4
			50 – 150	0.29
			>150	0.2
		Broadleaves	<75	0.46
			75–150	0.23
			>150	0.24
Boreal	Boreal coniferous forest, Boreal tundra woodland, Boreal mountain system	<75	0.39	
		>75	0.24	

[a]Used IPCC tropical mountain systems values.

Table 9.4 Interpolated root-to-shoot $R1$ and $R2$ values.

Climate domain	Ecological zone	$R1$ (tonnes $root_{DM}$ per tonne $shoot_{DM}$)	$R2$ ($tonne_{DM}^{-1}$)
Tropical	Tropical rainforest[a]	0.37	—
	Tropical moist deciduous forest	801	0.00029
	Tropical dry forest	33	0.01113
	Tropical shrubland[a]	0.40	—
	Tropical mountain system[a]	0.27	—
Subtropical	Subtropical humid forest	848	0.00028
	Subtropical dry forest	474	0.00126
	Subtropical steppe[a]	0.32	—
	Subtropical mountain system[a]	0.27	—
Temperate	Temperate oceanic forest, Temperate continental forest, Temperate mountain system	Conifers 25 Broadleaves 24	Conifers 0.02159 Broadleaves 0.02275
Boreal	Boreal coniferous forest, Boreal tundra woodland, Boreal mountain system	25	0.04466

[a]Used IPCC "root-to-shoot" value.

Maximal above-ground biomass stock and growth of natural forests have been respectively reported within the ranges of 15–300 $tonnes_{DM}$ per ha and 0.4–7.0 $tonnes_{DM}$ per ha per year. When including below-ground biomass, total biomass stock ranges from 23 $tonnes_{DM}$ per ha in cold ecological zones (*e.g.*, boreal tundra woodlands) to 410 $tonnes_{DM}$ per ha in warm and moist ecological zones (*e.g.*, tropical rainforests).

Example estimates are shown in Figure 9.8 for different ecological zones and tree species. Thinning operations are observed to increase the period required for forests to reach their maximum biomass stock. In tropical and subtropical mountain systems, characterised by high above-ground biomass stock and low above-ground biomass growth, the thinning frequency and intensity defined by the FMC impact considerably the above-ground biomass stock curve, *i.e.*, >200 years for the afforested curve *versus* 140 years for the curve of the reference system. This reflects the importance of selecting appropriate climate, tree species and adapted FMC to generate good forest growth and consequently provide CDR *via* AR.

9.3 Biogenic C (and CO_2) Sequestration Model

This chapter provides a detailed description of the biogenic C (and CO_2) sequestration model and its associated C pools.

9.3.1 Biogenic C Pools

Growing forests capture CO_2 from the atmosphere *via* photosynthesis. The sequestrated C – the CO_2 is sequestrated in the form of C – contained in the

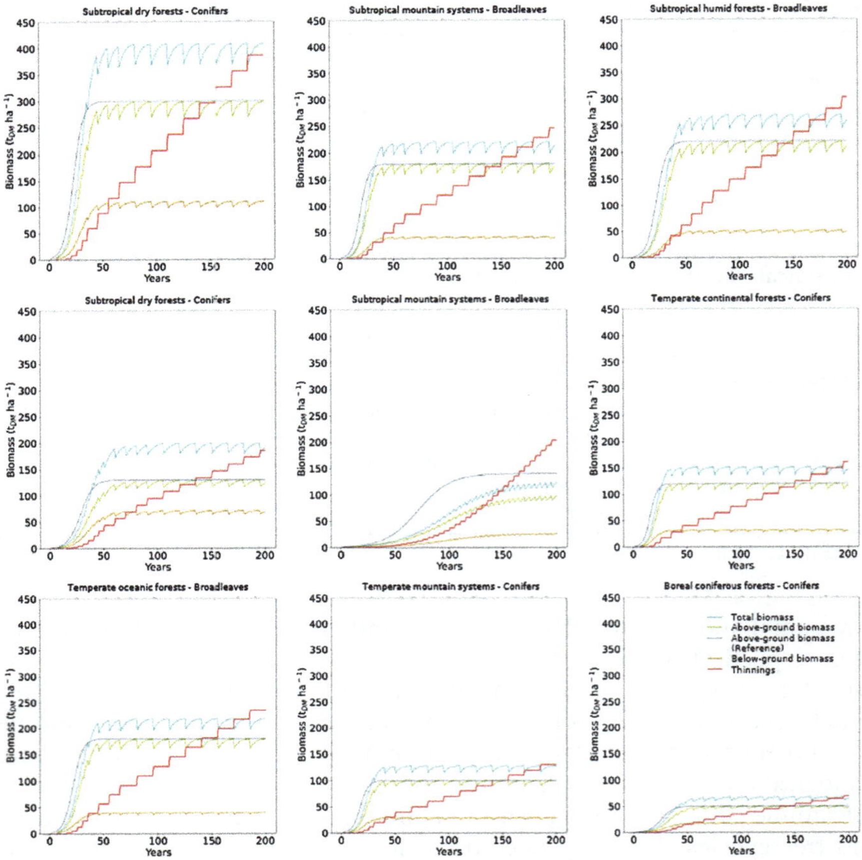

Figure 9.7 Examples of total, above-ground and below-ground biomass stocks and thinning stocks (tonnes$_{DM}$ per ha) for different ecological zones over 200 years.

above-ground biomass is then partially transferred to the below-ground biomass, dead organic matter and soil. During harvesting or thinning operations, timber and forest residues are extracted from the forest stands, and are considered as "harvested wood products". All together they constitute biogenic C pools.[26,27] Figure 9.7 illustrates the C flow into and out of the AR's whole system, as modelled here, as well as between the 5 aforementioned C pools.

9.3.2 Biomass C Pool

The above-ground biomass C pool is comprised of all C that is contained in the vegetation above the soil, such as stems, branches, foliage or bark, and the below-ground biomass C pool is comprised of the C contained in the roots. Together, they constitute the biomass C pool, of which the C stock

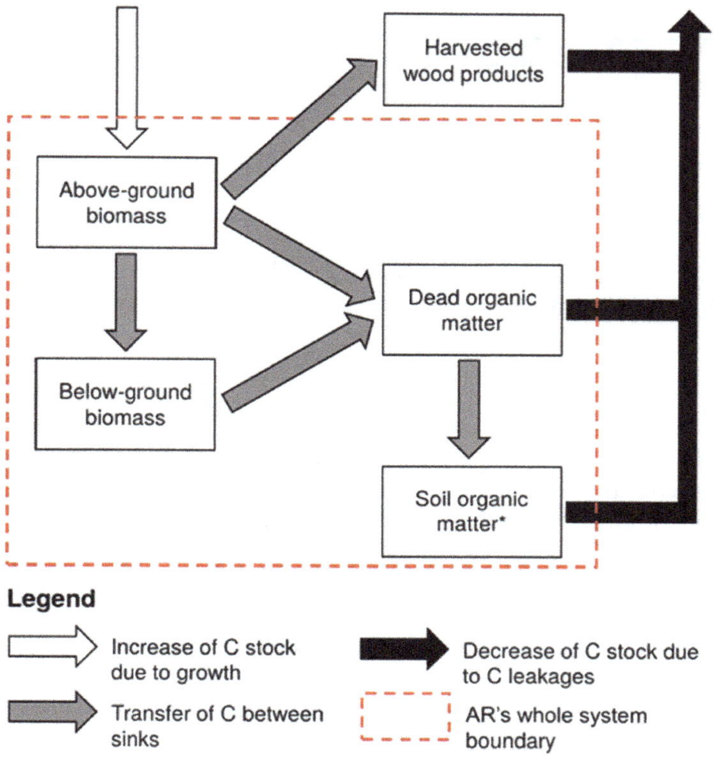

Figure 9.8 Illustration of the C flow into and out of the AR's whole system, as modelled here, as well as between the 5 main carbon pools: 1) the above-ground biomass, 2) the below-ground biomass, 3) the dead organic matter, 4) the soil organic matter and 5) the harvested wood products pools. Adapted from the 2006 IPCC Guidelines for National Greenhouse Gas Inventories.[27]

can be estimated from the biomass stock with the use of a carbon content factor C_f. C_f depends on climate, tree species, such as conifers or broadleaves, and tree characteristics, such as age, size or tree parts. Average values have been reported within the range of 0.43–0.55 tonnes C per tonnes_{DM}.[26] Above-ground and below-ground biomass carbon pools consist respectively of all carbon that is contained in the living biomass above the soil, and in the living roots.

In this chapter, the total biomass carbon stock (after thinning) C_{AT}^{Total} derives from the total biomass stock B^{Total} . C_{AT}^{Total} is defined for each ecological zone gez, each tree species sp and each year yr over a default period of 200 years and parametrised with the IPCC default values for C_f,[26] given in Table 9.5, as shown below:

$$C_{AT}^{Total}(yr, gez, sp) = B_{AT}^{Total}(yr, gez, sp) \times C_f(gez, sp)$$

Table 9.5 IPCC default values for carbon content, characterised by forest type.[26]

Climate domain		Carbon content (tonne C per tonne$_{DM}$)
	Broadleaves	Conifers
Tropical	0.47	0.47
Subtropical	0.47	0.47
Temperate	0.48	0.51
Boreal	0.48	0.51

Finally, the total biomass CO_2 stock CO_{2AT}^{Total} derives from the conversion of C to CO_2, based on the ratio of molecular weights (44/12):

$$CO_{2AT}^{Total}(yr, gez, sp) = C_{AT}^{Total}(yr, gez, sp) \times \frac{44}{12}$$

The estimated total biomass carbon stock range is 11–194 tonnes C per ha, equivalent to 40–709 tonnes CO_2 per ha. Similarly, to biomass stock, low bounds accounts for cold ecological zones (*e.g.*, boreal tundra woodlands) and upper bounds for warm and humid ecological zones (*e.g.*, tropical rainforests). Broadleaves/deciduous forests are seen to sequestrate 5–6% CO_2 less than conifers/evergreen forests in boreal and temperate climates, but to perform equally in subtropical and tropical climates. Subtropical steppes and tropical shrublands sequestrates 57–60% less than other ecological zones in a similar climate, and mountain systems 25–33% less, owing to their lower biomass density. Above-ground biomass contributes most to the forest biomass carbon stock, between 64–82%.

9.3.3 Dead Organic Matter C Pool

Dead organic matter is comprised of litter and dead wood. The dead organic matter C pool is highly variable, site-specific and time-evolving, depending specifically on forest characteristics, forest management and disturbance history, such as fire, hurricanes or pest. For instance, dead organic matter decay rates range from high in warm and moist climates to low in cold and dry climates.[27] Litter decay has also been observed to be much faster in deciduous (broadleaves) forests than in evergreen (conifers) forests, owing to humus formation, higher acidity of coniferous litter, and different canopy density resulting in different amount of light and water reaching the forest floor.[28]

A range value of 15–150 t CO_2 per ha, with an average value of 66 t CO_2 per ha can be found in the literature.[29] Although the C dynamics of the dead organic matter pool are qualitatively well understood, it is currently difficult to obtain a complete set of data at the national or regional scales. Therefore, we conservatively simplify the dead organic matter C pool by assuming that all C transferring from biomass to dead organic matter is directly emitted to the atmosphere.

9.3.4 Harvested Wood Product C Pool

The duration of the C contained in the "harvested wood products" pool varies with the product itself and its uses.[30] As a result, there are currently several different approaches to account for C (and CO_2) sequestration in wood products.

Here, all C contained in extracted above-ground biomass (resulting from thinning operations) – both thinnings and forest residues – is included in the "harvested wood products" C pool. Because thinning operations are carried out in the context of climate mitigation rather than timber production, the "harvested wood products" C pool is excluded from the AR's whole-system boundaries. All C contained in this pool can therefore be considered as a "C leakage". This is equivalent to assume conservatively that all C contained in the "harvested wood products" pool is directly emitted to the atmosphere, as is the case in the dead organic matter or soil organic matter pools.

Moreover, the below-ground biomass stock, proportional to the above-ground biomass stock, is also affected by thinning operations. Because all C contained in the affected below-ground biomass pool should be transferred to the dead organic matter pool, it is therefore simplified accordingly, and assumed to be directly emitted to the atmosphere.

The thinning C stock C^{Thinning} and CO_2 stock CO_2^{Thinning} are estimated with the use of the carbon content factor C_c and the ratio of molecular weights between C and CO_2. They are defined for each ecological zone gez, each tree species sp and each year yr over a default period of 200 years as follows:

$$C^{\text{Thinning}}(\text{yr}, \text{gez}, \text{sp}) = \sum_{1}^{\text{yr}} \text{AT}(\text{yr}, \text{gez}) \times C_c(\text{gez}, \text{sp})$$

$$CO_2^{\text{Thinning}}(\text{yr}, \text{gez}, \text{sp}) = C^{\text{Thinning}}(\text{yr}, \text{gez}, \text{sp}) \times \frac{44}{12}$$

9.3.5 Overview

Example estimates are shown in Figure 9.9 for different ecological zones and tree species, under the aforementioned FMC.

Two key points can be made. 1) Overall, CDR potential from AR ranges between 40–709 tonnes CO_2 per ha, when considering the above-ground and below-ground biomass carbon stocks only. Above-ground biomass contributes mostly to the total CO_2 stock, with 64–82%. Warm (*i.e.*, tropical and subtropical) and humid climates have the best CDR efficiency on a per hectare basis.

2) The HWP carbon stock (*i.e.*, thinnings and forest residues) created with AR is not negligible and can exceed the one of total biomass (after 150 years in this chapter). Whilst forest residues from AR do not contribute to carbon sequestration, they could be used as feedstocks for the bioenergy and biofuels sectors, and therefore reduce global GHG emissions reduction (*i.e.*, by providing/enabling avoided GHG emissions from fossil fuels). Even more, synergies between AR and BECCS could be unlocked, where forest residues would be used as a bioenergy feedstock for BECCS, and therefore enhance both CDR potentials (*i.e.*, by providing additional CDR).

There is thus considerable value in intelligent forest management to: 1) generate optimal forest growth (*i.e.*, higher forest growth than forest residues removal rate) and 2) provide biomass residues resulting from forest management

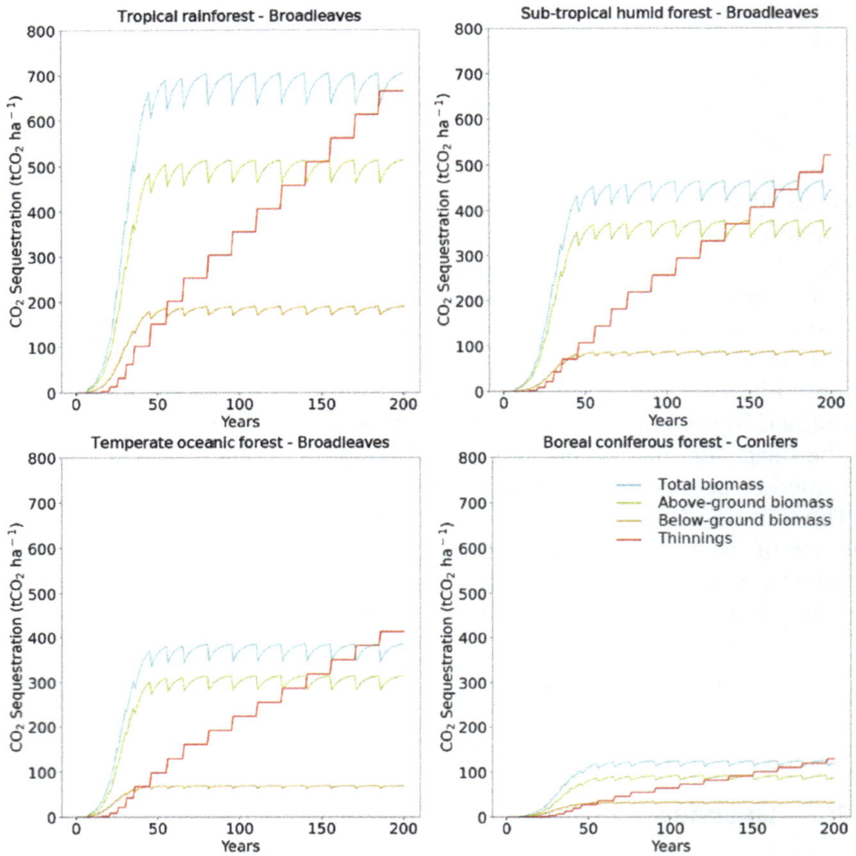

Figure 9.9 Examples of CO_2 sequestration (tonnes CO_2 per ha) in total, above-ground and below-ground biomass stocks and thinnings stocks for different ecological zones.

and contribute to additional GHG emissions reductions or CDR (*i.e.*, synergies between AF and BECCS), *via* feedstock for bio-energy and bio-fuels.

9.4 Risk-adjusted C Sequestration Model

This section discusses the permanence of CO_2 sequestration *via* AR, subject specifically to the risk of wildfires, and describes in detail the resulting fire-penalty model.

9.4.1 Permanence of Biogenic CO_2 Sequestration

Forests are vulnerable to natural and human disturbances such as drought, hurricanes, forest fires and pests, or to human-induced reversals, such as active deforestation. With climate change, this vulnerability could even be exacerbated.[31] Consequently, the permanence of biogenic CO_2 sequestration is less reliable than the one of geological CO_2 sequestration, such as in the cases of BECCS or DACCS.

Because the impact of natural disturbances on forest stands can be catastrophic, both in terms of biodiversity or financial losses – specifically in the context of timber production – how the risk of natural disturbances should be anticipated and integrated in forest management has been increasingly investigated. However, the focus has been predominantly set on maximising timber productivity and economic value, with or without carbon sequestration benefits, such as carbon price or carbon tax, and scarcely on minimising biomass and the resulting CO_2 sequestration losses.[32–34]

A few risk-accounting methods have been introduced, specifically for hurricanes or wildfires,[35–38] although most of the literature focuses on wildfires. In spite of the increasing widespread use of remote-sensing – the use of satellites to search for and collect geo-spatial data – these risk-accounting methods remain site-, region-, or at most, country-specific. For instance, the Landscape Fire and Resource Management Planning Tools (LandFire) Program in the USA has been providing national geo-spatial datasets (partially or completely based on remote-sensing) on vegetation distribution, fire regime and other fuel characteristics.[39] The NASA Land Use and Land Cover Program has also been releasing the MODIS Active Fire Products and Burned Area Products, ones of the most complete datasets at the global scale, but this chapter describes a model in which the risk of wildfires is accounted for in the form of a penalty coefficient in order to evaluate the impact of such wildfires on AR's CO_2 sequestration potential.

9.4.2 Wildfire-penalty Coefficient

The model introduced in this chapter is adapted from a risk-accounting methodology developed in Hurteau *et al.*[36] to define a wildfire-penalty coefficient R_{fire}, characterised by ecological zones, and therefore applicable at the global scale. R_{fire} is built upon wildfires' severity – the potential biomass loss given a fire occurrence – and periodicity – the probability of a fire event occurring during a specified time period.

As in Hurteau *et al.*,[36] wildfires' severity and periodicity are respectively quantified by the mean fire return interval (mFRI) – ranging from 0 years (very frequent) to 1000 years (very rare) – and the vegetation departure index (VDep) – ranging from 0% (zero potential biomass loss) to 100% (complete potential biomass loss).

This is written as follows:

$$R_{fire} = \begin{cases} \text{VDep} \times \left(1 - \dfrac{\text{mFRI}}{P}\right), & \mid \text{mFRI} \leq 100 \\ 0, & \mid \text{mFRI} > 100 \end{cases}$$

where P is permanence (years), which represents the length of time that biogenic carbon must be sequestered to be accounted as permanent here. P default value is 100 years.

Table 9.6 Spatial datasets.

Data layer/Spatial dataset	Source	Scale	Description
Shapefile Global Ecological Zones (GEZ), version 2010	FAO	World	GEZ represents the global ecological zoning distribution.
30m Raster Existing Vegetation Type (EVT), version LF 2.0.0	LANDFIRE	USA	EVT represents the distribution of the terrestrial ecological systems using the U.S. National Vegetation Classification System (NVCS).
30m Raster Vegetation Departure index (VDep), version LF 1.4.0	LANDFIRE	USA	VDep indicates how different current vegetation on a landscape is from estimated historical conditions, ranging from zero (no departure) to one (complete departure).
30m Raster Mean Fire Return Interval (mFRI), version LF 1.2.0	LANDFIRE	USA	mFRI is the average period between fires under the presumed historical fire regime.

mFRI and VDep are 2 geo-spatial datasets initially developed by the LandFire Program, and cover the US territory with a 30m geographic resolution.[39] Here, the mFRI and VDep datasets were aggregated at the ecological scale[§]. First, the existing vegetation type (EVT) dataset, also developed by the LANDFIRE project,[39] was used to restrict the mFRI and the VDep datasets to forest vegetation types. Then, the global ecological zone (GEZ) dataset, developed by the FAO[24] was used to aggregate the restricted mFRI and VDep datasets with ecological zones and finally extrapolate the aggregated and restricted mFRI and VDep datasets at the global scale. Because all ecological zones are not present on the US territory, values from "Tropical moist deciduous forest" were used for the "Tropical rainforest" ecological zone (closest ecological zone), and default values of 1000 years and 0 were assigned to mFRI and VDep, respectively, in boreal ecological zones (extremely low risk of wildfire in such cold and humid climate).

Table 9.6 gives an overview of the four datasets used to evaluate mFRI and VDep here.

The resulting values for mFRI, VDep and R_{fire} are provided in Table 9.7.

9.4.3 CO_2 Sequestration Potential

AR's CO_2 sequestration potential CO_{2AT}^{Seq} is defined for each ecological zone gez, each tree species sp and each year yr over a default period of 200 years and is estimated with the use of R_{fire}, characterised by ecological zone gez:

$$CO_{2AT}^{Seq}(\text{yr, gez, sp}) = (1 - R_{\text{fire}}(\text{gez})) \times CO_{2AT}^{Total}(\text{yr, gez, sp})$$

[§]mFRI and VDep are processed in ArcGIS 10.6 (ESRI).

Table 9.7 Mean fire return interval mFRI and vegetation departure index VDep values.

Climate domain	Ecological zone	mFRI (years)	VDep (–)
Tropical	Tropical rainforest[a]	46	65
	Tropical moist deciduous forest	46	65
	Tropical dry forest[a]	46	65
	Tropical shrubland[a]	46	65
	Tropical mountain system[a]	46	65
Sub-tropical	Subtropical humid forest	50	57
	Subtropical dry forest	24	30
	Subtropical steppe	52	47
	Subtropical mountain system	40	43
Temperate	Temperate oceanic forest	424	61
	Temperate continental forest	165	61
	Temperate mountain system	158	51
Boreal	Boreal coniferous forest[b]	1000	0
	Boreal tundra woodland[b]	1000	0
	Boreal mountain system[b]	1000	0

[a]Used tropical moist deciduous values for mFRI and VDep.
[b]Used default values of 1000 and 0 for mFRI and VDep, respectively.

AR CO_2 sequestration potential can be further aggregated at the regional scale:

$$CO_{2AT}^{Seq}(yr, sr) = \frac{\sum_{sp,gez} CO_{2AT}^{Seq}(yr, gez, sp) \times \%_{sp}(sp, sr) \times \%_{gez}(gez, sr)}{\sum_{sp,gez} \%_{sp}(sp, sr) \times \%_{gez}(gez, sr)}$$

where:

- CO_2^{Seq} is the AR CO_2 sequestration potential in sub-region *sr* and year *yr* (tonnes CO_2 captured per ha),
- $\%_{sp}$ is the share of each tree species sp in sub-region sr (%),
- and $\%_{gez}$ is the share of each ecological zone gez in sub-region sr (%).

The maximal CO_2 sequestration potential of AR in Brazil, China, the EU, India and the USA, at the regional scale is illustrated in Figure 9.10. AR's CO_2 sequestration potential is not affected by wildfires in boreal and temperate climates, whereas it decreases by 23–29% in subtropical climates, and up to 35% in tropical climates.

9.5 Forestry Operations Model

This chapter provides a detailed description of a forestry operations model. Specifically, AR requires the establishment and the on-going maintenance of the forest to maximise and maintain CDR. These include site establishment, forest roads construction and annual maintenance, and annual forestry (thinning) operations.[3,10,12]

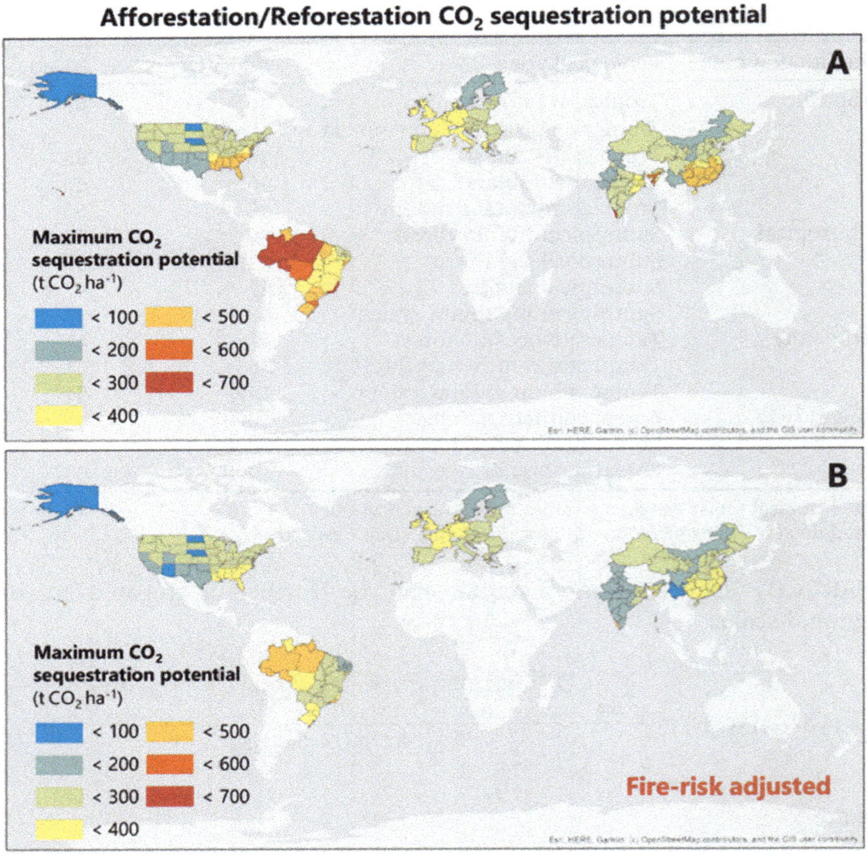

Figure 9.10 Maximum CO_2 sequestration potential: A) without considering the risk of fire, and B) with considering the risk of fire.

9.5.1 Site Establishment

The forest is established by land preparation and planting of new seedlings.

For land preparation, mounding is carried out by an excavator,[3,10,40] and herbicide and fertiliser are applied using a tractor.[3,10]

Tree seedlings are prepared in nurseries,[41] then planted by hand.[3,10,42]

9.5.2 Forest Road Construction and Maintenance

The access and the maintenance of forests requires forest roads, that are classified according to the frequency of their usage. Specifically, type A roads are the principal routes, used very frequently, and type B roads are secondary routes, only used during specific activities, such as thinning operations. Here, it is assumed that forest roads construction (for types A and B roads) involves spreading blasted rock on top of the soil, and then covering with a layer of finer, crushed aggregate.[3,10,40] Forest roads maintenance depends on the road type. Type A road maintenance involves the re-grading of the road every year, or the re-surfacing – the re-application of the top layer of aggregate – before thinning operations, whereas type B road maintenance only involves the re-grading and rolling of the remaining aggregate layer before thinning operations.[3,10,40] Mining and crushing of road rock and aggregate are also included,[10,40] as contributing to forest road construction and maintenance of indirect GHG emissions.

9.5.3 Annual Forestry Operations

As part of the FMC, a selection of trees is thinned using a cut-to-length logging system. This involves the felling and the extraction of trees from the forest site using a combination of harvesters and forwarders.[3,10,43,44] Here, it is assumed that this selection of thinned forest biomass is composed of 80% thinnings (whole tree thinnings and roundwood) and 20% forest residues, such as branches, foliage or bark.[12] Early whole tree thinnings involves tree felling by harvesters, followed by whole tree removal from the site to the roadside by forwarders. Harvesting roundwood requires the use of harvesters that cut and top the trees, leaving branches and other forest residues on the forest floor. The roundwood is then transported to the roadside by forwarders. Lastly, 35% of the residues are left in the forest for ecological reasons[12] – forest residues are left on the forest floor to maintain the nutrient and soil carbon balance – and the remainder is collected and extracted by forwarders that compress the forest residues into bundles. All thinned and extracted forest biomass are then stored at the roadside to allow for natural drying from 50% to 30% moisture content.[10,12]

Dry matter losses are also included at every step of the forestry operations – tree felling, harvesting, forwarding and storage – resulting in a total loss percentage of 11.6%. This value is consistent with the literature.[12,26] As a reference, the IPCC defaults values for harvest loss are 10% for broadleaves and 8% for conifers.[26] Figure 9.11 gives an overview of the annual forestry operations involving thinnings.

9.6 CO_2 and N_2O Balance Model

9.6.1 CO_2 and N_2O Emission Contributions

The total CO_2 and N_2O emissions of AR equal direct GHG emissions from the combustion of fuels,[45] and indirect GHG emissions due to the production of these fuels,[45] the generation of electricity[45,46] and the manufacture of materials or equipment, including agrochemicals,[11,47–49] seedlings,[11,47,48] or road rocks and aggregate.[10,40,45] Direct N_2O emissions arising from the application of nitrogen-based fertiliser during the forest establishment and from the use of ammonium nitrate-based explosive for road rocks extraction are also considered.[50–53] The energy system is also assumed to be subject to

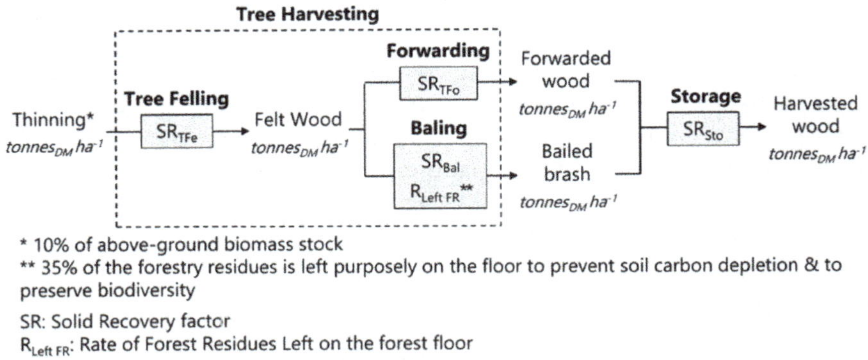

Figure 9.11 Overview of the annual forestry operations including tree (*i.e.*, thinnings and forest residues) felling, thinnings forwarding, forest residues baling, and storage.

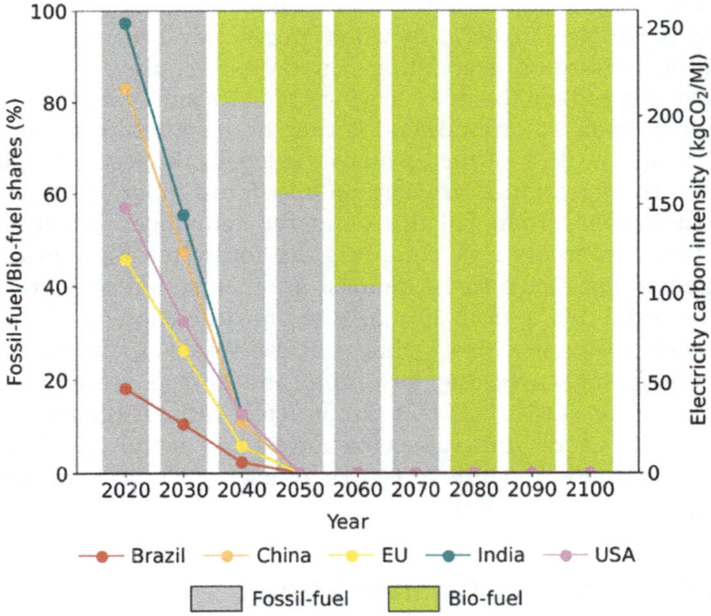

Figure 9.12 Decarbonisation of the energy mix: fossil/bio-fuels (left axis) and electricity (right axis).

decarbonisation (ESD), with the electricity carbon footprint following a decreasing carbon intensity as projected in the IPCC P2 mitigation scenario[54] and with fossil fuels being progressively replaced by 100% bio-fuels by 2080, as shown in Figure 9.12. Year 2020 represents the energy system without decarbonisation, and years 2040 and 2080 account respectively for the beginning and the end of the ESD.

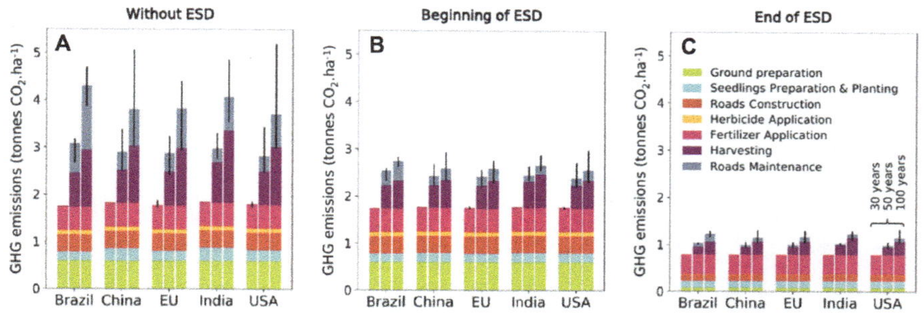

Figure 9.13 Breakdown of the GHG emissions of AR over 30, 50 and 100 years: (A) without ESD, (B) at the beginning of the ESD, and (C) at the end of the ESD.

Figure 9.13 shows the average contributions of different forestry and road operations to the final GHG emissions (per hectare) in different regions, over 10, 50 and 100 years, (a) without energy system decarbonisation (ESD), (b) at the beginning of the ESD, and (c) at the end of the ESD, within the uncertainty range.

All GHG emission contributions of AR are observed to decrease with the ESD (from 3.8–4.4 to 1.2 tonnes CO_2 per ha over 100 years), with the exception of the fertiliser application, that still emits N_2O emissions and represents the major GHG emissions contribution by the end of the ESD.

GHG emissions associated with harvesting and forest road maintenance are seen to be country-specific without ESD but to turn to be within the same range by the end of the ESD. This is because: 1) different countries have different forest growths (*i.e.*, different climates and with different tree species) resulting in different thinning frequencies and intensities, and therefore different fuel consumptions, and 2) different initial fuel carbon footprints, but the ESD fades these differences as all countries undergo through it.

The key point is that whilst total GHG emissions decrease with the ESD and are very slightly country-specific, they are, with or without ESD, negligible compared to the AR CO_2 sequestration potential, adjusted or not, by fire-risk. The little amount and spatial invariance of total GHG emissions from forestry operations, as early as at the beginning of the ESD, imply therefore that AR deployment should only be driven by its CO_2 sequestration potential, regardless of its total GHG emissions.

9.6.2 CO_2 Removal Efficiency

Overall, the AR total GHG balance, equivalently referred to as the CO_2 removal potential, is the difference between the risk-adjusted CO_2 sequestration from forest growth, and the total GHG emissions generated by all AR stages and result in a net CDR balance.

Figure 9.14 illustrates the impact of ESD on afforestation GHG removal efficiency (*i.e.*, ratio of CO_2 removal potential and CO_2 sequestration potential) in different archetypal regions.

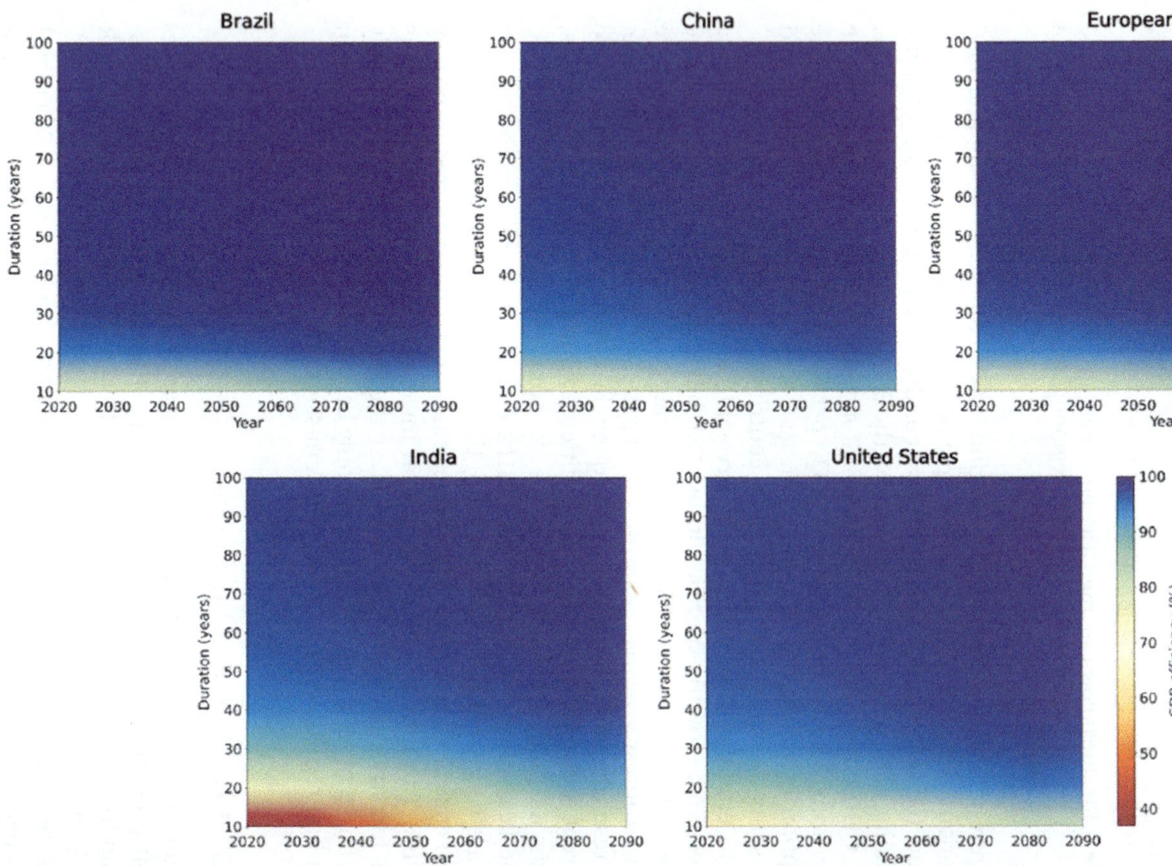

Figure 9.14 CDR efficiency of AR as a function of the year of establishment (horizontal axis) and the duration of the project (vertical axis).

The ESD is found to play a significant role in the first decades of AR deployment only. AR's CO_2 removal efficiency over a period length of 10 years increases indeed from 37.8–78.2% to 74.8–91.0 between 2020 (*i.e.*, without ESD) and 2080 (*i.e.*, end of ESD). Beyond 30 years of deployment, however, AR's CDR efficiency is seen to be as high as 88.8–95.8% over 30 years, regardless of the year of establishment, and higher than 90.0% beyond that. More specifically, AR's CDR efficiency reaches a maximum of 97.5-99.6% over a period length of 100 years, regardless of the year of establishment, making AR one of the most efficient CDR options. CDR efficiency is the lowest in India, likely due to the predominance of ecological zones characterised by slow forest growths: Tropical mountain systems and tropical shrublands account indeed for 48% of India total area and have an average growth rate of 1.7 tonne CO_2 per ha per year.

Whilst AR CDR efficiency in the mid/long-term is observed to be high and spatially invariant, AR initial carbon debt is not negligible and AR CDR efficiency in the short term is less high and spatially-specific. This could be of importance to policymakers or industrials with interest in CDR, with regards to an AR financial and business plan. If negative emissions were to be remunerated on an annual basis (*i.e.*, the annual amount of CO_2 removed is remunerated, in opposition to a cumulative basis, where the cumulative amount of CO_2 removed would be remunerated), *via* negative emissions credits (NECs) for instance, this could advantage regions with initial good CDR efficiency (*e.g.*, Brazil) and disadvantage regions with initial low CDR efficiency (*e.g.*, India), despite the fact that all regions reach similar CDR efficiency after 20 years. Conversely, cumulative NECs would remunerate CDR providers on the long term and on the cumulative amount of CDR achieved, and therefore incentivise AR projects everywhere. It is also worth underlining that, with a cumulative CDR target perspective, the ESD does not have any impact on its final CO_2 removal potential.

9.6.3 GHG Balance/CO_2 Removal Potential

Figure 9.15 shows AR CO_2 removal potential (per hectare) at the sub-regional level in Brazil, China, the European Union, India and the United States, over 100 years.

Southern regions are seen to be the best candidates for AR, with the exception of India, owing to low forest growth rates and consequently carbon sequestration rates.

9.7 Cost Balance Model

9.7.1 Cost Contributions

AR's total costs include:

- the cost of energy, such as fuels[55] and electricity;[56–60]

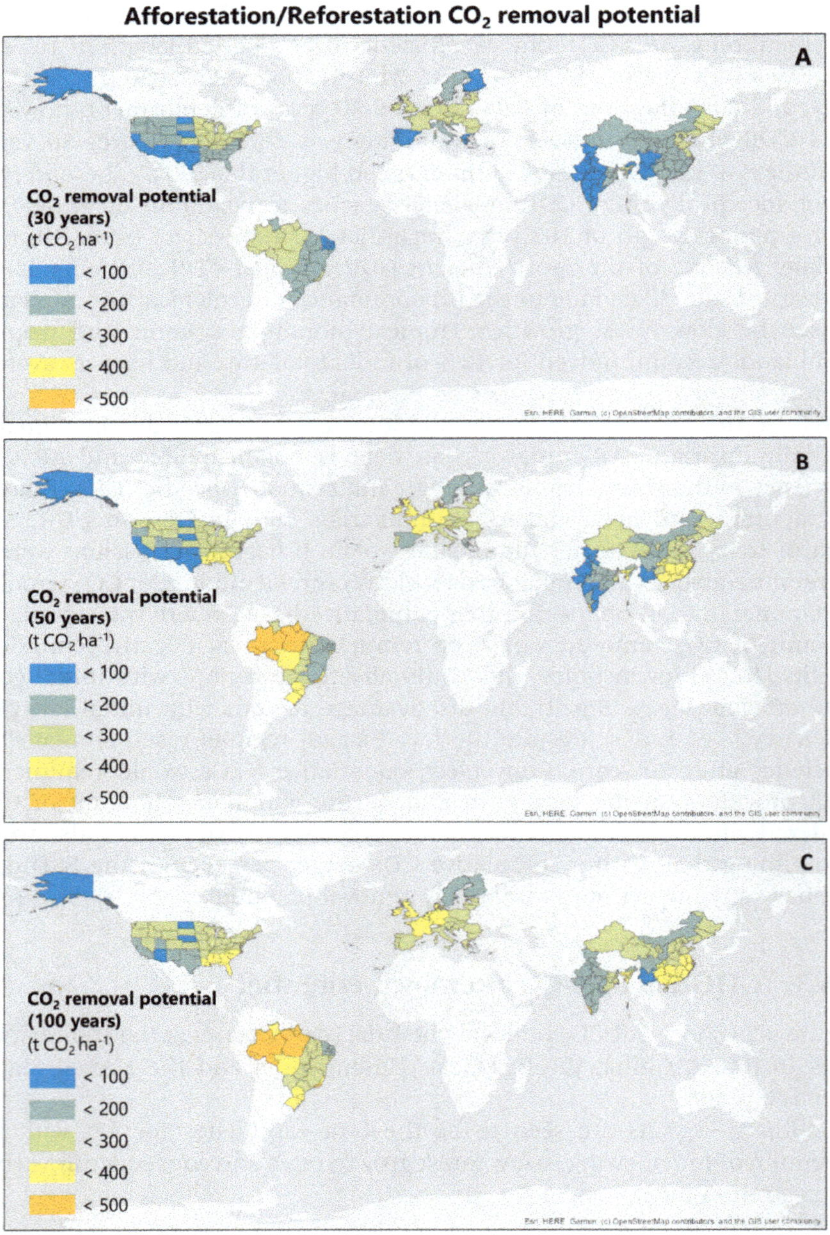

Figure 9.15 CO_2 removal potential: A) over 30 years, B) over 50 years, and C) over 100 years.

- the cost of machinery, such as trucks or excavator for land preparation, harvester and forwarder for harvesting operations, or other machinery for road construction and maintenance;[61,62]

- the cost of labour, such as ground worker, forest worker, road operator, *etc*;[63]
- and the cost of feedstocks, materials, such as agrochemicals,[64,65] seedlings[66] or road rocks.[67]

Here, costs are expressed in 2018 US $, but disaggregated at the national level.

When available in another currency or another year, costs of energy, machinery and feedstocks are converted with the use of exchange rate and inflation factors respectively.[68,69] When only available for one country (often the USA), costs of energy, machinery and feedstocks are converted with the use of purchasing power parity (PPP),[70] and costs of labour are converted proportionally to national hourly compensation costs from the Conference Board, as described previously.[71,72]

9.7.2 Cost Balance Over 100 Years

AR's initial investment, due to the establishment of the forest and the construction of forest roads, is levelised with the use of a capital recovery factor (CRF), itself calculated with an interest rate of 8% and a financial lifetime of 30 years. Other annual costs (*i.e.*, forest and forest roads maintenance) are summed up each year. Figure 9.16 shows the contributions of different forestry and forest road operations to the total levelised cost of AR per tonne of CO_2 removed, in different regions, over 10 to 100 years.

The contract length during which AR is deployed plays a crucial role in AR total levelized cost as it is seen to decrease significantly during the first fifty years and then level off. This is because of AR high initial investments that are only amortised when CO_2 sequestration increases and reaches saturation levels.

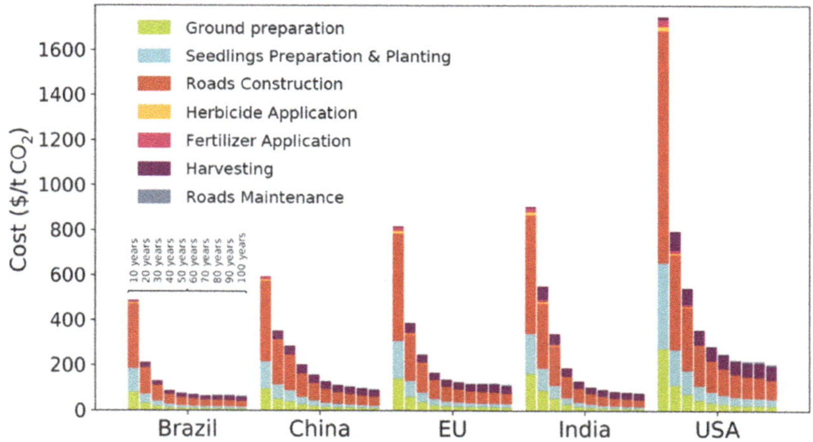

Figure 9.16 AR levelised cost breakdown of the contributions associated with different forestry and forest road operations in $ per tonne of CO_2 removed, shown for different regions and time-periods (10 to 100 years).

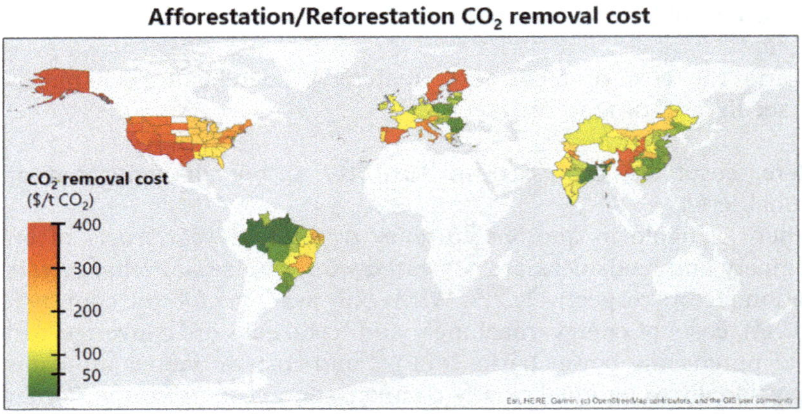

Figure 9.17 CO_2 removal costs for AR over a 100-year time-period in Brazil, China, the EU, India and the USA. With each of the 5 regions considered in MONET, differences between regions at a national scale is also accounted for.

AR is on average most expensive in the United States, with $1,749 per t CO_2 over 10 years, to $207 per t CO_2 over 100 years. It is on average cheapest in Brazil, with $487 per t CO_2 to $65 per t CO_2 over 10 and 100 years respectively. The construction of forest roads account for most of AR total levelised costs, and that throughout AR all lifetime. As thinning operations occur, they become the second highest contribution by the end of a century.

As shown in Figure 9.17, because of different economies across the world and different AR's CDR potentials within these economies, the CO_2 removal costs for AR varies at the global and national scale.

Overall, CO_2 removal *via* AR is the most expensive in the USA, ranging between $77–395 per t CO_2 whereas it is the cheapest in India, ranging between $22–139 per t CO_2, and in Brazil, ranging between $44–242 per t CO_2. Finally, AR's CO_2 removal costs are between $52–247 per t CO_2 in the EU, and between $59–265 per t CO_2 in China.

This is of importance, as although AR CDR has been identified as generally the most efficient in warm and humid regions, and usually the cheapest in southern countries such as Brazil or India, the cost of CDR is highly spatially-specific, and if intelligently managed and well-located, AR could be implemented in many more regions of the globe. If NECs were to be developed, some additional regions could decrease their cost and unlock additional CDR *via* AR, like the Eastern coast of the United States or the South of the European Union.

9.7.3 Cost Balance Over 1000 Years

Conversely to geo-engineered CDR options (*i.e.*, BECCS or DACCS) that store CO_2 into geological formations, and consequently can be practically and legally considered as permanent CDR options, immediately and over any period length, AR raises more concerns about the permanence of its CO_2 removal potential, due to the perpetual need to maintain the forest to keep the CO_2 sequestrated. Although the forest growth and its associated increase of CO_2 sequestration is found to drive predominantly AR total levelised cost per tonne of CO_2 removed down during the first century (Figure 9.16), annual costs, and GHG emissions, resulting from forest and forest roads maintenance will continue to increase whereas CO_2 sequestration will saturate. Consequently, AR total levelised cost per tonne of CO_2 removed will reach a minimal threshold, before increasing again, infinitely, in the absence of revenues, from timber and/or forest residues production for instance. For the CO_2 to be physically and legally considered as permanently removed from the atmosphere, a period length of 100 years seems inappropriate and some studies have thus been suggesting a period length of 300 years, or more instead.[73] Figure 9.18 shows more generally AR total levelised cost per tonne of CO_2 removed over 1000 years in different regions to evaluate the impact of this duration length on AR's CDR total levelized cost.

The AR total levelised cost is found to reach a minimal threshold between 40–60 years, and to increase afterwards. By 100 years, AR total levelised cost increases linearly, by as little as \$12 per tonne CO_2 every 100 years in Brazil and in India and by as much as \$26 per tonne CO_2 in the United States. In China and in the European Union, AR total levelised cost increases by \$17 per tonne CO_2 and \$19 per tonne CO_2 every 100 years, respectively.

These results indicate therefore that AR could be as low as \$43–97 per tonne CO_2 in the five archetypal nations considered in this study (*i.e.*, when AR total levelised cost threshold is reached), which is consistent with the current literature on AR cost. However, as explained above, it is unlikely that

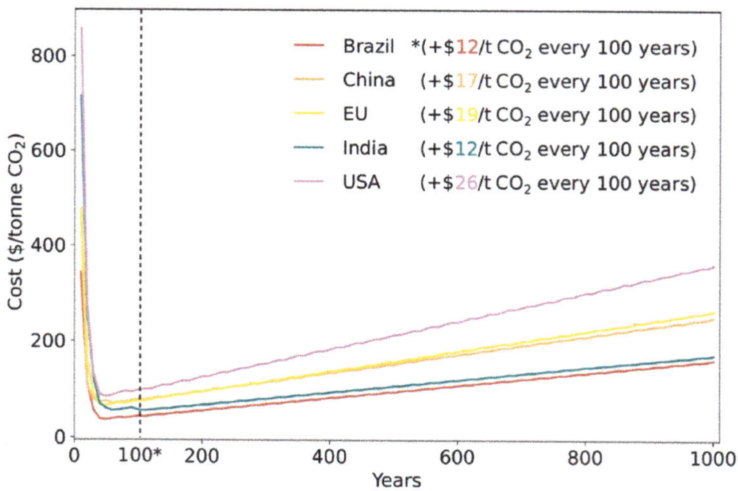

Figure 9.18 AR levelised total of CDR over 1000 years in Brazil, China, the European Union, India and the United States. After 100 years, AR total levelised costs increase linearly by \$12–26 per t CO_2 every 100 years.

AR should only be deployed over about 50 years to be considered a permanent CDR option (*i.e.*, the period length required to reach AR minimal total levelised cost). As a reference, over 300 years, AR total levelised cost is found to range between $68–153 per tonne CO_2, which (1) is not consistent with the current literature, and (2) comparable with other CDR options, BECCS particularly, of which permanence is more reliable than the one of AR.

Overall, these results thus demonstrate the high variability of AR cost, spatially and temporally, and urge for more clarity and accuracy in the current literature on AR, as it is likely to underestimate AR total levelised cost, and therefore overestimate AR potential and competitiveness with other CDR options. Lastly, and as for other nature-based CDR options (*e.g.*, biochar, soil carbon sequestration), there is a need to clarify the concept of permanence and to quantify, from a juridic and liable perspective, the amount of permanent CO_2 removal achievable *via* AR.

9.8 Water Footprint Model

9.8.1 Water Contributions

Water footprint WF is a water resource use indicator that has recently been developed for agricultural production (*i.e.*, crops for the production of food and bioenergy) to assess water use and water stress, and that can be extended to natural vegetation such as forests, shrubs or grasslands.[19,74] Usually expressed as the water volume used to produce a unit of product (m^3 per t_{DM}) or as the water volume per year of a delineated area (mm per day), it is comprised of three components: green, blue and grey water footprints:

$$WF = WF^{Green} + WF^{Blue} + WF^{Grey}$$

The green water footprint WF^{Green} is the volume of water from rainfall used during the vegetation growth, *via* evapotranspiration, and represents the major part of water used. The blue water footprint WF^{Blue} is the volume of additional fresh water required when the effective rainfall is insufficient. For agricultural production, the blue water is provided directly *via* irrigation from surface and/or groundwater whereas for natural vegetation, it is directly withdrawn from run-offs to surface and/or groundwater. At high levels, green and blue waters combined can cause water stress and water scarcity. Lastly, the grey water footprint WF^{Grey} refers to pollution and is the volume of fresh water required to assimilate the load of pollutants to natural background concentrations. For afforestation, this accounts for the application of nitrogen-based fertiliser during forest land establishment and the use of ammonium nitrate-based explosive during rocks extraction for forest roads construction and maintenance.

9.8.1.1 Vegetation Evapotranspiration

A key element in the evaluation of green and blue water footprints is the vegetation evapotranspiration ET (mm per day), which is the volume of water evaporated from soil, absorbed from roots and transpired by the vegetation during its growth. ET can be derived from ground measurements, eddy covariance stations (*i.e.*, remote-sensing), or calculated.

The evaluation of ET for agricultural production is well-established, and usually calculated following Allen *et al.*,[74] with the use the reference potential evapotranspiration ET_0 (mm per day), derived from Penman–Monteith equation and crop coefficients K_c (–), defined over the crop development stages:

$$\mathrm{ET} = K_c \times \mathrm{ET}_0$$

While ET_0 is only correlated to meteorological data climate (*e.g.*, relative humidity, wind speed, altitude), K_c is also determined by crop/vegetation characteristics and defined over time in accordance to the crop development/growth stage. For agricultural production, such crop coefficients can be found in reference tables. For natural vegetation, Allen *et al.* proposed the use of leaf area index LAI (–) and tree height h (m) as follows:[74]

$$K_C = K_C^{\mathrm{min}} + (K_C^{\mathrm{max}} - K_C^{\mathrm{min}})(1 - \mathrm{e}^{-0.7 \times \mathrm{LAI}})$$

$$K_C^{\mathrm{max}} = K_C^{\mathrm{h}} + [0.04(u_2 - 2) - 0.004(\mathrm{RH}_{\mathrm{min}} - 45)]\left(\frac{h}{3}\right)^{0.3}$$

$$K_C^{\mathrm{h}} = \begin{cases} 1 + 0.1h, & | \, h < 2\,\mathrm{m} \\ 1.2 & | \, h \geq 2\,\mathrm{m} \end{cases}$$

where:

- K_C^{min} is the minimal vegetation evapotranspiration of natural vegetation (–), whose default value for bare soil is 0.15.
- K_C^{max} is the maximal vegetation evapotranspiration of natural vegetation (–),
- u_2 is the mean value for wind speed at 2 m height (ms^{-1}),
- and $\mathrm{RH}_{\mathrm{min}}$ is the minimum daily relative humidity (–).

Similar to (seasonal) crops growing phases, natural vegetation and forests are also subject to an annual evapotranspiration cycle (*i.e.*, winter, spring, summer and fall). In this study, this is reflected in the LAI, which varies from 0 (*i.e.*, Winter) to 4.9 and 4.05 (*i.e.*, Summer) for deciduous and evergreen forest, respectively.[75]

In the context of AR, forests are also subject to growth (*i.e.*, from initial phase to old-growth phase). This is reflected in the tree height, that varies from 0 m (*i.e.*, planting of the seedlings) to average canopy height, obtained from the GLAS-LiDAR (NASA),[76] in accordance with the forest growth calculated in this study.

9.8.1.2 Green Water Footprint

The daily green water footprint $\mathrm{WF}_{\mathrm{d}}^{\mathrm{Green}}$ (m^3 per ha) is defined for each sub-region sr, tree species sp, ecological zone gez, day d, and year yr, over a default period of 200 years:[20]

$$\mathrm{WF}_{\mathrm{d}}^{\mathrm{Green}}(\mathrm{sr},\mathrm{sp},\mathrm{gez},\mathrm{d},\mathrm{yr}) = \begin{cases} P^{\mathrm{eff}}(\mathrm{sr},\mathrm{d}) \times 10, \mid P^{\mathrm{eff}}(\mathrm{sr},\mathrm{d}) < \mathrm{ET}(\mathrm{sr},\mathrm{sp},\mathrm{gez},\mathrm{d},\mathrm{yr}) \\ \mathrm{ET}(\mathrm{sr},\mathrm{sp},\mathrm{gez},\mathrm{d},\mathrm{yr}) \times 10, \mid P^{\mathrm{eff}}(\mathrm{sr},\mathrm{d}) \geq \mathrm{ET}(\mathrm{sr},\mathrm{sp},\mathrm{gez},\mathrm{d},\mathrm{yr}) \end{cases}$$

where:

- P^{eff} (sr, d) is the daily effective precipitation in sub-region sr and for day d (mm per day).
- and ET(sr, sp, d, yr) is the daily vegetation evapotranspiration in sub-region sr, for tree species sp, ecological zone gez, for day d and year yr (mm per day),

P^{eff} is usually calculated using the Soil Conservation Service Method developed by the US Department of Agriculture (USDA):

$$P^{\mathrm{eff}}(\mathrm{sr},\mathrm{d}) = \begin{cases} 0.6 \times P(\mathrm{sr}) - 10, \mid P(\mathrm{sr}) < 75 \\ 0.8 \times P(\mathrm{sr}) - 25, \mid P(\mathrm{sr}) \geq 75 \end{cases}$$

where P(sr, d) is the daily average precipitation in sub-region sr and for day d (mm per day).

9.8.1.3 Blue Water Footprint

The daily blue water footprint $\mathrm{WF}_{\mathrm{d}}^{\mathrm{Blue}}$ (m^3 per ha) is defined for each sub-region sr, tree species sp, ecological zone gez, day d, and year yr, over a default period of 200 years:[20]

$$\mathrm{WF}_{\mathrm{d}}^{\mathrm{Blue}}(\mathrm{sr},\ \mathrm{sp},\ \mathrm{gez},\ \mathrm{d},\ \mathrm{yr}) = \begin{cases} \mathrm{ET}(\mathrm{sr},\ \mathrm{sp},\ \mathrm{gez},\ \mathrm{d},\ \mathrm{yr}) \times 10, & \mid P^{\mathrm{eff}}(\mathrm{sr},\mathrm{d}) < \mathrm{ET}(\mathrm{sr},\ \mathrm{sp},\ \mathrm{gez},\ \mathrm{d},\ \mathrm{yr}) \\ 0, & \mid P^{\mathrm{eff}}(\mathrm{sr},\mathrm{d}) \geq \mathrm{ET}(\mathrm{sr},\ \mathrm{sp},\ \mathrm{gez},\ \mathrm{d},\ \mathrm{yr}) \end{cases}$$

9.8.1.4 Ecological Water Deficit

The ecological water deficit indicator represents the water stress exerted by AR through evapotranspiration. It is equal to the difference between ET and P_{eff}, also equivalent to the difference between $\mathrm{WF}_{\mathrm{Green}}$, $\mathrm{WF}_{\mathrm{Blue}}$ and P_{eff}. The daily ecological water deficit $\mathrm{EWD}_{\mathrm{d}}$ (m^3 per ha) is defined for each sub-region *sr*, ecological zone *gez*, tree species sp and year yr over a default period of 200 years as follows:

$$\begin{aligned} \mathrm{EWD}_{\mathrm{d}}(\mathrm{sr},\ \mathrm{sp},\ \mathrm{gez},\ \mathrm{d},\ \mathrm{yr}) &= \mathrm{WF}_{\mathrm{d}}^{\mathrm{Green}}(\mathrm{sr},\ \mathrm{sp},\ \mathrm{gez},\ \mathrm{d},\ \mathrm{yr}) + \mathrm{WF}_{\mathrm{d}}^{\mathrm{Blue}}(\mathrm{sr},\ \mathrm{sp},\ \mathrm{gez},\ \mathrm{d},\ \mathrm{yr}) - P^{\mathrm{eff}}(\mathrm{sr},\ \mathrm{d}) \\ &= \mathrm{ET}(\mathrm{sr},\ \mathrm{sp},\ \mathrm{gez},\ \mathrm{d},\ \mathrm{yr}) - P^{\mathrm{eff}}(\mathrm{sr},\ \mathrm{d}) \end{aligned}$$

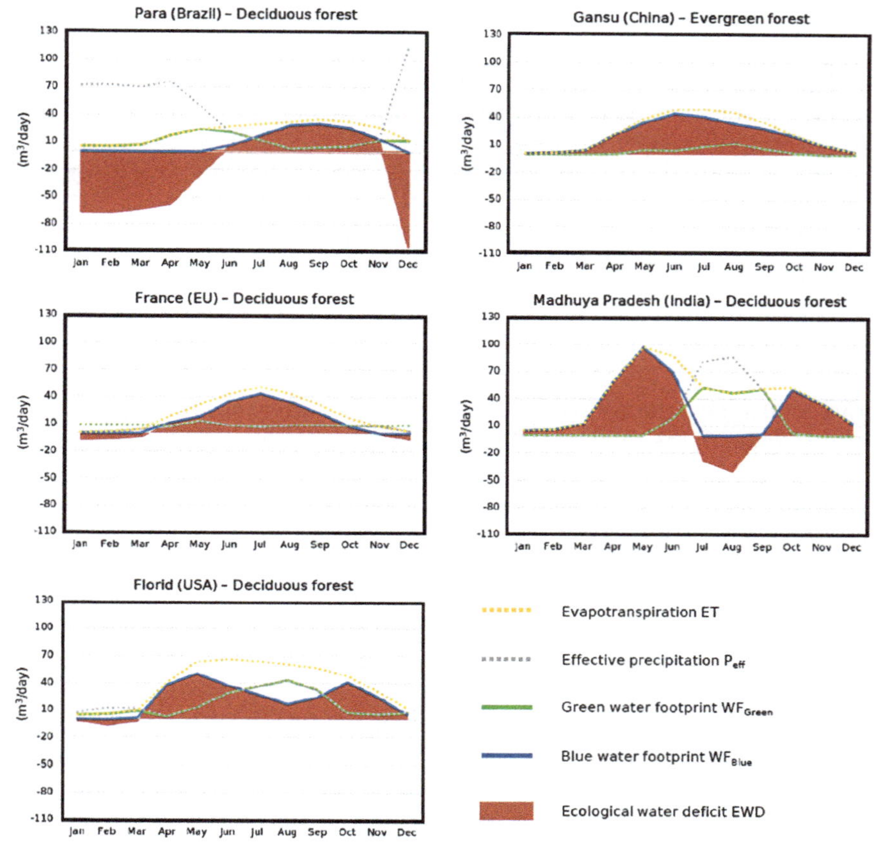

Figure 9.19 Evapotranspiration ET, effective precipitation P_{eff}, green and blue water footprints WF_{Green} and WF_{Blue}, and ecological water deficit EWD in five archetypal regions over a reference year.

Examples of green and blue water footprints, ecological water deficit, and evapotranspiration and precipitation in archetypes regions are illustrated in: Figure 9.19

9.8.1.5 Grey Water Footprint

The daily grey water footprint $\text{WF}_{\text{d}}^{\text{Grey}}$ (m^3 per ha) is defined for each subregion sr, tree species sp, ecological zone gez, day d, and year yr, over a default period of 200 years:[20]

$$\text{WF}^{\text{Grey}}(\text{sr}, \text{gez}, \text{sp}, \text{yr}) = \frac{R_{\text{N}}^{\text{FApp}} \times N_{\text{leach}} + \text{FRD}^{\text{RC}} \times R_{\text{N}}^{\text{RC}} + \text{FRD}^{\text{RM}} \times R_{\text{N}}^{\text{RM}}}{N_{\text{level}}}$$

where:

- N_{level} is the authorised levels of nitrate concentration ($\text{kg}\,\text{m}^{-3}$),

- N_{leach} is the direct emissions of nitrogen (N) leaching (kg N leaching per kg N applied),
- R_N^{FApp} is the rate of N fertiliser applied (kg N per ha),
- R_N^{RC} is the rate of N in explosive used during rock mining for road construction (kg N per km),
- R_N^{RM} is the rate of N in explosive used during rock mining for road maintenance (kg N per km),
- FRD^{RC} is the forest road under construction density (km/ha),
- and FRD^{RM} is the forest road under maintenance density (km/ha),

9.8.2 Water Balance

Figure 9.20 shows the contributions of different water footprints (*i.e.*, green, blue and grey) to the total water balance of AR per tonne of CO_2 removed, in different regions, over 30, 50 and 100 years.

In all regions, the grey water footprint is negligible compared to green and blue water footprints (≤0.5% of total water footprint). The total water footprint is the lowest in Brazil (from 2289 m^3 per t CO_2 over 30 years to 2289 m^3 per t CO_2 over 100 years), where climates are warm and humid, and the highest in India (from 15 434 m^3 per t CO_2 over 30 years to 8485 m^3 per t CO_2 over 100 years), where the climate is warm and dry. Humidity (*i.e.*, dry or humid) and temperature (*i.e.*, warm or cold) seem to be key in determining water footprint as the warmer and drier regions are, the higher the water footprint is. Green and blue water footprint contributions are highly spatially-specific, as blue water footprints account for the lowest in Brazil (44–46% of total water footprint), and the highest in the EU (74–78% of total water footprint).

Grey water footprint is also the highest in India (from 52 m^3 per t CO_2 over 30 years to 10 m^3 per t CO_2 over 100 years) and lowest in Brazil (from 10 m^3 per t CO_2 over 30 years to 4 m^3 per t CO_2 over 100 years).

The evapotranspiration of forests is a key element in determining the water footprint of AR. If high, AR in warm and dry regions, like India,

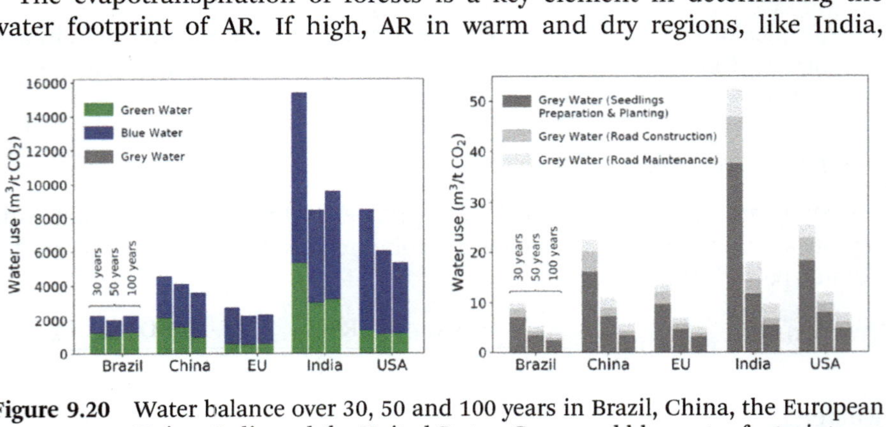

Figure 9.20 Water balance over 30, 50 and 100 years in Brazil, China, the European Union, India and the United States. Green and blue water footprints are found to be predominant.

requires a significant amount of water that precipitations alone cannot provide (*i.e.*, green water footprint), and is therefore more subject to water stress and water scarcity (*i.e.*, blue water footprint). A compromise between low forest evapotranspiration and sufficient precipitation is necessary to conserve an equilibrated water cycle. From a water use perspective, with concerns for water stress and water scarcity in mind, AR seems the most appropriate in regions like Brazil (*i.e.*, warm and very humid) or the EU (*i.e.*, temperate and humid).

9.9 Conclusion

Afforestation/Reforestation (AR) can achieve negative emissions through the growth of additional biomass capturing and sequestering CO_2 from the atmosphere *via* photosynthesis. The CO_2 sequestration differs from climates, tree species and forestry managements, with the highest potential in warm and humid regions of the world. In spite of the AR simplicity, notably compared with BECCS (Chapter 5) or DACCS (Chapter 6), and its co-benefits (*e.g.*, protecting biodiversity, increasing soil carbon and nutrients content), AR CO_2 sequestration is inherently saturated, and subject to natural (*e.g.*, fire, flood) or human-induced (*e.g.*, deforestation) disturbances, which can reduce, or even reverse its climate mitigation action. In this chapter, we have presented a whole-system analysis of AR, including notably a forest growth and associated CO_2 sequestration model, adjusted with a fire-penalty model. AR maximum CO_2 sequestration of 40–709 tonnes CO_2 per ha has been found to be reduced by as much as 35% when considering such risk.

AR for the purpose of negative emissions is not limited to forest growth, however, and a set of forestry operations have thus been considered to assess AR GHG balance, cost and water footprint. We have found that GHG emissions arising from AR forestry operations, both initial (*i.e.*, forest establishment and forest roads construction) and annual (*i.e.*, forest maintenance forest road maintenances) only accounts for 1.2–4.4 tonnes CO_2 per ha over a period length of 100 years, with forest roads construction contributing the most. AR has been found to be one of the most efficient CDR options, as high as 88.8–95.8% over 30 years, higher than 90.0% over 40 years, and as high as 97.5–99.6% over 100 years. However, depending on how much the energy system will be decarbonised, AR's initial carbon debt may not be negligible (*e.g.*, AR's CDR efficiency over the first decade could be as low as 37.8% in India). It might therefore be of interest to investigate a further financial and business plan for AR, based on AR's long-term CDR efficiency and consequently removal potential rather than on AR's effective and annual CDR potential.

Because AR requires the forest to be maintained regularly to achieve permanent CO_2 sequestration, AR's total levelised costs are found to be minimal over 30–50 years around \$43–97 per tonne CO_2, but to increase as much as \$12–26 per tonne CO_2 every 100 years afterwards. The cheapest AR has been found to be deployed in Brazil, as an archetypal country for Southern American countries, where a combination of good climatic conditions (*i.e.*, warm and humid) and low-cost labour is available. The most expensive AR has been seen to be deployed in the European Union and the United States, characteristic of temperate climates and more expensive labour. There is overall, a need to clarify: 1) the concept of permanence with regards to CO_2 removal and 2) the hypothesis of the current literature on AR, as the analysis in this chapter has indicated higher cost ranges than in the current literature, and as AR might consequently be less cost-efficient than other CDR options, such as BECCS. Alternatively, exploiting revenues from the production of timber and forest residues (from the forest management operations) could provide one way to compensate for AR increasing CDR costs over time.

Lastly, AR should be deployed with respect to other sustainability objectives, as defined in the United Nations' Sustainable Development Goals for instance. The (blue) water footprint of AR has been found to be the highest in warm and dry countries, such as India and the lowest in very humid countries, tropical such as in Brazil, or temperate such as in the European Union. The analysis presented in this chapter is, however, still at a high level of disaggregation, and further local research is necessary for better understanding, and to deploy AR without enhancing water scarcity, or conversely, with the purpose of enhancing water quality.

References

1. M. Amano, P. M. Fearnside, J. Frangi, P. Frumhoff, D. Goldberg, N. Higuchi, A. Janetos, M. Kirschbaum, R. Lasco, G. J. Nabuurs, R. Persson, W. Schle and R. Scholes, in *Land Use, Land-Use Change and Forestry*, 2000, ed. I. Noble, M. Apps, R. Houghton, D. Lashof, W. Makundi, D. Murdiyarso, B. Murray, W. Sombroek and R. Valentini.
2. T. A. R. Jenkins, E. D. Mackie, R. W. Matthews, G. Miller, T. J. Randle and M. E. White, *FC Woodland Carbon Code: Carbon Assessment Protocol (v2.0)*, 2018.
3. J. Morison, R. Matthews, G. Miller, M. Perks, T. Randle, E. Vanguelova, M. White and S. Yamulki, *Understanding the carbon and greenhouse gas balance of forests in Britain*, 2012.
4. USDA, Forest Vegetation Simulator (FVS), https://www.fs.fed.us/fvs/index.shtml, (accessed 14 December 2020).
5. British Columbia, Table Interpolation Program for Stand Yields (TIPSY), https://www2.gov.bc.ca/gov/content/industry/forestry/managing-our-forest-resources/forest-inventory/growth-and-yield-modelling/table-interpolation-program-for-stand-yields-tipsy, (accessed 14 December 2020).
6. IPCC, *Good Practice Guidance for Land Use, Land-Use Change and Forestry*, ed. J. Penman, M. Gytarsky, T. Hiraishi, T. Krug, D. Kruger, R. Pipatti, L. Buendia, K. Miwa, T. Ngara, K. Tanabe and F. Wagner, Intergovernmental Panel on Climate Change, 2003, https://www.ipcc-nggip.iges.or.jp/public/gpglulucf/gpglulucf_files/GPG_LULUCF_FULL.pdf.

7. MODIS and NASA, *Gross Primary Production (GPP)/Net Primary Production (NPP)*, https://modis.gsfc.nasa.gov/data/dataprod/mod17.php, (accessed 14 December 2020).
8. J. G. van Minnen, B. J. Strengers, B. Eickhout, R. J. Swart and R. Leemans, *Carbon Balance Manage.*, 2008, **3**, 3.
9. M. Gusti, *G4M Overview (global version)*, 2015.
10. C. Whittaker, N. Mortimer, R. Murphy and R. Matthews, *Biomass Bioenergy*, 2011, **35**, 4581–4594.
11. M. A. Elsayed, R. Matthews and N. D. Mortimer, *Carbon and energy balances for a range of biofuels options*, 2003.
12. M. Röder, C. Whittaker and P. Thornley, *Biomass Bioenergy*, 2015, **79**, 50–63.
13. L. Eriksson and L. Gustavsson, *Biomass Bioenergy*, 2010, **34**, 610–619.
14. M. Fiala and J. Bacenetti, *Biomass Bioenergy*, 2012, **42**, 107–113.
15. F. Zhang, D. M. Johnson, J. Wang and C. Yu, *Energy*, 2016, **114**, 1053–1062.
16. L. Eriksson and L. Gustavsson, *Biomass Bioenergy*, 2010, **34**, 82–90.
17. K. A. Farley, E. G. Jobbágy and R. B. Jackson, *Glob. Chang. Biol.*, 2005, **11**, 1565–1576.
18. A. I. J. M. van Dijk and R. J. Keenan, *For. Ecol. Manage.*, 2007, **251**, 1–9.
19. D. Lovarelli, J. Bacenetti and M. Fiala, *Sci. Total Environ*, 2016, **548–549**, 236–251.
20. A. Y. Hoekstra, A. K. Chapagain, M. M. Aldaya and M. M. Mekonnen, *The Water Footprint Assessment Manual: Setting the Global Standard*, 2011.
21. D. Chi, H. Wang, X. Li, H. Liu and X. Li, *Ecol. Indic.*, 2018, **92**, 141–150.
22. Z. Wang, D. Peng, D. Xu, X. Zhang and Y. Zhang, *J. Arid Environ.*, 2020, **182**, 104257.
23. C. Corbari, G. Ravazzani, M. Galvagno, E. Cremonese and M. Mancini, *Sensors*, 2017, **17**, 2664.
24. FAO, *Global Ecological Zoning for the Global Forest Resources Assessment 2000*, Rome, 2001.
25. FAO, *Global Ecological Zones for FAO Forest Reporting: 2010 Update*, Rome, 2012.
26. H. Aalde, P. Gonzalez, M. Gytarsky, T. Krug, W. A. Kurz, S. Ogle, J. Raison, D. Schoene, N. H. Ravindranath, N. G. Elhassan, L. S. Heath, N. Higuchi, S. Kainja, M. Matsumoto, M. J. S. Sánchez and Z. Somogyi, in *Volume 4: Agriculture, Forestry and Other Land Use - IPCC Guidelines for National Greenhouse Gas Inventories*, 2006.
27. H. Aalde, P. Gonzalez, M. Gytarsky, T. Krug, W. A. Kurz, R. D. Lasco, D. L. Martino, B. G. McConkey, S. Ogle, K. Paustian, J. Raison, N. H. Ravindranath, D. Schoene, P. Smith, Z. Somogyi, A. van Amstel and L. Verchot, in *Volume 4: Agriculture, Forestry and Other Land Use - IPCC Guidelines for National Greenhouse Gas Inventories*, 2006.
28. C. J. E. Schulp, G. J. Nabuurs, P. H. Verburg and R. W. de Waal, *For. Ecol. Manage.*, 2008, **256**, 482–490.
29. B. Berg, P. Gundersen, C. Akselsson, M. B. Johansson, Å. Nilsson and L. Vesterdal, in *Silva Fennica*, 2007, vol. 41, pp. 541–558.

30. K. Pingoud, K. E. Skog, D. L. Martino, M. Tonosaki and Z. Xiaoquan, in *Volume 4: Agriculture, Forestry and Other Land Use - IPCC Guidelines for National Greenhouse Gas Inventories*, 2006.
31. S. Fuss, W. F. Lamb, M. W. Callaghan, J. Hilaire, F. Creutzig, T. Amann, T. Beringer, W. De Oliveira Garcia, J. Hartmann, T. Khanna, G. Luderer, G. F. Nemet, J. Rogelj, P. Smith, J. V. Vicente, J. Wilcox, M. Del Mar Zamora Dominguez and J. C. Minx, *Environ. Res. Lett.*, 2018, **13**, 063002.
32. G. A. Stainback and J. R. R. A. Lavalapati, *Nat. Resour. Model.*, 2004, **17**, 299–317.
33. L. Hu, G. A. Stainback and X. Li, in *Southern Agricultural Economics Association's 2016 Annual Meeting*, San Antonio, Texas, USA, 2016.
34. S. Couture and A. Reynaud, *Ecol. Econ.*, 2011, **70**, 2002–2011.
35. S. G. McNulty, *Environ. Pollut.*, 2002, **116**, S17–S24.
36. M. D. Hurteau, B. A. Hungate and G. W. Koch, *Carbon Balance Manage.*, 2009, **4**, 1.
37. E. Chuvieco, I. Aguado, S. Jurdao, M. L. Pettinari, M. Yebra, J. Salas, S. Hantson, J. De La Riva, P. Ibarra, M. Rodrigues, M. Echeverría, D. Azqueta, M. V. Román, A. Bastarrika, S. Martínez, C. Recondo, E. Zapico and F. J. Martínez-Vega, *Int. J. Wildl. Fire*, 2014, **23**, 606–619.
38. D. M. Szpakowski and J. L. R. Jensen, *Remote Sens.*, 2019, **11**, 2638.
39. LANDFIRE Program: Data Products - Fire Regime, https://www.landfire.gov/fireregime.php, (accessed 15 October 2020).
40. C. L. Whittaker, N. D. Mortimer and R. W. Matthews, *Understanding the carbon footprint of timber transport in the United Kingdom*, 2010.
41. Y. Aldentun, *J. Clean. Prod.*, 2002, **10**, 47–55.
42. V. Timmermann and J. Dibdiakova, *Int. J. Life Cycle Assess.*, 2014, **19**, 1593–1606.
43. S. Berg and T. Karjalainen, *Forestry*, 2003, **76**, 271–284.
44. J. Kenney, T. Gallagher, M. Smidt, D. Mitchell and T. Mcdonald, in *37th Council on Forest Engineering Annual Meeting*, Moline, Illinois, USA, 2014.
45. BEIS, *MS Excel Spreadsheet - Greenhouse gas reporting - conversion factors 2020: full set (for advanced users)*, UK, 2020.
46. IEA, *CO_2 emissions from fuel combustion*, 2018.
47. G. G. T. Camargo, M. R. Ryan and T. L. Richard, *Bioscience*, 2013, **63**, 263–273.
48. I. Lewandowski, A. Kicherer and P. Vonier, *Biomass Bioenergy*, 1995, **8**, 81–90.
49. P. Nixon and M. Bullard, *Planting and Growing Miscanthus: Best Practice Guidelines for Applicants to DEFRA's Energy Crops Scheme*, 2001.
50. C. De Klein, R. S. A. Novoa, S. Ogle, K. A. Smith, P. Rochette, T. C. Wirth, B. G. McConkey, A. Mosier and K. Rypdal, in *Volume 4: Agriculture, Forestry and Other Land Use - IPCC Guidelines for National Greenhouse Gas Inventories*, 2006, pp. 1–54.
51. R. Parajuli, S. Løkke, P. A. Østergaard, M. T. Knudsen, J. H. Schmidt and T. Dalgaard, *Biomass Bioenergy*, 2014, **68**, 115–134.

52. F. Murphy, G. Devlin and K. McDonnell, *Renewable Sustainable Energy Rev.*, 2013, **23**, 412–420.
53. M. M. Mekonnen and A. Y. Hoekstra, *Hydrol. Earth Syst. Sci.*, 2011, **15**, 1577–1600.
54. IPCC, *Global Warming of 1.5 °C. An IPCC Special Report on the impacts of global warming of 1.5 °C above pre-industrial levels and related global greenhouse gas emission pathways, in the context of strengthening the global response to the threat of climate change*, Press, 2018.
55. World Bank, *Pump price for diesel fuel (US$ per liter)*, https://data.worldbank.org/indicator/ep.pmp.desl.cd, (accessed 15 February 2020).
56. Ministry of Mines and Energy*Monthly Energy Bulletin - December 2018*, 2018.
57. CEIC, Price of electricity, https://www.ceicdata.com/en/china/electricity-price%0A, (accessed 15 February 2020).
58. EUROSTAT, Electricity prices by type of user (EUR per kWh), https://ec.europa.eu/eurostat/databrowser/view/ten00117/default/table?lang=en, (accessed 15 February 2020).
59. Government of India, *Evaluation Report on Rajiv Gandhi Grameen Vidyutikaran Yojana*, New Delhi, India, 2014.
60. Energy Information Administration, *Average retail price of electricity (cents per kilowatthour)*, https://www.eia.gov/electricity/data/browser/#/topic/7?agg=1,0&geo=hfvvvvvvvvvo&endsec=2&linechart= ∼ ∼&columnchart=ELEC.PRICE.US-IND.A&map=ELEC.PRICE.US-IND.A&freq=A&ctype=linechart&ltype=pin&rtype=s&pin=&rse=0&maptype=0%0A, (accessed 15 February 2020).
61. R. W. Brinker, J. Kinard, R. Rummer and B. Lanford, *Machine rates for selected forest harvesting machines*, 2002.
62. J. Johansson, J.-E. Liss, T. Gullberg and R. Björheden, *Biomass Bioenergy*, 2006, **30**, 334–341.
63. US Bureau of Labor Statistics, Occupational Employment and Wage Statistics, https://www.bls.gov/oes/current/oes_stru.htm, (accessed 26 November 2019).
64. M. Duffy, *Estimated Costs for Production, Storage and Transportation of Switchgrass*, 2008.
65. M. de Wit and A. Faaij, *Biomass Bioenergy*, 2010, **34**, 188–202.
66. ALBA TREES, ALBA TREES - The UK's largest producer of cell grown plants, http://www.albatrees.co.uk/browse-products/, (accessed 16 February 2020).
67. HomeGuide, How Much Does Crushed Stone or Gravel Cost?, https://homeguide.com/costs/gravel-prices#grave, (accessed 16 February 2020).
68. World Bank, Official exchange rate (LCU per US$, period average), https://data.worldbank.org/indicator/PA.NUS.FCRF, (accessed 15 February 2020).
69. World Bank, inflation, GDP deflator (annual%), https://databank.worldbank.org/data/reports.aspx?source=2&series=NY.GDP.DEFL.KD.ZG, (accessed 15 February 2020).

70. World Bank, PPP conversion factor, GDP (LCU per international $), https://databank.worldbank.org/reports.aspx?source=2&series=NY.GDP.DEFL.KD.ZG, (accessed 15 February 2020).
71. M. Fajardy and N. Mac Dowell, *One Earth*, 2020, **3**, 214–225.
72. M. Fajardy, Imperial College London, 2020.
73. D. Archer, *J. Geophys. Res.: Oceans*, 2005, **110**, 1–6.
74. R. G. Allen, L. S. Pereira, D. Raes and M. Smith, *Crop evapotranspiration - Guidelines for computing crop water requirements*, 1998.
75. R. B. Myneni, S. Hoffman, Y. Knyazikhin, J. L. Privette, J. Glassy, Y. Tian, Y. Wang, X. Song, Y. Zhang, G. R. Smith, A. Lotsch, M. Friedl, J. T. Morisette, P. Votava, R. R. Nemani and S. W. Running, *Remote Sens. Environ.*, 2002, **83**, 214–231.
76. S. P. Healey, M. W. Hernandez, D. P. Edwards, M. A. Lefsky, J. E. Freeman, P. L. Patterson, E. J. Lindquist and A. J. Lister, *CMS: GLAS LiDAR-derived Global Estimates of Forest Canopy Height, 2004–2008*, 2015. https://daac.ornl.gov/CMS/guides/LIDAR_FOREST_CANOPY_HEIGHTS.html.

CHAPTER 10

Feasibility of Using Biologically-based Processes in the Open Ocean and Coastal Seas for Atmospheric CO_2 Removal

PHILLIP WILLIAMSON,*[a] PHILIP W. BOYD,[b] DANIEL P. HARRISON,[c] NICK REYNARD[d] AND ALI MASHAYEK[d]

[a] University of East Anglia, Norwich, UK; [b] University of Tasmania, Hobart, Australia; [c] Southern Cross University, Coffs Harbour, Australia; [d] Imperial College, London, UK
*Email: p.williamson@uea.ac.uk

This chapter assesses the feasibility of six ways to achieve greenhouse gas removal by using marine biological processes to increase CO_2 drawdown from the atmosphere. Four approaches involve increasing the supply of nutrients to the open ocean: adding iron directly; adding iron indirectly *via* the atmosphere; adding macronutrients; and by enhancing upwelling to increase natural nutrient recycling. The intention is to stimulate primary production and CO_2 uptake by marine phytoplankton, and hence deep-water carbon storage *via* the ocean biological carbon pump. Two coastal methods are also considered to increase CO_2 drawdown: restoring blue carbon ecosystems (mangroves, salt marsh and seagrass habitats) as a 'nature-based solution', and large-scale cultivation of seaweeds (macroalgae) and subsequent carbon storage, primarily by the marine equivalent of BECCS (bioenergy with carbon capture and storage). For each approach, information is

Energy and Environment Series No. 31
Greenhouse Gas Removal Technologies
Edited by Mai Bui and Niall Mac Dowell

Published by the Royal Society of Chemistry, www.rsc.org

given on scientific understanding of key processes, environmental consequences, cost-effectiveness, carbon accounting, governance and public acceptability. Major uncertainties are identified regarding the quantification of carbon removal, the permanence of storage, and the possible emissions of other greenhouse gases. Governance-related constraints are particularly important for the open ocean approaches. Marine BECCS using seaweed theoretically offers high potential, but its large-scale deployment has not been well investigated. Coastal ecosystem restoration is already being used for small-scale carbon offsetting, and can be considered a no-regrets mitigation option, on the basis of its non-climatic benefits. The scalability and effectiveness of this approach for increasing CO_2 removal are, however, uncertain, with many associated risks.

10.1 Introduction

The underlying mechanism for biologically-based greenhouse gas removal (GGR) in the ocean is the same as on land. Land-based GGR or negative emission technologies (NETs) are covered in Chapters 5, 8 and 9 of this volume. For biologically-based GGR, the greenhouse gas carbon dioxide (CO_2) is removed from the atmosphere through photosynthesis by plants, and a sufficiently high proportion of that carbon remains out of wider circulation for long enough to provide climate benefits. The scientific knowledge on using marine processes for CO_2 removal has recently been reviewed by the Joint Group of Experts on the Scientific Aspects of Marine Environmental Protection (GESAMP),[1] by the US National Academies of Sciences, Engineering and Medicine[368] and other comparative analyses,[2,3] and by two reports of the Intergovernmental Panel on Climate Change (IPCC).[4,369] These reviews mostly focus on the scale and effectiveness of initial drawdown of atmospheric CO_2.

Yet carbon removal from the air is only half the story. For biologically-based CO_2 removal using marine processes, an equally important component of effectiveness relates to the form and security of subsequent carbon storage;[5] both aspects may be highly variable in both time and space. In the absence of geological carbon sequestration (as would be used for BECCS; Chapter 5), high 'leakage' from natural storage is inevitable. That is because >99% of marine organic carbon will be eventually be remineralised by heterotrophs (microbes, fungi and animals) in the ocean carbon cycle and subsequently returned to the atmosphere, through geophysical processes occurring over months, years and millennia. While carbon storage over 100 years is commonly used as the standard for comparisons of sequestration efficiency,[6] information on shorter and longer time periods is also important for climate policy decisions.

Figure 10.1 summarises the main processes and timescales relevant to both natural re-cycling of carbon in the open ocean and artificially increased production through ocean fertilisation.

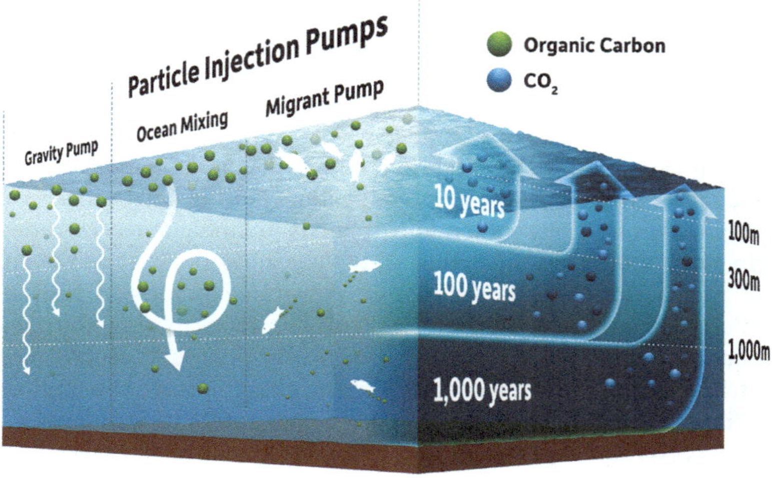

Figure 10.1 Key components of the ocean carbon cycle relevant to open ocean fertilisation. Left-hand side: the three main components of the biological carbon pump, comprising the gravity pump and two examples of particle injection pumps: ocean mixing and the migrant pump (driven by vertical animal movement). Right-hand side: the range of time periods for the return of biologically-fixed carbon to the atmosphere, depending on the depth at which particulate organic carbon is decomposed to CO_2. The relative importance of the pumps, of different water depths and of different return periods are determined by ocean physics, which varies greatly between different ocean regions. Image from ref. 7 courtesy of Michael Osadciw with permission from the University of Rochester.

Not all the carbon in marine organic matter will return to the atmosphere as CO_2; some may be released as methane (CH_4) and there may also be associated emissions of nitrous oxide (N_2O), with those pathways favoured by low oxygen conditions. Since N_2O has a global warming potential up to 300 times higher than CO_2, and CH_4 30–80 times higher (depending on whether releases are sustained or as pulses, and the time period under consideration),[8] the additional release of seemingly small amounts of these gases from increased marine carbon stores can have significant counteracting effects. Quantification of these 'multiplier' release rates of other greenhouse gases will therefore be as important as measuring CO_2 removal and release rates, particularly with regard to net-zero ambitions over the next 20–50 years. There may also be opposing climatic feedbacks operating through other biogenic gases, such as dimethyl sulphide (DMS).[9] Reliable estimation of future multi-gas release rates is, however, challenging, due to a lack of understanding of the factors that may cause spatial and temporal variability in biogeochemical processing, including the potential for non-linear responses, together with inherent uncertainties in societal behaviour (and hence the rate of climate change) over the next 50–100 years. Many environmental factors and human influences will alter over that period, with a real risk that such changes will make both natural and enhanced carbon storage in the ocean and marine sediments less secure.[4]

In this chapter, most attention is given to increasing marine productivity in the upper ocean using direct ocean iron fertilisation (OIF), achieved by the addition of soluble iron to near-surface waters. While this is a relatively well-studied approach, relevant research to date has primarily investigated natural processes rather than R&D for climate mitigation. For the latter purpose, OIF is politically and scientifically contentious,[10] and is unambiguously considered as geoengineering. Both the Convention on Biological Diversity[11] and the London Protocol[12,13] have introduced resolutions to constrain commercial development and implementation. While scientific research continues to be permitted under both frameworks, there have been no OIF field experiments led by scientific bodies, for either fundamental or applied research purposes, in the period 2009 to date (late 2021). International regulatory constraints were, however, side-stepped by a privately funded OIF experiment carried out off North West Canada in 2012, with the stated aim of fishery restoration. While of questionable legality[14] and with uncertain scientific outcomes,[15] this experiment (unintentionally) provided information on public and political responses[16] to the large-scale manipulation of ocean productivity, as discussed further in Section 10.2.1 below.

For biologically-based coastal CO_2 removal, the main approach covered here is the restoration of 'blue carbon' ecosystems, using rooted vegetation to increase the carbon content of their sediments. In contrast to OIF, interventions involving coastal blue carbon have high political and public acceptability, receiving wide international endorsement as a 'natural solution' to climate change[17–19] and included in the Nationally Determined Contributions to the Paris Agreement of over 50 countries.[20] Since carbon removal by vegetated coastal ecosystems is a natural process, rather than a 'technology', it is not always considered as a negative emissions technology (NET). However, additional CO_2 removal (considering current conditions as the baseline) can only be achieved by active management of such systems, through restoration, expansion or changing hydrological regimes, *i.e.*, comparable to reforestation/afforestation, and actions to enhance soil carbon. This wider usage of the term 'technology' is therefore consistent with its general definition as 'the practical application of scientific knowledge', rather than its alternative, more specific association with engineering and industry.

Directly-achieved ocean iron fertilisation and actions to increase the climatic benefits of coastal blue carbon are reviewed here with a focus on scientific understanding of key processes, wider environmental consequences, cost-effectiveness, carbon accounting, and governance and public acceptability. In addition, three other proposed methods for increasing open ocean productivity are considered in less detail: iron salt aerosol, to provide atmospheric delivery of iron; macronutrient additions; and enhanced upwelling, to increase *in situ* nutrient re-supply. In addition to the restoration of coastal blue carbon ecosystems, the large-scale cultivation of macroalgae (seaweed) is also considered as a coastal CO_2 removal process, although this technique could in theory also be deployed offshore.

10.2 Biologically-based CO_2 Removal in the Open Ocean

10.2.1 Direct Ocean Iron Fertilisation

10.2.1.1 Scientific Understanding of Key Processes

Photosynthesis by microscopic ocean plants (phytoplankton) accounts for around 45% of global carbon fixation and is responsible for a gross annual flux from the atmosphere to the ocean interior of around 40 Gt CO_2 *via* the 'biological carbon pump'.[21,22] Several different processes are involved in moving this carbon from the upper ocean to deeper waters[23] (Figure 10.1). A closely similar amount of CO_2 is returned annually to the atmosphere, arising from decomposition processes and subsequent outgassing in other ocean regions. Such recycling takes from years to millennia to achieve. Because of the delay in return processes, there would be a step-wise increase in the total carbon stored in the ocean (*i.e.*, net removal from the atmosphere) if downward carbon flux through the biological carbon pump could be substantively increased.

Trace amounts of iron are essential for photosynthesis. While there is rarely any shortage of this element on land, studies in the late 1980s showed that the combination of the low solubility of iron in seawater and its low external re-supply rates (*e.g.*, from wind-blown dust and rivers) limits marine productivity in many ocean regions, particularly in high latitudes.[24,25] Natural long-term variability in terrestrial iron supply to the ocean helped drive glacial cycles,[26,27] although its contribution was probably limited to an atmospheric change of 2–15 ppm CO_2,[28,29] with several other processes also involved.[30] See Section 10.2.2 for further discussion of these effects.

Twelve well-documented, patch-scale field experiments (Figure 10.2) were carried out between 1993 and 2009 to improve scientific understanding of the role of ocean iron supply in natural climate dynamics, particularly during glacial-interglacial cycles. Seven of these experiments were in the Southern Ocean, where near-surface levels of dissolved iron are usually extremely low, at sub-nanomolar levels, resulting in 'unused' macronutrients (primarily nitrate and phosphate), and hence where ocean iron fertilisation (OIF) might be most effective. Together, these experiments confirmed that iron additions could greatly stimulate the growth of phytoplankton in such high nutrient, low chlorophyll regions.[31–33] However, there was high variability between experiments: the iron-mediated increase in chlorophyll varied between 2–25 times, with 0–50% of the additional production (carbon within organic material) sinking to water depths greater than 250 m.[34–36] These experiments provided many other insights into the initial effects of OIF,[37,38] as well as the factors controlling the ocean carbon cycle. Nevertheless, because of their limited duration and relatively small spatial scale, additional information is needed to assess the effectiveness of this approach for climate mitigation purposes. This knowledge deficit has subsequently

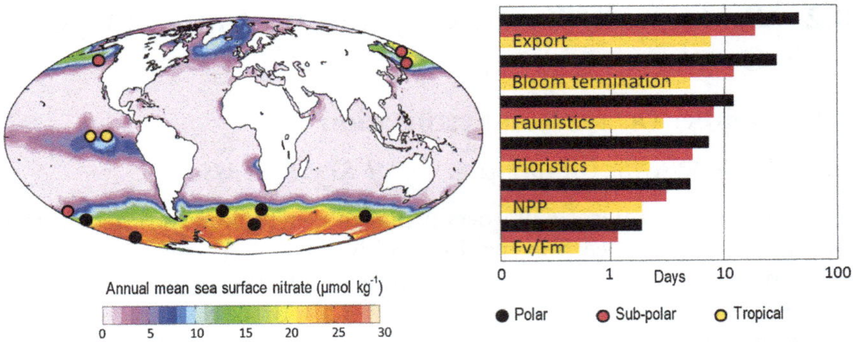

Figure 10.2 Left panel, locations of *in situ* ocean iron fertilisation experiments in polar (black circles), subpolar (red circles) and tropical (orange circles) waters, shown on the World Ocean Atlas map[39] of annual surface nitrate availability. Right, summary of timing of different biological responses to iron additions with the same colour coding, on a log scale. 'Export' represents period of increased downward flux of particles in the top 300 m; 'bloom termination', the time when plankton biomass begins to fall; 'faunistics and floristics', the times of greatest taxonomic changes; 'NPP', net primary production; and 'Fv/Fm' the initial stimulation of photosynthesis. Based on ref. 33 and 40.

been mostly addressed using ocean biogeochemical models of various complexity, with the more advanced including explicit representation of iron biogeochemistry in seawater, as discussed below.

Larger (*i.e.*, much greater than 1000 km^2), more ambitious field experiments[41] would be required to better-constrain data on the fate of carbon fixed by OIF and improve the representation of other relevant processes in computational models. There is a particular need to improve understanding of macro-scale nutrient interactions and their 'downstream' impacts,[42–44] and also the importance of physical features and trophic interactions[45] in carbon export from the upper ocean. While further field experiments have been planned by both publicly- and privately-funded research groups,[46,47] such studies now need to obtain international approval based on environmental impact assessments.[1,12]

In addition to the information from patch-scale field experiments, assessment of the potential for using OIF for climate mitigation requires knowledge of: (i) the natural dynamics of iron biogeochemistry in the ocean,[48–51] with relevant observations provided by the international GEOTRACES project;[38,52] and (ii) the many parameters affecting the export of organic particles from the upper ocean, their remineralisation to CO_2 in deeper water, and subsequent transport and transformations of their carbon in different depth strata of the global ocean[21–23,53–55] (Figure 10.1). Three factors are particularly important in determining the rate of return to the atmosphere, over the timescales of weeks to millennia: the sinking rate of particulate carbon, and hence the water depth where it is re-mineralised (affected by particle size and density, also by food-web processes); site-specific mesoscale hydrodynamic features affecting vertical water movement (eddy-induced upwelling and downwelling; winter mixing and island effects); and the large-scale, long-term patterns of lateral water movements within and between ocean basins. All these processes are likely to be sensitive to future climate change, through changes in planktonic community composition, mixing and stratification, and the strength of global ocean circulation.[4]

Biogeochemical models that take into account some (but not all) of these processes have been used to estimate the maximum amount of CO_2 that might be removed by OIF, *i.e.*, the additional (cumulative) ocean carbon drawdown if iron were no longer a limiting nutrient. In 1991, the potential total removal over 100 years was estimated to be at least 550 Gt CO_2, based on fertilising only the Southern Ocean.[56] That estimate was subsequently lowered to 257 Gt CO_2 and 95 Gt CO_2 in fully-global estimates made in 2006[43] and 2008[57] respectively. Differences between models mostly arise in later decades of the 100-year period, due to different rates of CO_2 return; all models gave the highest annual removal rates within 10–20 years of full OIF deployment. For the two more recent model estimates,[43,57] the maximum removal rates were 3–4 Gt CO_2 per year: that value could therefore, in theory, be achieved by 2050, assuming that the decision to use OIF is made in the very near future, and rapid implementation is achieved. Although politically highly unlikely (for reasons discussed below), full OIF implementation is not logistically impossible, requiring around 5 Mt Fe per year (less than 1% of the current global iron production[43]), that would need to be added in soluble form.

New model-based estimates of the potential OIF effectiveness are currently being developed, using improved understanding of the 3D complexity and heterogeneity of the biological carbon pump,[23] taking into account the carbon transport by ocean circulation and water masses, and also the 'downstream' effects of OIF macronutrient removal (in review).[58]

10.2.1.2 Environmental Consequences and Wider Context

For OIF to produce climatically-significant benefits (generally considered to be the sustained, long-term removal of more than 1 Gt CO_2 per year, *i.e.*, around 2.5% of current annual emissions), major increases in marine productivity would be needed at the global scale. These productivity increases will necessarily also cause changes in marine species abundance, biodiversity, ecosystem functioning and biogeochemistry throughout the water column, including seafloor ecosystems.[37,59] The wide range of these impacts would not be limited to the fertilised area (that would need to include polar and sub-polar waters, for gigaton-scale CO_2 removal), and could be either positive or negative from a human perspective. It is possible that fish biomass could increase and benthic invertebrates, such as crustacea and molluscs, also become more abundant, as currently occurs in the more productive regions of the Southern Ocean.[60] Beneficial effects on fish and fisheries are also indicated from other natural analogues, *e.g.*, the fertilising effect of the Kasatoshi volcanic eruption in the northeast Pacific.[61]

However, fishery co-benefits have yet to be demonstrated experimentally, and would be inherently difficult to verify[3] because of fish movements, population patchiness and natural variability in year-class success. Furthermore, local gains could be offset by reductions elsewhere, due to the non-uniform changes in primary production. These effects include far-field decreases due to the non-replenishment of other nutrients that become depleted at the site of iron additions, and therefore are no longer available to support productivity at other locations ('nutrient stealing').[37] Indeed, model simulations indicate that "the bulk of the changes in production do not occur at the fertilisation site or in the immediate vicinity but on larger spatial scales",[42] with major regional-scale productivity reductions (as well as increases) occurring in equatorial waters, also between 25–30°S and in the northwest Pacific.[43,62]

Additional biogeochemical perturbations arising from higher productivity under OIF are likely to accentuate three undesirable global trends: first, a decrease in midwater oxygen,[63] an issue of increasing global concern;[64] and second, an increase in the rate of acidification of the ocean interior,[65,66] that could extend throughout the water column in the Southern Ocean[57] if it were fully used for OIF. Any acceleration of anthropogenic ocean acidification (the fall in seawater pH due to increasing atmospheric CO_2) would be of particular concern in the Southern Ocean, since its waters are already close to undersaturation of aragonite (the form of $CaCO_3$ widely used by marine calcifiers), with potentially far-reaching ecosystem impacts.[67,68]

The third potential biogeochemical consequence of large-scale OIF is an increase in the production and release of two marine trace gases that might counteract the benefits of CO_2 removal: methane (CH_4) and nitrous oxide (N_2O). These gases are increasingly important for climate policy, since their rising atmospheric levels[69,70] jeopardise the fundamental UNFCCC goal of stabilising greenhouse gas concentrations in the atmosphere, and the associated aim of achieving a balance between emissions and removals under the Paris Agreement. For OIF, additional N_2O emissions are of greatest concern.[37,71–73] The limited N_2O measurements during OIF experiments showed relatively minor increases, offsetting CO_2 removal by $<12\%$.[74,75] However, most N_2O production is likely to occur over much longer timescales,[73] with scaling proportional to carbon export, thereby representing a major uncertainty in assessing OIF effectiveness.

10.2.1.3 Cost-effectiveness and Carbon Accounting

Reliable information on costs in relation to climatic benefits (and any associated non-climatic benefits and losses) is a prerequisite for making any GGR approach an actionable policy option. OIF was initially considered to be a low-cost option for counteracting climate change, with an early estimate of around US$ 2 per tonne of CO_2 "sequestered for 1000 to 2000 years".[76] That estimate assumed that the weight of carbon removed would be ~16 000 times higher than the weight of iron added. However, using values of the carbon exported from the upper ocean in five OIF field experiments, costs were subsequently re-estimated[77] to be at least US$ 7–90 per tonne CO_2 (converting published data from per t C to per t CO_2), not accounting for monitoring and verification costs, nor potential N_2O offsets.

A further, more comprehensive analysis[36] included the cost of nutrient manufacture (using the agricultural fertiliser, ferrous sulphate), transport and delivery; carbon conversion and export efficiencies; nutrient dynamics; N_2O release; and carbon 'leakage' over 100 years (Figure 10.3). The estimated average cost of OIF then increased to US$ 457 per t CO_2 removed, higher than for most other CO_2 removal approaches[78] and well above the cost currently considered competitive in an international carbon market.[79]

Caution is, however, needed with regard to the wider application of any single estimated value for OIF costs. As already noted, the field experiments showed high variability: not only was there a range of 0–50% in carbon export efficiencies for studies at different localities, but there was a five-fold difference in the biological response (based on the maximum increase in chlorophyll levels) when an OIF experiment was repeated at the same site three years later.[80] Based on pessimistic assumptions, costs could become infinitely high; on optimistic assumptions, a cost range of US$ 18–58 per tCO_2 might be achievable.[36] However, the lower cost range would require OIF implementation to be limited to the environmental conditions that favour strong 'particle injection pumps'[23] (Figure 10.1) to consistently

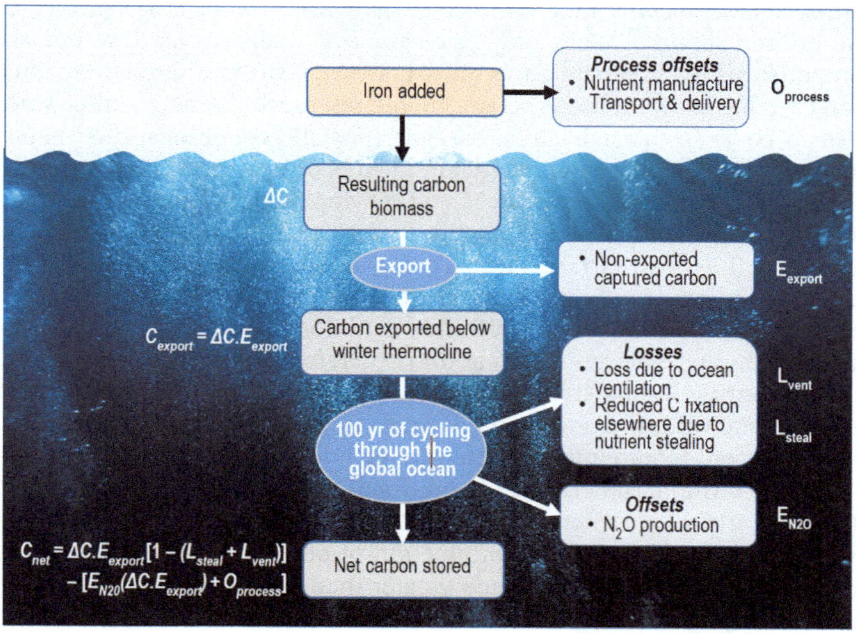

Figure 10.3 Framework for costing ocean iron fertilisation based on the efficiency of the main processes involved over a 100-year timescale. Modified from ref. 36.

deliver high proportions of the biologically-fixed carbon to below the depth of winter mixing. In practice, that would mean restricting fertilisation to times and regions where physically, chemically and biologically-favourable oceanographic conditions align. Such optimal combinations are not yet well-specified, and may not be easy to locate operationally. The relatively limited occurrence of such locations is also likely to constrain the scale of total benefits that might be obtained.

Nevertheless, costs might be decreased by using aircraft, rather than ships, to add the iron,[40] and/or by using iron chelates to provide slow release,[81] possibly combined with floating pellets, flakes[82] or other materials.[83] Such approaches have not yet been field-tested, and the ingestion of buoyant, synthetic iron-rich particles by fish, seabirds or sea mammals could introduce additional environmental risks. The potential for regional-scale atmospheric delivery *via* the iron salt aerosol method is considered separately in Section 10.2.2 below.

If climate-scale OIF were to be implemented, two technologies would be crucial for both pre-deployment site selection and post-deployment carbon accounting: satellite remote sensing and autonomous *in situ* water column observing. The former has already been extensively used during OIF experiments, with images directly confirming the areal extent of biological changes due to iron fertilisation (Figure 10.4), and distinguishing changes in biomass from those due to phytoplankton physiology.[84] The wider challenge is to quantify additional productivity compared to untreated areas where natural phytoplankton blooms may also occur, as a result of heterogeneity in the natural iron supply.[38] When only patch-size iron additions are involved, that comparison is relatively straightforward; however, if OIF treatments were multi-year, covering many thousand square kilometres, much greater attention would need to be given to establishing baseline (pre-OIF) conditions, their natural inter-annual variability and potential influences of long-term climate change.

Remote sensing can reliably show phytoplankton bloom occurrence (under cloud-free conditions), but is limited to the detection of spectral changes at the ocean surface. Furthermore, as already noted, iron-stimulated increases in upper ocean biomass are often poorly correlated with the magnitude of downward carbon flux and long-term storage.[88,89] Comprehensive water column information is therefore also needed for OIF carbon accounting. While ship-based observations and sampling have to date provided such information in the vicinity of OIF experiments, larger-scale treatments will require much greater use of autonomous *in situ* platforms in order to determine the 4D fate of additional new production, together with other potential biogeochemical impacts that will need to be monitored (deoxygenation, acidification and production of other greenhouse gases) over an area many orders of magnitude greater than was fertilised.

Vertically-profiling Argo floats have collected year-round data on temperature and salinity to depths of ~2000 m on a global basis since 2004,[90] and can now also include a wide range of bio-optical and biogeochemical sensors.[91,92] Yet it may take a further decade of improvements in the

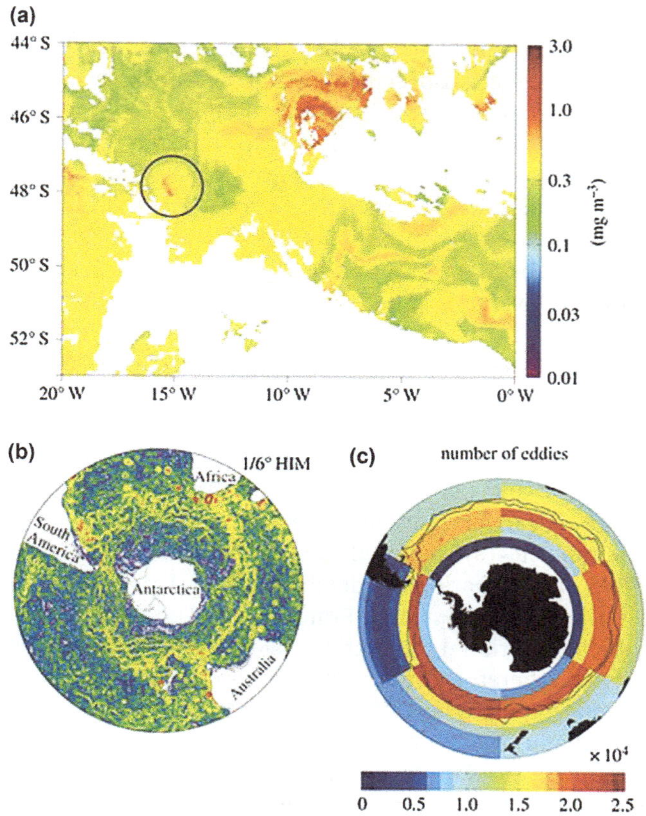

Figure 10.4 Ocean iron fertilisation will necessarily be superimposed on highly heterogeneous patterns of natural phytoplankton blooms, primarily driven by eddies, as shown here for the Southern Ocean. (a) MODIS satellite image of chlorophyll concentrations in part of the Atlantic sector of the Southern Ocean, showing the location (circled) of the 2009 LOHAFEX iron fertilisation study; (b) and (c) frequency and spatial pattern of eddies in the Southern Ocean based on high-resolution modelling and from satellite altimetry datasets, respectively. Adapted from ref. 85, 86 and 87 and reproduced from ref. 38 with permission from Royal Society Publishing, Copyright 2016.

reliability and stability of *in situ* iron sensors[93] before they can routinely provide the quantity and quality of data needed for reliable OIF monitoring and verification, closely linked to high-resolution model simulations.

10.2.1.4 Governance and Acceptability

As identified in previous assessments,[2,5,37,94] there are major governance-related difficulties in implementing large-scale OIF as a climate mitigation action. Such problems arise from the widely-held (but not unanimous) opinion of scientists, non-governmental organisations, intergovernmental bodies, governments and the wider public that such an approach is risky and undesirable. OIF was not considered to be a feasible approach to climate mitigation by either of two recent Special Reports of the Intergovernmental Panel on Climate Change;[4,95] its large-scale usage was described as "ineffective and potentially deleterious" by the leadership of the international Surface Ocean-Lower Atmosphere Study (SOLAS);[96] and ocean fertilisation, by iron or macronutrients, has been identified as the CO_2 removal method that presented the greatest threat to biodiversity, out of 15 approaches considered.[97] Similar views have been expressed by many other bodies[98–100] of varying scientific expertise, within a much wider debate regarding the ethics of climate geoengineering.[11,101,102]

At the intergovernmental level, such concerns resulted in a non-binding 'moratorium' on OIF by the Convention on Biological Diversity (Decision X/33 in 2010)[11] and the development of detailed regulation by the Convention on the Prevention of Marine Pollution by Dumping of Wastes and Other Matter (London Convention and London Protocol; LC-LP).[12] In 2008, Resolution LC-LP.1(1) prohibited ocean fertilisation except for "legitimate scientific research"; in 2013, Resolution LP.4(8) amended the London Protocol to require fertilisation experiments outside national jurisdictions to seek permission under a two-step Ocean Fertilisation Assessment Framework.[1] The LP amendment has yet to achieve sufficient national ratification (by two-thirds of parties) for it to be legally binding, and therefore ambiguities remain regarding the status of ocean fertilisation in international law.[44,103–105] Nevertheless, we are not aware of any current governmental interest to develop OIF for climate mitigation, and we understand that, to date (late 2021), no ocean fertilisation experiments have been nationally submitted and approved under LC/LP arrangements.

Additional regulatory constraints, established before climate-scale OIF was envisaged, apply in the Southern Ocean, where there is the greatest potential to increase productivity and associated long-term carbon uptake. In particular, the Convention on the Conservation of Antarctic Marine Living Resources (CCAMLR) provides increased ecosystem-level protection of 36 million km^2 of ocean south of the Antarctic convergence, while the Protocol on Environmental Protection to the Antarctic Treaty (Madrid Protocol)[106] further strengthens conservation south of 60°S. Although several OIF experiments were allowed to take place in CCAMLR-protected waters in the 1990s, such studies were relatively small-scale and subject to additional precautions and reporting procedures. Larger-scale interventions in the Southern Ocean seem unlikely to be permitted by these international bodies, regardless of other global regulations, due to their far-reaching potential for adverse environmental consequences and uncertain climatic benefits.

International law is not immutable, and 'rogue states' or the private sector can flout agreements. Nevertheless, formal changes in the policy positions of such a wide variety of decision-makers would be a lengthy process, requiring major shifts in the perceived balance between risks and benefits, and large-scale unilateral or commercial use of OIF for CO_2 removal would currently seem unrealistic, given the very low public acceptability of this approach and the long-term commitment necessary to achieve measurable climatic benefits. The 'real world' potential contribution of OIF to CO_2 removal by 2050 therefore seems extremely small.

10.2.2 Atmospheric Ocean Iron Fertilisation (Iron Salt Aerosol)

10.2.2.1 Scientific Understanding of Key Processes

Since aerial inputs are responsible for much of the iron naturally added to the upper ocean,[48] it has been proposed that atmospheric delivery would be advantageous for OIF; in particular, by using the 'iron salt aerosol' (ISA) method. This approach is currently only theoretical, covered in detail by only two peer-reviewed papers[107,108] and associated online information.[109,110] A major feature of the ISA approach is that the large concentration gradients initially arising from direct iron additions would be avoided, with a comparison made between "sprinkling salt onto a meal and being forced to eat a kilogram of salt".[110]

The ISA method envisages that iron compounds would initially be vaporised into a carrier gas through a combustion process, occurring either in ship engines or on land, or by using jet aircraft. If emissions are land-based, uplift may be assisted by towers, vortex generators or tethered balloons, to deliver an ISA plume to a height of around 1000 m, aiming to achieve distribution over thousands of square kilometres. Rain-out would subsequently directly deliver soluble iron compounds to the surface ocean. While many additional beneficial effects on climate processes and marine biogeochemistry are claimed in the main published description of the ISA method,[107] the focus here is whether it might be preferable to directly-based iron additions.

The local efficacy of ISA delivery to the ocean could be tested using aerosol-generating devices to generate ISA plumes in the lower marine boundary layer at a relatively small scale (tens of km^2), in ocean waters known to be iron-limited. Assuming funding and regulatory approval for such experiments (see Section 10.2.2.4 below), any resultant changes in productivity and carbon export to deeper water could then be measured in the treated area, together with other relevant hydrodynamic and biogeochemical parameters, using approaches previously deployed in 'conventional' OIF experiments.

Testing of the full ISA method (to include atmospheric transport and processing stages) would, however, be much more difficult, due to adverse signal-to-noise ratios. Thus, high-resolution measuring technologies, deployed over large areas and long timescales, would be needed to detect the effect of low-level additions (estimated to average $\sim$600 g km^{-2}),[110] in the context of high spatial and temporal heterogeneity in both the additions and the natural atmospheric and oceanic baselines. Comparisons with natural dust-driven fertilisation are only of limited value in this regard. Although effects of episodic dust events on marine productivity have been detected,[111] the oceanic and climatic consequences of natural variability (and human-driven changes[112]) in atmospheric iron inputs are now considered less important than variability in sedimentary and hydrothermal sources, and in the availability of iron-binding ligands and other internal cycling processes.[113,114]

Iron solubility in the ISA method can be expected to be high as a result of the proposed initial pyrogenic processing[115] and subsequent atmospheric reactions,[116,117] assisted by anthropogenically-driven increases in atmospheric acidity. Such acidity increases (caused by a range of industrial activities) have been estimated to have tripled the natural dissolved iron input to the global ocean.[118] Yet modelling studies indicate that the resulting climatic benefits have been modest,[112] reducing atmospheric CO_2 by <2 ppm since 1850, *i.e.*, two orders of magnitude less than the recent annual growth rate of atmospheric CO_2, at 2.3 ppm per year between 2009–2018.[119]

10.2.2.2 Environmental Consequences and Wider Context

If OIF using ISA were successful in greatly increasing marine productivity and long-term carbon removal, the risks of generally adverse side effects would be much the same as already discussed for directly-added iron. Nevertheless, many additional advantageous consequences, both climatic and non-climatic, are claimed (although untested) by ISA proponents.[107–110] These potential co-benefits include: an increase in cloud albedo; oxidation of methane and carbonaceous aerosols; reductions in tropospheric ozone; remediation of terrestrial iron deficiencies; and benefits to human health. The magnitude of such effects needs to be more rigorously explored by modelling, and/or small-scale field trials.[108] Further information would also seem necessary on the risk of toxic effects from impurities in the ISA precursor compounds, provisionally proposed[107] as either ferric chloride, $FeCl_3$, or ferrocene, $Fe(C_5H_5)_2$; and on the carbon/energy costs of the methods used for their bulk preparation.

10.2.2.3 Cost-effectiveness and Carbon Accounting

A cost of less than \$1 per t CO_2 removed has been estimated,[110] based on 150 000 tonnes of ISA iron removing 12 Gt CO_2 equivalent per year, around 50% directly and around 50% *via* removal of other greenhouse gases and other cooling effects. This would seem optimistic since many aspects are untested or highly uncertain. Most of the caveats regarding the cost-effectiveness of direct OIF would also apply, with additional uncertainties in monitoring and verification for the large ocean area receiving ISA-derived iron. The spatial heterogeneity in those inputs, and in natural productivity patterns (Figure 10.4), would make reliable carbon accounting particularly difficult. There is also a need to use full life cycle analyses to take account of CO_2 emissions arising from pyrogenic ISA release, proposed as fossil-fuel based (by burning oil or lignite).[107]

10.2.2.4 Governance and Acceptability

Regulation by the London Protocol would be expected, although not yet specifically covered. The wider constraints of the Convention on Biological

Diversity,[11] and regional environmental protection measures in the Southern Ocean[106] would also apply to ISA deployment techniques. Furthermore, unless safety can be demonstrated, the high likelihood of transboundary impacts would seem likely to jeopardise the general obligation of states "to ensure that activities within their jurisdiction or control respect the environment of other states or of areas beyond national control" (Principle 2 of the Rio Declaration).[120] The atmospheric delivery of ISA, with deliberate changes to air-quality and poorly-controlled outcomes likely to occur in highlypopulated areas, could reinforce the perception that such an approach is a technical fix that 'messes with nature',[121–123] rather than being a 'nature-based' solution'– despite the analogy that ISA mimics the influence of natural dust on climate.[108,110]

10.2.3 Macronutrient Addition

10.2.3.1 Scientific Understanding of Key Processes

The underlying rationale for using macronutrients (such as biologically-available nitrogen, phosphorus or both) for ocean-based CO_2 removal is the same as for OIF: the enhancement of upper ocean productivity that is expected to result in increased carbon cycling and storage in the ocean interior.[124] Many of the issues already considered for OIF are therefore also applicable; nevertheless, there are also important differences, relating both to the processes involved[125] and the information available.

There have been fewer studies on the potential for deliberate large-scale fertilisation using macronutrients than for using iron. However, there is a very extensive literature linking open ocean productivity to natural macronutrient availability and elemental ratios;[126–129] the role of ocean macronutrients in past climate perturbations;[130] and anthropogenic macronutrient enrichment in coastal waters.[131–133] The main research group that has investigated macronutrient additions for climate mitigation is closely linked to the Ocean Nourishment Corporation, which is seeking to commercialise the concept.[134] That association may have hindered, rather than helped, the perceived scientific credibility of the approach.

Outside polar regions, biologically-available nitrogen is mostly at very low levels in the upper ocean, due to either seasonal (temperate regions) or year-round stratification (sub-tropics and tropics). Adding nitrogen, in the form of manufactured urea, ammonium hydroxide or nitrate, might therefore seem near-certain to stimulate phytoplankton growth. However, such productivity increases will only occur where there is also 'excess' of other macronutrients, primarily phosphate.[129,135] Even if that is initially the case, any fertilising effects will end locally when other limitations develop within a growing season. Over longer periods, any unused nitrogen fertiliser could potentially contribute to enhanced productivity and carbon export at sites hundreds or thousands of km away from where it was added, although such effects would also be progressively subject to phosphorus limitation. As a result of these interactions and constraints, the maximum extra global carbon uptake from using nitrogen-only additions is expected to decline over a few years, from 13.2 Gt CO_2 per year to around 2.6 Gt CO_2 per year, assuming all nitrogen limitation can be alleviated under conditions of continued worldwide fertilisation.[136] If sufficient phosphate is also added, a continued annual removal of around 5.5 Gt CO_2 is considered theoretically possible.[136] These model-based values are maxima, involving treatment of ~60% of the global ocean. By focusing on the 16.5% of the ocean surface where effects are greatest the "once off" total for nitrogen-only carbon removal is estimated to be around 7 Gt CO_2.[136] This would be a very ambitious undertaking, involving treatment of ~60 million km^2.

Although greater amounts of macronutrients are needed to achieve productivity enhancements compared with OIF, the overall efficiency of the process may be much higher, estimated at 75% if urea or ammonium hydroxide is used.[137] This is because a high proportion of added macronutrients (much more than for iron) are recycled in soluble form between the deep and surface ocean, on timescales similar to carbon cycling. A proportion of the additional future CO_2 release to the atmosphere will therefore be matched by additional future CO_2 removal, due to nutrient reuse when reactive nitrogen is returned to the surface ocean. The occurrence of feedback processes is, however, likely to limit the scale of this effect.

For nitrogen additions, two such feedbacks have been identified. First, a decrease in the process of marine N_2 fixation from the atmosphere[138] by a wide range of 'diazotrophic' bacteria[139] that would lose their competitive advantage under conditions of increased nitrogen availability; and second, an increase in denitrification in the water column and sediments, enhanced by decreased oxygen availability.[140] The magnitude of these effects under conditions of macronutrient addition is currently uncertain, and has not been tested experimentally. Nevertheless, a modelling study[141] of the impact of more-than-doubled reactive nitrogen inputs to the global ocean since pre-industrial times (due to human activities) indicated that around half of the added nitrogen had been lost from the system, and that effects on net primary production were globally small, at ~1%. The geological record provides supporting evidence for such negative feedback, based on the long-term stability of the marine nitrogen cycle.[142,143] Due to this feedback, it is now widely recognised that nitrogen is only a limiting nutrient on relatively short time scales; phosphorus availability is instead the critical control, determining the strength of the biological pump on climate-relevant time scales.[144]

Other factors affecting the strength of the biological carbon pump,[21–23] as previously discussed in Section 10.2.1, will also affect the rate of nutrient recycling and the longevity of carbon sequestration, *i.e.*, whether carbon removal is short-term (weeks or months) or long-term (>100 years). Shifts in phytoplankton community composition[145] would seem a near-inevitable consequence of changes in the relative abundance of different macronutrients, with concern that such taxonomic changes may favour 'undesirable' species, either in terms of their toxicity[146] or their effectiveness in exporting carbon to deeper water. For the latter consideration, larger-celled, heavier and hence faster-sinking diatoms are climatically more desirable than smaller dinoflagellates or nanophytoplankton. However, the stimulation of diatom blooms would either require excess silicate to be already available, or to be included in the added macronutrients.[135,147] Species-specific sinking rates may also depend on nutrient availability.[148]

There have been two patch-scale field experiments for macronutrient additions.[149,150] Both used phosphorus enrichment in waters considered to be phosphorus-limited, in the Eastern Mediterranean and the subtropical North Atlantic. Contrary to expectations, neither study showed any significant increase in either primary production or chlorophyll concentrations. Take-up of the added phosphorus by nitrogen-fixing bacteria may have been involved,[151] showing that responses to changes in nutrient concentrations are not limited to algal phytoplankton. Indeed, if decomposition pathways are stimulated, and/or if there are changes in decomposition products (*e.g.*, refractory dissolved organic carbon) then macronutrient additions could be counterproductive, increasing CO_2 release rates. Such effects may explain ambiguities in CO2 flux changes in coastal seas in response to anthropogenic nutrient enrichment.[152,153]

10.2.3.2 Environmental Consequences and Wider Context

The major co-benefit claimed for macronutrient additions is increased fish biomass, with one study estimating that there could be an increase of up to 0.3 tonnes of fish (wet weight) from each tonne of nitrogen added.[154] Lower estimates are based on exploited fish mostly being at trophic level 3, with their biomass – and hence carbon and nitrogen – typically being around 1% of the biomass of phytoplankton[136] (although highly variable, as shown by the naturally-fluctuating abundances of many fish populations). It might be possible to directly demonstrate the conversion efficiencies between macronutrient additions, phytoplankton and different fish species by tracing the isotopic difference between oceanic and fertiliser nitrogen, with the latter being atmospherically-derived *via* the Haber–Bosch process.[155]

The three potential adverse biogeochemical impacts identified for OIF would also apply to macronutrient additions:[156] increased mid-water deoxygenation; increased ocean acidification in the ocean interior (although temporary decrease in near-surface waters, that will vary according to the chemical form of macronutrient added[136]); and an increase in ocean emissions of non-CO_2 greenhouse gases, primarily nitrous oxide. These effects have been investigated in models, with estimates that there might be a 17.5% global increase in anoxic water after 80 years of phosphorus-only fertilisation,[135] and that nitrous oxide emissions might reduce the climatic benefits of CO_2 removal by 20% for phosphorus-only[135] and by 8% for nitrogen-only[136] additions (all at maximum theoretical levels).

Two other areas of concern are specific to the macronutrient approach. First, issues arising from using phosphate rock to provide phosphorus (if used either in addition to nitrogen, or on its own). In particular, phosphate mining can be environmentally damaging; there may be trace metal contaminants; and there is only a finite global supply, which is likely to be increasingly prioritised for agricultural application.[157,158] While global phosphate resources are almost certainly greater than currently identified, their magnitude is contested,[159] and a high proportion of known reserves are in two countries with disputed borders (Morocco and Western Sahara). The second major concern relates to the possible stimulation of harmful algal blooms, as a consequence of the taxonomic changes already mentioned above. This problem already occurs as a result of (non-deliberate) anthropogenic nutrient enrichment in coastal waters[160] and there is the risk that it may be preferentially stimulated by urea additions.[146] However, controlling the quantity and quality of macronutrient additions could minimise or avoid such effects, without greatly reducing the target productivity increase.[125,136]

10.2.3.3 Cost-effectiveness and Carbon Accounting

A detailed scenario-based analysis[156] has estimated the cost of macronutrient additions to be around US$ 20 per t CO_2 removed for 100 years at 2010 prices, using ship-spraying of ammonium hydroxide within 500 km from the port where the nutrient is produced, and assuming adequate phosphate. This costing is limited to direct operational costs, using the pricing of long-term, large-supply contracts for natural gas (used to produce ammonia), with no allowance for monitoring or administration; it does not assume any fishery benefits, nor any adverse impacts that might need to be financially compensated. Costs for using urea (at market prices) as the nitrogen source are estimated to be higher, at around US$ 35 per t CO_2 removed, with a further near-doubling if phosphate is also included (D. Harrison; unpublished, using peer-reviewed methods[36,136]). These costings take into account estimated nitrogen losses through denitrification (N_2O formation, and its offset effects) but not the feedback operating through reduced marine N_2 fixation. As noted earlier, the latter could be significant.

Carbon accounting for macronutrient addition would need similar monitoring methods to those already discussed for OIF, based on remote sensing of ocean processes and automated *in situ* sensors, as well as direct ship-based observations and high-resolution modelling. Some simplifications would be possible, since the logistical difficulties of working in the Southern Ocean and the technical difficulties of trace iron measurements would not apply. Nevertheless, uncertainties would remain in quantifying the climate-mitigation benefits (and potential adverse impacts) of treatments against a background not only of natural spatial and temporal variability in marine productivity and CO_2 ocean-atmosphere fluxes, but also in climate-induced long-term changes in ocean conditions.[4] These changes include observed and projected trends in stratification,[161] primary production (as indicated by chlorophyll),[162] mid-water oxygenation,[63] and ocean acidification (throughout the water column),[163,164] and also changes in other anthropogenic nutrient inputs to the ocean, both river-based and atmospheric.[141,165]

The larger the ocean area treated with macronutrients for climate mitigation purposes, and the longer that such deployments are carried out, the more difficult it would become to reliably assess benefits and impacts compared to 'control' conditions. For this reason, (assuming international approval of the approach), increasing uncertainties in carbon accounting could constrain implementation benefits to 10–20% of the theoretical potential for CO_2 removal from the atmosphere.

10.2.3.4 Governance and Acceptability

Deliberate macronutrient additions for climate mitigation are considered as marine geoengineering under the London Protocol, and regulated under the specific provisions relating to ocean fertilisation.[1] The wider constraints of the Convention on Biological Diversity[11] would also apply, but not international regulations specific to the Southern Ocean (since there would be no need for that ocean region to be treated).

Although there have been scientific criticisms[96,146] of macronutrient-based approaches, the concept has attracted far less interest than OIF from the perspective of governance, ethics and public acceptability, probably because of the paucity of patch-scale field trials, with none to date using nitrogen additions. The continued release of large amounts of anthropogenic nutrients both directly to coastal seas (through river inputs) and indirectly to the open ocean (*via* shelf seas and atmospheric deposition) may also influence acceptability. Such action is, however, increasingly subject to national and regional pollution control. Sewage sludge dumping in a range of water depths was prevalent in the US and Europe in the late 20th century; that 'inadvertent' form of macro-nutrient addition has now ceased, due to the US Ocean Dumping Ban Act (1988) and policy decisions of the OSPAR Convention.[166,167]

10.2.4 Enhanced Upwelling to Increase Productivity

10.2.4.1 Scientific Understanding of Key Processes

Deep water regeneration of soluble macronutrients and micronutrients provide the main route for natural ocean fertilisation, supporting primary production in the upper, sunlit ocean. Upward transport for re-supply is naturally achieved through: diffusive fluxes across density and concentration gradients; seasonal or temporary breakdown of stratification by winter mixing or tropical storms; and by large-scale upwelling processes, driven by eddies or shelf-edge circulation systems. Use of artificially-enhanced upwelling[168,169] for CO_2 removal envisages the use of floating pipes or fixed structures to pump nutrient-rich water from 100–300 m depth to the near-surface. The increased supply of naturally-occurring nutrients (in proportions expected to provide a relatively good match to biological requirements) would stimulate CO_2 fixation by phytoplankton, and is also expected to increase subsequent long-term removal of carbon through the biological carbon pump.

A major drawback to artificial upwelling is that the deeper, nutrient-rich water will also have relatively high CO_2 concentrations, negating the intended climatic benefits.[170–172] Furthermore, modelling indicates that physical effects would be more important than biological ones, and although initially beneficial they would subsequently become counter-productive. Thus, simulations indicate that global warming would initially be reversed if very large amounts of cold, deep water could be brought to the ocean surface, yet such action is projected to *increase* warming within 50 years, with major regional climatic disruptions and an equilibrium surface temperature rise of >8 °C.[173] During the initial cooling phase, additional carbon storage (equivalent to 3.3 Gt CO_2 per year) is projected to occur; however, this would mostly be on land as a result of decreased soil respiration under lowered temperatures.[174]

Despite the apparent shortcomings of this approach, it continues to attract research interest, mostly in China[175–177] and Japan,[178] as a potential climate mitigation action that might also benefit fisheries[179] and aquaculture.[180] Pumping systems have been developed and trialled,[181–183] and theoretical efficiencies determined for different water conditions.[184] While significant net uptake of CO_2 may be possible under some circumstances, this has yet to be demonstrated under field conditions.

10.2.4.2 Environmental Consequences and Wider Context

If enhanced upwelling were to be used for large-scale enhancement of open ocean productivity, there is the risk of the same adverse impacts as already noted for direct and indirect OIF. Nevertheless, local increases in macro-nutrient availability could be advantageous for aquaculture,[180] and surface cooling effects could help protect coral reefs from bleaching,[185] providing that the associated changes in water column productivity and nutrient status do not result in other ecological stressors. A role in the suppression of hurricanes and tropical cyclones has also been proposed[186,187] since cooler surface waters could significantly reduce storm energy. Relatively large-scale deployment of upwelling devices would be required for that purpose.

Enhanced upwelling is also the physical process involved in Ocean Thermal Energy Conversion (OTEC), whereby electric power can be derived from the temperature difference between the upper ocean and deeper water[188,189] in tropical waters. The basic theory of OTEC is considered robust; its effectiveness has been demonstrated at the pilot scale; and OTEC undoubtedly has potential for climate mitigation by providing renewable energy. Yet existing OTEC schemes are not designed for CO_2 removal,[1] and the intention to date has generally been to minimise their perturbation of upper ocean nutrients and associated ecological impacts.[190] Net carbon removal using OTEC design modifications would seem more achievable using chemical rather than biological processes;[191] for that reason, such an approach is not considered further here.

10.2.4.3 Cost-effectiveness and Carbon Accounting

No cost estimates are known for using enhanced upwelling to provide long-term CO_2 removal through ocean fertilisation. Although local net removal may be possible under some conditions, large-scale development of this approach is unlikely to be climatically effective, precluding economic viability, for the reasons given above. If it were attempted, carbon accounting would present the same difficulties as for other approaches based on increasing the strength of the biological carbon pump (Sections 10.2.1 and 10.2.2). While the use of enhanced upwelling for other purposes (*e.g.*, OTEC) might be justified, its effectiveness and feasibility as a CO_2 removal technology have yet to be demonstrated.

10.2.4.4 Governance and Acceptability

Large-scale enhanced upwelling for the purpose of increasing ocean CO_2 uptake would meet the definition of marine geoengineering under the London Protocol, and therefore subject to its regulatory control.[1] The wider constraints of the Convention on Biological Diversity[11] would also apply. However, OTEC would seem exempt. No specific governance issues are known for enhanced upwelling trials carried out to date, nor for the OTEC pilot-scale power plant in Hawaii; larger-scale OTEC developments are currently being planned.[192] The public acceptability of using such approaches to achieve large-scale carbon removal does not seem to have been investigated.

10.3 Biologically-based CO_2 Removal in Coastal Seas

10.3.1 Coastal Blue Carbon Ecosystems

10.3.1.1 Scientific Understanding of Key Processes

Mangrove forests, salt marshes and seagrass meadows are vegetated coastal wetlands that together provide climatically-significant carbon removal and storage, primarily in their sediments.[4,17,18,193] They occur in every continent except Antarctica, as coastal 'blue carbon' ecosystems (CBCEs). The blue carbon concept has also been used more widely,[194,195] covering other productive marine ecosystems such as those structured by macroalgae (considered in Section 10.3.2), based on an early definition that potentially included all biotic components of the ocean carbon cycle.[196] Management of the three CBCEs considered here can contribute to climate mitigation in two ways: first, by habitat conservation to achieve emission reductions, preventing anthropogenically-driven CO_2 release from their carbon stores; and second, by actions to increase their CO_2 removal, *i.e.*, as a GGR technology, closely comparable to land-based afforestation (Chapter 9, this volume[197]). Estimates of the potential mitigation by the two approaches and three ecosystems have frequently been combined. Where separated for specific CBCEs, the estimated magnitudes of avoided emissions are generally greater than the potential for additional removals (particularly for seagrass and salt marsh ecosystems)[19] – with the implication that, for climate mitigation policy, conservation should be prioritised over restoration. This prioritisation would also seem desirable from a biodiversity perspective, with intact, natural ecosystems supporting a much greater range of species and ecosystem services than restored habitats.

CBCEs accumulate carbon in their soils and sediments, with net CO_2 removal from the atmosphere typically occurring at rates around an order of magnitude higher (on a per m^2 basis) than productive terrestrial ecosystems, such as forests, grasslands or peatland.[198,199] The below-ground carbon storage in CBCEs is a consequence of water saturation producing anaerobic conditions and therefore limiting microbial decomposition of organic material. The sources of the accumulated carbon may be local, within-ecosystem (autochthonous) or from elsewhere (allochthonous),[200,201] either marine or terrestrial.[202] External, allochthonous sources may provide 70–90% of the accumulated carbon for seagrass meadows in estuaries.[203]

Such lateral transport raises additionality issues, since a proportion of long-term sequestration (particularly for recalcitrant carbon[204]) might have naturally occurred at other storage sites, *e.g.*, offshore, in the absence of the CBCE. As a result, not all the carbon accumulating in CBCEs should be included in their negative emissions accounting.[205] Furthermore, for CBCEs to be considered as GGR, their carbon uptake and storage have to be deliberately increased by management action, with that increase reliably quantified. Although such an increase can be achieved by improving local environmental conditions (*e.g.*, by restoring natural hydrodynamics,[206] or reducing pollution), significant climatic benefits are only likely to arise from increasing the areal coverage of CBCEs. In particular, through actions to re-establish such habitats where they had previously been lost, through land-use change or other coastal developments, or by creating entirely new habitats, by planting mangroves or seagrasses in intertidal or subtidal sediments, or purposeful coastal flooding to stimulate salt marsh formation.

All such management actions are here considered as 'restoration', and many such initiatives have already been carried out for mangroves,[207] salt marshes[208] and seagrass meadows.[209] Consistent conclusions are that restoration success[210] critically depends on remediation of the factors (usually anthropogenic) that caused the original habitat loss.[211] In addition, larger-scale plantings,[209] natural regeneration and strong local stakeholder engagement all increase the likelihood of long-term survival for the restored system, with the need for long-term monitoring to demonstrate that such actions are having the desired effect.

Global estimates for maximum potential climatic benefits for CBCE restoration (for all three ecosystems combined) range between ~0.8 Gt CO_2 per year by 2030[212] with subsequent plateauing (assuming major policy action had been started in 2015, *i.e.*, achievable within 15 years) and 0.33 Gt CO_2 per year by 2050.[19] These estimates are theoretical maxima, based on full reinstatement of historical cover and therefore involving a near-doubling of current areal CBCE extent. Such action is unlikely to be feasible, because of the semi-permanent and ongoing nature of coastal land-use change, including human settlement, shoreline hardening and port development, in addition to conversion to agriculture and aquaculture.[4,213] Based on 40% restoration of historical cover (more realistic, yet still challenging), potential removal by this approach has been estimated as 0.20 Gt CO_2 per year by 2050.[19] Based on 'cost effective' restoration (<US$ 100 per t CO_2), global estimates are ~0.25 Gt CO_2 per year by 2030,[212] and 0.13 Gt CO_2 per year for an unspecified date.[5]

There have been major research efforts in the past decade to improve knowledge of CBCE carbon removal processes[214–216] and develop methodological and policy guidance for carbon accounting in the context of climate mitigation.[17,205,217–219] Nevertheless, the IPCC Special Report on the Ocean, Cryosphere and Climate Change[220] only gave 'medium confidence' to the effectiveness of this approach, on the basis of the many uncertainties regarding the quantification of climatic benefits, as summarised in Table 10.1. There are also socio-economic and local-to-national governance challenges[221] regarding the use of CBCEs for climatically-significant carbon removal, as discussed below. These uncertainties do not mean that CBCE restoration is not worthwhile; however, the wide range of co-benefits of such action[193,194] seem likely to have greater societal value than its contribution to CO_2 removal.

Three uncertainties (identified as factors 4, 5 and 6 in Table 10.1) potentially jeopardise the viability of using CBCEs for climate mitigation, since each could negate all or most of the intended climatic benefits. First, variability in the emissions of the greenhouse gases methane (CH_4) and nitrous oxide (N_2O), both of which have much higher global warming potentials than CO_2; second, the potential for calcification-driven CO_2 release (particularly from subtropical seagrass meadows);[232,233] and third, the vulnerability of CBCEs to future climatic conditions, that could greatly reduce their CO_2 removal role while also causing release of their carbon stores. Climatic vulnerabilities are considered in greater detail here; the potential emissions of CH_4 and N_2O are considered in Section 10.3.1.2 below.

The effects on CBCEs of future warming (including marine heat waves[241,242]), sea level rise and increased storminess have been assessed by IPCC[4,220] and elsewhere.[239] Seagrasses are considered the most sensitive to higher temperatures,[240] particularly episodic extremes, with their ecological structure and functioning already being subject to 'moderate' impacts, attributable to warming with medium confidence.[220] 'High' impacts, likely to involve the loss of their carbon stores,[234,243] are projected if global mean surface temperatures increase by more than ~2.3 °C relative to pre-industrial levels.[220] For salt marshes, the thresholds for moderate and high impacts are estimated[220] to be global temperature increases of ~1.2 °C and 3.1 °C, respectively; for mangroves, ~2.0 °C and 3.7 °C, respectively. Although mangroves are regarded as relatively resilient, warming-related mortalities of such ecosystems have already occurred in Australian coastal waters – linked

Table 10.1 Six factors affecting the reliable quantification of using coastal blue carbon ecosystems (CBCEs; mangrove forests, salt marshes and seagrass meadows) for climate mitigation through CO_2 removal, primarily through direct restoration. The uncertainties considered of greatest importance are discussed in detail in the text.

Factor	Summary of issues	Overall effect	Ref
1. Local variability in carbon burial rates	Multiple biological, chemical and physical factors affect carbon burial rates through their effects on primary production, sedimentation and preservation. For natural salt marshes, 600-fold difference occurs between the highest and lowest reported burial rates; for seagrasses, 76-fold, and for mangroves, 12-fold. Further differences likely between natural and restored ecosystems.	Global mean values unlikely to be locally applicable. May result in under- or over-estimation of actual climatic benefits.	198, 216
2. Errors in determining carbon burial rates	Most estimates of sedimentation and associated carbon burial rates are indirect, rather than using direct flux measurements or isotopic dating. Errors can arise by underestimating/disregarding bioturbation and surface remineralisation, or other invalid assumptions; these effects may be particularly important for seagrasses.	Most errors will result in over-estimation of carbon burial rates.	222, 223, 224
3. Lateral carbon transport	At some sites, a high proportion of the carbon buried can be from non-local sources (allochthonous) that might otherwise be removed from circulation by other processes. This would reduce the effectiveness of restored CBCEs in increasing total carbon storage. There may also be significant carbon export from CBCEs to deeper water, *e.g.*, as dissolved organic carbon.	Net carbon imports (particularly high in estuaries) would over-estimate climatic benefits; exports would have the opposite effect, but are difficult to quantify.	200, 201, 203, 204, 225

4. Fluxes of non-CO_2 greenhouse gases	CBCEs can be significant emitters of methane (CH_4) and nitrous oxide (N_2O). However, such emissions are highly variable, and associated processes are not well understood. Additional uncertainties relate to the timescales of CH_4 warming effects (Global Warming Potential) affecting contributions to relatively short term Nationally Determined Contributions to climate policy.	Likely over-estimation of climatic benefits (unless CH_4 and N_2O fluxes decrease as a result of restoration).	226, 227, 228
5. Carbonate formation (and dissolution)	The biogenic formation of calcium carbonate in seagrass and mangrove ecosystems, releasing CO_2, may partly or fully negate their effectiveness in climate mitigation. Net carbonate dissolution, increasing CO_2 uptake, is also possible for mangroves on carbonate-rich sediments.	Global effect could either be an over- or under-estimation of climatic benefits.	229, 230, 231, 232, 233
6. Vulnerability of CBCEs to future climate change	The functionality and survival of restored CBCEs are likely to be affected by future global warming (temperature increases $>1.5\ °C$), with seagrass ecosystems being the most vulnerable. Future increases in storminess and sea level rise under medium or high emission scenarios could greatly reduce the long-term effectiveness of CBCE carbon removal and integrity of carbon storage.	Over-estimation of climatic benefits if future CO_2 removal is reduced and/or stored carbon is released.	4, 234–240

to drought and other factors.[235] Interactions with other stressors are an important consideration in that regard:[239] the reduction of such stressors (through conservation action) seems likely to have increased seagrass coverage and abundance in Europe, despite recent warming.[244]

The risk of future climate-related impacts would be much reduced (but not completely averted), if the internationally-agreed commitments of the Paris Agreement are implemented, limiting global surface warming to below 2 °C. That outcome is currently considered unlikely[245,246] but not impossible.[370] Nevertheless, warming impacts could be minimised if climate mitigation efforts *via* CBCE restoration are focussed on the cooler (rather than warmer) parts of the distributional ranges of the species that structure such ecosystems, reflecting the well-established latitudinal influence of temperature changes on the distributions and abundances of marine (and terrestrial) species and communities, with associated poleward range shifts in response to warming.[247]

The implications of future changes in storm events, wave energy and sea level rise for CBCEs are less well understood, and are likely to show greater local and regional variability.[238,239,248] Thus, vulnerability – and resilience – will also be affected by sediment erodibility and sediment resupply, as well as by vegetation type, the frequency of extreme events, the rate of local sea level rise, and whether there is space for landward relocation.[239,249–251] Mangroves are generally unaffected by current rates of mean sea level rise (~3 mm per year),[252,253] maintaining their sediment accretion rates; however, they generally do not develop[254] where the relative sea level rise is greater than 7 mm per year. The latter rate is projected to be widely exceeded by mid-century under high emission scenarios (likely within the range 10–20 mm per year by 2100 under RCP 8.5).[252] Additional adverse impacts are also expected to arise from changes in precipitation and hurricane frequency.[255,256]

10.3.1.2 Environmental Consequences and Wider Context

The anaerobic conditions in CBCE soils and sediments that are responsible for long-term carbon storage favour[226] the production and emissions of methane (CH_4) and nitrous oxide (N_2O). These two potent greenhouse gases are of increasing climatic concern,[69,70,257,258] with significant (yet uncertain) marine sources.[259] It has been widely assumed that saline conditions in coastal sediments inhibit CH_4 production,[260] with the main risk of CH_4 release occurring when CBCEs are degraded or converted, *e.g.*, for agriculture or aquaculture.[261] However, CH_4 emissions also occur[262] in waters with salinity >18, with a recent synthesis[228] estimating that current global CH_4 emissions from natural mangroves and salt marshes together could exceed 5 million tonnes CH_4 per year. Because CH_4 has stronger warming effects, such emissions might fully counteract the climatic benefits of CO_2 removal by these ecosystems, estimated at 160 million tonnes CO_2 per year. Emissions of CH_4 from seagrass habitats were lower, with a net climate benefit indicated by that synthesis.[228] Statistical methods affect these estimates and comparisons, with lower global values for CH_4 emissions obtained when based on the median, rather than the mean, of site-specific studies.[263]

Further methodological issues relate to the timescales and assumptions used to estimate the global warming potential (GWP) of CH_4, for comparison of its warming effects to those of CO_2. Because of the relatively short atmospheric lifetime of CH_4, higher GWP values apply[264] when considering medium-term mitigation targets (*e.g.*, Nationally Determined Contributions by 2030, or national net zero by 2050) rather than contributions to long-term climate policy (*e.g.*, national inventories relating to global radiative flux budgets in 2100). For long-established natural CBCEs, their CH_4 emissions (even if high) can have a neutral radiative balance, since such emissions are counteracted by atmospheric removal processes, *i.e.*, they do not result in any further atmospheric accumulation.[265] A key issue for using CBCEs as GGR, affecting the cost-effectiveness of this approach,[266] is therefore whether actions to increase CO_2 removal might also increase CH_4 emissions (likely if the total area is increased), or might decrease them. The desired decrease could occur if pre-restoration conditions are themselves strongly methanogenic (*e.g.*, rice paddies[267] or shrimp ponds[268]) or might be achieved by changing water flows to existing habitat.[269]

Emissions of N_2O from CBCEs seem likely to be climatically less important than CH_4, and net uptake can also occur.[270,271] However, these N_2O fluxes are highly variable, with a 40-fold range between the upper and lower quartiles for the estimated global values for the three ecosystems combined.[272] While emissions are higher under anthropogenically-increased levels of coastal ammonium and nitrate (eutrophication),[272] many other biological, physico-chemical and anthropogenic factors affect the production, consumption, and transport pathways of N_2O for CBCEs.[226] Figure 10.5 summarises information on CH_4 and N_2O fluxes for the three ecosystems, and also gives the observed ranges of their carbon burial rates (on a per m^2 per day basis). Because of the very wide ranges for these removal and release rates, and our limited understanding of the causes of this variability, careful site-specific monitoring of CH_4, N_2O and CO_2 fluxes is needed for CBCE restoration actions carried out for climate mitigation purposes, both before and after such interventions. Ideally, such measurements should continue for 10–20 years,[226] as part of robust monitoring, reporting and verification (MRV) processes to provide confidence in carbon accounting for GGR purposes. This has been successfully done for a seagrass restoration project,[273] although with additional monitoring costs.

In addition to their role in climate mitigation, CBCEs also contribute to climate adaptation,[274] and provide many other ecosystem services,[193,194,275] as summarised in Table 10.2. The magnitude of these co-benefits could be increased by actions to increase their areal coverage for the purposes of CO_2 removal. Since the economic valuation of these other ecosystem services seems likely to be substantially higher than for their value for climate mitigation, the descriptor 'co-benefits' is only appropriate from a climatic perspective.

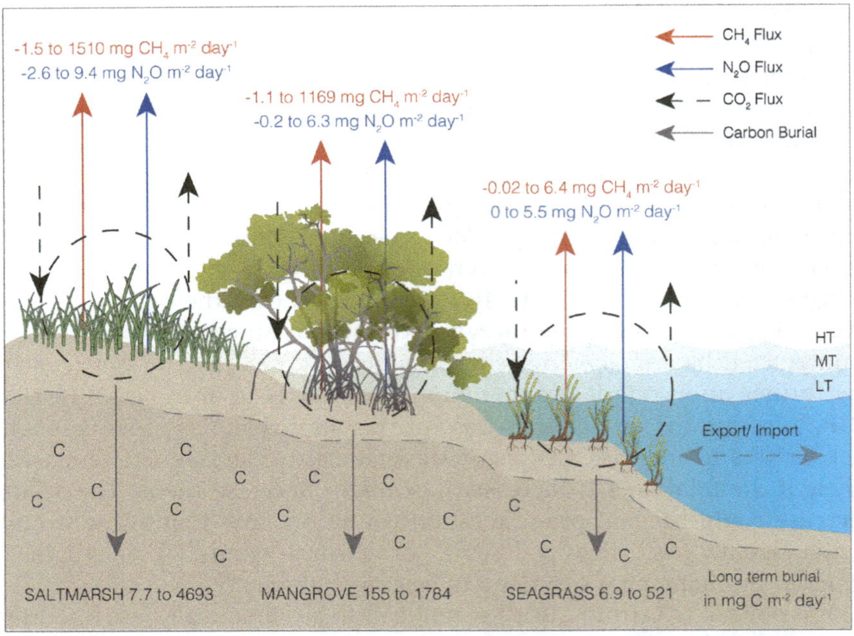

Figure 10.5 Summary of the high variability of carbon burial rates and associated methane (CH_4) and nitrous oxide (N_2O) fluxes in salt marsh, mangrove and seagrass ecosystems. Ranges show the lowest and highest literature values. Reproduced from ref. 226 with permission from John Wiley & Sons, Copyright © 2021 American Geophysical Union. All rights reserved.

The role of CBCEs in providing 'ecosystem-based adaptation'[252,277] through coastal protection and erosion control directly complements their potential use in climate mitigation. Incoming wave heights can be reduced by 35–71%;[278] wave energy reduced by around 50%;[279] and present-day surge-related flood damage reduced by more than 15% annually.[277] The additional advantage of vegetation-based coastal defences is their ability, under the right conditions, to self-maintain and regenerate after major storms,[280] with potential adjustment to future sea level rise under low emission scenarios (as discussed earlier). A recent global estimate[281] of the value of mangroves in preventing flood damage is US$ 82 billion per year, providing protection to around 18 million people.[274] The local effectiveness of such measures does, however, vary considerably, depending on storm and landscape factors, as well as vegetation biomass and taxonomic composition, making it difficult to extrapolate the physical and economic benefits across geographies.[252] It can also take several years before newly-established CBCEs reach maturity and become fully effective as natural sea defences.[282] While it is unlikely to be feasible to double the global area of CBCEs (*i.e.*, full restoration to their historical distributions), more modest increases would still provide major coastal protection benefits – in addition

Table 10.2 Potential co-benefits (regulating, provisioning, supporting and cultural ecosystem services) that could be enhanced by actions to increase CO_2 removal by coastal blue carbon ecosystems (CBCEs; here considered as mangrove forests, salt marshes and seagrass meadows). Relative valuations: •••, high; ••, medium; •, low; –, not estimated. Actual values (US$ per ha per year) are subject to many uncertainties, and are likely to reflect 'higher value' sites. Based on ref. 193, 275 and 276.

		Relative valuation of services with date-based actual values from ref. 276		
Ecosystem service	Ecosystem process or function	Mangroves	Salt marsh	Seagrasses
Regulating services (excluding climate mitigation through CO_2 removal)				
Natural hazards[a]	Storm and tsunami protection; attenuates and dissipates wave energy; flood control	•••US$ 9000–10 800 per ha per year (2007)	•••US$ 8200 per ha per year (2008)	••
Erosion control[a]	Provides sediment stabilisation and soil retention in vegetation root structure	•••US$ 3700 per ha per year (2001)	••	•
Pollution control	Provides uptake and retention of pollutants, including anthropogenic nutrients	•••	•••US$ 1940–37 000 per ha per year (1995)	•
Provisioning services				
Food	Harvest of fish, edible invertebrates and plants	•••US$ 480–600 per ha per year (2007)	••£15 per ha per year (1995)	•
Other raw materials and products	Timber and fuel-wood production (mangroves); other natural products and genetic material	•	•	–
Supporting services				
Maintaining fisheries; grazing for livestock	Reproductive habitat and nursery grounds for fish; food for domestic animals (salt marsh; not included in valuation)	••US$ 710–990 per ha (2007)	••US$ 2400–16 000 per ha (1997)	•
Providing habitat for biodiversity	Food and shelter for very many vertebrate and invertebrate species, including migratory and native birds; habitat for micro-organisms delivering a wide range of other services	–	–	–
Cultural services				
Recreational, spiritual and aesthetic	Tourism and other recreational activities; well-being	•	•••£33 per person (2007)	•

[a]Contributing to climate adaptation.

to enhancing their ecosystem services for biodiversity, fisheries, pollution control and cultural values, as identified in Table 10.2.

10.3.1.3 Cost-effectiveness

The overall cost-effectiveness of using CBCEs for CO_2 removal has been assessed as low,[3] although with many uncertainties. These uncertainties not only relate to the many factors affecting the estimation of CO_2 removal rates and storage longevity, and hence carbon accounting (Table 10.1), but also due to the very high variability in the cost of the management actions to increase CO_2 removal. Large differences in restoration costs occur both between and within the three vegetation types. Based on a synthesis of 91 studies for which detailed information was available and a relatively-wide definition of 'restoration' (both active and passive restoration and rehabilitation; as used here), total restoration costs for mangrove, salt marsh and seagrass habitats have median values of US$ 2500, US$ 151 100 and US$? 383 700 per ha (2010 prices), respectively.[283] The wide range of costs within each CBCE is shown by their mean values being 2–7 times higher than their medians, at US$ 15 000, US$ 1 042 100 and US$ 699 500 per ha, respectively.

A major cause of the variation for mangrove restoration costs (accounting for around a 30-fold difference in median values) was the difference between developed and developing countries. This was mostly due to the latter having much lower labour costs, with many CBCE restoration projects in developing countries being community- or volunteer-based. For salt marshes, there was also an 8-fold cost difference according to the planting method used. Further conclusions of the synthesis[283] were that the majority of restoration projects were short-lived; that they seldom reported monitoring costs; and that the data available may have been biased towards publishing successes rather than failures. A more recent review[266] found similar ranges in restoration costs per unit area for the three ecosystems, although with an even greater (>100-fold) difference in median annual costs for mangroves between developing and developed nations, at US$ 990 per ha and US$ 100 860 per ha, respectively.

The long-term survival of restored CBCEs is a crucial consideration for evaluating their effectiveness in sequestering carbon and hence their viability for climate mitigation. Low success rates of CBCE restoration projects typically occur due to poor site selection or poor choice of species introduced,[284,285] although a wide range of other factors can also be involved.[211,286,287] A key issue, as noted earlier, is whether the (human) factors causing original loss/degradation or land use conversion have been properly addressed.[4] The concept of social-ecological restoration[221] is applicable here, taking into account all relevant anthropogenic drivers and decision-making processes, from the local to international level. These include 'opportunity costs' (not necessarily included in the restoration costs quoted above); *i.e.*, the potential economic benefits that are not obtained when

CBCE restoration excludes other uses of the coastal land, for agriculture/aquaculture, industry or settlement.[252,288]

The above uncertainties and complexities mean that 'global average' cost estimates for using CBCEs as GGR must be regarded with extreme caution. While quantification can be improved by the further development of standardised carbon accounting methods,[205,206,218,219] few studies have implemented the necessary monitoring actions over sufficient time periods to provide reliable estimates of cost-effectiveness, including measurement of CH_4 and N_2O fluxes. The most comprehensive synthesis[266] of information to date, that takes assumed CH_4 emissions into account, estimated median removal costs of US$ 469 100 per t CO_{2e} for salt marsh restoration and US$ 560 per t CO_{2e} for mangrove restorations. That meta-analysis did not include estimates for seagrass restoration projects; a single long-term study[273] (measuring both CH_4 and N_2O) for that ecosystem gave a relatively 'favourable' removal cost of US$ 95 per t CO_{2e}. With the exception of that seagrass study, an earlier estimate[212] that around 25% of lost or degraded coastal wetlands could be restored at a cost less than US$ 100 per t CO_{2e} now seems optimistic. Nevertheless, the magnitude of potential co-benefits (Table 10.2) does not preclude the use of CBCEs for CO_2 removal as either a 'no-regrets' or 'low-regrets' option[289] on a site-specific basis, where other climatic and non-climatic ecosystem services can be delivered, particularly for mangroves.

10.3.1.4 Governance and Carbon Accounting

Since CBCE restoration necessarily occurs within national boundaries and near-shore waters, governance issues are inherently more straightforward than for carbon-removal interventions involving the open ocean. There is also strong support by many NGOs for such action, *e.g.*, *via* the Blue Carbon Initiative,[290] linked to long-standing international agreements on the desirability of coastland wetland protection and restoration, *e.g.*, through the Convention on Biological Diversity[291] and the Ramsar Convention.[292]

CBCE restoration is identified as a planned mitigation action in 23 initial Nationally Determined Contributions (NDCs) to the 2015 Paris Agreement, with 19 parties committing to mangrove restoration and 4 for seagrass restoration;[20] there are no mentions of salt marsh restoration. Quantitative targets are not stated in these NDCs, and so it is not clear how ambitious these targets might be, *i.e.*, whether they are intended to provide a substantive role in meeting national climate mitigation goals, and whether they are in addition to (rather than a substitute for) aggressive emission reductions.

The NDCs that include mangrove restoration are potentially supported by two UNFCCC carbon credit arrangements: Reducing Emissions from Deforestation and Degradation (REDD+) and the Clean Development Mechanism (CDM). The CDM, introduced in 2005, allows forestry-based carbon credits to be sold on the compliance market, with the intention of helping to meet international emission reduction commitments.[293] REDD+, which began in 2005, also creates financial value for the carbon stored in the forests of developing countries, with payments for reduced emissions from forested lands.[294] Both mechanisms therefore allow Annex I countries (developed countries) to implement mangrove restoration projects in Annex II countries (developing countries) and receive carbon offset credits in return, providing that registration and verification standards are met.[196] The effectiveness of offset mechanisms is debatable[295] with the risk of perverse incentives and wider issues regarding national reporting of afforestation and reforestation for climate mitigation[296] (as discussed further in Chapter 9).[197] Nevertheless, existing mechanisms do provide the basis for a more comprehensive valuation of coastal carbon removal and storage,[199] that might be extended to include other carbon sequestration processes in shelf seas.[297]

The CDM and REDD+ mechanisms are only feasible for relatively large CBCE restoration projects as they require a minimum of 5000 tonnes of carbon sold.[298] However, smaller CBCE restoration projects could use the voluntary carbon market through routes such as the Verified Carbon Standard (VCS)[210] or Plan Vivo, with the latter supporting the Mikoko Pamoja mangrove restoration project in Kenya.[299]

Whatever carbon accounting mechanism is used to validate CBCE restoration for climate mitigation purposes, it should also include clear criteria for active maintenance and long-term (50–100 years) protection to safeguard future continuation of CO_2 removal services and the security of carbon stores. Appropriate legal protection for coastal wetlands already exists in many countries, and has been successfully used to limit CBCE loss, *e.g.*, in Bangladesh.[199] Elsewhere, policy effort may need to be directed at the enforcement of existing national conservation regulations and international agreements,[291,292] as well as (where necessary) new legal mechanisms.[300,301]

10.3.2 Marine BECCS Using Seaweed (Macroalgae)

10.3.2.1 Scientific Understanding of Key Processes

Marine macroalgae, commonly known as seaweeds, belong to three main groups: brown, green and red algae. They mostly grow attached to hard substrates in intertidal and shallow waters, with the greatest abundance in temperate and sub-polar seas.

Unlike the coastal vegetation systems considered above, seaweeds do not directly transfer carbon into marine sediments *via* root systems. They are nevertheless considered as ‘blue carbon’ ecosystems by several authors,[302–304] with estimates of globally-high levels of carbon sequestration[305] through their biomass and breakdown products. These estimates are based on studies of the transfer of seaweed detritus to nearby sediments;[306] its decomposition to dissolved organic carbon (DOC);[307,308] and also its long-distance transport at the ocean surface.[308–311]

Yet caution is needed: a high proportion of organic carbon in shelf-sea sediments can be re-mineralised relatively rapidly,[312] with full decay of kelp biomass in less than a year under well-oxygenated conditions.[313] A key issue is therefore reliable measurement of the amount buried in anaerobic sediments or transferred to deep ocean waters. In this context, 'deep' needs to be below the winter mixing layer – much greater than ~50 m, as considered in one seaweed sequestration study.[306] Even in well-investigated regions (such as the North West European shelf), order of magnitude uncertainties remain regarding carbon transport processes in coastal waters, shelf seas and across the shelf edge, with these fluxes characterised by high spatial and temporal heterogeneity.[314]

There are also biogeochemical feedbacks that may arise from a relatively simplistic adoption of 'ocean afforestation' (growing more seaweed) as a climate change solution. There is likely to be calcification by encrusting marine life, releasing CO_2; nutrient reallocation, reducing the growth of other marine plants (and their carbon removal); and raised albedo, increasing global energy accumulation. These processes would all decrease the climatic benefits of the extra carbon drawdown by large-scale seaweed cultivations, potentially by as much as 100%.[315,371]

The use of seaweeds as a quantifiable GGR approach therefore requires several well-developed and integrated management actions, comparable to land-based BECCS (Chapter 5). In addition to increasing their productivity, and hence carbon fixation, a high proportion of their biomass needs to be harvested and processed.[316–320] Conversion to biofuels or other products[321] needs to be linked to carbon capture and storage,[322,323] or other mechanisms to ensure well-documented long-term removal, and hence valid carbon offsetting.[324]

Ongoing initiatives to investigate the feasibility of these integrated actions include the US$50 million MARINER (Macroalgae Research Inspiring Novel Energy Resources) program[325] supported by the US Department of Energy Advanced Research Projects Agency-Energy (ARPA-E), with 18 component projects.[326] Emerging results indicate that design and management improvements might achieve 3-to-5-fold increases in harvested yields for kelp species,[327] to >120 dry metric tons per ha per year, based on novel infrastructure developments, storm-resistant moorings, sustainable nutrient supply and with co-benefits of associated production of shellfish and finfish.

It has been estimated[328] that 20 million barrels per day of macroalgal biofuel might be produced (~20% of global liquid fuel production in 2020[329]) if such 'total ecosystem aquaculture' (TEA) approaches are used for growing seaweed over 11 million km^2 (the global area considered suitable for fish farming[330]). It is envisaged that the harvested biomass would be converted to biofuel using hydrothermal liquefaction,[331] and that biofuel subsequently used to produce electricity with high efficiency, *e.g.*, using the Allam Cycle process,[332] with carbon capture. It is estimated that such a system could potentially remove 28–38 Gt CO_2 per year, with associated production of up to 0.5 Gt per year of seafood. Many unverified assumptions underlie these estimates, and extensive R&D would be needed to test their validity. Nevertheless, proponents of this approach consider that "most components and technologies are proven in other forms of aquaculture",[328] also that hydrothermal liquefaction, advanced combustion and carbon capture technologies are now "nearly commercial", and that there is sufficient sequestration capacity for the CO_2 produced from large-scale seaweed-based BECCS.

An alternative to converting harvested seaweed biomass to biofuel (and subsequent CCS) would be to develop methods for either land-based storage, through seaweed-based biochar, or more direct disposal of the biomass at the deep seafloor. The relative merits of different biochar feedstocks are considered in Chapter 8 of this volume. While seaweed biochar may have some specific advantages for soil amelioration,[333] that option has not received much research attention, and its cost-effectiveness does not seem to have been determined. In contrast, the feasibility of deep-water disposal of seaweed biomass is being commercially investigated,[334] with conceptual similarities to the disposal of crop wastes in the deep ocean[335] (although without the adverse implications for soil quality[336] of the land-based approach). Ocean-based sequestration using crop residues[335] is not considered further here since that approach is not based on marine biology.

10.3.2.2 Environmental Consequences and Wider Context

The first detailed proposal[323] for large-scale use of macroalgae for CO_2 removal envisaged using 32 million km^2 (9% of the global ocean surface) for this purpose, in both shelf seas and the open ocean. Manipulation of the marine environment on that scale would inevitably involve major socio-economic trade-offs, since many existing activities (such as fishing, shipping and recreation) would be affected, as well as potentially severe disruptions to many marine species, including marine mammals and seabirds. While proposals arising from the MARINER study[328] are based on a lower scenario of 11 million km^2 and limited to a water depth less than 200 m (for moorings), that would still represent around 45% of global shelf seas,[337] excluding polar waters. That proportion is likely to be greater in temperate seas, since tropical waters are too warm for kelp,[338] the fast-growing seaweed generally favoured for biomass-to-biofuel production. Greater biogeographical constraints are likely to arise at low latitudes in future, as a consequence of global warming; however, polar seas in the Arctic may then become more available. Conflicts of interest regarding the multiple uses of suitable waters (see Section 10.3.2.4, below) would seem inevitable, particularly with regard to fishing and coastal conservation – notwithstanding that large-scale seaweed cultivation is also envisaged as providing high amounts of harvested shellfish and finfish as a co-benefit.[328]

A major environmental concern is the effect on nutrient cycling and availability, with multi-million km^2 deployments causing changes that impact marine ecosystem integrity and functioning[371]. Although there is generally nutrient over-supply[131] rather than under-supply in shelf seas, summer stratification can deplete near-surface nutrient availability to levels constraining seaweed growth in offshore waters. It is therefore recognised that the proposed very large-scale cultivation systems are likely to require managed nutrient recycling, and possibly additional supply (that might involve enhanced upwelling from deeper water[328]), affecting the seasonal cycles and abundances of natural phytoplankton, zooplankton and higher trophic levels.

With regards to non-CO_2 gas fluxes, the potential for increased macroalgal release of bromoform ($CHBr_3$) and other halomethanes[339,340] is an issue of concern, recognised as requiring additional research.[328] The natural marine sources of these gases[341] are currently estimated to be responsible for around 9% of stratospheric ozone loss, including depletion due to anthropogenic causes.[342] Significant changes to regional iodine cycling[343,344] could also arise from major expansion of seaweed cultivation, with implications for human health.[345]

The above consequences of large-scale seaweed cultivation for the purpose of climate mitigation would be more manageable if the target scale for full deployment were at least an order of magnitude lower, *e.g.*, coverage of 1 million km^2, focussed on those parts of the world where temperature and environmental conditions are most favourable for macroalgal growth.[346] If that were achievable, negative emissions could still be climatically worthwhile (at 3–4 Gt CO_2 per year), on the basis that all the proposed offshore and onshore component technologies can be deployed at their assumed efficiencies, and that deployment systems are sufficiently flexible to allow for species' sensitivity to future climate change.[346,347]

10.3.2.3 Cost-effectiveness and Carbon Accounting

Much of the cost of seaweed cultivation as a CO_2 removal approach will relate to offshore and onshore infrastructure, with unproven operational lifetimes for the novel arrangements proposed to provide increased yields and additional aquaculture benefits.[327,328] Until integrated demonstration plants are established for that purpose, their cost-effectiveness will be uncertain. Early estimates[323] were that a 100 km^2 demonstration unit could cost US$ 20–100 million, depending on water depth and the level of storm protection needed, with 15–20 years of development time for multiple units. More recent MARINER-based operational cost estimates[328] are US$ 11 GJ^{-1} for the production of macroalgal biocrude fuel, and as little as US$ 26 per t for removed CO_2. The latter estimate takes account of the income generated by electricity production (from the biofuel) and the value of associated fishery yields, but not any losses/compensation arising from effects on existing environmental uses.

A separate analysis[324] estimated a minimum cost of US$ 71 per t for CO_2 'sequestered' in harvested seaweed biomass. In that study, the costs of long-term carbon removal (*via* biofuel processing and CO_2 storage) were not included, nor were cost reductions arising from co-benefits. Very high variability was found between harvest-based costs for different seaweed species: median costs were US$ 543 per t CO_2 in biomass.

Carbon accounting for CO_2 removal and storage is relatively straightforward for marine BECCS, based on established CCS technologies. However, for a life cycle assessment,[348] CO_2 emissions from establishing and operating the seaweed cultivation units also need to be accounted for. These emissions could be substantial unless manufacturing, transport and operational energy needs are all fully de-carbonised. Co-location of large-scale seaweed cultivation with offshore renewable energy systems would seem beneficial in that regard, although introducing additional issues regarding the integration of different activities, operating under separate financial and regulatory regimes.

10.3.2.4 Governance and Acceptability

Seaweed cultivation is a well-established industry in many countries (predominantly China, South Korea and Japan), with a global annual harvest of around 8 million tonnes wet weight[316] from a total marine area of 1900 km^2 (ref. 324). Most of this is used for food. The thousand-fold global upscaling needed to achieve the climate mitigation benefits considered above would be in competition with the many other ongoing uses of shelf sea waters, nationally-regulated under exclusive economic zone (EEZ) arrangements. Marine spatial planning arrangements[349,350] have been developed in many countries to resolve competing interests; nevertheless, approval processes for using large-scale seaweed cultivation for climate mitigation are unlikely to be straightforward,[351] particularly where existing fisheries or marine protected areas are likely to be affected.

The political and public acceptability of aquaculture has generally not been a significant issue to date. However, there is some evidence (at least in Europe) of declining social acceptability of aquaculture activities,[352] and opposition to large-scale developments could be expected from affected groups. Governance considerations could include issues relating to the siting of onshore power plants (using the macroalgal biofuel) as well as the CCS arrangements.

If sequestration of seaweed carbon were to be achieved by large-scale direct deposition to the deep seafloor, as has been proposed as a commercial project,[334] such action would be subject to regulation by the London Convention/London Protocol, with assessment of its ecological consequences. Although not yet formally considered, it seems likely that such activity would be considered as marine geoengineering,[1,12] and therefore added to the new Annex 4 under the 2013 London Protocol amendments.

10.4 Conclusions

Six approaches are considered here that might increase CO_2 drawdown by marine biological processes, hence acting as GGR technologies. Table 10.3 provides a summary of the main issues considered to determine their overall feasibility for this role, identifying 'issues of concern'. These factors can also

Table 10.3 Major constraints regarding the feasibility of GGR approaches using marine biological processes for climatically-significant CO_2 removal. No shading, issue of low importance; light shading, medium importance; dark shading, high importance. "?" indicates large uncertainties due to high spatial and/or temporal variability; cost-effectiveness is based on published 'best estimates'.

	CO_2 removal		Environmental impacts		Cost effectiveness		Governance & acceptability	
	Carbon accounting	Storage security	Other emissions[a]	Other adverse impacts	Ignoring co-benefits	Including co-benefits	National	Inter-national
Open ocean								
Direct OIF	?	?	N_2O (& CH_4?)		?	?		
Atmospheric OIF	?	?	N_2O (& CH_4?)		?	?		
Macronutrients	?	?	N_2O (& CH_4?)		?	?		
Upwelling	?	?	N_2O (& CH_4?)		?	?		
Coastal ocean								
Blue carbon	?	?	CH_4 & N_2O					
Seaweed BECCS			$CHBr_3$					

[a] N_2O (nitrous oxide) and CH_4 (methane) are powerful greenhouse gases; $CHBr_3$ (bromoform) causes ozone-depletion.

be conceptually considered as barriers to implementation, or as weakest links in the chain of carbon processing and associated operational management, from the atmosphere to secure storage at climatically-significant levels. Major uncertainties are also indicated. As previously concluded for a wider range of ocean-based measures[1–4] (and for GGR and negative emission technologies more generally[5,78]), none of these approaches meet all the criteria that would provide an unambiguously-desirable policy response.

Major uncertainties relating to biogeochemical processes and their temporal and/or spatial variability are associated with all approaches except for marine BECCS. Since additional knowledge could reduce these uncertainties, the often-used conclusion could be made that 'further research is needed'.[353] While true in general terms, research is costly, not only when directly measured in dollars, pounds or euros but also in terms of the opportunity cost of skilled researchers foregoing other studies.[354] The desirability of keeping a wide portfolio of options open needs to be balanced with the political need to develop and implement cost-effective, large-scale CO_2 removal on decadal timescales.

There are also issues relating to climate change that are inherently uncertain and unpredictable: these relate to human behaviour (whether or not commitments to the Paris Agreement will be kept) as well as 'deep uncertainty'[355] relating to planetary thresholds and climatic instabilities.[356] Such scenario-based uncertainties affect the scale of future GGR deployment that will be needed; they also affect the operational effectiveness of many biologically-based carbon removal technologies over decadal to century timescales. Coastal blue carbon ecosystems may change from net CO_2 removers to CO_2 emitters if the increase in global surface temperature exceeds 2–3 °C, while changes to ocean circulation patterns under high emission scenarios may also compromise fertiliser-induced carbon sequestration in the ocean interior.

A focus on holistic feasibility[357,358] and identification of implementation barriers and weakest links[359] in deployment pathways is therefore appropriate, consistent with a recent framework for policy guidance[289] that distinguished three categories of climate mitigation actions: decisive, low regret and concept stage. Using that framework, none of the approaches considered here are decisive, requiring a combination of climatic effectiveness, cost-effectiveness and relatively-high global governability.

The restoration of coastal blue carbon ecosystems (mangroves, salt marsh and seagrass habitats) is here regarded as a low regret or no-regret option on the basis of its high co-benefits, consistent with a score-based, expert judgement[3,289] and other analyses.[193] However, as identified in this chapter, many constraints and uncertainties surround the climatic effectiveness, scalability, cost-effectiveness and storage security in using this approach as a GGR technology. Crucial concerns relate to the effect of future climate conditions on CO_2 'leakage', as well as the initial demonstration of quantified additionality. Site-specific factors can cause 1–2 orders of magnitude variability in CO_2 removal rates, and in the linked emissions of methane and nitrous oxide, potentially negating a high proportion of climatic benefits. The IPCC 6th Assessment Report (WG1)[360] has drawn specific attention to the increasing importance of methane in driving global warming, and therefore the need to reduce future emissions of this gas.

Several of the above concerns regarding carbon accounting and security of carbon storage can be considered as generic weaknesses of 'nature-based solutions'[361–364] as a policy response to climate change. Nevertheless, the associated uncertainties seem higher for marine ecosystems than for terrestrial ones. Their resolution requires additional effort on monitoring, reporting and verification, likely to incur significant additional costs – and hence further decreasing cost-effectiveness.

The four CO_2 removal approaches based on open ocean fertilisation considered here (*in situ* iron addition, OIF; iron salt aerosol, macronutrient addition and enhanced upwelling) are all regarded as conceptual, lacking strong justification for major R&D efforts for potential implementation in the near future. This assessment is consistent with many other policy-directed analyses (mostly focusing on OIF).[1–5,37,78,94,289] It is based on the fundamental scientific limitations and uncertainties relating to ocean fertilisation, as well as societal acceptance[365] and the political response to date toward such potential interventions, with the latter manifested through the regulatory controls and decision making of the Convention on Biological Diversity (CBD), the London Convention and London Protocol (LC-LP) and other international bodies. That does not mean there is no justification for additional research on the responses of marine ecosystems to nutrient manipulations and on the dynamic behaviour of the ocean carbon cycle. However, such research[368] should primarily be directed at the biogeochemical processes affecting ocean-climate interactions, and hence providing better understanding of the future behaviour of the climate system as a whole.

One other CO_2 removal method using marine biology is also currently considered to be conceptual: marine BECCS, involving the large-scale cultivation of seaweed (macroalgae) for use as a biofuel with subsequent CO_2 capture and storage. In this case, further R&D would seem justified, to determine the viability of the approach; in particular, whether current technological barriers can be overcome; potential co-benefits realised; scalability and governance issues resolved; and environmental impacts mitigated.

Technological Readiness Levels (TRLs) have not been assessed in this chapter, since they are not fully appropriate for biologically-based, rather than engineered, approaches. Nevertheless, it is considered that all but one of the CO_2 removal approaches considered here would require several decades of further R&D before they could potentially be implemented for climate mitigation purposes. The exception is blue carbon restoration, an approach that is already being developed with both private and public sector support, particularly for mangroves.[366,367] However, as discussed above, there are major challenges in quantifying the climatic benefits of this approach.

Acknowledgements and Competing Interests

We thank Chris Vivian, Valeria Guinder and Antonius Gagern for helpful comments. Funding support is acknowledged from: the Australian Research Council (Grant FL1601 00131) to P. W. B.; the Engineering and Physical Sciences Research Council (Centre for Doctoral Training in Sustainable Civil Engineering) to N. R.; and the Natural Environment Research Council (Grant NE/P018319/2) to A. M. None of the authors are in receipt of commercial funding to develop ocean-based climate mitigation using CO_2 removal processes.

References

1. GESAMP Working Group 41, *High Level Review of a Wide Range of Proposed Marine Geoengineering Techniques*, ed. P. W. Boyd and C. M. G. Vivian, Joint Group of Experts on the Scientific Aspects of Marine Environmental Protection, International Maritime Organization, London, 2019.
2. D. P. Keller in *Handbook on Marine Environment Protection: Science, Impacts and Sustainable Management*, ed. M. Salomon and T. Markus, Springer, Cham, 2018.
3. J. P. Gattuso, A. K. Magnan, L. Bopp, W. W. Cheung, C. M. Duarte, J. Hinkel, E. Mcleod, F. Micheli, A. Oschlies, P. Williamson and R. Bille, *Front. Mar. Sci.*, 2018, **5**, 337.
4. N. L. Bindoff, W. W. L. Cheung, J. G. Kairo, J. Arístegui, V. A. Guinder, R. Hallberg, N. Hilmi, N. Jiao, M. S. Karim, L. Levin, S. O'Donoghue, S. R. Purca Cuicapusa, B. Rinkevich, T. Suga, A. Tagliabue and P. Williamson, in *IPCC Special Report on the Ocean and Cryosphere in a Changing Climate*, ed. H.-O. Pörtner, D. C. Roberts, V. Masson-Delmotte, P. Zhai, M. Tignor, E. Poloczanska, K. Mintenbeck, A. Alegría, M. Nicolai, A. Okem, J. Petzold, B. Rama and N. M. Weyer, WMO/UNEP, 2019, ch. 5, pp. 447–587.
5. National Academies of Sciences, Engineering, and Medicine, *Negative Emissions Technologies and Reliable Sequestration: A Research Agenda*, National Academies Press, Washington DC, 2019.
6. P. M. Fearnside, *Mitigation Adapt. Strategies Global Change*, 2002, **7**, 19.
7. University of Rochester, New view of how ocean 'pumps' impact climate change, https://www.rochester.edu/newscenter/ocean-pumps-carbon-cycle-climate-change-377692, 2019, (accessed 10 August 2021).
8. S. C. Neubauer and J. P. S. Megonigal, *Ecosystems*, 2015, **18**, 1000.
9. J. Schwinger, J. Tjiputra, N. Goris, K. D. Six, A. Kirkevåg, Ø. Seland, C. Heinze and T. Ilyina, *Biogeosciences*, 2017, **14**, 3633.
10. A. Strong, S. Chisholm, C. Miller and J. Cullen, *Nature*, 2009, **461**, 347.
11. *Update on Climate Geoengineering in Relation to the Convention on Biological Diversity: Potential Impacts and Regulatory Framework*, ed.

P. Williamson and R. Bodle, Secretariat of the Convention on Biological Diversity, Montreal, 2016.

12. International Maritime Organization, Ocean Fertilization under the LC/LP, IMO, London; https://www.imo.org/en/OurWork/Environment/Pages/OceanFertilization-default.aspx, (accessed 10 August 2021).
13. G. H. Hong and Y. J. Lee in *Global Commons and the Law of the Sea*, 2018, ed. K. Zhou, Brill and Nijhoff.
14. G. Wilson, *Texas Int. Law J.*, 2014, 49, 507.
15. S. D. Batten and J. F. Gower, *J. Plankton Res.*, 2014, 36, 925.
16. Z. Horton, *Cultural Critique*, 2017, **97**, 128.
17. D. M. Alongi, *Blue Carbon: Coastal Sequestration for Climate Change Mitigation*, Springer, Cham, 2018.
18. L. Windham-Myers, S. Crooks and T. G. Troxler, *A Blue Carbon Primer: The State of Coastal Wetland Carbon Science, Practice and Policy*, CRC Press, Boca Raton, 2019.
19. *The Ocean as a Solution to Climate Change: Five Opportunities for Action*, ed. O. Hoegh-Guldberg, World Resources Institute, Washington DC, 2019.
20. N. D. Gallo, D. G. Victor and L. A. Levin, *Nat. Clim. Change*, 2017, **7**, 833.
21. H. W. Ducklow, D. K. Steinberg and K. O. Buesseler, *Oceanography*, 2001, **14**, 50.
22. T. DeVries, The oceanic anthropogenic CO_2 sink, *Global Biogeochem. Cycles*, 2014, **28**, 631.
23. P. W. Boyd, H. Claustre, M. Levy, D. A. Siegel and T. Weber, *Nature*, 2019, **568**, 327.
24. J. H. Martin and S. E. Fitzwater, *Nature*, 1988, **331**, 341.
25. J. H. Martin, S. E. Fitzwater and R. M. Gordon, *Global Biogeochem. Cycles*, 1990, **4**, 5.
26. J. H. Martin, *Paleoceanogr*, 1990, **5**, 2.
27. A. Martinez-Garcia, D. M. Sigman, H. Ren, R. F. Andersen, M. Straub, D. A. Hodell, S. L. Jaccard, T. L. Eglinton and G. H. Haug, *Science*, 2014, **343**, 1347.
28. K. E. Kohfeld and A. Ridgwell, *Geophys. Res. Ser.*, 2009, **182**, 251.
29. A. Tagliabue, O. Aumont and L. Bopp, *Geophys. Res. Lett.*, 2014, **41**, 920.
30. D. M. Sigman and E. A. Boyle, *Nature*, 2000, **407**, 859.
31. H. J. De Baar, P. W. Boyd, K. H. Coale, M. R. Landry, A. Tsuda, P. Assmy, D. C. Bakker, Y. Bozec, R. T. Barber, M. A. Brzezinski and K. O. Buesseler, *J. Geophys. Res.: Oceans*, 2005, **110**, C09S16.
32. P. W. Boyd, T. Jickells, C. S. Law, S. Blain, E. A. Boyle, K. O. Buesseler, K. H. Coale, J. J. Cullen, H. J. W. de Baar, M. Follows, M. Harvey, C. Lancelot, M. Levasseur, N. P. J. Owens, R. Pollard, R. B. Rivkin, J. Sarmiento, V. Schoemann, V. Smetacek, S. Takeda, A. Tsuda, S. Turner and A. J. Watson, *Science*, 2007, **315**, 612.
33. P. W. Boyd, D. C. E. Bakker and C. Chandler, *Oceanography*, 2012, **25**, 64.
34. P. W. Boyd, C. S. Law, C. S. Wong, Y. Noriji, A. Tsuda, M. Levasseur, S. Takeda, R. Rivkin, P. J. Harrison, R. Strzepek, J. Gower, R. M. McKay,

E. Abraham, M. Arychuk, J. Barwell-Clarke, W. Crawford, D. Crawford, M. Hale, K. Harada, K. Johnson, H. Klyosawa, I. Kudo, A. Marchetti, W. Miller, J. Needoba, J. Nishioka, H. Ogawa, J. Page, M. Robert, H. Saito, A. Sastri, N. Sherry, T. Soutar, N. Sutherland, Y. Taira, F. Witney, S.-K. E. Wong and T. Yoshimura, *Nature*, 2004, **428**, 549.

35. V. Smetacek, C. Klaas, V. H. Strass, P. Assmy, M. Montresor, B. Cisewski, N. Savoye, A. Webb, F. d'Ovidio, J. M. Arrieta, U. Bathmann, R. Bellerby, G. M. Berg, P. Croot, S. Gonzalez, J. Henjes, G. J. Herndl, L. J. Hoffmann, H. Leach, M. Losch, M. M. Mills, C. Neill, I. Peeken, R. Röttgers, O. Sachs, E. Sauter, M. M. Schmidt, J. Schwarz, A. Terbrüggen and D. Wolf-Gladrow, *Nature*, 2012, **487**, 313.
36. D. P. Harrison, *Int. J. Global Warming*, 2013, **5**, 231.
37. P. Williamson, D. W. R. Wallace, C. S. Law, P. W. Boyd, Y. Collos, P. Croot, K. Denman, U. Riebesell, S. Takeda and C. Vivian, *Proc. Saf. Environ. Prot.*, 2012, **90**, 475.
38. P. W. Boyd and M. Bressac, *Phil. Trans. R., Soc. A*, 2016, **374**, 20150299.
39. World Ocean Atlas, https://commons.wikimedia.org/wiki/File:WOA09_sea-surf_NO3_AYool.png, (accessed 10 August 2021).
40. E. R. Urban, A. R. Bowie, P. W. Boyd, K. N. Buck, M. C. Lohan, S. G. Sander, R. Schilitzer, A. Tagliabue and D. Turner, *Oceanography*, 2020, **33**(1), 11.
41. A. J. Watson, P. W. Boyd, S. M. Turner, T. D. Jickells and P. S. Liss, *Mar. Ecol. Progr. Ser.*, 2008, **364**, 303.
42. A. Gnanadesikan, J. L. Sarmiento and R. D. Slater, *Glob. Biogeochem. Cycles*, 2003, **17**, 1050.
43. O. Aumont and L. Bopp, *Glob. Biogeochem. Cycles*, 2006, **20**, GB2017.
44. D. Ianson, C. Völker, K. Denman, K. Kunze and N. Steiner, *Glob. Biogeochem. Cycles*, 2012, **26**, GB3002.
45. R. Kiko, D. Bianchi, C. Grenz, H. Hauss, M. Iversen, S. Kumar, A. Maas and C. Robinson, *Front. Mar. Sci*, 2020, **7**, 545.
46. J. E. Yoon, K. C. Yoo, A. M. Macdonald, Y. Ho-Il, K. T. Park, E. J. Yang, H. C. Kim, J. I. Lee, M. K. Lee, J. Jung and J. Park, *Biogeosci.*, 2018, **15**, 5847.
47. A. Vaughan, Controversial geoengineering scheme will dump iron in the sea; https://www.newscientist.com/article/2282188-controversial-geoengineering-scheme-will-dump-iron-in-the-sea/, 2021, (accessed 9 August 2021).
48. P. W. Boyd and M. Ellwood, *Nat. Geosci.*, 2010, **3**, 675.
49. A. Tagliabue, A. R. Bowie, P. W. Boyd, K. N. Buck, K. S. Johnson and M. A. Saito, *Nature*, 2017, **543**, 51.
50. M. Bressac, C. Guieu, M. J. Ellwood, A. Tagliabue, T. Wagener, E. C. Laurenceau-Cornec, H. Whitby, G. Sarthou and P. W. Boyd, *Nat. Geosci.*, 2018, **12**, 995.
51. P. J. Morris and M. A. Charette, *Deep Sea Res. II*, 2013, **90**, 147.
52. M. Lauderdale, R. Braakman, G. Forget, S. Dutkiewicz and M. J. Follows, *Proc. Natl. Acad. Sci. U. S. A.*, 2020, **117**, 4842.

53. S. Henson, F. Le Moigne and S. Giering, *Glob. Biogeochem. Cycles*, 2019, **33**, 891.
54. K. O. Buesseler, P. W. Boyd, E. E. Black and D. A. Siegel, *Proc. Natl. Acad. Sc. U. S. A.*, 2020, **117**, 9679.
55. J. Robinson, E. E. Popova, A. Yool, M. Srokosz, R. S. Lampitt and J. R. Blundell, *Geophys. Res. Lett.*, 2014, **41**, 2489.
56. F. Joos, J. L. Sarmiento and U. Siegenthaler, *Nature*, 1991, **349**, 772.
57. K. Zahariev, J. Christian and K. Denman, *Progr. Oceanogr.*, 2008, **77**, 56.
58. Project description at https://www.imas.utas.edu.au/home/home-features/arc-laureate-fellowship-geoengineering-the-southern-ocean-a-transdisciplinary-assessment. (Accessed 10 August 2021).
59. J. J. Cullen and P. W. Boyd, *Mar. Ecol.: Prog. Ser.*, 2008, **364**, 295.
60. G. A. Wolff, D. S. M. Billett, B. J. Bett, J. Holtvoeth, T. FitzGeorge-Balfour, E. H. Fisher, I. Cross, R. Shannon, I. Salter, B. Boorman, N. J. King, A. Jamieson and F. Chaillan, *PLoS One*, 2011, **6**, e20697.
61. T. Parsons and F. Whitney, *Fish. Oceanogr.*, 2014, **23**, 101.
62. K. Matsumoto, *J. Oceanogr.*, 2006, **62**, 887.
63. A. Oschlies, W. Koeve, W. Rickels and K. Rehdanz, *Biogeosci*, 2010, **7**, 4017.
64. L. A. Levin, *Ann. Rev. Mar. Sci.*, 2018, **10**, 229.
65. L. Cao and K. Caldeira, *Clim. Change*, 2010, **99**, 303.
66. P. Williamson and C. Turley, *Phil. Trans. R., Soc. A*, 2012, **370**, 4317.
67. C. Hauri, T. Friedrich and A. Timmermann, *Nat. Clim. Change*, 2016, **6**, 172.
68. B. I. McNeil and R. J. Matear, *Proc. Natl. Acad. Sci. U. S. A.*, 2008, **105**, 18860.
69. E. G. Nisbet, M. R. Manning, E. J. Dlugokencky, R. E. Fisher, D. Lowry, S. E. Michel, C. L. Myhre, S. M. Platt, G. Allen, P. Bousquet and R. Brownlow, *Global Biogeochem. Cycles*, 2019, **33**, 318.
70. R. L. Thompson, L. Lassaletta, P. K. Patra, C. Wilson, K. C. Wells, A. Gressent, E. N. Koffi, M. P. Chipperfield, W. Winiwarter, E. A. Davidson, H. Tian and J. G. Canadell, *Nat. Clim. Change*, 2019, **9**, 993.
71. J. A. Fuhrman and D. G. Capone, *Limnol. Oceanogr.*, 1991, **36**, 1951.
72. X. Jin and N. Gruber, *Geophys. Res. Lett.*, 2003, **30**, 2249.
73. C. S. Law, *Mar. Ecol.: Prog. Ser.*, 2008, **364**, 283.
74. C. S. Law and R. D. Ling, *Deep Sea Res. II: Top. Stud. Oceanogr.*, 2001, **48**, 2509.
75. S. Walter, I. Peeken, K. Lochte, A. Webb and H. W. Bange, *Geophys. Res.Lett.*, 2005, **32**, L23613.
76. M. Markels and R. T. Barber, in *Environmental Challenges and Greenhouse Gas Control for Fossil Fuel Utilization in the 21st Century*, ed. M. M. Maroto-Valer, C. Song and Y. Soong, Springer, Boston, 2002.
77. P. W. Boyd, *Mar. Ecol.: Prog. Ser.*, 2008, **364**, 213.
78. S. Fuss, W. F. Lamb, M. W. Callaghan, J. Hilaire, F. Creutzig, T. Amann, T. Beringer, W. de Oliveira Garcia, J. Hartmann, T. Khanna, G. Luderer, G. F. Nemet, J. Rogelj, P. Smith, J. L. V. Vicente, J. Wilcox, M. del Mar Zamora Dominguez and J. C. Minx, *Env. Res. Lett*, 2018, **13**, 063002.

79. W. Rickels, K. Rehdanz and A. Oschlies, *Res. Energy Econ.*, 2012, **34**, 129.
80. I. Kudo, Y. Noiri, W. P. Cochlan, K. Suzuki, T. Aramaki, T. Ono and Y. Nojiri, *Deep Sea Res., Part II*, 2009, **56**, 2755.
81. M. Markels, Method of increasing seafood production in the barren ocean with a fertilizer comprising chelated iron, *US Pat.* 6408792, 2002.
82. W. S. Clarke, Environmental Solutions via Buoyant Flake Fertilization; 2015. https://sustainabledevelopment.un.org/content/documents/5535 Buoyant_flake_fertilization_rev.pdf, (accessed 10 August 2021).
83. T.-J. Kim, *Open J. Mar. Sci.*, 2020, **10**, 149.
84. T. K. Westberry, M. J. Behrenfield, A. J. Milligan and S. C. Doney, *Deep Sea Res. Part I*, 2013, **73**, 1.
85. S. Thiele, B. M. Fuchs, N. Ramaiah and R. Amann, *Appl. Environ. Microbiol.*, 2012, **78**, 8803.
86. R. Hallberg and A. Gnanadesikan, *J. Phys. Oceanogr.*, 2006, **36**, 2232.
87. I. Frenger, N. Gruber, R. Knutti and M. Münnich, *Nat. Geosci.*, 2013, **6**, 608.
88. I. Schulz, M. Montreso, C. Klaas, P. Assmy, S. Wolzenburg, M. Gauns, A. Sarkar, S. Thiele, D. Wolf-Gladrow, W. Navqi and V. Smetacek, *Mar. Ecol. Progr. Ser.*, 2018, **601**, 77.
89. A. Gnanadesikan and I. Marinov, *Mar. Ecol.: Prog. Ser.*, 2008, **364**, 213.
90. D. Roemmich, G. C. Johnson, S. Riser, R. Davis, J. Gilson, W. B. Owens, S. L. Garzoli, C. Schmid and M. Ignaszewski, *Oceanography*, 2009, **22**, 34.
91. F. Chai, K. S. Johnson, H. Claustre, X. Xing, Y. Wang, E. Boss, S. Riser, K. Fennel, O. Schofield and A. Sutton, *Nat. Rev. Earth Environ.*, 2020, **1**, 315.
92. S. M. Bushinsky, A. R. Gray, K. S. Johnson and J. L. Sarmiento, *J. Geophys. Res. Oceans*, 2017, **122**, 8661.
93. M. Lin, X. Hu, D. Pan and H. Han, *Talanta*, 2018, **188**, 135.
94. The Royal Society and Royal Academy of Engineering, *Greenhouse Gas Removal*, RS/RAE, London, 2018.
95. H. de Coninck, A. Revi, M. Babiker, P. Bertoldi, M. Buckeridge, A. Cartwright, W. Dong, J. Ford, S. Fuss, J.-C. Hourcade, D. Ley, R. Mechler, P. Newman, A. Revokatova, S. Schultz, L. Steg and T. Sugiyama, Global Warming of 1.5 °C, in *An IPCC Special Report on the Impacts of Global Warming of 1.5 °C Above Pre-industrial Levels and Related Global Greenhouse Gas Emission Pathways, in the Context of Strengthening the Global Response to the Threat of Climate Change, Sustainable Development, and Efforts to Eradicate Poverty*, ed. V. Masson-Delmotte, P. Zhai, H.-O. Pörtner, D. Roberts, J. Skea, P. R. Shukla, A. Pirani, W. Moufouma-Okia, C. Péan, R. Pidcock, S. Connors, J. B. R. Matthews, Y. Chen, X. Zhou, M. I. Gomis, E. Lonnoy, T. Maycock, M. Tignor and T. Waterfield, WMO/UNEP, 2018, ch. 4.
96. SOLAS SSC Position Statement on Large-scale Ocean Fertilisation, https://web.whoi.edu/ocb-fert/wp-content/uploads/sites/100/2017/07/SOLASSSCPosition_statem_57503.pdf, 2017, (accessed 10 August 2021).

97. K. Dooley, E. Harrould-Kolieb and A. Talberg, *Global Policy*, 2021, **12**, 24.
98. M. Allsopp, D. Santillo and P. Johnston, A Scientific Critique of Oceanic Iron Fertilization as a Climate Change Mitigation Strategy, 2007, https://www.greenpeace.to/publications/iron_fertilisation_critique.pdf, (accessed 10 August 2021).
99. WWF Opposes Precarious Ocean Fertilization Project, 2009, https://wwf.panda.org/?155301/WWF-oppose, (accessed 10 August 2021).
100. K. J. Wetter and T. Zundel, The Big Bad Fix: The Case against Geoengineering, 2017, https://www.boell.de/en/2017/12/01/big-bad-fix-case-against-geoengineering, (accessed 9 Aug 2021).
101. C. Baatz, C. Heyward and H. Stelzer, *Environ. Values*, 2016, **25**, 1.
102. J. Anshelm and A. Hansson, *Energy Res. Social Sci.*, 2014, **2**, 135.
103. S. Gambardella, *Carbon Clim. Law Rev.*, 2019, **2019**(2), 122.
104. J. McGee, K. Brent and W. Burns, *Aust. J. Marit. Ocean Affairs*, 2018, **10**(1), 67.
105. L. Kovac, *Macqurie J. Int. Comp. Env. Law*, 2013, **9**(2), 39.
106. L. Cordonnery, A. D. Hemmings and L. Kriwoken, *Int. J. Mar. Coast. Law*, 2015, **30**, 727.
107. F. D. Oeste, R. D. Richter, T. Ming and S. Caillol, *Earth Syst. Dyn.*, 2017, **8**, 1.
108. T. Ming, R. de Richter, F. D. Oeste, R. Tulip and S. Calliol, *Atmos. Poll. Res.*, 2021, **2**(5), 101035.
109. Presentation by C. Elsworth and F. Oeste, 3 July 2020, https://restoreourclimate.earth/2020/07/isa-presentation/, (accessed 10 August 2021).
110. Iron Salt Aerosol: Climate Restoration with Iron Salt Aerosol (ISA), https://climategamechangers.org/game-changers/iron-salt-aerosol/, (accessed 10 August 2021).
111. K. Bali, A. K. Mishra, S. Singh, S. Chandra and Y. Lehahn, *Environ. Sci. Pollut. Res.*, 2019, **26**, 11940.
112. A. Krishnamurthy, J. K. Moore, N. Mahowald, C. Luo, S. C. Doney, K. Lindsay and C. S. Zender, *Global Biogeochem. Cycles*, 2009, **23**, GB3016.
113. C. Völker and A. Tagliabue, *Mar. Chem.*, 2015, **173**, 67.
114. A. Tagliabue, A. R. Bowie, T. DeVries, M. J. Ellwood, W. M. Landing, A. Milne, D. C. Ohnemus, B. S. Twining and P. W. Boyd, *Nat. Commun.*, 2019, **10**, 1.
115. A. Ito, S. Myriokefalitakis, M. Kanakidou, N. M. Mahowald, R. A. Scanza, D. S. Hamilton, A. R. Baker, T. Jickells, M. Sarin, S. Bikkina and Y. Gao, *Sci. Adv.*, 2019, **5**, eaau7671.
116. A. Ito, *Environ. Sci. Technol. Lett.*, 2015, **2**, 70.
117. Z. Shi, S. Bonneville, M. D. Krom, K. S. Carslaw, T. D. Jickells, A. R. Baker and L. G. Benning, *Atmos. Chem. Phys.*, 2011, **11**, 995.
118. S. Myriokefalitakis, N. Daskalakis, N. Mihalopoulos, A. Baker, A. Nenes and M. Kanakidou, *Biogeoscience*, 2015, **12**, 3973.

119. NOAA, Climate Change: Atmospheric Carbon Dioxide, 2020, https://www.climate.gov/news-features/understanding-climate/climate-change-atmospheric-carbon-dioxide, (accessed 10 August 2021).
120. R. Bodle, *Tulsa Law Rev*, 2010, **46**, 305.
121. A. Corner, K. Parkhill, N. Pidgeon and N. E. Vaughan, *Global Environ. Change*, 2013, **23**, 938.
122. A. Corner and N. Pidgeon, *Clim. Change*, 2015, **130**, 425.
123. R. Bellamy, J. Lezaun and J. Palmer, *Global Environ Change*, 2017, **45**, 194.
124. I. S. F. Jones, *Energy Convers. Manage.*, 1996, **37**, 1049.
125. I. S. F. Jones, *Mar. Ecol.: Prog. Ser.*, 2011, **425**, 281.
126. K. R. Arrigo, *Nature*, 2005, **437**, 349.
127. G. Shaffer, *J. Geophys. Res. Oceans*, 1996, **101**(C2), 3723.
128. T. M. Lenton and A. J. Watson, *Global Biogeochem.Cycles*, 2000, **14**, 225.
129. A. C. Martiny, M. W. Lomas, W. Fu, P. W. Boyd, L. C. Yuh-ling, G. A. Cutter, M. J. Ellwood, K. Furuya, F. Hashihama, J. Kanda and D. M. Karl, *Sci. Adv.*, 2019, **5**, eaax0341.
130. N. Gruber, in *The Ocean Carbon Cycle and Climate*, ed. M. Follows and T. Oguz, Springer, Dordrecht, 2004.
131. S. W. Nixon, *Ophelia*, 1995, **41**, 199.
132. S. P. Seitzinger, E. Mayorga, A. F. Bouwman, C. Kroeze, A. H. W. Beusen, G. Billen, G. Van Drecht, E. Dumont, B. M. Fekete, J. Garnier and J. A. Harrison, *Global Biogeochem. Cycles*, 2010, **24**, GB0A08.
133. D. F. Boesch, *Estuaries*, 2002, **25**, 886.
134. Ocean Nourishment Corporation: http://www.oceannourishment.com/, (accessed 10 August 2021).
135. R. J. Matear and B. Elliott, *J. Geophys. Res.*, 2004, **109**, C04001.
136. D. P. Harrison, *Environ. Res. Lett.*, 2017, **12**, 035001.
137. M. W. Lawrence, *Int. J. Global Warming*, 2014, **6**, 15.
138. A. Yool, A. P. Martin, C. Fernández and D. R. Clark, *Nature*, 2007, **447**, 999.
139. J. P. Zehr, E. J. Carpenter and T. A. Villareal, *Trends Microbiol.*, 2000, **8**, 68.
140. D. Bianchi, J. P. Dunne, J. L. Sarmiento and E. D. Galbraith, *Global Biogeochem. Cycles*, 2012, **26**, GB2009.
141. S. Yang and N. Gruber, *Global Biogeochem. Cycles*, 2016, **30**, 1418.
142. C. Deutsch, *Global Biogeochem. Cycles*, 2004, **18**, GB4012.
143. D. E. Canfield, A. N. Glazer and P. G. Falkowski, *Science*, 2010, **330**, 192.
144. T. Tyrrell, *Nature*, 1999, **400**, 525.
145. C. A. Heil, M. Revilla, P. M. Glibert and S. Murasko, *Limnol. Oceanogr.*, 2007, **52**, 1067.
146. P. M. Glibert, R. Azanza, M. Burford, K. Furuya, E. Abal, A. Al-Azri, F. Al-Yamani, P. Andersen, D. M. Anderson, J. Beardall, G. M. Berg, L. Brand, D. Bronk, J. Brookes, J. M. Burkholder, A. Cembella, W. P. Cochlan, J. L. Collier, Y. Collos, R. Diaz, M. Doblin, T. Drennen, S. Dyhrman, Y. Fukuyo, M. Furnas, J. Galloway, E. Graneli, D. V. Ha, G. Hallegraeff, J. Harrison, P. J. Harrison, C. A. Heil, K. Heimasn,

R. Howarth, C. Jauzein, A. A. Kana, T. M. Kana, H. Kim, R. Kudela, C. Legrand, M. Mallin, M. Mulholland, S. Murray, J. O'Neil, G. Pitcher, Y. Qi, N. Rabelais, R. Raine, S. Seitzinger, P. S. Salomon, C. Solomon, D. K. Stoecker, G. Usup, J. Wilson, K. Yin, M. Zhou and M. Zhu, *Mar. Poll. Bull.*, 2008, **56**, 1049.

147. J. K. Egge and D. L. Aksnes, *Mar. Ecol.: Prog. Ser.*, 1992, **83**, 281.
148. P. K. Bienfang and P. J. Harrison, *Mar. Biol.*, 1984, **83**, 293.
149. T. F. Thingstad, M. D. Krom, R. F. C. Mantoura, G. F. Flaten, S. Groom, B. Herut, N. Kress, C. S. Law, A. Pasternak, P. Pitta and S. Psarra, *Science*, 2005, **309**, 1068.
150. L. Dixon, *Environ. Chem.*, 2008, **5**, 135.
151. A. P. Rees, C. S. Law and E. M. S. Woodward, *Geophys. Res. Lett.*, 2006, **33**, L10607.
152. C. T. A. Chen and A. V. Borges, *Deep Sea Res.II: Top. Stud. Oceanogr.*, 2009, **56**, 578.
153. A. Roobaert, G. G. Laruelle, P. Landschützer, N. Gruber, L. Chou and P. Regnier, *Global Biogeochem. Cycles*, 2019, **33**, 1693.
154. I. S. F. Jones and M. Renilson, *J. Ocean Technol.*, 2011, **6**, 30.
155. M. Voss, B. Deutsch, R. Elmgren, C. Humboirg, P. Kuuppo, M. Pastuszak, C. Rolff and U. Shulte, *Biogeosci.*, 2006, **3**, 663.
156. I. S. F. Jones, *Int. J. Clim. Change Strategies Manage.*, 2014, **6**(4), 391.
157. D. Cordell, J. Drangert and S. White, *Global Environ. Change*, 2009, **19**, 292.
158. T.-S. S. Neset and D. Cordell, *J. Sci. Food Agric.*, 2012, **92**, 2.
159. J. D. Edixhoven, J. Gupta and H. H. G. Savenije, *Earth Syst. Dyn.*, 2014, **5**, 49.
160. P. M. Glibert, R. Maranger, D. J. Sobota and L. Bouwman, *Environ. Res. Lett.*, 2014, **9**, 105001.
161. G. Li, L. Cheng, J. Zhu, K. E. Trenberth, M. E. Mann and J. P. Abraham, *Nat. Clim. Change*, 2020, **10**, 1116.
162. M. L. Hammond, C. Beaulieu, S. A. Henson and S. K. Sahu, *Sci. Rep.*, 2020, **10**, 1.
163. Y. W. Watanabe, B. F. Li and M. Wakita, *Geophys. Res. Lett.*, 2018, **45**, 9106.
164. C. T. A. Chen, H. K. Lui, C. H. Hsieh, T. Yanagi, N. Kosugi, M. Ishii and G. C. Gong, *Nat. Clim. Change*, 2017, 7, 890.
165. T. D. Jickells, E. Buitenhuis, K. Altieri, A. R. Baker, D. Capone, R. A. Duce, F. Dentener, K. Fennel, M. Kanakidou, J. LaRoche and K. Lee, *Global Biogeochem. Cycles*, 2017, **31**, 289.
166. R. L. Swanson, M. L. Bortman, T. P. O'Connor and H. M. Stanford, *Mar. Pollut. Bull.*, 2004, **49**, 679.
167. N. Kress, E. Shoham-Frider and B. S. Galil, *Mar. Pollut. Bull.*, 2016, **110**, 99.
168. J. E. Lovelock and C. G. Rapley, *Nature*, 2007, **449**, 781.
169. W. H. Calvin, Using the oceans to remove CO_2 from the atmosphere, http://geo-engineering.blogspot.com/2013/03/using-the-oceans-to-remove-co2-from-the-atmosphere.html 2013, (accessed 10 August 2021).

170. A. Yool, J. G. Shepherd, H. L. Bryden and A. Oschlies, *J. Geophys. Res.*, 2009, **114**, C08009.
171. S. Dutreuil, L. Bopp and A. Tagliabue, *Biogeosciences*, 2009, **6**, 901.
172. S. J. Bauman, M. T. Costa, M. B. Fong, B. M. House, E. M. Perez, M. H. Tan, A. E. Thornton and P. J. S. Franks, *Oceanography*, 2014, **27**(3), 17.
173. L. Kwiatkowski, K. L. Ricke and K. Caldeira, *Environ. Res. Lett.*, 2015, **10**, 034016.
174. A. Oschlies, M. Pahlow, A. Yool and R. J. Matear, *Geophys. Res. Lett.*, 2010, **37**, L04701.
175. Y. Pan, W. Fan, D. Zhang, J. Chen, H. Huang, S. Liu, Z. Jiang, Y. Di, M. Tong and Y. Chen, *Sci. China Earth Sci.*, 2016, **59**, 236.
176. Y. Pan, W. Fan, T. H. Huang, S. L. Wang and C. T. A. Chen, *Sci. Total Environ.*, 2015, **511**, 692.
177. Y. Zhang, M. Zhao, Q. Cui, W. Fan, J. Qi, Y. Chen, Y. Zhang, K. Gao, J. Fan, G. Wang and C. Yan*Sci. China Earth Sci.*, 2017, 60, 809.
178. M. Magi, B. E. Casareto, T. Suzuki, Y. Honda, Y. Suzuki and M. P. Niraula, *Greenhouse Gas Control Technol.*, 2015, 7, 791.
179. B. Kirke, *Ocean Coast. Manage.*, 2003, **46**, 901.
180. W. Fan, Z. Zhang, Z. Yao, C. Xiao, Y. Zhang, Y. Zhang, J. Liu, Y. Di, Y. Chen and Y. Pan, *Appl. Ocean Res.*, 2020, **101**, 102260.
181. Y. Pan, Y. Li, W. Fan, D. Zhang, Y. Qiang, Z. P. Jiang and Y. Chen, *J. Atmos. Ocean. Technol*, 2019, **36**, 2191.
182. W. Fan, Y. Pan, D. Zhang, C. Xu, Y. Qiang and Y. Chen, *Ocean Eng.*, 2016, **113**, 191.
183. B. E. Casareto, M. P. Niraula and Y. Suzuki, *J. Exp. Mar. Biol. Ecol.*, 2017, **487**, 1.
184. Y. Pan, L. You, Y. Li, W. Fan, C. T. A. Chen, B. J. Wang and Y. Chen, *Sustainability*, 2018, **10**, 664.
185. Y. Sawall, M. Harris, M. Lebrato, M. Wall and E. Y. Feng, *Front. Mar. Sci*, 2020, 7, 720.
186. K. Klima, N. Lin, K. Emanuel, M. G. Morgan and I. Grossmann, *Env. Sci. Technol*, 2012, **46**, 636.
187. R. D. Blum, B. J. Blum, D. P. Duston and G. M. Hagerman, Method and Apparatus for Reducing the Intensity of Hurricanes at Sea by Deep-water Upwelling, *U.S. Pat.*, 09/994,860, 2009.
188. W. H. Avery and C. Wu, *Renewable Energy from the Ocean: a Guide to OTEC*. Oxford University Press, Oxford, U.K, 1994.
189. R. Fujita, A. C. Markham, J. E. D. Diaz, J. R. M. Garcia, C. Scarborough, P. Greenfield, P. Black and S. E. Aguilera, *Mar. Policy*, 2012, **36**, 463.
190. P. E. Grandelli, G. Rocheleau, J. Hamrick and M. Church, Modeling the Physical and Biochemical Influence of Ocean Thermal Energy Conversion Plant Discharges into their Adjacent Waters (DOE/EE0003638-1), Makai Ocean Engineering Inc., 2012.
191. G. H. Rau and J. R. Baird, *Renewable Sustainable Energy Rev.*, 2018, **95**, 265.

192. Makai's Ocean Thermal Energy Conversion (OTEC) Power Plant, Hawaii. https://www.power-technology.com/projects/makais-ocean-thermal-energy-conversion-otec-power-plant-hawaii/, (accessed 10 August 2021).
193. N. Reynard, E. Ellison, A. Wilson, P. Williamson, J. O'Niles, E. Ransome and A. Mashayek, *The Contribution of Coastal Blue Carbon Ecosystems to Climate Change Mitigation and Adaptation*, Grantham Institute, Imperial College London, 2020.
194. D. Laffoley, *Protecting and Effectively Managing Blue Carbon Ecosystems to Realise the Full Value to Society – A Sea of Opportunities*. Opinion piece for WWF-UK, Woking, 2020. https://www.wwf.org.uk/sites/default/files/2020-11/WWF_blue_carbon020.pdf, (accessed 2 December 2020).
195. N. Jiao, H. Wang, G. Xu and S. Aricò, *Natl. Sci. Rev.*, 2018, **5**, 464.
196. C. Nellemann, E. Corcoran, C. Duarte, L. Valdes, C. Young, L. Fonseca and G. Grimsditch, Blue Carbon: The Role of Healthy Oceans in Binding Carbon, United Nations Environment Programme, GRID-Arendal, 2009, https://www.grida.no/publications/145, (accessed 10 August 2021).
197. S. Chiquier, *Afforestation, in Greenhouse Gas Removal Technologies*, ed. M. Bui and N. Mac Dowell, Royal Society of Chemistry, London, 2022, ch. 9.
198. E. Mcleod, G. L. Chmura, S. Bouillon, R. Salm, M. Björk, C. M. Duarte, C. E. Lovelock, W. H. Schlesinger and B. R. Silliman, *Front. Ecol. Environ.*, 2011, **9**, 552.
199. P. Taillardat, D. A. Friess and M. Lupascu, *Biol. Lett.*, 2018, **14**, 20180251.
200. H. Kennedy, J. Beggins, C. M. Duarte, J. W. Fourqurean, M. Holmer, N. Marbà and J. J. Middelburg, *Global Biogeochem. Cycles*, 2010, **24**, GB4026.
201. S. Bouillon, F. Dahdouh-Guebas, A. Rao, N. Koedam and F. Dehairs, *Hydrobiology*, 2003, **495**, 33.
202. M. F. Adame, S. F. Wright, A. Grinham, K. Lobb, C. E. Reymond and C. E. Lovelock, *Limnol. Oceanogr.*, 2012, **57**, 1492.
203. A. M. Ricart, P. H. York, C. V. Bryant, M. A. Rasheed, D. Ierodiaconou and P. I. Macreadie, *Sci. Rep.*, 2020, **10**, 5865.
204. S. T. Chew and J. B. Gallagher, *Sci. Rep.*, 2018, **8**(1), 2553.
205. Verra: Verified Carbon Standard, VM0033 Methodology for Tidal Wetland and Seagrass Restoration, https://verra.org/methodology/vm0033-methodology-for-tidal-wetland-and-seagrass-restoration-v1-0/, 2015, (accessed 10 August 2021).
206. S. Crooks, A. E. Sutton-Grier, T. G. Troxler, N. Herold, B. Bernal, L. Schile-Beers and T. Wirth, *Nat. Clim. Change*, 2018, **8**, 1109.
207. J. López-Portillo, R. R. Lewis, P. Saenger, A. Rovai, N. Koedam, F. Dahdouh-Guebas, C. Agraz-Hernández and V. H. Rivera-Monroy, in *Mangrove Ecosystems: A Global Biogeographic Perspective*, ed. V. H. Rivera-Monroy, A. Y. Lee, E. Kristensen and R. R. Twilley, Springer, Cham., 2017, p 301–345.
208. P. Adam, in *Coastal Wetlands*, ed. G. M. E. Perillo, E. Wolanski, D. R. Cahoon and C. S. Hopkinson, Elsevier, Amsterdam, 2019, pp. 817–861.

209. M. M. van Katwijk, A. Thorhaug, N. Marbà, R. J. Orth, C. M. Duarte, G. A. Kendrick, I. H. Althuizen, E. Balestri, G. Bernard, M. L. Cambridge, A. Cunha, C. Durance, W. Giesen, Q. Han, S. Hosokawa, W. Kiswara, T. Komatsu, C. Lardicci, K.-S. Lee, A. Meinesz, M. Nakaoka, K. R. O. Brien, E. I. Paling, C. Pickerell, A. M. A. Ransijn and J. J. Verduin, *J. Appl. Ecol.*, 2016, **53**, 567.
210. Q. Zhao, J. Bai, L. Huang, B. Gu, Q. Lu and Z. Gao, *Ecol. Indic.*, 2016, **60**, 442.
211. L. Wylie, A. E. Sutton-Grier and A. Moore, *Mar. Policy*, 2016, **65**, 76.
212. B. W. Griscom, J. Adams, P. W. Ellis, R. A. Houghton, G. Lomax, D. A. Miteva, W. H. Schlesinger, D. Shoch, J. V. Siikamäki, P. Smith and P. Woodbury, *Proc. Natl. Acad. Sci.*, 2017, **114**, 11645.
213. X. Li, R. Bellerby, C. Craft and S. E. Widney, *Anthropocene Coasts*, 2018, **1**, 1.
214. S. Crooks, L. Windham-Myers and T. G. Troxler, in *A Blue Carbon Primer: The State of Coastal Wetland Carbon Science, Practice and Policy*, ed. L. Windham-Myers, S. Crooks and T. G. Troxler, CRC Press, Boca Raton, 2019, ch. 1, pp. 1–8.
215. C. E. Lovelock and C. M. Duarte, *Biol. Lett.*, 2019, **15**, 20180781.
216. P. I. Macreadie, A. Anton, J. A. Raven, N. Beaumont, R. M. Connolly, D. A. Friess, J. J. Kelleway, H. Kennedy, T. Kuwae, P. S. Lavery, C. E. Lovelock, D. A. Smale, E. T. Apostolaki, T. B. Atwood, J. Baldock, T. S. Bianchi, G. L. Chmura, B. D. Eyre, J. W. Fourqurean, J. M. Hall-Spencer, M. Huxham, I. E. Hendriks, D. Krause-Jensen, D. Laffoley, T. Luisetti, N. Marbà, P. Masque, K. J. McGlathery, J. P. Megonigal, D. Murdiyarso, B. D. Russell, R. Santos, O. Serrano, B. R. Silliman, K. Watanabe and C. M. Duarte, *Nat. Commun.*, 2019, **10**, 3998.
217. T. Hiraishi, T. Krug, K. Tanabe, N. Srivastava, J. Baasansuren, M. Fukuda and T. G. Troxler, *2013 Supplement to the 2006 IPCC Guidelines for National Greenhouse Gas Inventories: Wetlands*. IPCC, Geneva, 2014.
218. T. G. Troxler, H. A. Kennedy, S. Crooks and A. E. Sutton-Grier, in *A Blue Carbon Primer: The State of Coastal Wetland Carbon Science, Practice and Policy*, ed. L. Windham-Myers, S. Crooks and T. G. Troxler, CRC Press, Boca Raton, 2019, ch. 16, pp. 217–234.
219. B. A. Needelman, I. M. Emmer, M. P. J. Oreska and J. P. Megonigal, in *A Blue Carbon Primer: The State of Coastal Wetland Carbon Science, Practice and Policy*, ed. L. Windham-Myers, S. Crooks and T. G. Troxler, CRC Press, Boca Raton, 2019, ch. 20, pp. 283–292.
220. IPCC (Intergovernmental Panel on Climate Change). Summary for Policymakers, in *IPCC Special Report on the Ocean and Cryosphere in a Changing Climate*, ed. H.-O. Pörtner, D. C. Roberts, V. Masson-Delmotte, P. Zhai, M. Tignor, E. Poloczanska, K. Mintenbeck, A. Alegría, M. Nicolai, A. Okem, J. Petzold, B. Rama and N. M. Weyer, WMO/UNEP, 2019.
221. A. Abelson, D. C. Reed, G. J. Edgar, C. S. Smith, G. A. Kendrick, R. J. Orth, L. Airoldi, B. Silliman, M. W. Beck, G. Krause, N. Shashar, N. Stambler and P. Nelson, *Front. Mar. Sci.*, 2020, 7, 544105.

222. S. C. Johannessen and R. W. Macdonald, *Environ. Res. Lett.*, 2016, **11**(11), 113001.
223. A. Arias-Ortiz, P. Masqué, J. Garcia-Orellana, O. Serrano, I. Mazarrasa, N. Marbà, C. E. Lovelock, P. S. Lavery and C. M. Duarte, *Biogeosci.*, 2018, **15**, 6791.
224. J. C. Callaway, in *A Blue Carbon Primer: The State of Coastal Wetland Carbon Science, Practice and Policy*, ed. L. Windham-Myers, S. Crooks and T. G. Troxler, CRC Press, Boca Raton, 2019, ch. 7, pp. 81–92.
225. I. R. Santos, D. J. Burdige, T. C. Jennerjahn, S. Bouillon, A. Cabral, O. Serrano, T. Wernberg, K. Filbee-Dexter, J. A. Guimond and J. J. Tamborski, *Estuarine Coast. Shelf Sci.*, 2021, **255**, 107361.
226. J. A. Rosentreter, A. N. Al-Haj, R. W. Fulweiler and P. Williamson, *Glob. Biogeochem. Cycles*, 2021, **35**(2), e2020GB006858.
227. J. K. Keller, in *A Blue Carbon Primer: The State of Coastal Wetland Carbon Science, Practice and Policy*, ed. L. Windham-Myers, S. Crooks and T. G. Troxler, CRC Press, Boca Raton, 2019, ch. 8, pp. 93–106.
228. A. N. Al-Haj and R. W. Fulweiler, *Global Change Biol.*, 2020, **26**, 2988.
229. H. Kennedy, J. W. Fourqurean and S. Papadimitriou, in *A Blue Carbon Primer: The State of Coastal Wetland Carbon Science, Practice and Policy*, ed. L. Windham-Myers, S. Crooks and T. G. Troxler, CRC Press, Boca Raton, 2019, ch. 9, pp. 107–119.
230. V. Saderne, N. R. Geraldi, P. I. Macreadie, D. T. Maher, J. J. Middelburg, O. Serrano, H. Almahasheer, A. Arias-Ortiz, M. Cusack, B. D. Eyre, J. W. Fourqurean, H. Kennedy, D. Krause-Jensen, T. Kuwae, P. S. Lavery, C. E. Lovelock, N. Marba, P. Masqué, M. A. Mateo, I. Mazarra, K. J. McGlathery, M. P. J. Oreska, C. J. Sanders, I. R. Santos, J. M. Smoak, Y. Tanaya, K. Watanabe and C. M. Duarte, *Nat. Commun.*, 2019, **10**, 1106.
231. V. Saderne, M. Fusi, T. Thomson, A. Dunne, F. Mahmud, F. Roth, S. Carvalho and C. M. Duarte, *Limnol. Oceanogr. Lett.*, 2021, **6**, 61.
232. B. R. Van Dam, P. Polsenaere, A. Barreras-Apodaca, C. Lopes, Z. Sanchez-Mejia, T. Tokoro, T. Kuwae, L. G. Loza, A. Rutgersson, J. Fourqurean and H. Thomas, *Global Biogeochem. Cycles*, 2021, **35**, e20290GB006848.
233. B. R. Van Dam, M. A. Zeller, C. Lopes, A. R. Smyth, M. E. Böttcher, C. L. Osburn, T. Zimmerman, D. Pröfrock, J. W. Fourqurean and H. Thomas, *Sci. Adv.*, 2021, 7, eabj1372.
234. R. D. Ward, D. A. Friess, R. H. Day and R. A. MacKenzie, *Ecosyst. Health Sustainability*, 2016, **2**, 01211.
235. N. C. Duke, J. M. Kovacs, A. D. Griffiths, L. Preece, D. J. Hill, P. Van Oosterzee, J. Mackenzie, H. S. Morning and D. Burrows, *Mar. Freshwater Res.*, 2017, **68**, 1816.
236. S. Strydom, K. Murray, S. Wilson, B. Huntley, M. Rule, M. Heithaus, C. Bessey, G. A. Kendrick, D. Burkholder, M. W. Fraser and K. Zdunic, *Global Change Biol.*, 2020, **26**, 3525.
237. M. L. Kirwan and J. P. Megonigal, *Nature*, 2013, **504**, 53.

238. C. E. Lovelock and R. Reef, *One Earth*, 2020, **3**, 195–211.
239. P. Williamson and V. A. Guinder, in *The Impacts of Climate Change: A Comprehensive Study of Physical, Biophysical, Social and Political Issues*, ed. T. M. Letcher, Elsevier, Amsterdam, 2021, ch. 5.
240. H. M. Nguyen, P. J. Ralph, L. Marín-Guirao, M. Pernice and G. Procaccini, *Biol. Rev.*, 2021, **96**, 2009.
241. C. Laufkötter, J. Zscheischler and T. L. Frölicher, *Science*, 2020, **369**, 1621.
242. D. A. Smale, T. Wernberg, E. C. Oliver, M. Thomsen, B. P. Harvey, S. C. Straub, M. T. Burrows, L. V. Alexandre, J. A. Benthuysem, M. G. Donat, M. Feng, A. J. Hobday, N. J. Holbrook, S. E. Perkins-Kirkpatrick, H. A. Scannell, A. S. Gupta, B. L. Payne and P. J. Moore, *Nat. Clim. Change*, 2019, **9**, 306.
243. A. Arias-Ortiz, O. Serrano, P. Masqué, P. S. Lavery, U. Mueller, G. A. Kendrick, M. Rozaimi, A. Esteban, J. W. Fourqurean, N. J. N. C. C. Marbà and M. A. Mateo, *Nat. Clim. Change*, 2018, **8**, 338.
244. C. B. de los Santos, D. Krause-Jensen, T. Alcoverro, N. Marbà, C. M. Duarte, M. M. van Katwijk, M. Pérez, J. L. Sánchez-Lizaso, G. Roca, E. Jankowska, J. L. Pérez-Lloréns, J. Fournier, M. Montefalcone, G. Pergent, J. M. Ruiz, S. Cabaço, K. Cook, R. J. Wilkes, F. E. Moy, G. M.-R. Trayter, X. S. Arañó, D. J. de Jong, Y. Fernández-Torquemada, I. Auby, J. J. Vergara and R. Santos, *Nat. Commun.*, 2019, **10**, 3356.
245. A. E. Raftery, A. Zimmer, D. M. W. Frierson, R. Startz and P. Liu, *Nat. Clim. Change*, 2017, 7, 637.
246. UNEP (United Nations Environment Programme), Emissions Gap Report 2020, UNEP, Nairobi, https://www.unenvironment.org/emissions-gap-report-2020, (accessed 10 August 2021).
247. E. S. Poloczanska, M. T. Burrows, C. J. Brown, J. García Molinos, B. S. Halpern, O. Hoegh-Guldberg, C. V. Kappel, P. J. Moore, A. J. Richardson, D. S. Schoeman and W. J. Sydeman, *Front. Mar. Sci.*, 2016, **3**, 62.
248. K. Rogers, J. J. Kelleway, N. Saintilan, J. P. Megonigal, J. B. Adams, J. R. Holmquist, M. Lu, L. Schile-Beers, A. Zawadzki, D. Mazumder and C. D. Woodroffe, *Nature*, 2019, **567**, 91.
249. M. E. Hanley, T. J. Bouma and H. L. Mossman, *Ann. Bot.*, 2020 **125**, 197.
250. I. Valiela, J. Lloret, T. Bowyer, S. Miner, D. Remsen, E. Elmstrom, C. Cogswell and E. R. Thieler, *Sci. Total Environ.*, 2018, **640**, 1148.
251. M. Schuerch, T. Spencer, S. Temmerman, M. L. Kirwan, C. Wolff, D. Lincke, C. J. McOwen, M. D. Pickering, R. Reef, A. T. Vafeidis, J. Hinkel, R. J. Nicholls and S. Brown, *Nature*, 2018, **561**, 231.
252. M. Oppenheimer, B. C. Glavovic , J. Hinkel, R. van de Wal, A. K. Magnan, A. Abd-Elgawad, R. Cai, M. Cifuentes-Jara, R. M. DeConto, T. Ghosh, J. Hay, F. Isla, B. Marzeion, B. Meyssignac and Z. Sebesvari, in *IPCC Special Report on the Ocean and Cryosphere in a Changing Climate*, ed. H.-O. Pörtner, D. C. Roberts, V. Masson-Delmotte, P. Zhai, M. Tignor, E. Poloczanska,

K. Mintenbeck, A. Alegría, M. Nicolai, A. Okem, J. Petzold, B. Rama and N. M. Weyer, WMO/UNEP, Geneva, 2019, ch. 4, pp. 321–445.

253. S. D. Sasmito, D. Murdiyarso, D. A. Friess and S. Kurnianto, *Wetlands Ecol. Manage.*, 2016, **24**, 263.

254. N. Saintilan, N. S. Khan, E. Ashe, J. J. Kelleway, K. Rogers, C. D. Woodroffe and B. P. Horton, *Science*, 2020, **368**, 1118.

255. M. Simard, L. Fatoyinbo, C. Smetanka, V. H. Rivera-Monroy, E. Castañeda-Moya, N. Thomas and T. Van der Stocken, *Nat. Geosci.*, 2019, **12**, 40.

256. M. Collins, M. Sutherland, L. Bouwer, S.-M. Cheong, T. Frölicher, H. Jacot Des Combes, M. Koll Roxy, I. Losada, K. McInnes, B. Ratter, E. Rivera-Arriaga, R. D. Susanto, D. Swingedouw and L. Tibig, in *IPCC Special Report on the Ocean and Cryosphere in a Changing Climate*, ed. H.-O. Pörtner, D. C. Roberts, V. Masson-Delmotte, P. Zhai, M. Tignor, E. Poloczanska, K. Mintenbeck, A. Alegría, M. Nicolai, A. Okem, J. Petzold, B. Rama and N. M. Weyer, WMO/UNEP, Geneva, 2019, ch. 6, pp. 589–655.

257. M. Saunois, A. R. Stavert, B. Poulter, P. Bousquet, J. G. Canadell, R. B. Jackson, P. A. Raymond, E. J. Dlugokencky, S. Houweling, P. K. Patra, P. Ciais, V. K. Arora, D. Bastviken, P. Bergamaschi, D. R. Blake, G. Brailsford, L. Bruhwiler, K. M. Carlson, M. Carrol, S. Castaldi, N. Chandra, C. Crevoisier, P. M. Crill, K. Covey, C. L. Curry, G. Etiope, C. Frankenberg, N. Gedney, M. I. Hegglin, L. Höglund-Isaksson, G. Hugelius, M. Ishizawa, A. Ito, G. Janssens-Maenhout, K. M. Jensen, F. Joos, T. Kleinen, P. B. Krummel, R. L. Langenfelds, G. G. Laruelle, L. Liu, T. Machida, S. Maksyutov, K. C. McDonald, J. McNorton, P. A. Miller, J. R. Melton, I. Morino, J. Müller, F. Murguia-Flores, V. Naik, Y. Niwa, S. Noce, S. O'Doherty, R. J. Parker, C. Peng, S. Peng, G. P. Peters, C. Prigent, R. Prinn, M. Ramonet, P. Regnier, W. J. Riley, J. A. Rosentreter, A. Segers, I. J. Simpson, H. Shi, S. J. Smith, L. P. Steele, B. F. Thornton, H. Tian, Y. Tohjima, F. N. Tubiello, A. Tsuruta, N. Viovy, A. Voulgarakis, T. S. Weber, M. van Weele, G. R. van der Werf, R. F. Weiss, D. Worthy, D. Wunch, Y. Yin, Y. Yoshida, W. Zhang, Z. Zhang, Y. Zhao, B. Zheng, Q. Zhu, Q. Zhu and Q. Zhuang, *Earth Syst. Sci. Data*, 2020, 12 (3), 1561.

258. H. Tian, R. Xu, J. G. Canadell, R. L. Thompson, W. Winiwarter, P. Suntharalingam, E. A. Davidson, P. Ciais, R. B. Jackson, G. Janssens-Maenhout, M. J. Prather, P. Regnier, N. Pan, S. Pan, G. P. Peters, H. Shi, F. N. Tubiello, S. Zaehle, F. Zhou, A. Arneth, G. Battaglia, S. Berthet, L. Bopp, A. F. Bouwman, E. T. Buitenhuis, J. Chang, M. P. Chipperfield, S. R. S. Dangal, E. Dlugokencky, J. W. Elkins, B. D. Eyre, B. Fu, B. Hall, A. Ito, F. Joos, P. B. Krummel, A. Landolfi, G. G. Laruelle, R. Lauerwald, W. Li, S. Lienert, T. Maavara, M. MacLeod, D. B. Millet, S. Olin, P. K. Patra, R. G. Prinn, P. A. Raymond, D. J. Ruiz, G. R. van der Werf, N. Vuichard, J. Wang, R. F. Weiss, K. C. Wells, C. Wilson, J. Yang and Y. Yao, *Nature*, 2020, **586**, 248.

259. S. T. Wilson, A. N. Al-Haj, A. Bourbonnais, C. Frey, R. W. Fulweiler, J. D. Kessler, H. K. Marchant, J. Milucka, N. E. Ray, P. Suntharalingam, B. F. Thornton, R. C. Upstill-Goddard, T. S. Weber, D. L. Arévalo-Martínez, H. W. Bange, H. M. Benway, D. Bianchi, A. V. Borges, B. X. Chang, P. M. Crill, D. A. del Valle, L. Farías, S. B. Joye, A. Kock, J. Labidi, C. C. Manning, J. W. Pohlman, G. Rehder, K. J. Sparrow, P. D. Tortell, T. Treude, D. L. Valentine, B. B. Ward, S. Yang and L. N. Yurganov, *Biogeosciences*, 2020, **17**, 5809.
260. H. J. Poffenbarger, B. A. Needelman and J. P. Megonigal, *Wetlands*, 2011, **31**, 831.
261. C. E. Lovelock, D. A. Freiss, J. B. Kauffman and J. W. Fourqurean, in *A Blue Carbon Primer: The State of Coastal Wetland Carbon Science, Practice and Policy*, ed. L. Windham-Myers, S. Crooks and T. G. Troxler, CRC Press, Boca Raton, 2019, ch. 3, pp. 17–24.
262. J. A. Rosentreter, D. T. Maher, D. V. Erler, R. H. Murray and B. D. Eyre, *Sci. Adv.*, 2018, **4**, eaao4985.
263. J. A. Rosentreter and P. Williamson, *Global Change Biol.*, 2020, **26**, 5351.
264. S. C. Neubauer and J. P. Megonigal, *Ecosystems*, 2015, **18**, 1000.
265. S. C. Neubauer and J. T. A. Verhoeven, in *Wetlands: Ecosystem Services, Restoration and Wise Use*, ed. S. An and J. T. A. Verhoeven, Ecological Studies 28, Springer Nature, Switzerland, 2019, ch. 3, pp. 39–610.
266. P. Taillardat, B. S. Thompson, M. Garneau, K. Trottier and D. A. Friess, *Interface Focus*, 2020, **10**, 20190129.
267. R. Chauhan, A. Datta, A. L. Ramanathan and T. K. Adhya, *Agric., Ecosyst. Environ.*, 2017, **246**, 38.
268. P. Yang, D. Bastviken, D. Y. F. Lai, B. S. Lin, X. J. Mou, C. Tong and Y. Yao, *Estuarine Coastal Shelf Sci.*, 2017, **199**, 125.
269. K. D. Kroeger, S. Crooks, S. Moseman-Valtierra and J. Tang, *Sci. Rep.*, 2017, 7, 11914.
270. S. Q. Foster and R. W. Fulweiler, *Front. Mar. Sci*, 2016, **3**, 40.
271. D. T. Maher, J. Z. Sippo, D. R. Tait, C. Holloway and I. R. Santos, *Sci. Rep.*, 2016, **6**, 25701.
272. R. H. Murray, D. V. Erler and B. D. Eyre, *Global Change Biol.*, 2015, **21**, 3219.
273. M. P. J. Oreska, K. J. McGlathery, L. R. Aoki, A. C. Berger, P. Berg and L. Mullins, *Sci. Rep.*, 2020, **10**, 7325.
274. Global Commission on Adaptation, Adapt Now: A Global Call for Leadership on Climate Resilience, World Resources Institute, Washington DC, https://gca.org/reports/adapt-now-a-global-call-for-leadership-on-climate-resilience/, 2019, (accessed 10 August 2021).
275. T. Vegh, L. Pendleton, B. Murray, T. Troxler, K. Zhang, E. Castañeda-Moya, G. Guannel and A. E. Sutton-Grier, in *A Blue Carbon Primer: The State of Coastal Wetland Carbon Science, Practice and Policy*, ed. L. Windham-Myers, S. Crooks and T. G. Troxler, CRC Press, Boca Raton, 2019, ch. 18, pp. 249–266.

276. E. B. Barbier, S. D. Hacker, C. Kennedy, E. K. Koch, A. C. Stierand and B. R. Silliman, *Ecol. Monogr.*, 2011, **81**, 169.
277. A. E. Sutton-Grier, R. K. Gittman, K. K. Arkema, R. O. Bennett, J. Benoit, S. Blitch, K. A. Burks-Copes, A. Colden, A. Dausman, B. M. DeAngelis, A. R. Hughes, S. B. Scyphers and J. H. Grabowski, *Sustainability*, 2018, **10**, 523.
278. S. Narayan, M. W. Beck, B. G. Reguero, I. J. Losada, B. Van Wesenbeeck, N. Pontee, J. N. Sanchirico, J. C. Ingram, G. M. Lange and K. A. Burks-Copes, *PLoS One*, 2016, **11**, e0154735.
279. B. M. John, K. G. Shirlal and S. Rao, *Procedia Eng.*, 2015, **116**, 600.
280. K. B. Gedan, M. L. Kirwan, E. Wolanski, E. B. Barbier and B. R. Silliman, *Clim. Change*, 2011, **106**, 7.
281. T. Worthington and M. Spalding, Mangrove Restoration Potential: A Global Map Highlighting a Critical Opportunity, https://doi.org/10.17863/CAM.39153, 2018, (accessed 10 August 2021).
282. Y. Mazda, M. Magi, M. Kogo and P. N. Hong, *Mangroves Salt Marshes*, 1997, **1**, 127.
283. E. Bayraktarov, M. I. Saunders, S. Abdullah, M. Mills, J. Beher, H. P. Possingham, P. J. Mumby and C. E. Lovelock, *Ecol. Appl.*, 2015, **26**, 1055.
284. R. A. Konisky and D. M. Burdick, *Wetlands*, 2004, **24**, 434.
285. J. Beck and D. J. Gustafson, *Southeastern Nat.*, 2012, **11**, 747.
286. K. A. S. Kodikara, N. Mukherjee, L. P. Jayatissa, F. Dahdouh-Guebas and N. Koedam, *Restor. Ecol.*, 2017, **25**, 705.
287. M. M. van Katwijk, A. Thorhaug, N. Marbà, R. J. Orth, C. M. Duarte, G. A. Kendrick, I. H. Althuizen, E. Balestri, G. Bernard, M. L. Cambridge, A. Cunha, C. Durance, W. Giesen, Q. Han, S. Hosokawa, W. Kiswara, T. Komatsu, C. Lardicci, K.-S. Lee, A. Meinesz, M. Nakaoka, K. R. O. Brien, E. I. Paling, C. Pickerell, A. M. A. Ransijn and J. J. Verduin, *J. Appl. Ecol.*, 2016, **53**, 567.
288. R. Stewart, T. Noyce and H. Possingham, *Mar. Ecol.: Prog. Ser.*, 2003, **253**, 25.
289. J.-P. Gattuso, P. Williamson, C. M. Duarte and A. K. Magnan, *Front. Clim.*, 2021, **2**, 575716.
290. Blue Carbon Initiative, Mitigating Climate Change Through Coastal Ecosystem Management; https://www.thebluecarboninitiative.org/, (accessed 10 August 2021).
291. CBD (Convention on Biological Diversity), Wetlands and the Aichi Biodiversity Targets, CBD Press Brief, Montreal, Canada, https://www.cbd.int/waters/doc/wwd2015/wwd-2015-press-brief-aichi-en.pdf, 2015, (accessed 10 August 2021).
292. *Wetlands, Biodiversity and the Ramsar Convention: The Role of the Convention on Wetlands in the Conservation and Wise Use of Biodiversity*, ed. A. J. Halls, Ramsar Convention Bureau, Gland, Switzerland, 1997.
293. K. H. Olsen, *Clim. Change*, 2007, **84**, 59.
294. E. Corbera and H. Schroeder, *Environ. Sci. Pol.*, 2011, **14**, 89.

295. J. Correa, R. van der Hoff and R. Rajão, *Forests*, 2019, **10**, 272.
296. G. Grassi, J. House, W. A. Kurz, A. Cescatti, R. A. Houghton, G. P. Peters, M. J. Sanz, R. A. Viñas, R. Alkama, A. Arneth, A. Bondeau, F. Dentener, M. Fader, S. Frederici, P. Friedlingstein, A. K. Jain, E. Kato, C. D. Koven, D. Lee, J. E. M. S. Nabel, A. A. Nassikas, L. Perugini, S. Rossi, S. Sitch, N. Viovy, A. Wiltshire and S. Zaehle, *Nat. Clim. Change*, 2018, **8**, 914.
297. T. Luisetti, S. Ferrini, G. Grilli, T. D. Jickells, H. Kennedy, S. Kröger, I. Lorenzoni, B. Milligan, J. Van Der Molen, R. Parker, T. Pryce, R. K. Turner and E. Tyllianakis, *Nat. Commun.*, 2020, **11**, 4599.
298. H. C. Lovell, *WIREs Clim. Change*, 2010, **1**, 353.
299. J. G. Kairo, A. J. Hamza and C. Wanjiru, in A Blue Carbon Primer: The State of Coastal Wetland Carbon *Science, Practice and Policy*, ed. L. Windham-Myers, S. Crooks and T. G. Troxler, CRC Press, Boca Raton, 2019, ch. 24, pp. 341–351.
300. D. A. Miteva, B. C. Murray and S. K. Pattanayak, *Ecol. Econ.*, 2015, **119**, 127.
301. D. Herr, M. Unger, D. Laffoley and A. McGivern, *Aquat. Conserv. Mar. Freshwater Ecosyst*, 2017, **27**(S1), 116.
302. J. Raven, *Biol. Lett.*, 2018, **14**, 20180336.
303. D. Krause-Jensen, P. Lavery, O. Serrano, N. Marba, P. Masque and C. M. Duarte, *Biol. Lett.*, 2018, **14**, 20180236.
304. K. Filbee-Dexter and T. Wernberg, *Sci. Rep.*, 2020, **10**, 12341.
305. D. Krause-Jensen and C. M. Duarte, *Nat. Geosci.*, 2016, **9**, 737.
306. A. M. Queirós, N. Stephens, S. Widdicombe, K. Tait, S. J. McCoy, J. Ingels, S. Rühl, R. Airs, A. Beesley, G. Carnovale, P. Cazenave, S. Dashfield, E. Hua, M. Jones, P. Lindeque, C. L. McNeill, J. Nunes, H. Parry, C. Pascoe, C. Widdicombe, T. Smyth, A. Atkinson, D. Krause-Jensen and P. J. Somerfield, *Ecol. Monogr.*, 2019, **89**, e01366.
307. S. Wada, M. N. Aoki, A. Mikami, T. Komatsu, Y. Tsuchiya, T. Sato, H. Shinagawa and T. Hama, *Mar. Ecol.: Prog. Ser.*, 2008, **370**, 33.
308. K. Watanabe, G. Yoshida, M. Hori, Y. Umezawa, H. Moki and T. Kuwae, *Biogeosciences*, 2020, **17**, 2425.
309. Y. Kokubu, E. Rothäusler, J.-B. Filippi, E. D. H. Durieux and T. Komatsu, *Sci. Rep.*, 2019, **9**, 4331.
310. M. Wang, C. Hu, B. B. Barnes, G. Mitchum, B. Lapointe and J. P. Montoya, *Science*, 2019, **365**, 83.
311. A. Ortega, N. R. Geraldi, I. Alam, A. A. Kamau, S. G. Acinas, R. Logares, J. M. Gasol, R. Massana, D. Krause-Jensen and C. M. Duarte, *Nat. Geosci.*, 2019, **12**, 748.
312. P. V. R. Snelgrove, K. Soetart, M. Solan, S. Thrush, C.-L. Wei, R. Danovaro, R. W. Fulweiler, H. Kitazato, B. Ingole, A. Norkko, R. J. Parkes and N. Volkenborn, *Trends Ecol.*, 2018, **33**, 96.
313. M. F. Pedersen, K. Filbee-Dexter, N. L. Frisk, Z. Sárossy and T. Wernberg, *Mar. Ecol.: Prog. Ser.*, 2021, **660**, 53.
314. O. Legge, M. Johnson, N. Hicks, T. Jickells, M. Diesing, J. Aldridge, J. Andrews, Y. Artioli, D. C. Bakker, M. T. Burrows, N. Carr, G. Cripps,

S. L. Felgate, L. Fernand, N. Greenwood, S. Hartman, S. Kröger, G. Lessin, C. Mahaffey, D. J. Mayor, R. Parker, A. M. Queirós, J. D. Shutler, T. Silva, H. Stahl, J. Tinker, G. J. C. Underwood, J. Van der Molen, S. Wakelin, K. Weston and P. Williamson, *Front. Mar. Sci.*, 2020, 7, 43.

315. L. T. Bach, V. Tamsitt, J. Gower, C. L. Hurd, J. A. Raven and P. W. Boyd, *Nat. Commun.*, 2021, **12**, 2556.
316. K. Chung, J. Beardall, S. Mehta, D. Sahoo and S. Stojkovic, *J. Appl. Phycol.*, 2011, **23**, 877.
317. J. Wu, H. Zhang, Y. Pan, D. Krause-Jensen, Z. He, W. Fan, X. Xiao, I. Chung, N. Marbà, O. Serrano and R. B. Rivkin, *Ocean Coast. Manage.*, 2020, **194**, 105241.
318. C. F. Sondak, P. O. Ang, J. Beardall, A. Bellgrove, S. M. Boo, G. S. Gerung, C. D. Hepburn, D. D. Hong, Z. Hu, H. Kawai and D. Largo, *J. Appl. Phycol.*, 2017, **29**, 2363.
319. I. K. Chung, J. H. Oak, J. A. Lee, J. A. Shin, J. G. Kim and K. S. Park, *ICES J. Mar. Sci*, 2013, **70**, 1038.
320. C. M. Duarte, J. Wu, X. Xiao, A. Bruhn and D. Krause-Jensen, *Front. Mar. Sci.*, 2017, **4**, 100.
321. L. M. Laurens and R. S. Nelson, in *Sustainable Seaweed Technologies*, ed. M. D. Torres, S. Kraan and H. Dominguez. Elsevier, Amsterdam, 2020, p. 643.
322. A. D. Hughes, K. D. Black, I. Campbell, K. Davidson, M. S. Kelly and M. S. Stanley, *Greenhouse Gas Sci. Technol.*, 2012, **2**, 402.
323. A. R. N'Yeurt, D. P. Chynoweth, M. E. Capron, J. R. Stewart and M. A. Hasan, *Proc. Saf. Environ. Prot.*, 2012, **90**, 467.
324. H. E. Froehlich, J. C. Afflerbach, M. Frazier and B. S. Halpern, *Current Biol.*, 2019, **29**, 3087.
325. Macroalgae Research Inspiring Novel Energy Resources (MARINER) Program Overview. https://arpa-e.energy.gov/sites/default/files/documents/files/MARINER_ProgramOverview_FINAL.pdf (accessed 10 August 2021).
326. Macroalgae Research Inspiring Novel Energy Resources (MARINER) Project Descriptions https://arpa-e.energy.gov/sites/default/files/documents/files/MARINER-Project-Descriptions-92117.pdf, (accessed 10 August 2021).
327. M. E. Capron, R. Blaylock, K. Lucas, M. D. Chambers, J. R. Stewart, S. F. DiMarco, K. Whilden, B. Wang, M. H. Kim, Z. Moscicki, C. Sullivan, I. Tsukrov, M. R. Swift, S. C. James, M. Brooks S. Howden, S. Fredericq, S. A. Krueger-Hadfield, A. D. R. N'Yeurt and C. Webb, Ocean forests: Breakthrough yields for macroalgae, in *Proc. OCEANS 2018 MTS/IEEE*, Charleston SC, USA, 2018, pp. 22–25.
328. M. E. Capron, J. R. Stewart, A. R. N'Yeurt, M. D. Chambers, J. K. Kim, C. Yarish, A. T. Jones, R. B. Blaylock, S. C. James, R. Fuhrman and M. T. Sherman, *Energies*, 2020, **13**, 4972.

329. U.S. Energy Information Administration: Short Term Energy Outlook, 2021, https://www.eia.gov/outlooks/steo/report/global_oil.php, (accessed 30 January 2021).
330. R. R. Gentry, H. E. Froehlich, D. Grimm, P. Kareiva, M. Parke, M. Rust, S. D. Gaines and B. S. Halpern, *Nat. Ecol. Evol.*, 2017, **1**, 1317.
331. J.-L. Jiao, F. Wang, P.-G. Duan, Y.-P. Xu and W. H. Yan, *Energy Sources, Part A*, 2017, **39**, 851.
332. R. Allam, S. Martin, B. Forrest, J. Fetvedt, X. Lu, D. Freed, G. W. Brown, T. Sasaki, M. Itoh and J. Manning, *Energy Procedia*, 2017 **114**, 5948.
333. D. A. Roberts, N. A. Paul, S. A. Dworjanyn, M. I. Bird and R. de Nys, *Sci. Rep.*, 2015, **5**, 9665.
334. Fast Company, Forget planting trees: This company is making carbon offsets by putting seaweed on the ocean floor, 2020, https://www.fastcompany.com/90548820/forget-planting-trees-this-company-is-making-carbon-offsets-by-putting-seaweed-on-the-ocean-floor, (accessed 10 August 2021).
335. S. E. Strand and G. Benford, *Environ. Sci. Technol.*, 2009, **43**, 1000.
336. D. L. Karlen, R. Lal, R. F. Follett, J. M. Kimble, J. L. Hatfield, J. M. Miranowski, C. A. Cambardella, A. Manale, R. P. Anex and C. W. Rice, *Environ. Sci. Technol.*, 2009, **43**, 8011.
337. P. T. Harris, M. Macmillan-Lawler, J. Rupp and E. K. Baker, *Mar. Geol.*, 2014, **352**, 4.
338. H. Teagle, S. J. Hawkins, P. J. Moore and D. A. Smale, *J. Exp. Mar. Biol. Ecol.*, 2017, **492**, 81.
339. L. J. Carpenter and P. S. Liss, *J. Geophys. Res.: Atmos.*, 2000, **105**, 20539.
340. F. S.-L. Keng, S.-M. Phang, N. A. Rahman, E. C. Leedham Elvidge, G. Malin and W. T. Sturges, *J. Appl. Phycol.*, 2020, **32**, 1377.
341. M. Mehlmann, B. Quack, E. Atlas, H. Hepach and S. Tegtmeier, *Environ. Sci. Proc. Impacts*, 2020, **22**, 679.
342. S. Tegtmeier, F. Ziska, I. Pisso, B. Quack, G. J. M. Velders, X. Yang and K. Krüger, *Atmos. Chem. Phys.*, 2015, **15**, 13647.
343. R. von Glasow, *Environ. Chem.*, 2005, **2**, 243.
344. C. Leblanc, C. Colin, A. Cosse, L. Delage, S. La Barre, P. Morin, B. Fiévet, C. Voiseux, Y. Ambroise, E. Verhaeghe, D. Amouroux, O. Donard, E. Tessier and P. Potin, *Biochimie*, 2006, **88**, 1773.
345. P. P. A. Smyth, R. Burns, R. J. Huang, T. Hoffman, K. Mullan, U. Graham, K. Seitz, U. Platt and C. O'Dowd, *Environ. Geochem. Health*, 2011, **33**, 389.
346. I. Bartsch, C. Wiencke and T. Laepple, in *Seaweed Biology*, ed. I. Bartsch, C. Wiencke and T. Laepple, Springer, Heidelberg, 2012, p. 383.
347. I. K. Chung, C. F. Sondak and J. Beardall, *Eur. J. Phycol.*, 2017 **52**, 495.
348. P. Goglio, A. G. Williams, N. Balta-Ozkan, N. R. Harris, P. Williamson, D. Huisingh, Z. Zhang and M. Tavoni, *J. Cleaner Prod.*, 2020, **244**, 118896.
349. F. Douvere, *Mar. Policy*, 2008, **32**, 762.

350. S. E. Lester, J. M. Stevens, R. R. Gentry, C. V. Kappel, T. W. Bell, C. J. Costello, S. D. Gaines, D. A. Kiefer, C. C. Maue, J. E. Rensel, R. D. Simons, L. Washburn and C. White, *Nat. Commun.*, 2018, **9**, 945.
351. I. Galparsoro, A. Murillas, K. Pinarbasi, A. M. M. Sequeira, V. Stelzenmüller, Á. Borja, A. M. O'Hagan, A. Boyd, S. Bricker, J. M. Garmendia, A. Gimpel, A. Gangnery, S.-L. Billing, Ø. Bergh, Ø. Strand, L. Hiu, B. Fragoso, J. Icely, J. Ren, N. Papageorgiou, J. Grant, D. Brigolin, R. Pastres and P. Tett, *Rev. Aquacult.*, 2020, **12**, 2061.
352. S.-L. Billing, J. Rostan, P. Tett and A. Macleod, *Aquaculture*, 2021, **534**, 736203.
353. K. Güssow, A. Proelss, A. Oschlies, K. Rehdanz and W. Rickels, *Mar. Policy*, 2010, **34**, 911.
354. C. V. Phillips, *Int. J. Epidemiol.*, 2001, **30**(4), 771.
355. M. Kandlikar, J. Risbey and S. Dessai, *C. R. Geosci.*, 2005, **337**(4), 443.
356. W. Steffen, J. Rockström, K. Richardson, T. M. Lenton, C. Folke, D. Liverman, C. P. Summerhayes, A. D. Barnosky, S. E. Cornell, M. Crucifix, J. F. Donges, I. Fetzer, S. J. Lade, M. Scheffer, R. Winkelmann and H. J. Schellnhuber, *Proc. Natl. Acad. Sci.*, 2018, **115**, 8252.
357. N. E. Vaughan and C. Gough, *Environ. Res. Lett.*, 2016, **11**, 095003.
358. T. Thoni, S. Beck, M. Borchers, J. Förster, K. Görl, A. Hahn, N. Mengis, A. Stevenson and D. Thrän, *Front. Clim.*, 2020, **2**, 590305.
359. A. Hamilton, *Proc. Inst. Civil Eng. Munic. Eng.*, 2003, **156**, 263.
360. IPCC (Intergovernmental Panel on Climate Change), Summary for Policymakers, in *Climate Change 2021: The Physical Science Basis. Contribution of Working Group I to the Sixth Assessment Report of the Intergovernmental Panel on Climate Change* ed. V. Masson-Delmotte, P. Zhai, A. Pirani, S. L. Connors, C. Péan, S. Berger, N. Caud, Y. Chen, L. Goldfarb, M. I. Gomis, M. Huang, K. Leitzell, E. Lonnoy, J. B. R. Matthews, T. K. Maycock, T. Waterfield, O. Yelekçi, R. Yu and B. Zhou, Cambridge University Press, 2021, pp. 3–32.
361. C. Nesshöver, T. Assmuth, K. N. Irvine, G. M. Rusch, K. A. Waylen, B. Delbaere, D. Haase, L. Jones-Walters, H. Keune, E. Kovacs and K. Krauze, *Sci. Total. Environ.*, 2017, **579**, 1215.
362. N. Seddon, A. Chausson, P. Berry, C. A. Girardin, A. Smith and B. Turner, *Philos. Trans. R., Soc. B*, 2020, **375**, 20190120.
363. N. Seddon, A. Smith, P. Smith, I. Key, A. Chausson, X. C. Girardin, J. House, S. Srivastava and B. Turner, *Global. Change Biol.*, 2021, **27**(8), 1518.
364. E. Cox, M. Boettcher, E. Spence and R. Bellamy, *Front. Clim.*, 2021, **3**, 576294.
365. H. Keith, M. Vardon, C. Obst, V. Young, R. A. Houghton and B. Mackey, *Sci. Total. Environ.*, 2021, **769**, 144341.
366. S. Raghav, K. Siman, A. Gross, A. Wu, Y. Zeng, M. Comstock, J. Zhang, J.-R. Yoshioka, M. Kanakasabai, L. Prakash, J. Howard, A. Roopsind, T. V. Sarira, L. R. Carrasco, K. Koenig, A. Schoenberg, K. Tabor, B. Griscom, A. Uraguchi and L. P. Koh, *The Business Case for Natural*

Climate Solutions: Insights and Opportunities for Southeast Asia, Temasek, 2020.

367. Y. Zeng, D. A. Friess, T. V. Sarira, K. Siman and L. P. Koh, *Curr. Biol.*, 2021, **31**, 1.
368. National Academies of Sciences, Engineering and Medicine, *A Research Strategy for Ocean-based Carbon Dioxide Removal and Sequestration*, The National Academies Press, Washington DC, 2022.
369. M. Babiker, G. Berndes, K. Blok, B. Cohen, A. Cowie, O. Geden, V. Ginzburg, A. Leip, P. Smith, M. Sugiyama and F. Yamba, *Climate Change 2022, Mitigation of Climate Change, Working Group III Contribution to the Sixth Assessment Report of the Intergovernmental Panel on Climate Change*, WMO/UNEP, 2022, ch.12.
370. M. Meinshausen, J. Lewis, C. McGlade, J. Gutshow, Z. Nicholls, R. Burdon, L. Cozzi and B. Hackmann, *Nature*, 2022, **604**, 304.
371. J. B. Gallagher, V. Shelamoff and C. Layton, *ICES J. Mar. Sci*, 2022, **79**, 585.

CHAPTER 11

Comparison of Technologies and Practices for Removing Carbon Dioxide from the Atmosphere

SABINE FUSS[a,b]

[a] Mercator Research Institute on Global Commons and Climate Change, Berlin, Germany; [b] Institute of Geography, Humboldt University of Berlin, Germany
Email: fuss@mcc-berlin.net

This chapter presents a synthesis of the technology information provided in Chapters 5–10 and compares the technologies and practices covered in this volume with respect to not only their costs and potentials, but also their side effects and general level of readiness. This comparison includes technology-specific implementation challenges, which will be addressed in the second half of this volume and conducts a gap analysis of technological knowledge.

11.1 Introduction

2020 marked the year in which the Covid-19 pandemic forced many economies into a slowdown. First estimates show that for reasonable scenarios of post-lockdown recovery, the resulting contraction in economic recovery will lead to about the same amount of emissions reductions that would be needed each year for the coming decade to move toward a climate stabilisation pathway aligned with the ambitious temperature goals agreed upon

Energy and Environment Series No. 31
Greenhouse Gas Removal Technologies
Edited by Mai Bui and Niall Mac Dowell

Published by the Royal Society of Chemistry, www.rsc.org

in Paris in 2015.[1†] It is difficult to disagree that these emissions reductions have come at a high price and that, therefore, emissions pathways compatible with the 1.5 °C goal that try to comply with a dwindling carbon budget without tapping into the potentials of Carbon Dioxide Removal (CDR)[2] look politically no more feasible than their CDR-based counterparts. Adding to this, difficult-to-decarbonise sectors will need offsetting in the short to medium term.[3–5] So, without saying that CDR potentials will need to be fully exploited, it looks like at least some removal of previously emitted CO_2 will be necessary. Previous chapters have introduced an array of technologies and practices to do so and where they currently stand in practice (Chapter 2) and have provided framing conditions and physical background for their effectiveness (Chapters 3 and 4).

Indeed, there is growing knowledge on CDR, as indicated by an exponential increase of peer-reviewed publications[6] and the fact that the topic is more and more comprehensively taken up in IPCC assessment reports (Chapter 2, Minx *et al.*[6]). In addition, many of the chapters describing specific CDR technologies and practices (Chapters 5–10) point to first pilots and existing programs. In particular, notwithstanding the remaining uncertainties and disagreement at the different ends of the estimates, there is by now a better understanding of CDR potentials, costs and side effects,[7–11] indicating that significant potentials would be available for exploitation by mid-century in a cost range from slightly negative (mostly changes in agricultural practices enhancing soil carbon sequestration that pay off economically without additional incentives[12]) to several hundred US$ per ton of CO_2 stored (in the case of DACCS[13]). In this chapter, the assessments of the CDR technologies and practices with respect to potentials, costs and side effects of deployment at a gigaton-scale will be put into perspective, focusing on new insights that this volume contributes rather than iterating previous, comprehensive assessments.

Furthermore, despite a growing evidence base on the supply side of CDR, demand-side aspects are still subject to considerable knowledge gaps when it comes to innovation, policy, governance, and public acceptability.[9,13,14] In this chapter, a new framework will be proposed for the assessment of the readiness of CDR technologies and practices. Importantly, this concept of readiness will go beyond the notion of technological readiness and include drivers of and barriers to deployment such as policy, commercial incentives, availability of business models, public acceptability, legal and regulatory frameworks and the political economy. The readiness framework and how it is operationalised here will be introduced in more detail in Section 11.2 and it will be applied to the different technologies and practices in Section 11.3. Section 11.4 provides the synthesis (Table 11.1), allowing for a tentative comparison of CDR approaches,

---

[†]Naturally, these estimates are subject to a number of uncertainties. For example, assumptions have to be made on the length of confinement in different countries and the type and extent of the corresponding restrictions. Varying degrees would lead to emissions reductions between 2% and 13% for 2020, with government actions and post-crisis economic incentives potentially impacting emissions pathways for decades – for better or worse.

Table 11.1 Comparison of technologies and practices to remove CO_2 from the atmosphere.

	Cost (US$)	Potential (Gt CO_2 per year)	Side effects	TRL	Business model	Public acceptability	Legal and regulatory frameworks	Political economy
BECCS	< 200	0.5–40	Mostly related to land use change (direct and iLUC), except if based on algae (see Section 11.3.6)	Commercial	Voluntary offsetting, tax credits	Geographically diverse, very problematic in some European countries	Geographically diverse, often regulatory barriers to storage, barriers to transport partially alleviated with London Protocol amendment; regional regulation concerning biomass imports	Depending on residual emissions & industrial structure
DACCS (weak base sorbent, solid adsorption)	600–1000	> 5	Weak evidence, but potential implications from materials use, including land	Demonstration, modular approach allowing for maximum learning by doing	CCU, offsetting (especially from wealthy high-tech companies), LCFS, 45Q	Geographically diverse attitudes towards CCS	Geographically diverse, often regulatory barriers to storage	Depending on residual emissions, energy mix, airlines interested in DACCS-fuels
DACCS (strong base sorbent, liquid absorption)	94–232 (based on ref. 42, subject to optimistic assumptions on energy requirements, see Chapter 6)	> 5	Weak evidence, but potential implications from materials use, including land; higher water requirements than the other DAC system	Prototype, but likely to benefit from economies of scale	CCU, offsetting (especially from wealthy high-tech companies), EOR, LCFS, 45Q	Controversies over combining with EOR and continued use of fossil fuels	Geographically diverse, often regulatory barriers to storage	Depending on residual emissions, energy mix, airlines interested in DACCS-fuels

Table 11.1 (*Continued*)

	Cost (US$)	Potential (Gt CO_2 per year)	Side effects	TRL	Business model	Public acceptability	Legal and regulatory frameworks	Political economy
Enhanced weathering	50–200	2–4	Soil improvement, through increased phosphate and other nutrients, reduced soil acidification, increased soil carbon, reduced heavy metal toxicity in crops, increased CEC improving nutrient retention; adverse impacts of mining activities, potentially heavy metal release, health risks of fine grained material	Prototype	Only small-scale, incentives possible through co-benefits for food and soil security	Limited evidence, but potential lessons learned from acceptability problems associated with upscaling other mining activities	Limited evidence on legal and regulatory issues of applying minerals; governance of mining activity regulated at national level	Depending on residual emissions, sectoral composition and mining industry
Alkaline materials	Lower cost than Enhanced Weathering	2.9–8.5[a]	See Enhanced Weathering, without the adverse side effects of mining; potential of hazardous waste stabilisation	Prototype	Incentives possible through co-benefits for food and soil security; potential opportunities for using wastes from cement and steel industry for sustainable construction	No evidence	Regulation under national governance	Diverse, decentralised interest groups in industry and agriculture.
	− 45–10	7		Commercial				

Soil carbon sequestration			Impacts on land use change, albedo and energy use dependent on practice; positive impacts on all soil functions, water retention and nutrients.		Only for negative cost potential or in combination with other co-benefits, *e.g.*, restoration programmes	Limited evidence on barriers	Existing regulations and ag support may not hold incentives to enhance soil carbon, but no evidence of direct barriers for most SCS practices and new trends (*e.g.*, EU Farm-to-Fork strategy) integrate CO_2.	Many diverse, decentralised interest groups, partially with strong lobbies
Biochar	30–120	0.3–2	Potential N_2O decreases, soil carbon and fertility potential yield increases, need for biomass feedstock, reductions in soil contamination, biomass waste recycling, co-production of energy, possible albedo reduction, possible methanogenisis	Demonstration	Not yet in the context of CO_2 removal, but conceivable in the context of the co-benefits.	Limited evidence on barriers	Limited evidence on barriers	Many small, decentralised actors
Afforestation	5–50	0.5–7	Positive impact on water retention, hazard protection, biodiversity/habitat, soil, local climate; negative impact on albedo in the North, increased water use, high land footprint	Commercial	PES, offsets, additional income sources from forestry	Higher acceptance than BECCS	Limited evidence on barriers, except regions with high land tenure uncertainty	Perceived trade-offs between economic development/ livelihoods and climate change mitigation, opportunity

Table 11.1 (*Continued*)

	Cost (US$)	Potential (Gt CO_2 per year)	Side effects	TRL	Business model	Public acceptability	Legal and regulatory frameworks	Political economy
								costs needing compensation
Blue carbon from enhanced coastal vegetation	95–6600 (incl. CH_4, lower end for seagrass[69] and higher end for mangroves[70])	0.2 (2050 estimate for restoring 40% of historical cover[71])	Potentially high co-benefits for adaptation and for other ecosystem services; potential increase in CH_4 and N_2O	Prototype	Payments for Ecosystems Services, non-carbon benefits; CDR and REDD + for larger-scale projects, voluntary carbon markets for smaller	Strong support from NGOs, CBD	Regulation under national governance	No evident barriers, but also not much literature
Marine-biomass-fuelled BECCS	26–71 (with species-dependent harvesting costs possibly increasing this by an order of magnitude, see Chapter10)	3–4	At large scale currently discussed: potentially-severe disruptions to marine species at large scale, potentially adverse effect on nutrient cycling and availability, increased macroalgal release of bromoform, potentials' adverse effect on human health through changes in regional iodine cycling.	Concept for large-scale seaweed cultivation, Commercial for CCS	Voluntary offsetting, tax credits	Some evidence on declining public acceptance of aquacultures, potential opposition to onshore plant siting and CCS	Potential barriers in approval processes for using large-scale seaweed cultivation as CDR; potential regulatory barriers for CCS	Competition with other uses of shelf sea waters, competing interests particularly concerning existing fisheries or marine protected areas; see above for CCS and removal in general.

[a]2100 estimate.

which will be critically reflected upon in Section 11.5, which also serves to distil more systematically the knowledge gaps and indicate which of them will be addressed in the following chapters (*e.g.*, social issues in Chapter 13, policy in Chapter 14). Finally, Section 11.6 will condense the main conclusions.

11.2 Readiness Framework for CDR

Previous assessments that tentatively compared CDR practices and technologies have often focused on costs, potentials and side effects of upscaling CO_2 removals to a gigaton-scale,[8,10,13,15] while others have assessed technologies with respect to their deployment limitations comparing their levels of technological readiness.[16] Clearly, however, there are a number of factors beyond technological readiness, which will influence whether a CDR approach can be rolled out in the near or medium term or not.

The IPCC has used the notion of "feasibility" to assess the readiness of mitigation and adaptation approaches more broadly.[13] While the term is otherwise used in a binary way (feasible or not feasible) or as contingent on certain conditions, *e.g.*, feasible in some countries, but too costly, or otherwise inapplicable in others, the IPCC introduced a more nuanced definition: "the degree to which climate goals and response options are considered possible and/or desirable".[17] This is in principle sufficiently broad to capture all potential barriers to and drivers of CDR deployment, yet the synthesis is difficult because of the above-mentioned context-specificity of feasibility and also because different dimensions[‡] will interact with each other, which is a challenge to capture in a clean way when aggregating indicators. Finally, the varying extent and quality of the underlying literature does not lend itself easily to synthesis either.

The community around Carbon Capture and Storage (CCS) in tandem with the UK government has been engaged in the development of a readiness index for more than a decade.[18] The assessment logic is converse in this case, where CCS is at the centre of the assessment with different criteria, which are then explored in different national contexts. These criteria encompass storage (*i.e.*, all geological and technical aspects that could impact the ability to inject and store CO_2), the country's legal and regulatory frameworks, the policy landscape including also indirect support for CCS, and CCS inherent interest (mostly measured by global shares of fossil fuel production and consumption).[19] While impractical to conduct as a comparative assessment of an array of partially very differently structured technologies and practices, it points to important factors largely unexplored in other assessments such as the underlying political economy.

In this chapter, a hybrid framework will be used that draws on the strengths of these other assessment approaches without trying to come up with a quantitative ranking of technologies. Rather, the goal is to highlight

[‡]In the case of the IPCC, these dimensions are economic, technological, institutional, socio-cultural, geophysical, and environmental, all them approximated by a number of indicators.

the performance across different sets of criteria in order to elicit possible trade-offs and risks, but also entry points, without forestalling any weighting.§

The readiness of a CDR technology or practice will be explored along three dimensions: first of all, their costs and potentials, second, the impact of the upscaling on non-climate outcomes, and third, the drivers and barriers of deployment including policy, commercial incentives, availability of innovative business models, acceptance, legal and regulatory frameworks, and the political economy. Working technology by technology, the relevance of the context of each of these dimensions will be discussed without trying to preempt a more precise geographical distribution (see Chapter 12 on this).

11.3 Comparison of CDR Technologies and Practices

11.3.1 Bioenergy with Carbon Capture and Storage (BECCS)

Surveying all the literature until and including 2017, BECCS costs span a range of US$15–400 per t CO_2.[14] The large range reflects differences in assumptions about distance to feedstock and storage sites, types of processes and feedstock, availability of land, sustainability constraints, social preferences, projection horizon and many other parameters. Taking the different assumptions into account and assessing the confidence and robustness of cost estimates across the literature supplemented by expert opinion, Fuss *et al.*[14] narrow this range down to US$100–200 and estimate potentials of 0.5–5 Gigatonnes CO_2 per year (Gt CO_2 per year) by mid-century, which is far below the potentials cited in the literature that extends up to 85 Gt CO_2 per year and which typically focus on technical potentials with optimistic assumptions about land availability and yield increases. More recent assessments have put the potentials in the range of 3.5 to 15 Gt CO_2 per year as a function of land availability.[10] The latest estimates of purely biophysical potentials point to up to 40 Gt CO_2 per year when accounting for emissions over 80 years,[20] noting several bottlenecks that might limit this potential in reality.

The discussion of potentials already indicates that a large-scale rollout of BECCS bears the risk of certain side effects, mostly associated with the increased demand for land for growing the bioenergy feedstock[21,22] – land that is also needed to feed a growing population and safeguard terrestrial biodiversity.[23] Other side effects discussed in the literature are related to the processes of producing biomass, *i.e.,* increased use of scarce resources like water and fertiliser, also associated with higher non-CO_2 Greenhouse Gases (GHGs)[24] and energy, *i.e.*, the corresponding air pollutants.[25] On the geophysical side, researchers have suggested that if biomass plantations replace reflective snow cover in higher latitudes, albedo effects could partially offset the gains in terms of temperature reduction.[26] Furthermore, the Carbon Capture and Storage (CCS) portion of the BECCS supply chain has been scrutinised for risks of

§Even assigning equal weights to different criteria contains value judgements.

leakage during transport and storage, the mitigation of which is now a mature research field of its own.[27] Positive side effects are much less prominent in the current literature. While some local actors may stand to gain from scaling bioenergy, many will start from a disadvantaging institutional setting, thus implying risks for local livelihoods.[28] Others have pointed to BECCS presenting an opportunity to save jobs in deep decarbonisation transformation pathways.[29]

Related to many of the negative side effects, but also concerns about carbon removal in general, such as ethical considerations around hubris or moral hazards,[30] public acceptability of BECCS has fought with concerns related to both the bioenergy and the CCS portion of the BECCS supply chain[31–33] – with partially mixed evidence in different contexts which is indicative of a dire need for more systematic research into the topic. However, in some European countries, the debate around CCS has been so controversial[34] that it has even culminated in quite grave impacts on legal and regulatory frameworks, *e.g.*, the inclusion of a clause in the 2012 German CCS law[¶] allowing the German Länder to forbid geological storage of CO_2 on their respective territories, which they subsequently made use of. Scandinavian countries are less restrictive in this respect, which has resulted in a massive effort to invest in overcapacities of CO_2 storage by the Norwegian private sector, for example.[‖] While legal and regulatory settings thus differ across countries as demonstrated by these European examples, some issues are also of a transnational nature. Related to CCS, this affects mostly the transportation of CO_2 to the injection site. Resorting again to the example of Europe, the transport of CO_2 as regulated by the London Protocol was not allowed across borders in the past. Only in 2019 the contracting parties to the London Protocol approved an amendment – proposed by Norway and the Netherlands – to remove the barrier to cross-border CO_2 transport. Related to bioenergy, regulation has hitherto been very clear, and in many countries also conducive to expansion, as part of efforts to increase the share of renewable energy within their economies. However, the above-mentioned concerns around competition for land and risks of indirect Land Use Change (iLUC) and biodiversity loss have required additional regulation, in Europe in the form of the sustainability safeguards of the revised Renewable Energy Directive (EU) 2018/2001, adopted in 2018 by the European Parliament and by the Council of Ministers of the European Union. Even this additional regulation has been deemed inadequate, owing to, *inter alia* its focus on prohibiting biomass extraction from primary forest only.[35]

Furthermore, a decisive factor in ramping up BECCS infrastructure will be the political economy. The Global CCS Institute tracks progress on CCS and its enabling factors and barriers to deployment by monitoring a variety of indicators, one of them being inherent CCS interest, which mainly builds on

---

[¶]Kohlendioxid-Speicherungsgesetz (KSpG), Art. 1 G. v. 17.08.2012 BGBl. I S. 1726 (No. 38).
[‖]Equinor (2019). Northern Lights Project Concept Report. Available at: https://northernlightsccs.com/assets/documents/Northern-Lights-Project-Concept-report.pdf.

global shares of fossil fuel production and consumption.[19] CCS has been moving up the agenda in the US and Canada, while the leaders in Europe are typically the ones rich in natural gas.[36] This growing interest in CCS[3–5] – together with other residual emissions, which are difficult to drive down in the short to medium term, such as in the agricultural sector[37] – is a relevant variable in the context of BECCS. Existing industries relying on biomass will also play a role – both where established firms have an interest to solidify and expand and where competition between industries could be exacerbated. At the EU level, for example, Kallio *et al.* (2017) found that policy choices will have strong impacts on the allocation of biomass use between heat and power production and the production of liquid biofuels.[38]

Finally, despite the fact that both bioenergy and post-combustion CCS have reached commercial-level technological maturity (TRL 9[36]) with a number of pilots combining the two having emerged around the world (Chapter 2), there is currently no business model across the whole supply chain that would allow for rapid scaling of BECCS. Without integration into carbon markets or other government support, utilisation of CO_2 or voluntary offsetting seem the only, limited route in many countries, even though some have stepped up incentives for parts of the supply chain, *e.g.*, in the form of tax credits for CO_2 capture, such as 45Q in the US, or bioenergy mandates. Others have first considerations for a more structured support to BECCS, *e.g.*, in terms of contracts for difference or reverse auctioning.[39]

11.3.2 Direct Air Carbon Capture and Storage (DACCS)

Direct Air Capture is no longer a nascent technology these days and several plants are operating at demonstration and semi-commercial scale. Bui *et al.* (2017) estimate a level of technological readiness (TRL) of 7.[36] However, there is so far only one installation which truly removes CO_2 from the atmosphere on a net basis (see Chapters 2 and 6). Furthermore, a TRL of 7 refers to DAC systems based on solid adsorption with weak base sorbent like the Climeworks technique, while systems working with liquid absorption and strong base sorbent like Carbon Engineering extrapolate from smaller-scale pilots at this stage, thus being situated rather around TRL 4–5 and featuring higher uncertainties in cost estimates. The two DAC routes are therefore assessed separately from each other in Table 11.1.

How much CO_2 could potentially be removed from the atmosphere by means of DACCS is technically only constrained by the capacity of storage, given that sufficient amounts of carbon-free energy could be provided. While there is large agreement that there should be sufficient potential available globally to geologically store amounts of CO_2 comparable to those removed in many 1.5 °C and 2 °C scenarios,[8,13,40] there could be storage bottlenecks in specific regions.[40,41] Global estimates of total geological storage potential range from 320 to 50 000 Gt CO_2, with much of the variation coming from estimates of potentials in aquifers, where the lower end considers only aquifers with structural traps, while allowing for the integration of other trapping mechanisms can dramatically increase the potential. With these massive geological storage potentials some assessments do not limit the removal potential of DACCS, but Fuss *et al.* (2018) do so on account of upscaling and economic challenges and put a maximum of 5 Gt CO_2 per year as a mid-century estimate, alongside costs of US$100–300 per ton of CO_2.[8] This range has subsequently been reconsidered, especially in the light of new research from industry indicating the possibility of more rapid and deeper cost reductions to below US$100 per ton of CO_2.[42] In this volume, the previously assessed cost range and subsequent suggestions of even lower costs have been critically scrutinised (Chapter 6). In particular, Chapter 6 studies the process design for the DAC processes based on chemical absorption and adsorption and assesses estimates for DAC costs going forward. The results indicate clearly that US$100–300 per t CO_2 might not be reached mid-century – maybe not at all. The suggested cost range for DAC in 2030 is estimated in the range of US$600–1000 per ton of CO_2 removed.

Notwithstanding this assessment, other research has shown that DACCS has certain advantages when it comes to containing some of the side effects that other CDR approaches have. The technology has been commended for integrating well into renewable energy systems, for example, and is also less dependent on suitable areas of land as is BECCS.[43]** While this integration may materialise over the longer term, it is important to keep in mind some economic and technical hurdles that need to be overcome, *e.g.*, that DAC systems will need to be run continuously to avoid stranding expensive capital, which implies a need for expensive storage of renewable energy. Also, while DAC systems relying on weak base sorbent for solid adsorption already work with renewable energy, liquid absorption techniques with strong base sorbent (such as Carbon Engineering) operate the calciner at 900 °C, which is not yet compatible with renewable energy.

As for public acceptance, legal and regulatory barriers, and political economy considerations, much of what applies to the CCS portion of the BECCS supply chain is also relevant for DACCS. While DACCS is not exposed to the bioenergy issues raised in the context of BECCS, it does however face criticism in the public debate in the cases where the DAC energy source is fossil, and the storage is associated with enhancing oil recovery (EOR) pointing to oil companies as an important interest group – as is the case with Carbon Engineering (Chapter 6).

The absence of a high carbon price is also the main reasons why business models for DACCS are scant. Partially justifying the need for learning, companies like Climeworks initially did not inject the CO_2, but used it in

**As a caveat, depending on the source of energy, significant amounts of land might still be needed, as demonstrated in a recent assessment of possible CDR futures in California.[83] Also, Chapter 6 points out that there is considerable uncertainty on the size of land area the scrubbers need to be spread over. Unlike wind farms, which can coexist with agriculture, it is unclear what type of multiple use DAC farms will allow. This is an important gap for future research to specify.

greenhouses or to produce synfuels. Enhanced oil recovery has already been mentioned as another route of making DACCS profitable. In spite of a general absence of high carbon prices, some countries do have policy instruments in place that foster CCS, the US being a prime example. As Chapter 6 points out for the context of DACCS, the oil produced from DAC-based EOR would indeed qualify for California's low carbon fuel standard credits and the 45Q tax credits, resulting in even higher revenues per ton of CO_2. Finally, an increasing demand by companies finding it hard to decarbonise their operations such as airlines starts representing a modest market for offsets from DACCS. At the same time, there is also continued interest in DAC fuels to substitute for kerosene, noting that this would not fall under CDR, but merely avoided emissions, contingent on a decarbonised energy source for both the DAC and the CCU process.

11.3.3 Mineral Carbonation and Alkaline Materials

Another method for removing CO_2 from the atmosphere is through the enhancement of natural weathering processes. The processes summarised under this label are actually multi-fold, including *ex situ* and *in situ* mineral carbonation, enhanced rock weathering (2050 potential of 2–4 Gt CO_2 per year at US$50–200 per t CO_2[8††]), and absorbing CO_2 through reaction with alkaline materials (see Chapter 7 and the National Academies of Sciences, Engineering and Medicine report[10] for an in-depth overview of the mechanics). To the extent that these processes compete for the same resources, their removal potentials will not be additive. However, there is general agreement in the scientific community that the necessary rock material is abundant, leading to high technical potentials.

These will be difficult to exploit without considering the required infrastructure, both in terms of rock extraction and processing, which is an energy-intensive process.[10] These activities are associated with potentially harmful side effects on the environment and society, especially if the energy source is not completely decarbonised. The use of industrial alkaline by-products or even waste would get around the problem to a large extent and could significantly reduce the cost, benefit from higher levels of reactivity, thus reducing the need for pre-treatment, temperature and pressure, and get rid of the need to mine the materials (Chapter 7). By 2100, this could result in the removal of up to 8.5 Gt CO_2 per year, mostly depending on assumptions about projected sectoral growth,[44] thus representing a relevant contribution to a possible CDR portfolio for net negative emissions. Next to the adverse side effects associated with high non-decarbonised energy use and scaling mining activity, a potential of heavy metal release and health

[††]Extraction of natural rock could enable end-of-the-century removals of 1-10 Gt CO_2 per year based on expectations of the global aggregate extraction industry to grow by 25Gt per by mid-century, assuming that 1Gt CO_2 removal requires 2-6 Gt rock and positing that an additional 2–60 Gt of rock extraction per year is possible to realise over 50–80 years. (45) estimate a potential of 0.5–4Gt CO2/yr by 2100 using 10–30t/ha/yr.

risks from the finely ground material have been reported in the literature.[8] There is also more and more evidence on the positive side, however, where applying the minerals can positively impact soil and thereby secure food production,[46] alkalinity from weathering, if reaching the ocean, may help to mitigate the impact of ocean acidification,[47] and heavy metal toxicity in crops will be reduced as silica acts as a competitive inhibitor (see Chapter 7). Using alkaline materials from industry can furthermore play a role in stabilising hazardous wastes (see Chapter 7, Section 7.15).

When it comes to implementation, the component processes for Enhanced Weathering are technologically mature, although piecing together these components in a coherent system for CO_2 removal is immature. However, current business models are largely inexistent, with incentives mostly at the small scale, *e.g.*, through co-benefits for food and soil security.[45,46] For alkaline materials, there are potential business opportunities for using wastes from the cement and steel industries for sustainable construction projects.

There is relatively limited evidence concerning public acceptability, regulation and legal frameworks and the political economy, but potential lessons learned might emerge from the literature on acceptability and regulatory barriers associated with upscaling other mining activities and maybe application of fertilisers and pesticides in agriculture.

11.3.4 Soil Carbon Sequestration and Biochar

Soil Carbon Sequestration (SCS) would remove CO_2 through enhancing soil carbon – Table 8.8 in Chapter 8 of this volume presents an overview of suitable practices to do so, ranging from reduced tillage and improved nutrient management to agroforestry and restoration of degraded land areas. The attractive feature of SCS as a CDR approach is its potentially negative cost of −45 US$ per ton of CO_2 for tillage and residue management at the lower end of the spectrum and about 10 US$ per ton of CO_2 for managing cropland and grazing land to enhance soil carbon.[12] It is noteworthy that approximately 20% of the SCS potential can be realised at negative cost indicating it as already economically profitable (see Table 8.1 in Chapter 8 for a detailed overview of the potential across practices). Still, adherence to existing practices and traditions, scattered knowledge with a disconnect to local context-specific needs, the challenge of coordinating a vast set of diverse actors, political economy considerations in a landscape of existing regulation that is not necessarily targeted at sustainability, and short-term lack of finance before benefits materialise, all represent barriers to reaping the technical SCS potential that Chapter 8 puts at 7 Gt CO_2 per year globally, which is well in the range of the current literature, which extends to up to 11 Gt CO_2 in 2050.[8] However, these potentials have to be considered in the light of saturation levels, *i.e.*, they are larger where soils are poorer and smaller where soil carbon is already close to saturation and Chapter 8 therefore emphasises that SCS is a potent, but rather short-term CDR approach, which will likely complement a broader portfolio of CDR methods.

In contributing to a CDR strategy, the ancillary benefits associated with SCS are helpful when it comes to implementation. It is important to note, however, that side effects of SCS implementation vary across soil types and climate regions and, most importantly, management practices. While cropland management, for example, does not require additional land areas, restoration programmes will do so. Albedo could be affected if light materials like straw would be used as cover crops.[46] Additional energy might be necessary for some practices (*e.g.*, involving irrigation), while other approaches may lead to energy savings.[46] Chapter 8's assessment also shows positive impacts of SCS on all soil functions. Both water retention and nutrients are enhanced by SCS.[12,24,46]

While evidence is limited when it comes to the barriers to SCS implementation due to public acceptance issues or regulation, the wide endorsement of the "4 per mille" initiative described in Chapter 8 is at least an indication that this will not constitute a major bottleneck.

Another way of removing CO_2 through amendment of soils is through the application of biochar, the global technical potential of which has been estimated to range up to 35 Gt CO_2 per year, but Fuss *et al.*[8] lower this to a maximum of 2 Gt CO_2 in 2050 given the absence of large-scale field trials and limited amount of available biomass. Costs range from US$30–120 per ton of CO_2 according to the same assessment.

Even though some studies indicate that there could be short-term decreases in non-CO_2 GHGs, notably of N_2O,[48,49] there is not enough evidence to allow for any robust quantifications, but there is high agreement that soil carbon will be enhanced by biochar[50] despite substantial uncertainty concerning the magnitude (Chapter 8). Whether the enhanced plant growth from biochar application will serve as a CO_2 removal mechanism obviously depends on whether a yield increment materialises in the first place[51] and whether it will subsequently be used for food or feed, in which case there is no removal.[12] Even though the application itself of biochar would not involve land use change, biochar needs biomass as a feedstock, which could imply quite high land requirements that could partially be reduced by additional bio-oil sequestration, for example.[52] Positive impacts are further related to reductions in soil contamination and toxic substances, the possibility to use biomass waste for the pyrolysis, and the co-production of energy, while some doubts have been raised to the large-scale application of biochar because of the possible reduction in albedo (see Chapter 8 for an in-depth overview).

Similar to SCS, there is limited evidence indicating barriers to the application of biochar from the perspective of public acceptance, regulation and legal frameworks or the political economy. In the absence of large-scale field trials, there is currently also not a business model for biochar application as a CDR method, but given its in principle technological maturity, the co-benefits (potential yield increases, purification qualities, *etc.*) could provide entry points to achieve some scaling at least, maybe in combination with SCS strategies.[46] Also, ongoing certification processes could unlock access to other business opportunities including carbon markets.[‡‡]

11.3.5 Afforestation

Afforestation is a near-universally popular option to remove CO_2 from the atmosphere, with high-profile initiatives such as the "trillion-tree" pledge[53] being supported by a range of NGOs. Acceptance has therefore been higher than for other CDR approaches with many contributions to the literature emphasising the ancillary benefits in terms of temporary and permanent habitats for a large variety of species and, consequently, a positive impact on biodiversity,[46] which does of course critically depend on the mode of implementation, where large-scale monocultures evidently do not harbour synergies with biodiversity strategies. Locally, afforestation and restoration have been shown to improve air quality, but large-scale afforestation is also exposed to impacts of ongoing climate change and wildfires, for example, they can quickly release the sequestered CO_2 and cause additional air pollution.

In the literature, the CDR potentials arising from afforestation are often conflated with restoration and even forest management techniques (see Chapter 9), making for wide ranges of potentials of 3 to 18 Gt CO_2 per year.[8,11] This is narrowed to 0.5 to 7 Gt CO_2 per year in 2050, but land use competition may further lower the upper bound of the range. Griscom *et al.* (2017)[54] estimate the cost-effective mitigation potential for reforestation assuming a global ambition to hold warming to <2 °C of about 3 Gt CO_2 per year in 2030.[54]

In general, the cost estimates also reflect the wide ranges of practices. However, all assessments locate the afforestation cost at the lower end of the CDR spectrum, with the US$5–50 range for 2050[8] hiding the differentiation of costs across the full potential, which the Royal Society demarcates at 1.2 Gt CO_2 for under US$30 per t CO_2 and 0.4 Gt CO_2 at less than US$3 per t CO_2; future cost estimates for afforestation (including reforestation) are assessed at US$15–30 per t CO_2 in 2100.[11]

Despite the relative attractiveness of afforestation as a CDR approach when considering costs and acceptance, the land footprint is higher when compared to BECCS and lower permanence and saturation discount its value as a silver bullet.[13] Furthermore, large-scale afforestation in Northern latitudes has been suggested to reduce the albedo, thereby partially offsetting climate benefits.[55]

On the positive side, afforestation can regulate local temperature and precipitation. In general, impacts on non-CO_2 environmental variables like water consumption are extremely dependent on local context (see also Section 9.8 in

---

[‡‡]EBI. (2020). Mit Pflanzenkohle basierten Kohlenstoffsenken dem Klimawandel entgegenwirken. EBI Whitepaper. http://www.biochar-industry.com/wp-%09content/uploads/2020/09/Whitepaper_Pflanzenkohle2020.pdf http://www.biochar-industry.com/wp-content/uploads/2020/09/Whitepaper_Pflanzenkohle2020.pdf

Chapter 9) and may, for example, improve water balances by decreasing runoff, and improving water retention and filtering capacity,[46] but afforestation can also demand more water than the previous land cover[24] and can change local streamflow to the extent of completely drying out streams.[56] A lower-scale rollout would coincide with co-benefits for soil formation and quality, nutrients, and soil organic carbon. Also, afforestation and reforestation play a role in climate change adaptation due to their protective impact vis-à-vis flooding and erosion[57]. Smith *et al.* (2019)[46] also emphasise the fact that afforestation provides food and feed for many people, including forest fruits, mushrooms, game meet, and medicinal plants, and biomass for *e.g.*, energy purposes when combined with sustainable forest management.[46]

While public acceptance appears to be higher for afforestation than for other CDR approaches (see above), the political economy often looks differently, with opportunity costs of agricultural production playing an important role. Business models therefore often rely on additional sources of income such as those derived from sustainable forestry or Payments for Ecosystems Services (PES) such as under Reduced Emissions from Deforestation and Degradation (REDD+), which also covers reforestation and restoration, or carbon offsets on voluntary markets or project-based offsetting for individuals and companies. In comparison with some of the other CDR options, afforestation can be categorised as technologically mature, with regulatory and legal barriers in regions with land tenure uncertainty and overlapping policy objectives.

11.3.6 Marine Biologically-based CDR

There are several options to enhance marine carbon uptake, but in this chapter the focus is mainly on two of the possible pathways because they are currently considered more promising to gain traction in the deliberation of possible CDR portfolios.[§§] One pathway is targeted at coastal vegetation systems, the enhancement of which would lead to a transfer of carbon into marine sediments *via* root systems. This is often addressed as blue carbon in the recent literature.[58,59] The literature is still too sparse to allow for a robust assessment of global Blue Carbon potentials, where the current Blue Carbon sink is tentatively being put at just below 1% of global anthropogenic emissions[60] with indications that this is probably a too optimistic estimate and subject to a risk of reversal, which is of course true for most nature-based approaches.

Compared to other terrestrial CDR approaches, restoring coastal vegetation systems is relatively expensive in many of the regions with high potentials and thus many projects have been failing and individual technologies have had trouble scaling.[61] In addition, costs will critically hinge on the climate change mitigation effectiveness of blue carbon approaches to CDR, which is a significant factor of uncertainty for restoration of coastal ecosystems. According to Chapter 10, variability in the emissions of methane and nitrous oxide, the

[§§]See Chapter 10 for an assessment of ocean fertilisation for completeness.

potential for calcification-driven CO_2 release and the vulnerability of carbon in coastal ecosystems to future climatic conditions could lead to a negation of the CO_2 removal effect and thus needs to be factored into studies, which should in addition be subject to long-term monitoring. The cost estimates in Table 11.1 do consider CH_4, but should of course still be interpreted with caution.

Still, the primarily positively evaluated side effects (Chapter 10) might represent a promising entry point independently from the carbon storage potential. In particular, Blue Carbon has been found to have synergies with adaptation,[62] *e.g.*, by reducing human and environmental impacts on coastal ecosystems and people living in the area.[60] Chapter 10 demonstrates in its assessment that the value of potential co-benefits arising from the restoration of blue carbon ecosystems (mangroves, saltmarsh and seagrass) can be several orders of magnitude higher than for CO_2 removal benefits; notwithstanding high uncertainties and site-specificity. More research would be needed to better understand the underlying biogeochemical processes and their temporal and spatial variability. This is an obvious disadvantage of most blue carbon options compared to the other route assessed (marine BECCS).

Though the literature on public acceptability, legal and regulatory frameworks and the political economy is scarce, much should be transferable from the lessons of REDD+, a PES mechanism to compensate tropical nations for their opportunity costs when reducing deforestation and increasing restoration. Also, Chapter 10 points to the support for blue carbon approaches coming from NGOs and the Convention on Biodiversity. Larger projects would work both under CDM-like mechanisms and REDD+, whereas voluntary carbon markets would be a better fit for the smaller ones.

The other pathway assessed in this volume in the area of marine biologically-based CDR is based on the cultivation of macroalgae, which can then be used as biomass feedstock for BECCS or be buried for carbon removal. The differences of marine-biomass-based BECCS compared to other BECCS pathways mainly relate to the feedstock, which on the one hand has the advantage of not competing with other land uses. On the other hand, the cultivation and processing of the algae are more costly than other biomass feedstocks, even though there could be synergies when substituting algae for high-protein soy.[63] The assessment of the CCS part of the algae-based BECCS chain is the same as for conventional BECCS, see Section 11.3.1. Gattuso *et al.* (2021)[64] assess marine-biomass-fuelled BECCS as one of the most cost-effective routes they look at, with removals at US$26 per ton of CO_2 and possible improvements in the future.[64] However, that assessment rests on a single estimate and Chapter 10 also points to various factors that could increase costs substantially. In addition,[64] the low technological maturity is stressed as a major barrier to scale-up in the short to medium run: Some of the key components of the approach have not been demonstrated at scale, so their estimate is of at least a decade to close the necessary knowledge gaps. Keeping those uncertainties in mind, a tentative estimate for yearly CO_2 removals through marine-biomass-fuelled BECCS is put between 3 and 4 Gt CO_2, given that all the necessary on- and offshore infrastructure can be rolled out accordingly. However, at the large scales currently discussed, Chapter 7 also assesses the environmental side effects to be potentially severe, including disruptions to marine species at a large scale, potentially adverse effects on nutrient cycling and availability, increased macroalgal release of bromoform, and the possibility of adverse effects on human health through changes in regional iodine cycling. This leads the authors to argue for lower-scale deployment in order to contain environmental risks, which would also help circumventing some of the barriers to be expected in the context of competition for shelf sea waters that are currently being used by many different actors, most notably for fisheries and conservation. Still, a rollout would be associated with challenging approval processes for using large-scale seaweed cultivation as CDR and the same regulatory barriers for the CCS component discussed in the context of traditional (terrestrial-biomass-based) BECCS would apply.

11.4 Synthesis

The assessment of technologies and practices to remove CO_2 from the atmosphere in Section 11.3 clearly shows that all of them feature relevant potentials, yet also come with a unique set of trade-offs and opportunities, which corroborates earlier findings that no single technology or practice will be shouldering removals at double-digit Gigaton-scale, but that a portfolio of CDR options is the most robust and sustainable solution.[7,9] Despite the commonly held belief that CDR methods are all in their infancy, we furthermore see that most of the technologies and practices are actually technologically mature, with their components at relatively advanced levels of technological readiness, even if there are few installations incorporating all elements to date.[65] Barriers to upscaling seem to exist more along the dimensions of public acceptability and a pervasive lack of business models, but these two fields also feature vast knowledge gaps, which have been highlighted in systematic reviews before.[14] It is noteworthy that public acceptance can be an issue at two different levels: at the technology level, there is some heterogeneity, where practices associated with planting trees and restoring nature enjoy larger public support than technologies involving bioenergy and CCS,[13] while at a broader level, the very concept of enabling CDR is rejected because of fears of moral hazard,[30] delaying emissions reductions and thus jeopardising the achievement of ambitious mitigation goals.[66] There is also evidence that the choice of policy instrument for incentivising removals has an impact on the level of acceptance. In particular, payments for removal are found to be preferred over price guarantees for producers who are selling energy derived from BECCS.[67] Similarly, the framing is identified as a critical factor in how society responds to CCS[31] and CDR technologies.[68]

Legal and regulatory frameworks can provide barriers to deployment, *e.g.*, if restricting local storage opportunities, but also important safeguards for sustainable upscaling, *e.g.*, EU regulation concerning the import of biomass designed to reduce the risk of iLUC.

Finally, political economy considerations will play into national CDR strategies. Similar to the CCS readiness index's dimension of CCS inherent interest as measured by global shares of fossil fuel production and consumption,[19] the role of CDR in the overall mitigation strategy will *inter alia* depend on a country's residual emissions, and sectoral composition will be a determining factor in the portfolio of CDR options. While technologies like DACCS will depend on industry to a large extent, most of the nature-based approaches, like SCS, enhancing coastal vegetation, and applying biochar, require the coordination of manys small and decentralised actors. In as far as resources like land are concerned, there will also be regional differences in the political economy, *e.g.*, tropical reforestation will come at the opportunity cost of not being able to expand agriculture or grant palm oil concessions, for example – activities associated with economic development and improvements in rural livelihoods. All in all, there is a large knowledge gap when it comes to the political economy of CDR, however. Clearly, different extents of CDR and different portfolios of CDR will have very different distributional implications, which have to be understood in order to develop successful roadmaps to reach net-zero emissions in time.

11.5 Discussion

This chapter has provided a synthesis of the previous chapters on individual technologies and practices for removing CO_2 from the atmosphere and put them into perspective of the wider literature, going beyond their costs, potentials and side effects, but also discussing what we know and do not know on other aspects of their implementation, such as their acceptance, political economy, technological readiness, legal and regulatory frameworks, and existing business models. This assessment framework leans on earlier efforts to evaluate the readiness of CCS,[18,19] the feasibility assessment in the IPCC Special Report on Global Warming of 1.5 °C[13,17] and a broad review of the literature (see Section 11.2).

This synthesis enables, on the one hand, more straightforward comparisons when planning for net-zero targets, weighing different aspects of technologies against each other when composing environmentally and socially sustainable and politically and economically feasible CDR portfolios across technologies, geographies and time. On the other hand, it allows for the identification of concrete knowledge gaps that in themselves present barriers to moving forward with implementation. In particular, Section 11.4 very clearly finds that despite not being rolled out to any significant extent at the moment, many of the assessed CDR technologies and practices are relatively technologically mature at this stage and that the main barriers to upscaling seem to exist more along the dimensions of public acceptability and lack of business models, where also most of the knowledge gaps can be located.

Business models will of course critically hinge on developments in climate policy, which need to work on incentivising large-scale, rapid deployment rather than small-scale demonstration.[72] Yet to date, it is unclear how CDR will be integrated into net zero strategies, *e.g.*, whether and how removals would be accounted for in emissions trading,[73] whether reverse auctions would be used to achieve removal targets[74] even in a situation where residual positive emissions would no longer provide sufficient revenues for funding negative ones, and whether concerns about undermining impacts on conventional mitigation could be addressed by adjusted designs or separate targets.[75] Chapter 13 of this volume will explore the advantages and challenges of different policy frameworks in more detail, also discussing some of the challenges associated with the views of stakeholders, *e.g.*, moral hazards and social-economic issues.

Other knowledge gaps, which will gain importance as implementation becomes more imminent include distributional aspects, which will influence the political economy at a national scale, but may also result in a re-opening of burden-sharing discussions at an international scale.[76,77] Chapter 14 will discuss the priorities for policy design which can potentially overcome challenges associated with the deployment of CDR to address the former set of questions, while Chapter 12 will unpack region-specific potentials in order to shed more light on the latter, the international dimension.

Finally, global pathway analysis will not be sufficient to map out national net-zero strategies: Without wanting to deny the value of the literature derived from the pathways, which has focused on identifying sustainability trade-offs associated with double-digit Gigaton removals through BECCS, it would for the near-term be more constructive to determine country-level potentials and deployment strategies that could also drive synergies with other SDGs.[29,78] Unfortunately, this part of the literature is still somewhat underdeveloped, and Chapter 15 will provide a more in-depth exploration of the topic and therefore promising entry points for CDR rollout, specifically for socially equitable energy transitions. This chapter can therefore be understood as building the bridge between the technology assessments of Chapters 5–10 and Chapters 12–15 on the issues relevant for implementation.

The analysis in this chapter is limited in its scope, as it has focused on those CDR technologies and practices that have been assessed in this volume and which feature prominently in the current discussion. However, this leaves out some other technology pathways, which are currently gaining more traction. For example, as of the Special Report on Global Warming of 1.5 °C,[79] there has been a new definition of Carbon Dioxide Removal, which extends not only to geological and ocean storage of CO_2, but also to storage in products.¶¶ Carbon Capture and Utilisation (CCU), which has been briefly mentioned in the context of DACCS in Section 11.3.2, provides an attractive narrative of recycling and re-use, with proponents arguing that net costs of emissions reductions or removals can be reduced in this way and that CCS technologies can be facilitated through early learning despite the absence of high carbon prices. However, a ton of CO_2 utilised does not necessarily lead to a ton mitigated or even

¶¶"Anthropogenic activities removing CO_2 from the atmosphere and durably storing it in geological, terrestrial, or ocean reservoirs, or in products. It includes existing and potential anthropogenic enhancement of biological or geochemical sinks and direct air capture and storage, but excludes natural CO_2 uptake not directly caused by human activities."[84]

removed, *e.g.*, emissions could even increase if the CO_2 is captured from fossil fuel combustion, transformed into synthetic fuels using non-decarbonised energy, and CO_2 subsequently released when using the synfuel. The atmospheric carbon stock will still increase. It is therefore important to incentivise mitigation and removal and not utilisation *per se*[80] and to closely account for the duration of the CO_2 storage and CCU's lifecycle emissions. Indeed, early research into global potentials and business models like EOR finds that due to the rate and scale of CO_2 production compared to that of utilisation, it is unlikely that chemical CO_2 conversion will deliver more than 1% of the mitigation challenge, and even scaled-up EOR will likely only account for 4–8%.[81] On the other hand, some literature also points to CO_2 storage in building materials as a promising entry point in the face of high demands to expand construction in many parts of the world – this does not even have to be chemically converted CO_2, but could come in the form of wooden structures.[82] It is probably too early to draw robust conclusions and the CCU pathways are too diverse to be summarised into one category, but it is clear that there is vast research demand and plenty of caveats against setting perverse incentives that may increase emissions rather than removing them.

Abbreviations

AR5	Fifth Assessment Report of the Intergovernmental Panel on Climate Change
BECCS	Bioenergy with Carbon Capture and Storage
CCS	Carbon Capture and Storage
CDR	Carbon Dioxide Removal
CCU	Carbon Capture and Utilisation
DAC	Direct Air Capture
DACCS	Direct Air Carbon Capture and Storage
EOR	Enhanced Oil Recovery
GHG	Greenhouse Gas
iLUC	indirect Land Use Change
IPCC	Intergovernmental Panel on Climate Change
LCFS	Low-Carbon Fuel Standards
MRV	Monitoring, Reporting and Verification
NCP	Nature's Contribution to People
PES	Payments for Ecosystem Services
SCS	Soil Carbon Sequestration
SDG	Sustainable Development Goal
SR1.5	Special Report of the Intergovernmental Panel on Climate Change on Global Warming of 1.5 °C

Acknowledgements

S.F. acknowledges funding by the RESTORE + project (http://www.restoreplus.org/), part of the International Climate Initiative, supported by the Federal

Ministry for the Environment, Nature Conservation, and Nuclear Safety (BMU) on the basis of a decision adopted by the German Bundestag, and from the Natural Environment Research Council of the UK in the frame of the project 'Comparative assessment and region-specific optimisation of GGR' under grant reference NE/P019900/1. She is indebted to the other chapter authors of this volume for information on specific CDR technologies and practices.

References

1. C. Le Quéré, R. B. Jackson, M. W. Jones, A. J. P. Smith, S. Abernethy, R. M. Andrew, A. J. De-Gol, D. R. Willis, Y. Shan, J. G. Canadell, P. Friedlingstein, F. Creutzig and G. P. Peters, *Nat. Clim. Change*, 2020, **10**, 647–653.
2. J. Rogelj, D. Shindell, K. Jiang, S. Fifita, P. Forster, V. Ginzburg, C. Handa, H. Kheshgi, S. Kobayashi, E. Kriegler, L. Mundaca, R. Séférian and M. Vilariño, *Global Warming of 1.5 °C an IPCC Special Report on the Impacts of Global Warming of 1.5 °C Above Pre-industrial Levels and Related Global Greenhouse Gas Emission Pathways, in the Context of Strengthening the Global Response to the Threat of Climate Change*, 2018.
3. D. Tong, Q. Zhang, Y. Zheng, K. Caldeira, C. Shearer, C. Hong, Y. Qin and S. J. Davis, *Nature*, 2019, **572**, 373–377.
4. G. Luderer, Z. Vrontisi, C. Bertram, O. Y. Edelenbosch, R. C. Pietzcker, J. Rogelj, H. S. De Boer, L. Drouet, J. Emmerling, O. Fricko, S. Fujimori, P. Havlík, G. Iyer, K. Keramidas, A. Kitous, M. Pehl, V. Krey, K. Riahi, B. Saveyn, M. Tavoni, D. P. Van Vuuren and E. Kriegler, *Nat. Clim. Change*, 2018, **8**, 626–633.
5. S. J. Davis, N. S. Lewis, M. Shaner, S. Aggarwal, D. Arent, I. L. Azevedo, S. M. Benson, T. Bradley, J. Brouwer, Y.-M. Chiang, C. T. M. Clack, A. Cohen, S. Doig, J. Edmonds, P. Fennell, C. B. Field, B. Hannegan, B.-M. Hodge, M. I. Hoffert, E. Ingersoll, P. Jaramillo, K. S. Lackner, K. J. Mach, M. Mastrandrea, J. Ogden, P. F. Peterson, D. L. Sanchez, D. Sperling, J. Stagner, J. E. Trancik, C.-J. Yang and K. Caldeira, *Science* , 2018, **360**, eaas9793.
6. J. Minx, W. F. Lamb, M. W. Callaghan, L. Bornmann and S. Fuss, *Environ. Res. Lett.*, 2017, **12**, 035007.
7. J. C. Minx, W. F. Lamb, M. W. Callaghan, S. Fuss, J. Hilaire, F. Creutzig, T. Amann, T. Beringer, W. de Oliveira Garcia, J. Hartmann, T. Khanna, D. Lenzi, G. Luderer, G. F. Nemet, J. Rogelj, P. Smith, J. L. Vicente Vicente, J. Wilcox and M. del Mar Zamora Dominguez, *Environ. Res. Lett.*, 2018, **13**, 063001.
8. S. Fuss, W. F. Lamb, M. W. Callaghan, J. Hilaire, F. Creutzig, T. Amann, T. Beringer, W. De Oliveira Garcia, J. Hartmann, T. Khanna, G. Luderer, G. F. Nemet, J. Rogelj, P. Smith, J. V. Vicente, J. Wilcox, M. Del Mar Zamora Dominguez and J. C. Minx, *Environ. Res. Lett.*, 2018, **13**, 063002.

9. S. Fuss, J. G. Canadell, P. Ciais, R. B. Jackson, C. D. Jones, A. Lyngfelt, G. P. Peters and D. P. Van Vuuren, *One Earth*, 2020, **3**, 145–149.
10. National Academies of Sciences Engineering and Medicine, *Negative Emissions Technologies and Reliable Sequestration*, Washington, DC, 2019.
11. Royal Society & Royal Academy of Engineering, *Greenhouse Gas Removal*, London, 2018.
12. P. Smith, *Global Change. Biol.*, 2016, **22**, 1315–1324.
13. H. de Coninck, A. Revi, M. Babiker, P. Bertoldi, M. Buckeridge, A. Cartwright, W. Dong, J. Ford, S. Fuss and J.-C. Hourcade, *in Global Warming of 1.5 C an IPCC special report on the impacts of global warming of 1.5 C above pre-industrial levels and related global greenhouse gas emission pathways, in the context of strengthening the global response to the threat of climate change*, Intergovernmental Panel on Climate Change, 2018.
14. G. F. Nemet, M. W. Callaghan, F. Creutzig, S. Fuss, J. Hartmann, J. Hilaire, W. F. Lamb, J. C. Minx, S. Rogers and P. Smith, *Environ. Res. Lett.*, 2018, **13**, 063003.
15. UNEP, The Emissions Gap Report 2017 - A UN Environment Synthesis Report, 2017.
16. D. McLaren, *Process Saf. Environ. Prot.*, 2012, **90**, 489–500.
17. C. Singh, J. Ford, D. Ley, A. Bazaz and A. Revi, *Clim. Change*, 2020, **162**, 255–277.
18. N. Markusson and S. Haszeldine, *Energy Procedia*, 2009, **1**, 4625–4632.
19. C. P. Consoli, I. Havercroft and L. Irlam, *Energy Procedia*, 2017, **114**, 7348–7355.
20. S. V Hanssen, V. Daioglou, Z. J. N. Steinmann, J. C. Doelman, D. P. Van Vuuren and M. A. J. Huijbregts, *Nat. Clim. Change*, 2020, **10**, 1023–1029.
21. F. Creutzig, N. H. Ravindranath, G. Berndes, S. Bolwig, R. Bright, F. Cherubini, H. Chum, E. Corbera, M. Delucchi, A. Faaij, J. Fargione, H. Haberl, G. Heath, O. Lucon, R. Plevin, A. Popp, C. Robledo-Abad, S. Rose, P. Smith, A. Stromman, S. Suh and O. Masera, *GCB Bioenergy*, 2015, 7, 916–944.
22. C. Robledo-Abad, H. J. Althaus, G. Berndes, S. Bolwig, E. Corbera, F. Creutzig, J. Garcia-Ulloa, A. Geddes, J. S. Gregg, H. Haberl, S. Hanger, R. J. Harper, C. Hunsberger, R. K. Larsen, C. Lauk, S. Leitner, J. Lilliestam, H. Lotze-Campen, B. Muys, M. Nordborg, M. Ölund, B. Orlowsky, A. Popp, J. Portugal-Pereira, J. Reinhard, L. Scheiffle and P. Smith, *GCB Bioenergy*, 2017, **9**, 541–556.
23. H. Haberl, T. Beringer, S. C. Bhattacharya, K.-H. Erb and M. Hoogwijk, *Curr. Opin. Environ. Sustainability*, 2010, **2**, 394–403.
24. P. Smith, S. J. Davis, F. Creutzig, S. Fuss, J. Minx, B. Gabrielle, E. Kato, R. B. Jackson, A. Cowie, E. Kriegler, D. P. Van Vuuren, J. Rogelj, P. Ciais, J. Milne, J. G. Canadell, D. McCollum, G. Peters, R. Andrew, V. Krey, G. Shrestha, P. Friedlingstein, T. Gasser, A. Grübler, W. K. Heidug, M. Jonas, C. D. Jones, F. Kraxner, E. Littleton, J. Lowe, J. R. Moreira, N. Nakicenovic, M. Obersteiner, A. Patwardhan, M. Rogner, E. Rubin,

A. Sharifi, A. Torvanger, Y. Yamagata, J. Edmonds and C. Yongsung, *Nat. Clim. Change*, 2016, **6**, 42–50.
25. C. Robledo-Abad, H.-J. Althaus, G. Berndes, S. Bolwig, E. Corbera, F. Creutzig, J. Garcia-Ulloa, A. Geddes, J. S. Gregg, H. Haberl, S. Hanger, R. J. Harper, C. Hunsberger, R. K. Larsen, C. Lauk, S. Leitner, J. Lilliestam, H. Lotze-Campen, B. Muys, M. Nordborg, M. Ölund, B. Orlowsky, A. Popp, J. Portugal-Pereira, J. Reinhard, L. Scheiffle and P. Smith, *GCB Bioenergy*, 2017, **9**, 541–556.
26. R. M. Bright, F. Zhao, R. B. Jackson and F. Cherubini, *Global Change Biol.*, 2015, **21**, 3246.
27. M. Bui, C. S. Adjiman, A. Bardow, E. J. Anthony, A. Boston, S. Brown, P. Fennell, S. Fuss, A. Galindo, L. A. Hacket, H. J. Herzog, G. Jackson, J. Kemper, S. C. M. Krevor, G. C. Maitland, M. Matuszewski, C. Petit, G. Puxty, J. Reimer, D. M. Reiner, E. S. Rubin, A. Scott Stewart, N. Shah, B. Smit, J. P. M. Trusler, P. Webley, J. Wilcox and N. Mac Dowell, *Energy Environ. Sci.*, 2018, **11**, 1062–1176.
28. C. Hunsberger, S. Bolwig, E. Corbera and F. Creutzig, *Geoforum*, 2014, **54**, 248–260.
29. P. Patrizio, S. Leduc, F. Kraxner, S. Fuss, G. Kindermann, S. Mesfun, K. Spokas, A. Mendoza, N. Mac Dowell, E. Wetterlund, J. Lundgren, E. Dotzauer, P. Yowargana and M. Obersteiner, *Joule*, 2018, **2**(12), 2633–2648.
30. D. Lenzi, *Global Sustainability*, Cambridge University Press, 2018, vol. 1, p. e7.
31. L. Whitmarsh, D. Xenias and C. R. Jones, *Palgrave Commun.*, 2019, **5**, 17.
32. C. Cummings, S. Lin and B. Trump, *Clim. Res.*, 2017, **73**, 247–264.
33. D. Fytili and A. Zabaniotou, *Curr. Opin. Green Sustainability Chem.*, 2017, **8**, 5–9.
34. C. Braun, C. Merk, G. Pönitzsch, K. Rehdanz and U. Schmidt, *Clim. Policy*, 2018, **18**, 1–14.
35. T. D. Searchinger, T. Beringer, B. Holtsmark, D. M. Kammen, E. F. Lambin, W. Lucht, P. Raven and J.-P. van Ypersele, *Nat. Commun.*, 2018, **9**, 3741.
36. M. Bui, C. S. Adjiman, A. Bardow, E. J. Anthony, A. Boston, S. Brown, P. S. Fennell, S. Fuss, A. Galindo, L. A. Hackett, J. P. Hallett, H. J. Herzog, G. Jackson, J. Kemper, S. Krevor, G. C. Maitland, M. Matuszewski, I. S. Metcalfe, C. Petit, G. Puxty, J. Reimer, D. M. Reiner, E. S. Rubin, S. A. Scott, N. Shah, B. Smit, J. P. M. Trusler, P. Webley, J. Wilcox and N. Mac Dowell, *Energy Environ. Sci.*, 2018, **11**, 1062–1176.
37. S. Frank, P. Havlík, E. Stehfest, H. van Meijl, P. Witzke, I. Pérez-Domínguez, M. van Dijk, J. C. Doelman, T. Fellmann, J. F. L. Koopman, A. Tabeau and H. Valin, *Nat. Clim. Change*, 2019, **9**, 66–72.
38. A. M. I. Kallio, R. Chudy and B. Solberg, *Biomass Bioenergy*, 2018, **108**, 415–425.

39. Swedish Environmental Protection Agency, *Det Klimatpolitiska Ramverket*, 2017.
40. J. J. Dooley, *Energy Procedia*, 2013, **37**, 5141–5150.
41. K. Calvin, J. Edmonds, B. Bond-Lamberty, L. Clarke, S. H. Kim, P. Kyle, S. J. Smith, A. Thomson and M. Wise, *Energy Econ.*, 2009, **31**, S107–S120.
42. D. W. Keith, G. Holmes, D. St Angelo and K. Heidel, *Joule*, 2018, **2**, 1 573–1594.
43. C. Breyer, M. Fasihi, C. Bajamundi and F. Creutzig, *Joule*, 2019, **3**, 2053–2057.
44. P. Renforth, *Nat. Commun.*, 2019, **10**, 1401.
45. D. J. Beerling, E. P. Kantzas, M. R. Lomas, P. Wade, R. M. Eufrasio, P. Renforth, B. Sarkar, M. G. Andrews, R. H. James, C. R. Pearce, J.-F. Mercure, H. Pollitt, P. B. Holden, N. R. Edwards, M. Khanna, L. Koh, S. Quegan, N. F. Pidgeon, I. A. Janssens, J. Hansen and S. A. Banwart, *Nature*, 2020, **583**, 242–248.
46. P. Smith, J. Adams, D. J. Beerling, T. Beringer, K. V Calvin, S. Fuss, B. Griscom, N. Hagemann, C. Kammann, F. Kraxner, J. C. Minx, A. Popp, P. Renforth, J. L. Vicente Vicente and S. Keesstra, *Annu. Rev. Environ. Resour.*, 2019, **44**, 255–286.
47. P. Renforth and G. Henderson, *Rev. Geophys.*, 2017, **55**, 636–674.
48. Y. Yuan, H. Chen, W. Yuan, D. Williams, J. T. Walker and W. Shi, *Soil Biol. Biochem.*, 2017, **113**, 14–25.
49. N. Borchard, M. Schirrmann, M. L. Cayuela, C. Kammann, N. Wrage-Mönnig, J. M. Estavillo, T. Fuertes-Mendizábal, G. Sigua, K. Spokas, J. A. Ippolito and J. Novak, *Sci. Total Environ*, 2019, **651**, 2354–2364.
50. X. Bai, Y. Huang, W. Ren, M. Coyne, P.-A. Jacinthe, B. Tao, D. Hui, J. Yang and C. Matocha, *Glob. Change Biol.*, 2019, **25**, 2591–2606.
51. X. Liu, J. Zhou, Z. Chi, J. Zheng, L. Li, X. Zhang, J. Zheng, K. Cheng, R. Bian and G. Pan, *CATENA*, 2019, **179**, 20–28.
52. C. Werner, H.-P. Schmidt, D. Gerten, W. Lucht and C. Kammann, *Environ. Res. Lett.*, 2018, **13**, 44036.
53. Trillion Trees, Partnership of BirdLife International, Wildlife Conservation Society and the World Wide Fund for Nature (WWF), https://trilliontrees.org/home.
54. B. W. Griscom, J. Adams, P. W. Ellis, R. A. Houghton, G. Lomax, D. A. Miteva, W. H. Schlesinger, D. Shoch, J. V Siikamäki, P. Smith, P. Woodbury, C. Zganjar, A. Blackman, J. Campari, R. T. Conant, C. Delgado, P. Elias, T. Gopalakrishna, M. R. Hamsik, M. Herrero, J. Kiesecker, E. Landis, L. Laestadius, S. M. Leavitt, S. Minnemeyer, S. Polasky, P. Potapov, F. E. Putz, J. Sanderman, M. Silvius, E. Wollenberg and J. Fargione, *Proc. Natl. Acad. Sci.*, 2017, **114**, 11645–11650.
55. Y. Pan, R. A. Birdsey, J. Fang, R. Houghton, P. E. Kauppi, W. A. Kurz, O. L. Phillips, A. Shvidenko, S. L. Lewis, J. G. Canadell, P. Ciais, R. B. Jackson, S. W. Pacala, A. D. McGuire, S. Piao, A. Rautiainen, S. Sitch and D. Hayes, *Science*, 2011, **333**, 988–993.

56. R. B. Jackson, E. G. Jobbagy, R. Avissar, S. B. Roy, D. J. Barrett, C. W. Cook, K. A. Farley, D. C. le Maitre, B. A. McCarl and B. C. Murray, *Science*, 2005, **310**, 1944–1947.
57. N. Seddon, E. Daniels, R. Davis, A. Chausson, R. Harris, X. Hou-Jones, S. Huq, V. Kapos, G. M. Mace, A. R. Rizvi, H. Reid, D. Roe, B. Turner and S. Wicander, *Glob. Sustainability*, 2020, 1–12.
58. P. I. Macreadie, D. A. Nielsen, J. J. Kelleway, T. B. Atwood, J. R. Seymour, K. Petrou, R. M. Connolly, A. C. G. Thomson, S. M. Trevathan-Tackett and P. J. Ralph, *Front. Ecol. Environ.*, 2017, **15**, 206–213.
59. J. Siikamäki, J. N. Sanchirico, S. Jardine, D. McLaughlin and D. Morris, *Environ. Sci. Policy Sustainability Dev.*, 2013, **55**, 14–29.
60. S. C. Johannessen and R. W. Macdonald, *Environ. Res. Lett.*, 2016, **11**, 113001.
61. E. Bayraktarov, M. I. Saunders, S. Abdullah, M. Mills, J. Beher, H. P. Possingham, P. J. Mumby and C. E. Lovelock, *Ecol. Appl.*, 2016, **26**, 1055–1074.
62. A. Himes-Cornell, L. Pendleton and P. Atiyah, *Ecosyst. Serv.*, 2018, **30**, 36–48.
63. C. M. Beal, I. Archibald, M. E. Huntley, C. H. Greene and Z. I. Johnson, *Earth's Future*, 2018, **6**, 524.
64. J.-P. Gattuso, P. Williamson, C. M. Duarte and A. K. Magnan, *Front. Clim.*, 2021, **2**, 37.
65. L. Waller, T. Rayner, J. Chilvers, C. A. Gough, I. Lorenzoni, A. Jordan and N. Vaughan, *WIREs Clim. Change*, 2020, **11**, e649.
66. D. McLaren, K. A. Parkhill, A. Corner, N. E. Vaughan and N. F. Pidgeon, *Global Environ. Change*, 2016, **41**, 64–73.
67. R. Bellamy, J. Lezaun and J. Palmer, *Nat. Commun.*, 2019, **10**, 743.
68. C. Gough and S. Mander, *Curr. Sustainability Energy Rep.*, 2019, **6**, 116–123.
69. M. P. J. Oreska, K. J. McGlathery, L. R. Aoki, A. C. Berger, P. Berg and L. Mullins, *Sci. Rep.*, 2020, **10**, 7325.
70. P. Taillardat, B. S. Thompson, M. Garneau, K. Trottier and D. A. Friess, *Interface Focus*, 2020, **10**, 20190129.
71. O. Hoegh-Guldberg, K. Caldeira, T. Chopin, S. Gaines, P. Haugan, M. Hemer, J. Howard, M. Konar, D. Krause-Jensen, E. Lindstad, C. E. Lovelock, M. Michelin, F. Gunnar Nielsen, E. Northrop, R. Parker, J. Roy, T. Smith, S. Some and P. Tyedmers, *The Ocean as a Solution to Climate Change: Five Opportunities for Action*, Washington, DC, 2019.
72. G. P. Peters and O. Geden, *Nat. Clim. Change*, 2017, 7, 619–621.
73. W. Rickels, A. Proelß, O. Geden, J. Burhenne and M. Fridahl, *The Future of (Negative) Emissions Trading in the European Union*, Kiel, Germany, 2020.
74. SOU, Vägen till en klimatpositiv framtid: Statens Offentliga Utredningar (Governmental inquiry), Stockholm, 2020.
75. D. P. McLaren, D. P. Tyfield, R. Willis, B. Szerszynski and N. O. Markusson, *Front. Clim.*, 2019, **1**, 4.

76. C. Pozo, Á. Galán-Martín, D. M. Reiner, N. Mac Dowell and G. Guillén-Gosálbez, *Nat. Clim. Change*, 2020, **10**, 640–646.
77. C. L. Fyson, S. Baur, M. Gidden and C.-F. Schleussner, *Nat. Clim. Change*, 2020, **10**, 836–841.
78. S. Fuss and F. Johnsson, *Front. Energy Res.*, 2021, **8**, 553400.
79. IPCC, Global warming of 1.5 °C. An IPCC special report on the impacts of global warming of 1.5 °C above pre-industrial levels and related global greenhouse gas emission pathways, in the context of strengthening the global response to the threat of climate change, 2018.
80. C. Hepburn, E. Adlen, J. Beddington, E. A. Carter, S. Fuss, N. Mac Dowell, J. C. Minx, P. Smith and C. K. Williams, *Nature*, 2019, **575**, 87–97.
81. N. Mac Dowell, P. Fennell, N. Shah and G. C. Maitland, *Nat. Clim. Change*, 2017, **7**, 243.
82. G. Churkina, A. Organschi, C. P. O. Reyer, A. Ruff, K. Vinke, Z. Liu, B. K. Reck, T. E. Graedel and H. J. Schellnhuber, *Nat. Sustainability*, 2020, **3**, 269–276.
83. S. E. Baker, J. K. Stolaroff, G. Peridas, S. H. Pang, H. M. Goldstein, F. R. Lucci, W. Li, E. W. Slessarev, J. Pett-Ridge, F. J. Ryerson, J. L. Wagoner, W. Kirkendall, R. D. Aines, L. Sanchez, Daniel, B. Cabiyo, J. Baker, S. McCoy, R. Runnebaum, J. Wilcox, P. C. Psarras, H. Pilorgé, N. McQueen, D. Maynard and C. McCormick, *Getting to Neutral: Options for Negative Emissions in California*, 2020.
84. IPCC, Annex I: Glossary. in *Global Warming of 1.5 °C. An IPCC Special Report on the impacts of global warming of 1.5 °C*, 2018, ed. T. W. Masson-Delmotte, V., P. Zhai, H.-O. Pörtner, D. Roberts, J. Skea, P.R. Shukla, A. Pirani, W. Moufouma-Okia, C. Péan, R. Pidcock, S. Connors, J.B.R. Matthews, Y. Chen, X. Zhou, M.I. Gomis, E. Lonnoy, T. Maycock and M. Tignor, In Press.

CHAPTER 12

Global Distribution of Negative Emissions

CARLOS POZO,*[a] ÁNGEL GALÁN-MARTÍN,[b,c] DAVID M. REINER,[d] GONZALO GUILLÉN-GOSÁLBEZ,[e] MAI BUI[f,g] AND NIALL MAC DOWELL[f,g]

[a] Department of Chemical Engineering, Universitat Politècnica de Catalunya, EEBE, Av. Eduard Maristany 16, 08019 Barcelona, Spain; [b] Department of Chemical, Environmental and Materials Engineering, Universidad de Jaén, Campus Las Lagunillas s/n, 23071 Jaén, Spain; [c] Center for Advanced Studies in Earth Sciences, Energy and Environment. Universidad de Jaén, Campus Las Lagunillas s/n, 23071 Jaén, Spain; [d] Energy Policy Research Group, Judge Business School, University of Cambridge, Cambridge, UK; [e] Institute for Chemical and Bioengineering, Department of Chemistry and Applied Biosciences, ETH Zürich, Zürich, Switzerland; [f] Centre for Environmental Policy, Imperial College London, Exhibition Road, London SW7 1NA, UK; [g] Centre for Process Systems Engineering, Imperial College London, Exhibition Road, London SW7 2AZ, UK
*Email: carlos.pozo.fernandez@upc.edu

Meeting our climate goals will require large-scale deployment of carbon dioxide removal (CDR) options to compensate for both ongoing emissions and residual emissions from hard-to-abate sectors. Acknowledging that different regions are unequally endowed with natural capacities to deliver CDR, ensuring an equitable distribution of burdens and efforts, and promoting collaborative actions emerge as critical issues to guarantee fair contributions and effective actions to meet the CDR goals. This chapter

Energy and Environment Series No. 31
Greenhouse Gas Removal Technologies
Edited by Mai Bui and Niall Mac Dowell

Published by the Royal Society of Chemistry, www.rsc.org

departs from country-by-country CDR quotas based on equity principles and investigates the broader implications of adopting two different political approaches – *isolationist vs. cooperative* – for meeting such quotas. Cooperation is the preferred approach as it allows the most economically appealing biogeophysical resources to be exploited. However, incentivising such cooperation between actors and regions showing different interests will require designing new policy instruments and incentives to engage with the large-scale deployment of CDR options. Major barriers need to be overcome for governing CDR at global scales, such as the difficulty of simultaneously considering all the facets of the CDR problem, the necessity to integrate removal actions in frameworks for conventional mitigation, the need to deal with the transboundary aspects of international policies and the present inability to elucidate a coherent roadmap for CDR options deployment resulting from the lack of harmony in scientific contributions.

12.1 Introduction

Deploying carbon dioxide (CO_2) removal (CDR) options at the scale needed to achieve our climate ambitions by the end of the century requires a global response. With the recent widespread adoption of net-zero and climate neutrality targets in many countries, CDR has emerged as a key element for compensating for residual emissions from difficult-to-decarbonise sectors but also to compensate for emissions taking place in the coming decades that will exceed the carbon budget. This momentum poses significant challenges while opening up new opportunities to deliver the CDR required following a fair collaborative approach. Failing to do so may lead to a deadlock thereby jeopardising global action and putting climate goals out of reach.

The United Nations Framework Convention on Climate Change (UNFCCC) proposes distributing the burden of combatting climate change based on the principle of "equity and common but differentiated responsibilities and respective capabilities, in the light of different national circumstances".[1] This is particularly relevant in the context of CDR because, unlike some conventional mitigation options such as switching to less carbon-intensive resources (*e.g.*, from coal to natural gas or renewables), all CDR options are, to at least some extent, linked to biogeophysical resources. These domestic constraints prevent countries from being equally endowed to deliver CDR at scale. For example, even if two countries can grow the same crop, one might need higher irrigation rates or greater fertiliser application, which will lead to differential costs, water consumption, and social acceptance issues. Hence, the ability to deliver CDR using either technological approaches, *e.g.*, bioenergy with carbon capture and storage, or natural-based solutions, *e.g.*, afforestation, is intrinsically linked to region-specific factors and particularities.

This chapter investigates the fair distribution of the CDR burden, focusing on (i) the consequences of the limited ability of countries to deliver CDR, (ii) how cooperation can reduce overall costs by redistributing efforts more effectively, (iii) mechanisms that could spur greater cooperation among countries, and (iv) key aspects to be considered in any regulatory framework aimed at getting the ball rolling.

12.2 Definition of Fair CDR Quotas

For years, defining targets on emissions reductions at different scales – international, national, local, and organisational levels – has been a recurrent policy instrument to put the world on the path to fight against climate change. The Kyoto Protocol's failure in 1998 with a top-down approach, defining legally binding emissions reduction targets, led to the new approach followed in the Paris Agreement adopted in 2015. Unlike the Kyoto Protocol, the Paris accord provides flexibility to countries to undertake their own nationally determined contributions to the global response and maintain successive and more ambitious efforts with the hope that future commitments will align climate actions with goals.

In a context where the deployment of CDR options is key to achieving the Paris objectives, the definition of separate targeting for CDR is gaining attention to promote its early consideration in the climate agendas and to monitor CDR within its own framework.[2,3] Several studies used equity approaches—instead of the traditional cost-optimal approaches—to allocate CDR efforts across countries and regions.[4–6] For illustrative purposes, we refer here to the CDR quotas based on equity principles derived by Pozo *et al.*[6] (see Table 12.1). These CDR quotas result from the allocation of 687 Gt CO_2 to be removed cumulatively by 2100 (as in the IPCC SR1.5 'middle-of-the-road' P3 scenario[1]) over 176 of the 197 UNFCCC countries following three equity principles of *Responsibility*, *Capability* and *Equality*. *Responsibility* concerns the historical liability on the climate problem; therefore, giving higher quotas to countries with higher accumulated past emissions. In contrast, *Capability* considers the capacity to pay for mitigation actions, therefore, defining heavier weights for richer countries. Finally, *Equality* assigns equal rights per person, therefore, equal per capita CDR. This initial effort-sharing exercise yields three different yet arguably fair quotas for each country, one per principle, which are then merged into a single quota that captures the three fairness principles in a standalone value. This multicriteria quota is obtained by weighing the quotas of the different principles according to a hypothetical voting system where each country would advocate for the principle that allocates the least CDR to it, however in practice, this could be the result of international negotiations.

Acknowledging there might be opportunities for important contributions in other regions, we focus our discussion on the EU28 for a variety of reasons, namely its longstanding tradition in sharing climate obligations, the submission of a joint Nationally Determined Contribution (NDC), and the fact that market-based mechanisms for CO_2 emissions are already in place in this region. In addition, the EU28 committed in 2019 to reaching net-zero (see European Green Deal, Climate Law and EU climate action

Table 12.1 Quotas and gross removal and storage potentials for EU28 (2020–2100 cumulative Gt CO_2).[6]

	Cumulative CDR quotas by 2100 [Gt CO_2]	Gross removal and storage potentials by 2100 [Gt CO_2]		
		Reforestation	Biomass	Onshore storage
AUT	6.99	0.33	2.77	0.00
BEL	7.11	0.16	0.56	0.20
BGR	2.22	0.99	1.96	2.12
CYP	2.74	0.02	0.02	0.00
CZE	5.35	2.10	2.63	0.85
DEU	10.18	1.33	11.41	17.08
DNK	7.33	0.12	0.83	2.76
ESP	6.27	4.85	21.69	14.18
EST	6.19	0.16	2.72	0.00
FIN	6.10	1.16	5.74	0.00
FRA	8.27	5.44	11.33	8.69
GBR	9.17	3.13	3.92	14.40
GRC	4.95	2.43	2.38	0.25
HUN	3.12	0.48	2.17	0.62
IRL	5.40	1.49	2.66	0.00
ITA	7.50	2.38	5.25	6.55
LTU	2.73	0.21	3.31	0.04
LUX	10.87	0.01	0.06	0.00
LVA	2.64	0.29	3.34	0.40
MLT	3.02	0.00	0.00	0.00
NLD	7.99	0.25	0.35	2.34
POL	4.41	1.21	9.98	2.94
PRT	4.26	1.58	3.60	0.00
ROU	2.54	1.49	4.60	9.00
HRV	2.82	0.51	1.01	2.90
SVK	3.70	0.23	1.30	1.72
SVN	3.85	0.09	0.67	0.09
SWE	6.56	0.69	8.00	0.00
Total	154.29	33.13	114.25	87.13

proposal[7]), which will arguably require CDR efforts to compensate emissions from hard-to-abate sectors such as agriculture. Quotas for EU countries are provided in Table 12.1, which includes the UK as the data are for 2019.

12.3 Two Possible Pathways: Isolationist and Cooperative

As an illustrative exercise, we evaluate the capability of countries to meet these quotas using their own natural resources which, for the sake of simplicity, are limited here to reforestation, biomass for bioenergy with carbon capture and storage (BECCS), direct air CO_2 capture and storage (DACCS), with CO_2 stored in geological reservoirs (see Table 12.1). This analysis could be further extended to embrace other CDR technologies or options once they become sufficiently mature.[8] The gross CO_2 removal potentials for BECCS for the 28 EU countries are also taken from Pozo *et al.*[6] by aggregating the potentials for the distinct biomass sources therein. Analogously, the storage capacity for each EU country in onshore sites is derived by aggregating all the reservoir types described in the same ref. 6.

In this context, the countries have a set of CDR options at hand to meet their quotas. From all these technological and geographical options, the implications of two extreme scenarios for the EU28 are compared here to illustrate opposite political approaches (Figure 12.1).

In the first scenario, countries would act in an *isolationist* manner, with no or minimal interaction with other countries (Figure 12.1a). In this scenario, each EU Member State will attempt first to meet its quota using all of its natural domestic resources, mainly *via* reforestation and BECCS relying on domestic biomass. If additional removal is required, then the country will resort to direct air carbon capture and storage (DACCS), which is generally agreed to be more expensive than reforestation or BECCS (see Chapter 11). In this scenario, the CO_2 captured at BECCS or DACCS facilities will preferably be stored in domestic onshore sites, and only when these sites are exhausted, the captured CO_2 would then be shipped to offshore reservoirs, possibly located overseas. Note that reforestation offers both removal and storage services since CO_2 absorbed during tree growth is fixed in forest biomes as biogenic carbon.

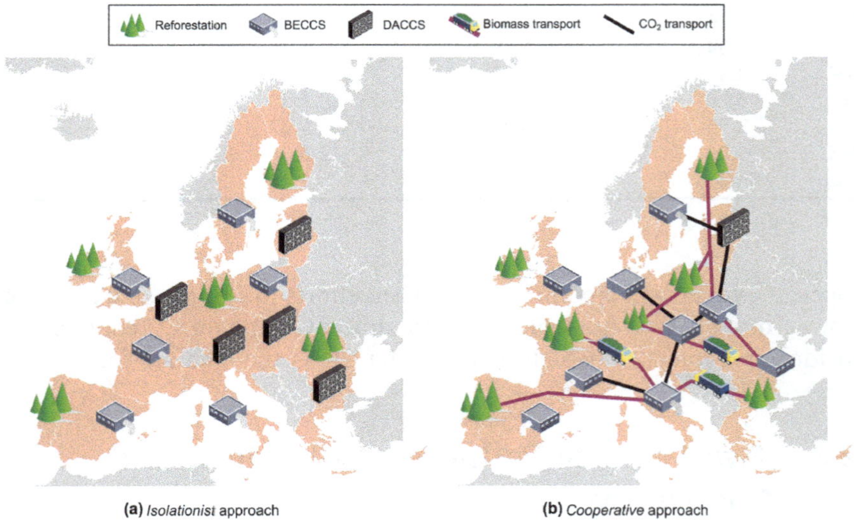

Figure 12.1 Hypothetical opposite political approaches for CDR. Subplot (a) illustrates an *isolationist* approach, where countries would meet their quotas using their domestic resources. In contrast, subplot (b) depicts a *cooperative* approach whereby biogeophysical resources would be shared among cooperating countries for the common benefit. In both cases, the CDR options depicted on the countries are for illustrative purposes only and do not intend to represent any realistic solution.

The comparison of country quotas with potentials reveals inefficiencies in the *isolationist* strategy. On the one hand, Member States with very limited reforestation and biomass potentials (*e.g.*, Luxembourg or Netherlands, with 0.07 and 0.60 Gt CO_2, respectively) would need to resort to expensive DACCS to meet their quotas of 10.87 and 7.99 Gt CO_2. Meanwhile, countries well-endowed with agricultural practices, *e.g.*, Spain or France, or with availability of residues from their abundant forests, *e.g.*, Germany, Poland or Sweden, would leave substantial BECCS resources unexploited. The amount of DACCS that could be globally displaced by exploiting these spare natural resources will depend on the final supply chains deployed, as life cycle emissions would need to be discounted from the gross CDR of each alternative before net negative emissions can be quantified and compared to country quotas.[9] Regardless, it is clear that a certain degree of cooperation between EU countries could be highly beneficial, sharing biomass resources and alleviating the need to deploy more costly technologies. The same observation is also valid for CO_2 storage, since unused onshore geological formations could reduce the need for more costly offshore or overseas storage, bringing the global costs down. As an example of this, the spare storage capacity in Germany, Spain and Great Britain, calculated as the storage potential less the difference between quotas and reforestation potentials, would be worth 8.23, 12.76 and 8.36 Gt CO_2, respectively.

By contrast, in a *cooperative* scenario (Figure 12.1b), EU countries will no longer try to meet their individual quota, but rather share all their natural resources to meet the global EU quota at minimum global cost. In this scenario, countries with poor domestic storage capacity for BECCS could direct captured CO_2 to an appropriate storage location in another country. This could be, for instance, the case of Poland, with a gross biomass potential of 9.98 Gt CO_2 but only 2.94 Gt CO_2 of proven onshore storage capacity. Otherwise, if the country has spare storage capacity but is short of biomass, it could import biomass pellets from other countries for use in BECCS plants before relying on more costly options like DACCS. Such a strategy could be implemented in Germany, where the storage capacity significantly exceeds gross biomass potential (17.08 Gt CO_2 compared to 11.41 Gt CO_2).

Regardless of the final supply chains deployed, the comparison of global quotas with global removal and storage potentials reveals that natural carbon sources and sinks within the EU28 are insufficient to meet the CDR goals. In this context, the remaining efforts – perhaps in the form of DACCS for removal and offshore reservoirs for storage – could be shared equally among the 28 Member States. In this context, it is important to note that offshore reservoirs in the Nordic region alone, with a capacity of 209 Gt CO_2,[10] would be more than enough to meet the EU CDR quota of 154 Gt CO_2.

In summary, the strength of the *cooperative* strategy relies on avoiding CDR in the most disfavoured placements, allowing countries to better exploit their natural biogeophysical resources. From a global perspective, this will result in lower costs since, for instance, imported biomass pellets constitute a cheaper carbon source compared to domestic DACCS (Chapter 11), while a large pipeline network for CO_2 transportation, deployed all over Europe, will likely attain lower marginal costs than isolated installations in selected countries. Note, however, that the asymmetric distribution of natural resources across countries, would likely lead to differences in national contributions towards the global CDR. Countries such as Spain, France, Germany or Great Britain, with abundant removal and storage potentials, might lead the CDR efforts, while others such as Luxembourg or Malta would contribute only marginally. In turn, this would translate into larger costs (*i.e.*, CDR burden) incurred by the former groups of countries, while the remaining Members would benefit economically by cooperation. These asymmetries in the total cost might act as deterrent agents, hampering the willingness of some countries to engage in a collective response to the climatic problem.

12.4 Encouraging Cooperation: United We Stand

Since the *cooperative* approach relies upon some countries delivering much larger quantities of CDR relative to their actual goals, it will be necessary to compensate them for this service. Such an incentive could, for example, be derived by calculating the difference between a reference cost, *i.e.*, the one in the *isolationist* approach, and the real cost incurred by each country in cooperation.[11] Importantly, one would anticipate a CDR supply curve emerging, with various countries able to offer a range of CDR options at different costs. A central authority, *e.g.*, the European Commission in this context, could collect payments from countries on whose behalf CDR services are being provided, and use those funds to compensate countries for the provision of CDR services (Figure 12.2). As a result, all countries will end up with a net cost that is lower than the cost of meeting their target individually by the same proportion as the total cost is reduced in cooperation.

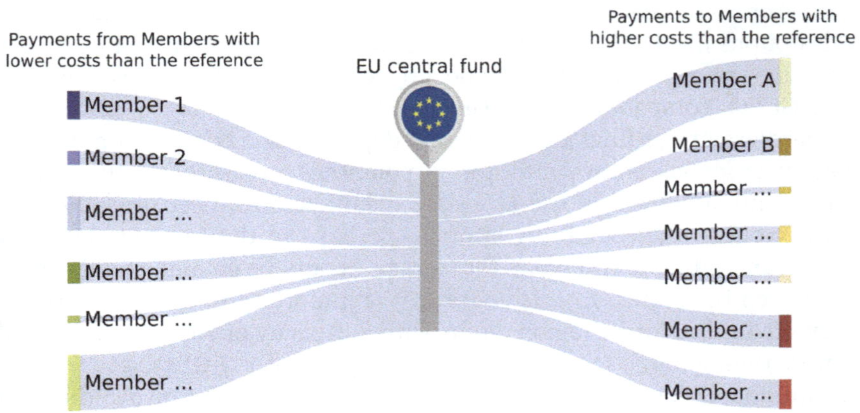

Figure 12.2 Illustrative compensation mechanism to encourage cooperation. Members on the left would contribute to a central fund, *e.g.*, administered by the EU, which in turn, will redistribute this money to the Members on the right to compensate them for their additional efforts.

12.5 Effective Policies to Remove Barriers

If cooperation is clearly appealing from an economic viewpoint,[11] what is deterring countries from adopting more ambitious and collaborative agreements for CDR? Likely, the main reason is the lack of reliable policy instruments to provide a safe environment for public and private investors and incentivise the initial demonstration of CDR options to facilitate their scale-up.

We next discuss four main barriers that we identified as hindering the development of such regulations, starting with the need to consider and assign value not only to CDR and associated costs, but also to related aspects of CDR. At this point, we should note that our discussion has been constrained to economic considerations and should be further expanded to consider other key collateral effects such as local pollution damage to soil, water or air, changes in albedo and ecosystems, biodiversity loss[28] or food security issues. CDR is complex and its multiple facets should be considered and integrated into regulatory frameworks. Along the same lines, some CDR alternatives are recognised to provide ancillary benefits beyond carbon sequestration. Afforestation and reforestation have the potential to offer ecosystem services such as protection of watersheds from erosion, nutrient retention, good water quality, wildlife habitat and diversity, recreational opportunities and additional social benefits.[12,13] Enhanced weathering expedites natural processes that neutralise CO_2 acidity.[14] BECCS can deliver negative-carbon energy services. Furthermore, employment opportunities or changes in socio-economic conditions of local communities remain largely overlooked,[15] but initial analysis implies that positive outcomes are possible.[16] This latter effect might be particularly relevant in developing countries where communities surrounding and benefiting from unregistered lands might be affected by CDR projects that involve land use (*e.g.*, BECCS or afforestation/reforestation).[17]

A second difficulty arises from the necessity of managing mitigation and CDR concurrently. CDR should not be viewed as an alternative to deep cuts in emissions in the near term,[18] as unabated emissions would have numerous associated side effects, and would increase the risk of failing to meet the climate targets.[19] In turn, there are additional accounting challenges owing to spatial and temporal issues. For instance, carbon removals from DACCS should in principle be straightforward to measure directly, yet the same does not hold for other CDR options such as biochar or enhanced weathering, which poses additional challenges for monitoring, reporting, and verifying carbon removal.[20] One key aspect here is the integration of CDR into existing regulatory mechanisms for carbon emissions, such as the ongoing EU Emission Trading System (EU ETS). Discussions about the eligibility of carbon capture and storage (CCS) under an international regulatory framework were initiated under the Clean Development Mechanism,[20] but have yet to crystallise. The advent of the Circular Economy Action Plan emerges as a promising stepping stone in the pathway towards a comprehensive policy, since robust and transparent carbon accounting for monitoring and verifying CDR has been flagged as one of the key actions to come in 2023.[21]

The third challenge refers to the international dimension of such policies. International law requires flexible frameworks of governance and regulation to address the use of CDR alternatives successfully. There are general principles, such as the duty to avoid deleterious transboundary effects, recognised in many treaty instruments (the UNFCCC, the United Nations Convention on the Law of the Sea or the Convention on Biological Diversity, among others). Nevertheless, international cooperation will be necessary to agree on specific regulatory aspects, including risk management, direction, ownership and control of CDR supply chains.[20] From a temporal viewpoint, some CDR strategies such as BECCS suffer from an initial carbon debt. This debt refers to the land-use change that may delay carbon removal, preventing the deployment of selected CDR options unless appropriate carbon accounting is in place.[9] Similarly, the technical and socio-political challenges of building and governing an infrastructure for the transportation of hundreds of Gt CO_2 over the century should not be overlooked.

Finally, we identified the lack of comprehensive and detailed frameworks to better inform policy-makers as a key barrier delaying effective policies. Today, CDR assessments are, in many cases, incomplete and highly fragmented. There is a wide range of models, which are often based on different assumptions and focus on a subset of CDR options. This lack of harmonisation leads to results that are hard to reconcile and compare in practice, preventing any agreement on the preferred roadmap for the deployment of CDR strategies.

12.6 Conclusions

This chapter provided an overview of the main distributive issues associated with the multinational deployment of multiple CDR options, emphasising the differences between burdens and efforts and outlining the potential mechanisms that aim to reconcile them. To this end, we discussed the implications of meeting a set of multicriteria quotas for CDR by implementing different technological (CDR alternatives) and political (*isolationist vs cooperative*) strategies.

First and foremost, figures suggest that no single CDR option will enable meeting our climate goals alone due to the limited capacity of natural endowments.[6,22] Indeed, natural climate solutions such as afforestation/reforestation might be complemented by engineering solutions such as BECCS and DACCS, shouldering additional carbon removals.

A similar conclusion can also be drawn for countries: we need all players to contribute to the best of their abilities to meet our climate goals. In this context, the EU offers a promising environment for collaboration and the development of international policy frameworks that could serve as models for more comprehensive efforts.

We have identified at least four critical aspects in developing regulatory frameworks for the certification of CDR.[21]

First, CDR is a multi-faceted problem with links to the economy but also to society and the environment, all of which need to be considered in any realistic policy. These span from co-benefits (*i.e.*, BECCS generation of energy) to potential collateral damage on ecosystems and the population (*e.g.*, BECCS harm on biodiversity). Establishing sustainability standards to regulate the deployment of CDR alternatives will be crucial to preserve the planet, avoid social conflicts and engage countries in a collective action perceived as fair. This will require cooperation among different panels of experts (*e.g.*, UNFCCC, Convention on Biological Diversity) to effectively address the problem's many dimensions.

The second challenge is to combine emissions reduction with carbon removal effectively under a common framework. Independent yet concurrent, quotas for conventional mitigation and CDR could be established to reduce the risk of delaying climate action forever.[2,23] In practice, the lower cost of conventional mitigation should be considered by regulations aiming to promote CDR. Some authors have reported that meeting our climate goals might require carbon prices in the range \$10–50 per t CO_2 in the short term but rising to \$500–2500 per t CO_2 by the end of the century.[24,25] Hence, encouraging early mitigation actions followed by CDR for later in the century could be a way forward. Continuing with and strengthening conventional mitigation will enhance the likeliness of meeting our goals, while reducing the dependence on more costly options. If we fail to implement these early actions on time, we will need to deploy CDR at a great rate, an issue not to be taken lightly. CDR alternatives such as BECCS or DACCS require access to CO_2 transport and storage infrastructure, while enhanced weathering and biochar need intense mining, agriculture and transportation infrastructures[26] at scales dictated by the extent to which actions are delayed. Reforestation is not exempt of risks either, since the carbon stored in the forest biomes could be eventually released owing to, *e.g.*, fire or land clearance for agriculture.

Policy-makers will also need to deal with the international aspects of these frameworks. Splitting CDR alternatives and related aids and incentives between domestic and trans-boundary,[27] might facilitate the early implementation of domestic policies laying the foundations for future, more geographically-comprehensive regulations. Alternatively, technology transfers might help to harmonise domestic regulatory frameworks by decoupling efforts and burdens spatially and policy-wise. Dedicated funds to this end, such as those under the Global Environment Facility, have so far focused only on conventional mitigation while overlooking projects for CDR or CCS.[15]

Finally, we raise concerns on the lack of harmonisation across scientific contributions, which ultimately fragments the information available for policymakers to base their decisions on. We call for greater cooperation and coordination of scientific efforts to share high-quality data and agree on standards to perform sound analyses and assessments. Ideally, these should capture the complexity of the CDR problem on a global scale, while providing sufficient detail at the local level. In this context, mathematical programming emerges as a promising analytical tool to model region-specific interactions and include ancillary side-effects on the environment and the society. The whole portfolio of CDR strategies must be well represented in such models to exploit regional opportunities without compromising sustainable development.

References

1. UNFCCC, *Adoption of the Paris Agreement*, UNFCCC, Bonn, Germany, 2015.
2. D. P. McLaren, D. P. Tyfield, R. Willis, B. Szerszynski and N. O. Markusson, *Front. Clim.*, 2019, **1**, 4.
3. O. Geden, G. P. Peters and V. Scott, *Clim. policy*, 2019, **19**, 487–494.
4. O. Geden and F. Schenuit, *Unconventional mitigation: Carbon Dioxide removal as a new approach in EU climate policy*, Stiftung Wissenschaft und Politik – SWP – Deutsches Institut für Internationale Politik und Sicherheit, Berlin, 2020.
5. C. L. Fyson, S. Baur, M. Gidden and C.-F. Schleussner, *Nat. Clim. Change*, 2020, **10**, 836–841.
6. C. Pozo, Á. Galán-Martín, D. M. Reiner, N. Mac Dowell and G. Guillén-Gosálbez, *Nat. Clim. Change*, 2020, **10**, 640–646.
7. European Commission, A European Green Deal: Striving to be the first climate-neutral continent.
8. S. Fuss, W. F. Lamb, M. W. Callaghan, J. Hilaire, F. Creutzig, T. Amann, T. Beringer, W. de Oliveira Garcia, J. Hartmann and T. Khanna, *Environ. Res. Lett.*, 2018, **13**, 63002.
9. M. Fajardy and N. Mac Dowell, *Energy Environ. Sci.*, 2017, **10**, 1389–1426.
10. K. L. Anthonsen, P. Aagaard, P. E. S. Bergmo, M. Erlström, J. I. Fareide, S. R. Gislason, G. M. Mortensen and S. Ó. Snæbjörnsdottir, *Energy Procedia*, 2013, **37**, 5080–5092.
11. A. Galán-Martín, C. Pozo, A. Azapagic, I. E. Grossmann, N. Mac Dowell and G. Guillén-Gosálbez, *Energy Environ. Sci.*, 2018, **11**, 572–581.
12. R. Watson, *Living Beyond our Means: Natural Assets and Human Well-being: Statement from the Board*, 2005.
13. A. J. Plantinga and J. Wu, *Land Econ.*, 2003, **79**, 74–85.
14. H. S. Kheshgi, *Energy*, 1995, **20**, 915–922.
15. N. Mac Dowell, M. Fajardy, H. Daggash and P. Patrizio, *Front. Clim.*, 2019, **1**, 6.
16. P. Patrizio, S. Leduc, F. Kraxner, S. Fuss, G. Kindermann, S. Mesfun, K. Spokas, A. Mendoza, N. Mac Dowell, E. Wetterlund, J. Lundgren, E. Dotzauer, P. Yowargana and M. Obersteiner, *Joule*, 2018, **2**(12), 2633–2648.
17. M. Greve, B. Reyers, A. M. Lykke and J.-C. Svenning, *Nat. Commun.*, 2013, **4**, 1–9.
18. S. Fuss, J. G. Canadell, G. P. Peters, M. Tavoni, R. M. Andrew, P. Ciais, R. B. Jackson, C. D. Jones, F. Kraxner and N. Nakicenovic, *Nat. Clim. Change*, 2014, **4**, 850–853.
19. S. Fuss, C. D. Jones, F. Kraxner, G. P. Peters, P. Smith, M. Tavoni, D. P. van Vuuren, J. G. Canadell, R. B. Jackson and J. Milne, *Environ. Res. Lett.*, 2016, **11**, 115007.
20. J. G. Shepherd, *Geoengineering the Climate: Science, Governance and Uncertainty*, Royal Society, 2009.

21. European Commission, Communication from the Comissions to the European Parliament, the Council, the European and Social Committee of the Regions. A new Circular Economy Action Plan. For a cleaner more Competitive Europe., Brussels, 2020.
22. J. C. Minx, W. F. Lamb, M. W. Callaghan, S. Fuss, J. Hilaire, F. Creutzig, T. Amann, T. Beringer, W. de Oliveira Garcia, J. Hartmann, T. Khanna, D. Lenzi, G. Luderer, G. F. Nemet, J. Rogelj, P. Smith, J. L. Vicente Vicente, J. Wilcox and M. del Mar Zamora Dominguez, *Environ. Res. Lett.*, 2018, **13**, 063001.
23. G. P. Peters and O. Geden, *Nat. Clim. Change*, 2017, 7, 619–621.
24. E. Kriegler, O. Edenhofer, L. Reuster, G. Luderer and D. Klein, *Clim. Change*, 2013, **118**, 45–57.
25. J. Edmonds, P. Luckow, K. Calvin, M. Wise, J. Dooley, P. Kyle, S. H. Kim, P. Patel and L. Clarke, *Clim. Change*, 2013, **118**, 29–43.
26. G. F. Nemet, M. W. Callaghan, F. Creutzig, S. Fuss, J. Hartmann, J. Hilaire, W. F. Lamb, J. C. Minx, S. Rogers and P. Smith, *Environ. Res. Lett.*, 2018, **13**, 63003.
27. O. Boucher, P. M. Forster, N. Gruber, M. Ha-Duong, M. G. Lawrence, T. M. Lenton, A. Maas and N. E. Vaughan, *Wiley Interdiscip. Rev. Clim. Change*, 2014, **5**, 23–35.
28. S. Cobo, Á. Galán-Martín, V. Tulus, V. M. A. J. Huijbregts and G. Guillén-Gosálbez, *Nat. Commun.*, 2022, **13**, 2535.

CHAPTER 13

Political and Socio-economic Challenges of Greenhouse Gas Removal Technologies

ZEYNEP CLULOW* AND DAVID M. REINER

Energy Policy Research Group, Judge Business School, University of Cambridge, Cambridge, UK
*Email: z.clulow@jbs.cam.ac.uk

While large-scale deployment of greenhouse gas removal (GGR) technologies is widely regarded as a prerequisite for meeting ambitious climate targets, upscaling efforts are obstructed by several socio-economic and political challenges. To a large extent, these issues are specific to each GGR approach, with the more mature options such as afforestation/reforestation (AR) and soil carbon sequestration (SCS) facing challenges relating to local socio-economic effects, resource scarcities and accounting issues and the more nascent options such as direct air capture (DAC), enhanced weathering (EW) and ocean fertilisation (OF) also impeded by technological problems and concerns over sustainability. Stakeholder engagement is not uniform across the different GGR options and perspectives vary widely across national governments, private interests, NGOs and publics. While land-based options are an obvious focus for agriculture, energy, utilities, transport and aviation have tended to concentrate on CCS-based options for reducing sectoral emissions as well as land-based options for delivering offsets. Ultimately, the ability to upscale GGR depends on the emergence of an effective governance system that can coordinate the interests of the diverse stakeholders involved,

Energy and Environment Series No. 31
Greenhouse Gas Removal Technologies
Edited by Mai Bui and Niall Mac Dowell

Published by the Royal Society of Chemistry, www.rsc.org

which rests on the ability to reach an agreement on a number of important sociocultural and political questions that are raised by the regulation of GGR.

13.1 Introduction

In 2015, the international community adopted the Paris Agreement, which set out to keep global warming well below 2 °C and 'to pursue efforts' to limit the temperature rise to 1.5 °C. Yet even if all countries were to honour their mitigation pledges as set out in their Nationally Determined Contributions (NDCs) to the international climate regime, median warming is expected to be between 2.6 to 3.1 degrees.[1] The inadequacy of mitigation efforts to date underlies the rationale for greenhouse gas removal (GGR) and negative emission technologies (NETs). Meeting the Paris target implies either: (i) mitigation significantly increases over the next decade, or (ii) global greenhouse gas (GHG) emissions become net negative in the second half of the century. Given the formidable obstacles that have delayed mitigation so far, it is unlikely that the first scenario will come to pass. In contrast, the benefit of the second scenario – and potential appeal of GGR – is that it allows more time for countries (and sectors) that face high mitigation costs to find ways of incentivising decarbonisation by removing CO_2 from the atmosphere to compensate for near-term emissions.

Most scenarios in the Intergovernmental Panel on Climate Change (IPCC) Fifth Assessment Report do not manage to keep global warming within 2 °C without large-scale GGR,[2] while a 1.5 °C target seems virtually impossible without them.[3] Interest in GGR technologies is growing around the world as an increasing number of countries and private stakeholders are committing to achieving carbon neutrality and net-zero as, for example, demonstrated by the 73 countries that signed the 2019 Climate Ambition Alliance.

Yet despite the strong rationale for GGR, deployment to date has been rather limited. To get an idea of how far we are lagging, it is helpful to consider the gap between actual negative emission levels and those projected as necessary to be Paris-compliant. Until recently, the GGR options that have received the most attention in integrated assessment models (IAMs) were afforestation and reforestation (AR) – the planting of trees that absorb CO_2 in previously deforested lands or growth of new forest lands where they previously did not exist – and bioenergy with carbon capture and storage (BECCS). To be compatible with 2 degrees, most IAMs recommend that AR needs to be deployed at an average scale of 5 Gt CO_2 and BECCS at 12 Gt CO_2 per year by mid-century through to 2100. At present, AR removes less than 1 Gt CO_2 from the atmosphere, and BECCS is close to zero annual negative emissions, *i.e.*, zero CO_2 removal.[4]

Of course, GGR encompass a wide range of options – any activity that involves anthropogenic efforts to permanently remove GHG, namely CO_2, from the atmosphere. Hence, some GGR options are also referred to as carbon dioxide removal (CDR) technologies – the main ones being AR, BECCS, direct air capture (DAC), enhanced weathering (EW), soil carbon sequestration (SCS) and ocean fertilisation (OF). These options vary widely in terms of technological readiness and deployment practices, but none of them are close to being deployed at climatically-relevant scales.

While AR has been practiced for centuries and often at a large scale, there is a need for substantial upscaling, which has spawned the emergence of several international initiatives to promote AR globally such as the Bonn Challenge and the New York Declaration on Forests. By contrast, while there are around 40 carbon capture and storage (CCS) projects at various stages of development globally, most are aimed at capturing CO_2 from fossil fuels and only five of these involve biomass as the energy source and can therefore be classed as BECCS.[5]

Partly in response to growing concerns over the resource limits to upscaling AR and BECCS, there has recently been a rise in interest in alternative GGR approaches. One such example is DAC, which involves direct removal of CO_2 from ambient air by chemical reaction and then storage of CO_2. The CO_2 can then be stored underground in saline formations or depleted oil and gas reservoirs similar to CCS or *via* other methods such as the Carbfix project in Iceland using basalt. Though still largely an experimental option, DAC is appealing because, theoretically at least, there is unlimited potential for upscaling at least on the capture if not the storage side. Currently, there are only a handful of prototypes in existence – each using a different technology to target different markets.[6] For example, Climeworks, a commercial DAC plant in Switzerland, supplies atmospheric CO_2 for use in mineral water production while Climate Engineering based in North America uses DAC to produce low-carbon fuel for aviation.

EW – the grinding of rocks that naturally absorb CO_2 to increase their surface area and absorption and spreading over land or oceans – is another option that avoids the resource challenges associated with BECCS and AR. It is also a nascent technology, with ongoing experiments on the weathering process and storage medium.[7,8]

SCS – enhancing the sequestration of carbon in the soil by increasing inputs or reducing outputs – provides another avenue for negative emissions. While interest in soils is mature compared to other GGR options, most existing soil management is practiced at the individual farm level.[9,10]

OF involves the adding of iron or macronutrients (such as urea or phosphate) to the ocean to stimulate ocean phytoplankton to absorb more CO_2 through photosynthesis and then potentially sink to the deep ocean and sequester after they die. OF comprises another set of experimental options, with existing implementations such as the GEOTRACES survey starting to approach relevant scales.[11,12] Notably, OF is often categorised as geoengineering and is opposed outright but many NGOs.[13]

A key research objective, therefore, is to explain why the rollout of GGR technologies has been slower than expected. All options involve political and socio-economic challenges to differing degrees (see Table 13.1). For more

Table 13.1 Current and potential deployment scales, major projects and core challenges associated with major NETs.

NET	Current Scale (Potential)[a]	Major initiatives and projects[b]	Core challenges
AR	<1 Gt (0.5–3.6 Gt)	At scale in the tropics (*e.g.*, 0.61 Gt in the Congo river basin, 0.1 Gt in the Amazon) Initiatives: Bonn Challenge, AFR100 and the New York Declaration on Forests	Resource needs and implications for biodiversity and food security, accounting, disruption to host communities
BECCS	1.5 Mt (0.5–5 Gt)	Large-scale: Illinois Industrial CCS (US) (1 Mt) Small-scale: Kansas Arkalon (US) (290 t) Bonanza CCS (US) (100 t) Husky Energy (Canada) (90 t) Drax (UK) (330 t)	Resource needs for biomass supply, implications for biodiversity and food security, storage risks, integration with existing infrastructures
DAC	9000 t (0.5–5 Gt, but some authors predict as high as 30 Gt)	Various prototypes targeting different markets *e.g.*, Climeworks (Switzerland) (900 t), Carbon Engineering (North America) (3500 t)	High energy needs and operating costs, integration with zero-carbon energy source
EW	1.1 Gt (from natural oceanic weathering) (2–4 Gt)	Mainly laboratory-scale experiments; testing materials and interaction with soil and oceans	High energy needs and operating costs, availability of sustainable materials, side-effects
SCS	~3 Gt stored in global soils (2–8.6 Gt additional storage)	Variable scale and permanence (*e.g.*, long-term experiments at Rothamsted Research UK capture 3.4 t). Initiatives: 4 per 1000 Initiative	Accounting, incentivising local stakeholders (*e.g.*, farmers)
OF	Extremely limited	Only small-scale experimental studies *e.g.*, GEOTRACES survey <100 t	Unknown effectiveness, sustainability

[a]Potential scales are from Fuss *et al.* (2018).[23]
[b]Data from: Global Climate Capture and Storage Institute, Bioenergy and Carbon Capture and Storage (2019),[5] Harris *et al.*, Global maps of twenty-first century forest carbon fluxes (2021),[219] Beuttler et al., The role of direct air capture in mitigation of anthropogenic greenhouse gas emissions (2019),[38] Poutlon *et al.*, Major limitations to achieving '4 per 1000' increases in soil organic carbon stock in temperature regions: Evidence from long-term experiments at Rothamsted Research, United Kingdom (2017),[220] Boyd and Bressac, Developing a test-bed for robust research governance of geoengineering: the contribution of ocean iron biogeochemistry (2016),[12] and Fuss *et al.*, Negative emissions – Part 2: Costs, potentials and side-effects (2018).[23]

nascent GGR approaches such as DAC, EW and OF, the challenges are also technological. DAC, for example, is energy-intensive and costly to operate as huge volumes of air need to be processed to capture large amounts of CO_2 from ambient air. Fundamentally, it needs to be powered by low-carbon energy if it is not to result in positive emissions.[14] Like DAC, EW requires large amounts of energy for grinding rocks and is therefore costly to run. There is also likely to be strong opposition to new rock mining, although recent research suggests that it may be possible to obviate the need for new mining by using excess industrial silicate materials from cement and steel production.[15] With OF, the major challenges relate to unknown risks to marine health. As noted, OF is perhaps the most controversial of the options considered here.[16] The NGO WWF, which tends to be more inclined to consider the role for GGR than many other NGOs, describes OF (as well as EW in the oceans) as 'not suitable for use at this time since, at present, the ecological and social uncertainties far outweigh the known benefits'.[17]

On the other end of the spectrum, GGR options such as AR and SCS have been practiced for centuries and are obstructed by more practical challenges such as resource needs, side-effects, and accounting issues. Major obstacles to AR include the disruptive effects on local communities, resource scarcities and the difficulty of monitoring carbon stored in biological sinks.[18,19] A major advantage of SCS is that it is not expected to put stress on land or water resources despite having a similar yield potential as AR.[20] Yet these options face similar accounting challenges and difficulties in incentivising farmers to implement soil techniques. The 4 Per 1000 Initiative, for example, seeks to address these issues by supporting local stakeholders to adopt agroecological practices that increase the organic carbon content of global soils by 0.4% per year and promote AR.[21]

There are important trade-offs involved with all options – both among themselves as GGR technologies to compete for limited financial, geophysical and infrastructural resources – and more broadly as large-scale deployment stands to compromise certain aspects of sustainable development and, counterintuitively, mitigation.[22] From a sustainability perspective, OF is particularly problematic because of its unknown side-effects on ocean acidity, making it unlikely to become the GGR of choice given the existence of other options that are associated with fewer side-effects.[23,24] Some options involve more specific trade-offs as they are likely to compete with other priorities for the same resources. For example, large-scale BECCS and AR are expected to put major pressure on land, water and fertiliser resources, which would be needed in significant volumes to feed biomass.[25,26] These resource scarcities are also expected to lead to higher food prices globally.[27] The high energy needs of DAC and EW raise concerns over their compatibility with existing energy infrastructures and energy security.[28] These trade-offs are becoming more pertinent as it emerges that a diverse GGR portfolio is likely to be more effective and less risky than relying on a limited number of technologies.[29,30]

GGR technologies also diverge in terms of how they are perceived by different stakeholders. For example, while environmentalists are generally sceptical of BECCS, DAC, EW and especially OF, they are more supportive of AR and SCS, although they are unlikely to support even these latter options if they are deployed at scale and threaten other priorities. GGR technologies also have different cost structures; most options involve significant start-up investments, making them more economic to deploy at scale whereas traditional options such as AR and SCS are initially relatively inexpensive, but face rising costs as resources become scarce at scale.

There are also some broader challenges which apply to all GGR options, to varying degrees. Compared to other mitigation options such as renewables and nuclear energy, most GGR technologies are unfamiliar to national publics and policymakers, with many policymakers and potential stakeholders only becoming interested following the adoption of the Paris Agreement in 2015. Huge uncertainties relating to efficiency, side-effects, carbon prices, stakeholder views, public engagement and governance challenges deter potential stakeholders from funding what are typically substantial investments, however, such funding is needed to make GGR both commercially and climatically viable.[31]

13.2 Economic Challenges

Since GGR options are generally much more expensive than mitigation options, incentives are crucial for attracting investment in R&D, demonstration and, ultimately, large-scale, sustainable deployment.[32] Yet there are several disincentives to deter potential stakeholders. Private actors are generally reluctant to invest because third parties are likely to benefit from the results of successful innovation,[33] as well as the expectation that only a relatively small share of innovations will be successful.[34] The most obvious disincentive, perhaps, is the sheer scale of investment needed to upscale most GGR technologies. For example, 5–10 full size demonstrations of BECCS would require investment in the range of $5–10 billion.[35] Understandably, therefore, the private sector is unlikely to provide the required investment without public intervention.[36] Yet governments are also hesitant to become involved because, in addition to the preceding disincentives, they do not want to appear to be picking winners (and losers), especially since investments are expected to result in failed projects rather than successes.

13.2.1 Cost Structures

In some regard, the disincentives are specific to each technology as GGR approaches vary widely in terms of their process components, technological maturity, resource needs, infrastructural requirements, operating costs, side-effects, and co-benefits. Most options entail a multi-stage process to remove CO_2 emissions from the atmosphere. In general, the more complex the process, the greater the number of moving parts such as installation requirements, resource inputs, transportation lines, storage, monitoring and liability, and associated costs, and, therefore, the greater the potential sources of uncertainty. From a process perspective, BECCS is arguably the most complicated CDR option as it involves multiple interlinked steps relating to biomass production, processing into fuel or electric power and CO_2 storage, with potential costs accruing from investment in the power plant or biofuel production facility, CCS infrastructure, general operating and maintenance costs, costs of biomass production and transport of biomass to the plant as well as transport of CO_2 to a sequestration site.[37] DAC involves similar types of costs, in addition to high energy inputs, but has the benefit that it can, in theory, be deployed proximate to storage facilities and renewable energy sites, which would significantly reduce transport costs.[38] BECCS will, of course, also contribute to the energy system by generating electricity (or fuels) whereas the other options serve only to provide carbon dioxide removal services although some, such as DAC, may explicitly be paired with energy production such as plans to link with aviation fuels.[39]

While involving fewer components, the major costs associated with EW relate to the technology chosen for rock grinding, material transport and rock source, while OF costs centre around nutrient production and its delivery to the application area.[40] GGR options that involve the fewest steps such as AR and SCS are associated with fewer, but different types of costs, which are generally better understood due to the existence of past experience. AR costs, for example, mainly reflect land prices, yield rates and accounting, while SCS costs are also related to labour prices.[41]

In general, technological maturity – particularly entering the deployment phase of the innovation cycle – tends to reduce cost uncertainty by demonstrating how cost trajectories behave in practice.[42] Indeed, the predicted costs of GGR technologies that are ahead in deployment such as AR, SCS and biochar tend to be lower with a narrower uncertainty range (from 5 to 120 US\$ per t CO_2) compared to more experimental options such as DAC and EW (from 50 to >300 US\$ per t CO_2, with DAC reaching ~1000 US\$ per t CO_2).[43–45]

Infrastructure needs (*e.g.*, for installing plants, transportation lines, integration with existing systems and mining) also have a bearing on the cost curves and optimal scales of deployment. Options that require high initial investments tend to have declining marginal costs and economies of scale, favouring large-scale deployment. DAC is probably the clearest example of this cost structure as its autonomy from geophysical inputs means that the potential scale for capture (if not storage) is essentially unlimited, assuming, of course, that its high energy needs can be met with an appropriately low carbon source.[46] BECCS, EW and OF also have high capital needs and are initially associated with economies of scale, however, as we discuss below, marginal costs have the potential to rise after intermediate levels of deployment as resource scarcities begin to inhibit upscaling.[47,48] By contrast, GGR options with lower capital needs such as AR and SCS are initially more economic to deploy at small-to-intermediate scales, but are expected to face increasing costs as upscaling beyond this point encounters saturation limits, resource scarcities and rising carbon prices.[49] More broadly, it is also worth noting that most cost forecasts are based on the characteristics of individual GGR options and do not take into account the interaction effects of deploying multiple technologies simultaneously such as, for example, competing demand for limited resources and trade-offs with other priorities such as food and energy security.[50]

The costs of individual GGR technologies also vary dramatically across countries and regions due to geophysical characteristics as well as socio-political and institutional differences, as discussed below. Fundamentally, GGR options have different natural resource needs – the main ones being land (BECCS, AR and, to a less extent, SCS and biochar) and water (BECCS and AR) for feeding natural carbon removal, storage sites including deep saline formations, depleted hydrocarbon fields and coal fields (BECCS and DAC), minerals suitable for weathering (EW), and coastal and ocean sites for OF. In general, GGR options are expected to cost less in contexts where geophysical (as well as social, legal, and political) conditions are conducive to deployment.[51,52] If we assume for a moment that GGR technologies are to be implemented from start to finish within a single national jurisdiction, then geophysical conditions appear to rule out certain options for certain countries. For example, DAC and BECCS are essentially off the table for countries that do not possess adequate geological storage for captured CO_2, whereas BECCS is not viable for countries that lack domestic biomass supply, and so forth. In such a scenario devoid of international cooperation, individual countries would only have the option of resorting to more expensive GGR options due to geophysical constraints.[53] For example, inadequate indigenous biomass supply could drive the UK to deploy DAC instead of BECCS, which would be significantly more energy-intensive and costly than importing biomass for BECCS to deliver the same amount of negative emissions.[54] Yet, as we discuss below, although international cooperation has the potential to deliver an optimal global GGR portfolio, it raises several obstacles for potential stakeholders such as, for example, the need for reliable transboundary transport lines, monitoring capacity and enforceable long-term agreements.

13.2.2 Compatibility with Other Priorities

To the extent that upscaling is successful, GGR technologies are likely to compete with other sectors for the same resources. Until recently, the main focus of these concerns has revolved around BECCS and AR, which, if deployed at scale, would jointly require 2–4 times the amount of land that is currently identified as abandoned or marginal.[55] Such changes in land use would significantly limit the amount of available land for food or other bioenergy production and put upward pressures on global food prices. At scale, BECCS and AR would also require around 5% of global freshwater, which is problematic in the context of rising population trends.[56] In contrast, while DAC and EW do not require land resources, they are hugely energy intensive. The energy required for DAC at scale approaches the same order of magnitude as current global natural gas demand.[57] It is easy to imagine how such options raise concerns over energy security, especially in developing countries where energy demand is rising, and energy transmission systems are often fragile.[58]

There are also concerns about the compatibility of GGR technologies with broader sustainability goals, especially those relating to socio-economic development, human, environmental and ecosystem health, the availability of clean water and reducing inequity.[59] Options involving CCS (BECCS and DAC) give rise to concerns over the risks associated with storage leakage.[60] AR projects under the CDM and initiatives to prevent deforestation within the REDD+ framework have been found to be particularly destructive for small farmers who have often lost their livelihoods due to foregone deforestation.[61] Since SCS has relatively low resource needs compared to other GGR options, the major side-effects of this option relate to changes in soil quality and potential inequalities stemming from changes in land use.[62] There are also concerns over the unknown consequences of OF and EW on ecosystems and marine and soil properties, as well as the well-known adverse effects of surface mining.[63] In some sense, OF is a step behind other GGR approaches as the absence of clear co-benefits and relatively long list of potential incompatibilities with sustainability make it difficult to justify the investment given the existence of other options with co-benefits and fewer side-effects.[64] Given this manifest complexity, it is imperative to ensure that GGR approaches are delivered in a sustainable manner, and that unintended outcomes are avoided.

13.2.3 Incentives

It is widely expected that the co-benefits associated with GGR will play a critical role in driving deployment by creating additional incentives for stakeholders. BECCS, for example, generates carbon-free electricity as a by-product of negative emissions, which could potentially provide an additional revenue stream alongside other beneficial services such as waste-processing.[65] Of course, not all co-benefits are financial or can be easily monetised. For example, when practiced on marginal land, AR could provide local benefits such as (mainly seasonal) livelihoods and environmental goods. Similarly, in certain configurations, EW and SCS could improve nutrient quality in the soil and increase land productivity.[66] Furthermore, given that some GGR approaches have high resource needs (most notably BECCS, AR and land, DAC and EW energy), the ability to avoid the resource pressures associated with other GGR options is increasingly becoming regarded as a co-benefit in its own right.

Stringent demand reduction is needed for high-carbon industries (aviation, heavy-duty transport, agriculture, energy and energy-intensive industries) to reduce sectoral emissions.[67] Since this is likely to be opposed both by consumers and the sector-specific interest groups themselves, the potential for GGR to offset emissions from these industries is an important incentive for private stakeholders. A number of national governments have created voluntary carbon markets to boost demand for local offsets such as, for example, the UK's Woodland Carbon Code, France's Voluntary Carbon Land Certification Initiative and Colombia's Voluntary Carbon Market.[68] GGR options that involve CCS (BECCS and DAC) are particularly attractive for their potential to draw revenues through carbon capture and utilisation (CCU), that is, by utilising the captured CO_2 in a range of commercially viable end-use industries such as, for example, carbonated drinks, synthetic fuels, enhanced oil recovery (EOR) and enhanced gas recovery (EGR). The latter option has proven popular as a revenue stream in early CCS, with around 30 operational projects in the US, Canada, China and UAE,[69] all of which have an abundance of natural resources and strong fossil fuel industries. This emphasises the importance of conducting a thorough LCA on any GGR project to ensure that it does, in fact, result in the net removal of CO_2 from the atmosphere.

Considering that these countries are among the most reluctant in the global climate regime, their apparent enthusiasm for EOR is more likely a reflection of the potential for CCU to improve the efficiency of fossil fuel extraction and provide a pathway for fossil fuel activity to continue rather than reflecting a concern for mitigation.[70] Although CCU achieves an important feat by creating what are apparently effective incentives for CCS, as well as providing early opportunities for learning, their long-term feasibility is limited as the utilisation of captured CO_2 is associated with increased fossil fuel supply and net-positive emissions.[71] Consequently, CCU only has the potential to make a limited contribution (EOR 4–8% and chemical conversion 1%) to the mitigation challenge.[72]

13.3 Stakeholder Views

While most stakeholders and laypeople are unfamiliar with GGR and NETs, interest among governments and potential stakeholders is growing as it becomes inevitable that our ability to meet ambitious climate targets largely depends on rapid upscaling. Engagement is not uniform across the different GGR technologies and perspectives vary widely across governments, private interests, NGOs and publics. Most projects have been driven by public–private sector collaborations, with national governments typically assuming a leadership role and commercial interests being the more reluctant party.

Awareness about nascent options such as DAC, EW and OF is largely limited to scientists, engineers and policymakers who are directly involved in the development of the technology, and is almost non-existent among the general public. Compared to other GGR options, the state has generally assumed a more limited – primarily financial – role in developing these technologies, although various governments in the Northern hemisphere have recently become more publicly interested in response to emerging evidence that these options may be deployable at substantially lower costs than previously thought. For example, in mid-2020, the UK government announced a competition worth £70 million to fund innovation on DAC alongside other GGR technologies[73] and as of 2019, the US federal government has invested $11 million in DAC, $44 million in OF and $25 in EW.[74] Several national governments have recently introduced schemes to incentivise CCS; notably the 45Q federal tax credit in the US as well as state funding for CCS retrofits on coal plants in India and China, which might help support DAC and BECCS projects (although the current track record suggests that EOR projects are likely to be the main beneficiaries).

By contrast, stakeholders and publics are relatively familiar with 'tried and tested' options (AR and SCS), although generally using different terminologies such as 'reforestation and forest degradation' (REDD+) projects and 'land management'.[75] Governments have tended to play a more proactive role in overseeing land-based GGR approaches, albeit usually for purposes other than carbon storage. Most countries are party to at least some of the existing multilateral forest agreements such as the UN REDD+, WB Forest Carbon Partnership Facility and Forest Investment Programme. However, market conditions have generally limited private involvement in these GGR programmes, leaving the government to assume a long-term leadership role.

From a stakeholder perspective, BECCS can be regarded as a leading GGR technology since interest from both governments and private stakeholders has been more forthcoming in comparison to other options. Governments have tended to invest more in BECCS than any other GGR option,[76] with policymakers preferring this technology over nuclear energy or fossil CCS for energy transition.[77] BECCS has attracted interest from commercial stakeholders spanning diverse sectors such as, for example, energy, agriculture, food and transport, with which national governments have collaborated on relatively advanced-stage projects. Although public awareness of BECCS is generally low,[78] interested governments and private actors are in the process of launching public engagement programs.

13.3.1 Governments

Given their necessity for meeting ambitious mitigation targets as well as their potential to significantly offset emissions from sectors that are difficult to decarbonise, one could expect that governments would be strongly supportive of GGR. While there have been some promising recent developments such as the growing number of countries to have invested substantial public funds in R&D or introduced tax returns or subsidies for GGR technologies, the historical track record of government involvement is rather different. As Pozo *et al.* point out, none of the NDCs even mention GGR or NETs, while only a handful mention CCS.[79] So far, governments have generally been reluctant to initiate the substantive policy changes that are needed for rapid upscaling.

There are several reasons for governments to be cautious about GGR technologies, although these vary by option and country. GGR approaches inevitably require a balancing act between different actors (private stakeholders, national and local publics and NGOs), sectors, current and future priorities and other mitigation measures. It is sometimes asserted that policymakers have weak incentives for implementing effective climate measures because of their potential to spark opposition.[80] It is easy to imagine how GGR technologies might be affected by a similar dynamic as the benefits of climatically relevant scales of deployment will also require significant changes and time to materialise.

Research suggests that, compared to technical experts and policymakers, lay publics are more likely to prioritise potential risks over payoffs when evaluating new technologies.[81] Therefore, the long list of unknown risks and potential adverse side-effects associated with EW and OF is likely to fuel strong public opposition.[82] By contrast, for BECCS and DAC, the major risks relate to CCS activities – most notably storage, transport and integration with existing infrastructure.[83] On the other hand, public dissent about more familiar options such as AR and SCS is likely to be driven by actual experience with previous projects rather than novel physical risks associated with the negative emissions activity.[84,85]

Governments are also likely deterred by the expectation that most GGR technology investments are likely to result in (commercial) dead-ends in the sense that they will not generate enough revenue to justify the initial investment, although there is a strong case to be made that the ultimate payoffs of these investments relate to learning rather than revenue.[86] For example, it was initially intended that REDD would eventually be financed by private investment, but since carbon markets do not deliver sufficient funds to cover the opportunity cost of forest conservation, the state has ended up playing a more central role in overseeing and maintaining the land use changes (LUCs) that are required for ongoing AR projects.[87] The first ever BECCS demonstration, in Russell, Kansas, provides a different kind of example. Funded by the US Department of Energy in 2005, it later came to be regarded as a 'failure' because it indicated that the use of CO_2 injection only had limited potential to increase oil recovery from EOR.[88] In some sense, capital-intensive options that involve large start-up costs (BECCS, DAC, OF and EW) might appear riskier than more familiar options, which require incremental funding and can therefore be aborted without investing the complete budget for the project. The latter options also benefit from the existence of suitable infrastructure and state institutions for overseeing deployment.[89]

Government perceptions are also influenced by deployment experiences of GGR and component or analogous technologies.[90] Although it has been only a decade since the first BECCS plant began operating in Illinois in 2010, BECCS has been stymied by numerous cancellations, failures and aborted projects, which is likely to decrease the credibility of future projects.[91] Similarly, past AR projects under REDD+ have resulted in loss of livelihoods for small farmers, creating pockets of strong resistance in host communities and among the agricultural sector more broadly.[92] Accounting difficulties have often impeded the flow of benefits for land-based negative emissions to developing countries, decreasing the appeal of these options for governments and host communities.[93] Policymakers' expectations about public engagement with DAC and BECCS are likely to be influenced by past experience associated with projects involving storage risks such as CCS and nuclear power.[94,95] Similarly, the public in countries that generate large amounts of bioenergy are presumably more likely to be supportive of BECCS than countries that have restricted bioenergy generation due to concerns over food security or adverse side-effects such as land scarcity and environmental damage. Similarly, countries that already host CCS projects may look more positively on BECCS because they will have the infrastructure and technical knowledge in place and, therefore, possess a smoother potential pathway for bringing in the biomass component by switching from fossil-fuel feedstocks to biomass.[96,97]

More broadly, because GGR approaches touch on several issues of national importance, they potentially fall under the remit of multiple government departments, such as, for example, the Ministries of Environment, Food, Rural Affairs, Economy, Transport, Agriculture, Business, Energy and Industry. Given the relatively recent entry of GGR technologies to the policy agenda, many governments do not yet possess the institutional organs that are needed to coordinate the involved departments. Therefore, policy initiatives may be obstructed by certain branches of government that are more sceptical of GGR. For example, because of the close links of options involving CCS with the fossil fuel industry, responsibility for BECCS and DAC generally tends to be with energy ministries. However, as there are usually cheaper ways of reducing emissions, energy ministries tend to prefer other methods, which frequently creates tensions with ministries that are more supportive of GGR (typically those overseeing the environment).[98]

13.3.2 Private Sector Views

In parallel with growing commitments by national governments to net-zero, commercial stakeholders, particularly in high-carbon sectors, are facing increasing pressure from national governments and consumers to reduce their own carbon footprints. As a result, an increasing number of private actors – especially multinational companies – have pledged to contribute to mitigation efforts, with many adopting their own net-zero targets. In 2020, the Coalition for Negative Emissions, consisting of 11 companies from climate engineering, agriculture and aviation wrote a letter calling on the UK government to adopt clear policies to support viable NETs projects over the next decade. While many governments, NGOs and publics would hold commercial interests responsible for reducing emissions within their own industry, stakeholders in high-carbon sectors demonstrate a keen interest in offsetting emissions through 'nature-based' GGR options such as AR and SCS. Moreover, because GGR approaches vary widely in terms of resource needs, mitigation potential, costs, side effects and technological readiness, stakeholders in different sectors of the economy often hold divergent attitudes towards the same technology.

We focus on three key sectors where relatively coherent stances towards the role of GGR technologies in reducing their carbon footprints have emerged, namely: land use, energy and transport, and aviation (see Table 13.2).

Table 13.2 Key NETs, incentives and challenges for deploying NETs in land use, energy and utility and transport and aviation.

Key NETs	Incentives/Co-benefits	Challenges
Land use		
SCS	Subsidies or rewards for SCS practices	Developing a support system for stakeholders to access training, machinery and inputs
	Policies to change demand for fertiliser/tillage practices	
	Remote technologies for monitoring soil carbon flows	
BECCS	Biomass subsidies	Sensitivity to volatile international markets, concerns over sustainability
	The position of BECCS in changing electricity markets	Adverse experience of biocrop boom
AR	Strong experience and knowledge of forestry	Risks to small and medium farmers
	Remote technologies for monitoring soil carbon flows	Adverse effects of previous AR projects under REDD + and CDM
	Higher returns offered by voluntary offset markets	Land rights
		Accounting problems delay benefits to host community
		Lack of competitiveness with other land uses *e.g.*, agriculture
Energy and utilities		
BECCS and DAC	CCU can provide an additional revenue stream, but must be balanced against positive emissions downstream	CCS increases costs
	CDR and CCS schemes in the energy sector *e.g.*, Capex and CfD/Opex support	
DAC	Location-independent sourcing of CO_2 can significantly reduce transport costs	High energy needs are likely to strain energy infrastructure, especially in developing countries
BECCS	Biomass electricity generation can provide an important revenue stream	High transport costs and infrastructure needs
	Choice of high-yield crops *e.g.*, sugarcane	
AR (offsets)	Ability to delay/avoid making substantial changes to own sector	Ethical and social acceptability of avoided sectoral mitigation
Transport and aviation		
AR (offsets)	Ability to delay/avoid making substantial changes to own sector	Ethical and social acceptability of avoided sectoral mitigation
BECCS and DAC	Ability to utilise captured CO_2 to produce low-carbon fuels to reduce sectoral emissions	Positive emissions downstream released from utilising synthetic fuels
		Competition for same resources to produce biofuel could increase costs for both sectors

13.3.2.1 Agriculture

With the exception of DAC and OF, all the major GGR technologies involve land-based components. Therefore, the agricultural sector and other land-based industries have the potential to play a central role in delivering negative emissions. For most countries to deploy large-scale BECCS, they either need to redeploy some of their agricultural land to biomass production or import biomass – although the latter simply outsources to the exporting country the switching of agricultural land to biomass.[99] Similarly, significant amounts of agricultural land would need to be diverted to forestation if AR is to be practiced at scale.[100] Agricultural land is also the main focus for carbon storage in SCS[101] (Smith *et al.* 2016), and, to a lesser extent, EW, which can also be deposited in water systems.

The agriculture sector is a priority for many governments and is generally well-supported through state subsidies, especially in food-exporting countries. Farmers usually have strong, well-entrenched lobbies and wield significant political influence, often obstructing policies deemed incompatible with sectoral interests. As a result, it is widely expected that farmers are more likely to engage with GGR approaches if they decide to do so on their own accord (with the enticement of incentives such as subsidies) rather than through top-down regulation.[102] A key challenge is ascertaining the factors that underlie farmers' attitudes towards GGR technologies and identifying potential strategies for eliciting support. Research suggests that farmers behave like economic actors when deciding whether to become involved in GGR projects.[103] Partly as a result of past experience of biofuel crop booms, they are often reluctant to commit to growing energy crops for BECCS as they are perceived to be vulnerable to international markets.[104] Furthermore, biomass production has been known to have severe adverse ecological and social impacts on host communities, creating additional grounds for caution. Yet recent research on farmer attitudes in the US and UK suggests that compared to other GGR approaches – particularly the more nascent options – farmers are more receptive to the prospect of biomass production due to their familiarity with the agricultural technology involved as well as expectations about growing market demand relative to other agricultural products such as dairy.[105] In the UK, for example, the National Farmers' Union has been supportive of putting the sector onto a path to net zero by 2040.[106]

SCS and EW, on the other hand, may appeal to farmers because of their potential to improve soil fertility and, therefore, increase returns from a variety of agricultural activities (including GGR options that rely on biomass production such as BECCS or AR).[107] Several major agricultural producers have already committed to reducing their carbon footprint by deploying SCS. For example, Danone and Indigo Agriculture have pledged to remove carbon *via* SCS and sustainable agriculture.[108] Various policies could be used to incentivise these options for farmers such as subsidies or low-cost loans to reduce the costs of purchasing new machinery and inputs. In particular, it has been suggested that the application of EW on agricultural land could reduce the need for industrial fertiliser, which would be appealing for nutrient-poor soils in the tropics.[109] Without significant state funding, however, small and medium farmers are unlikely to be able to access the necessary training, machinery and inputs for deploying these technologies, so the provision of adequate funds appears to be a critical facilitating factor.

Unlike other GGR technologies, farmers have extensive experience in land-use practices including managing forests and soils. As with biomass production, however, farmers' views are influenced by adverse effects of previous international forestation projects under the REDD+ and the CDM, which are infamously associated with loss of livelihoods. In developing countries, around 40% of land for forestation has been claimed from small farmers, who have generally not been compensated for foregone benefits such as crops, fuel, wood and livestock incomes.[110] Even for larger, wealthier farmers, the notable lack of competitive revenues for forestation renders AR less appealing compared to other land uses such as agriculture. More broadly, accounting for carbon flows in land-based GGR is notoriously difficult, which casts doubts for farmers over the likelihood that their involvement in deploying these GGR approaches will be remunerated. For some, the challenge of carbon accounting presents an opportunity for the agroindustry to make an important contribution as several companies that manufacture agricultural equipment have begun developing technologies for remotely sensing soil organic carbon stock to monitor land-based negative emissions.[111]

While international and state-led forestry projects are not usually associated with lucrative returns, AR has been able to attract higher revenues from offset purchases in voluntary carbon markets, which might be regarded as one avenue for overcoming some of the financial disincentives. Efforts to appeal to an increasingly green consumer base have prompted a growing number of private sector actors – from oil companies such as Shell to artists like Pearl Jam – to voluntarily pay above market prices for forestry projects to reduce their carbon footprint.[112] While the supply of AR-offsets has generally outpaced demand and consequently resulted in low carbon prices, the prioritisation of sustainability in the recovery from the Covid-19 pandemic appears to have increased demand for offsets from forestry, which is likely to render AR more appealing to farmers and other potential project developers going forward.[113]

13.3.2.2 Energy and Utilities

Most recent interest from energy and utility companies has focused on CCS-based options, which can be incorporated into the energy system – either by installing new, purpose-built plants, or through integration with existing facilities (*e.g.*, CCS retrofits to fossil fuel plants and DAC combined with renewable energy facilities). Electric utilities have tended to focus on renewables and other mitigation options after a brief foray into fossil CCS in the early 2000s,[114] whereas the oil and gas industry has maintained a more active involvement in CCS (primarily fossil) as well as DAC investments through their venture arms.[115]

From an energy sector perspective, involving CCS increases costs, but this can be mitigated in various ways depending on the characteristics of the chosen technology. A major advantage of BECCS is that biomass electricity can provide firm power generation in a system increasingly characterised by a high penetration of variable renewables and thus an additional revenue stream. Other BECCS approaches also generate multiple sources of value. For example, sugarcane processing, which involves converting primary energy into ethanol *via* fermentation, has received attention for its potential to reduce the cost of carbon capture by providing a pure CO_2 stream. Such projects are of particular interest to developing countries which already specialise in sugarcane processing, such as Brazil. Recent research suggests that the involvement of CCS does not have to result in significantly higher prices for end-users as sharing the costs between ethanol fuel and bioelectricity would result in small increases in market prices (2% for bioelectricity and 3.5% for ethanol).[116]

By contrast, while DAC is not associated with additional revenue streams, its ability potentially to offer location-independent sourcing of CO_2 has the potential to offset CCS costs by significantly reducing transport and energy-input costs if DAC facilities are located proximate to renewable energy sources.[117] Since BECCS involves a rather long transport chain (for biomass and CO_2), this cost-saving measure could increase the appeal of DAC over BECCS for some stakeholders. Energy firms that were to invest in BECCS or DAC could accrue additional revenues by selling a portion of their captured CO_2 for utilisation in a range of other industries such as synthetic fuels, horticulture, carbonated drinks and, most prominently, EOR/EGR. The largest BECCS plant, the Illinois Industrial, for example, utilises captured CO_2 in corn syrup production and food processing while the other four operational BECCS facilities supply CO_2 for use in EOR. Yet, as discussed above, the emissions released through the utilisation process as well as the use of CCS for supporting fossil energy generation counteract the negative emissions.

Energy sector views towards CCS also depend on national conditions. GGR technologies are likely to be more appealing if government initiatives are in place to promote certain technologies. For example, several major jurisdictions have regulations to encourage biomass production contingent on complying with sustainability targets (*e.g.*, EU Renewable Energy Directive), which could potentially incentivise BECCS. Furthermore, both BECCS and DAC are eligible for a growing number of government schemes that promote CCS and CDR. However, such policies are only likely to engage the energy sector if concerns relating to the impacts of resource changes on competing sectors are addressed in the eyes of the public and relevant stakeholders. For BECCS, the major hurdle is ensuring that diverting agricultural land to biomass production would not result in socially unacceptable impacts.[118] Another challenge is balancing the adverse effects of CCS on energy systems with other political priorities. In this regard, developing countries are likely to face significant hurdles as they often have fragile energy infrastructures that are already under pressure due to growing energy demand and the need to improve energy access. It is likely that energy security and cost would be prioritised over climate mitigation, with DAC appearing especially problematic due to its high energy needs.[119]

In addition to CCS-based options, various multinational energy firms have expressed an interest in purchasing offsets, mainly from AR projects, to reduce the carbon footprint of fossil energy. For example, in 2019, Shell announced its plans to reduce its net carbon footprint by 2–3% over the next three years by investing in reforestation projects in the Netherlands, Spain, Australia and Malaysia.[120] Similarly, the same year, BP invested $5 million in Finite Resources, a US forest carbon management company, to offset its emissions through reforestation projects in North America.[121]

13.3.2.3 Transport and Aviation

Road transport, shipping and aviation collectively account for around a quarter of global CO_2 emissions, often comprising a larger share of emissions in industrialised countries.[122] At present, only domestic flights count toward a country's emissions, therefore, light-duty vehicles, heavy goods vehicles, and shipping comprise the bulk of transport emissions, with aviation comprising a smaller share (*e.g.*, 1.2% in the UK).[123] The focus on domestic aviation ignores how the growing demand for air travel would mean that the aviation sector overall could eventually account for perhaps 5% of anthropogenic climate change.[124]

In parallel with the increasing number of countries that have adopted net-zero commitments, transport sectors around the world are looking to GGR technologies as possible options for reducing their carbon footprints. Similarly, while the onus for curbing emissions from aviation is currently more on airline companies than national governments, the aviation sector is also exploring GGR options following the 2009 pledge by the International Air Transport Association to halve emissions by 2050 compared to 2005 levels.

The aviation sector is often criticised for assuming that most of its mitigation efforts can be realised by purchasing offsets from other sectors rather than decarbonising their own operations.[125] In 2016, the International Civil Aviation Organization (ICAO) established the Carbon Offsetting and Reduction Scheme for International Aviation (CORSIA). Although participation is voluntary until 2021, several airlines have already committed, with some companies such as British Airways providing an option for passengers to offset (international) flights by paying extra to support forestry projects. Similarly, several companies – many of them from outside the transport sector – such as Shell, Lyft, Etsy and Anglo-American have chosen to offset their emissions from freight transport by investing in AR projects.[126] Yet, as critics point out, offsets alone are unlikely to be able to deliver the volume of negative emissions required to bring transport and aviation to net zero.[127]

More promisingly, stakeholders in transport and aviation are also considering how GGR could be used to directly reduce emissions within the industry. Most efforts focus on using CCS-based technologies to produce low-carbon fuels for air and road transport or offset sectoral emissions by sequestering CO_2 in permanent storage. Several transport firms and airlines have already collaborated with CCS-based companies to develop facilities for delivering these options. For example, United Airlines recently announced its plans to become net-zero by 2050 by investing in the 1Point5 project, which would be the first DAC plant in the US to permanently sequester millions of tonnes of CO_2 per year.[128] Similarly, in early 2020, IAG, the parent company of BA, invested $400 million in building a plant for converting household waste into sustainable fuel for aviation with renewable fuel companies Velocys and Shell.[129] DAC companies such as Carbon Engineering and ClimeWorks are explicitly targeting stakeholders in transport and aviation with technological options that could facilitate utilising the captured CO_2 for producing low-carbon synthetic fuels.[130,131] Both BECCS and DAC offer cost advantages which can be passed on to end-users; low-carbon biofuels produced *via* BECCS would be eligible for existing biofuel incentives which could translate into lower end-prices, while DAC could be deployed in locations that have the lowest renewable energy prices. Yet, bioenergy policies in energy and transport may be inversely related as competition over resources for producing biofuel in one sector could drive up prices in the other sector. Furthermore, increasing the use of biofuels in transport could retard electrification of road transport, which is the policy of choice for many governments.[132] Moreover, there is a more fundamental risk that releasing back the captured CO_2 that is utilised in biofuel production will result in net increases rather than reductions.[133] Most leading commercial DAC start-ups such as Carbon Engineering and Global Thermostat, for example, are linked to EOR, suggesting that initial DAC projects – which lack storage – do not qualify as genuine GGR.

13.3.3 NGOs

Environmental NGOs have tended to perceive GGR technologies as a dangerous distraction from what they regard as proper climate mitigation strategies such as renewable energy, efficiency improvements and energy saving, with some conceding a limited role for 'no-regrets' options – AR and SCS.[134] At its core, the environmentalist perspective is premised on moral hazard concerns in the sense that the mere existence of GGR as a measure to reduce the risk of dangerous global warming increases the probability of climate risks. Several NGOs cite the literature on mitigation deterrence[135,136] to set out how large-scale GGR could obstruct mitigation and 'lock-in' fossil fuel energy.[137–139] A report by Friends of the Earth, for example, argues that governments are tempted to resort to GGR technologies to avoid making politically risky decisions.[140] For the same reason, NGOs are also critical of private stakeholders who opt to invest in offsets rather than decarbonise their own industry. They also draw attention to the risk that carbon traders could use GGR to sell 'phantom' offsets that might not materialise.[141] A related concern is that the anticipation of large-scale GGR is likely to result in a larger CO_2 overshoot – one commentary in *Nature* claimed that this could be some 9 Gt above the predicted levels without GGR.[142] Furthermore, most GGR technologies are substantially more expensive than emissions reduction themselves.[143] However, as discussed above, deployment costs are expected to fall steeply at scale.

The importance of meeting net zero and taking a 1.5 °C target seriously does mean that NGOs have begun to pay more attention to GGR. One of the more nuanced reports on the subject was commissioned by WWF Scotland, which supported the need to 'Fund demonstration of land-based GGR at scale to establish CO_2 removal monitoring processes, and understand environmental risks. To unlock land-based GGR as a viable option to achieve net zero, demonstration is required now."[144] In general though, NGOs tend to side with pessimistic forecasts about the ability to make GGR technologies commercially and climatically viable in time for meeting an ambitious temperature target. Adding to these concerns are uncertainties about the technical feasibility, effectiveness, safety and permanence of carbon sinks, as well as counter-beneficial carbon flows from CCU, which is, rather unaptly, predominantly used to extract fossil energy.

NGOs also have a number of more specific concerns associated with different GGR approaches. For the more nascent technologies, the main criticisms relate to risks to ecosystems, the environment and human health and conflicts with other priorities that require the same resources. For DAC and BECCS, the main issues are energy requirements, storage capacity, leakage risks, and, when CCS is used for biofuel production, the sustainability of low-carbon fuels. Due to its land requirements, BECCS tends to attract attention from a broader range of NGOs relating to land rights and equity in addition to the environment. One Earth (2020), for example, describes BECCS as a 'false saviour' due to its adverse implications for agricultural land use and food prices.[145] EW is criticised for its potential to encourage rock mining as well as its associated adverse effects on environmental and human health, while OF is opposed due to its uncertain effects on marine ecosystems.

By contrast, AR and SCS are generally perceived to be more acceptable, albeit at limited scales, due to their association with fewer unknown environmental risks and widespread acceptance as natural activities. That said, these options are not unproblematic as even NGOs that are relatively supportive of their large-scale deployment are highly attuned to the adverse effects of previous land management projects on host communities. For example, while outlining a path of 'natural' reforestation to restore 75% of original global forest cover, the Climate Land Ambition and Rights Alliance goes to great lengths to spell out practical measures that should be taken to ensure that AR does not result in land grabbing or dispossession of livelihoods as in past REDD projects.[146] Similarly, forestry-related NGOs such as the Global Forest Coalition (2017) make a strong case that monoculture tree plantations, which are becoming increasingly common under offset projects, are incompatible with broader sustainability goals.[147] For some NGOs such as the Climate Institute, concerns over adverse effects justify limiting these GGR technologies to small-scale projects.[148]

13.3.4 Public Engagement

13.3.4.1 Public Attitudes

Since public approval is widely regarded as a precondition for a 'social license to operate',[149] it is increasingly construed as a critical factor in determining the scalability of GGR technologies.[150] While public awareness of GGR is generally low, research involving attempts to inform publics about GGR technologies as well as opinions among the minority who have heard about GGR/NETs suggest that attitudes are largely negative. In general, GGR technologies are viewed less favourably than emissions reduction, which is not helped by the fact that the leading deployments to date – offsets and EOR/EGR – have clear linkages to mitigation deterrence.

That said, there are important distinctions between public perceptions of familiar GGR technologies, which tend to be perceived as more natural and acceptable, and nascent options, which are generally seen as negative futuristic solutions.[151] Most people are familiar with AR and SCS. Publics in industrialised countries, which have typically funded AR projects in developing countries *via* international schemes, tend to perceive AR as natural and sustainable. Such positive views are reinforced by environmental NGOs and comprise a key reason behind the ability of corporate AR offsets to elicit support from Western publics. Perceptions tend to be more negative in developing countries where previous AR projects have caused disruption to host communities, often resulting in fierce opposition to new projects.[152] It is becoming more common for industrialised countries to host their own AR and SCS projects to contribute towards national net-zero commitments. As in developing countries, such initiatives are often met with local opposition, especially from rural communities and farmers, due to concerns that soil management is likely to compromise related priorities such as water management, soil erosion and food security.[153]

Public attitudes towards these GGR options are also influenced by the nature of land rights under national legislation. Eliciting local support for AR is usually more straightforward in contexts where forested land is mainly private property as financial compensation is generally an effective mechanism for legitimising land acquisition. Yet a different set of challenges emerge when private owners do not agree to part with their land as, compared to the state, they usually have less incentive to invest in negative emissions which are associated with long-term benefits. In the US, for example, only 2–5% of private cropland owners receive funds under programmes relating to soil health.[154] Land legislation poses greater challenges in developing countries where forestland is mostly owned by national governments and land rights are ambiguous. In such contexts, people who use the land will usually not be ones making the decision over LUCs, therefore, projects often encounter unrealistic conditions on the ground.[155]

Public awareness of most nascent GGR technologies (DAC, EW and OF) is almost non-existent. While knowledge has sometimes been shown to increase support,[156] more information often raises further questions relating to potential risks and adverse consequences.[157] In general, research suggests that when provided with sufficient time and resources, laypeople formulate thoughtful, sophisticated judgements about GGR technologies, which often mirror the main themes in the literature about risks, side-effects and moral hazard concerns.[158]

There is a rich literature on public perceptions of CCS[159,160] which has several implications for BECCS and DAC. First, publics are concerned about storage risks – both for their potential to compromise the effectiveness of CCS in reducing atmospheric CO_2 and because of the risks to local ecosystems, environments and human health. As far as storage concerns are rooted in NIMBYism rather than moral hazard concerns, offshore storage might provide a pragmatic solution as survey studies suggest that people are generally more supportive of CCS when it is stored offshore rather than onshore.[161] Second, attitudes tend to be more positive when CCS projects are expected to deliver economic benefits. Whitmarsh *et al.* found evidence of a 'Yes in my back yard' (YIMBY) effect whereby host communities who live proximate to CCS are more supportive than the wider public because of anticipation of local economic benefits.[162] Expectations about economic benefits are linked to private sector views, particularly from agriculture and transport. For example, declining demand for meat consumption and arable vehicles due to the electrification of transport is expected to increase demand for biofuel and has been shown to increase support for BECCS.[163] Third, and relatedly, attitudes are also influenced by whether the captured CO_2 is utilised in another industry, and therefore, linked to revenue streams, rather than stored. Yet the nature of this relationship appears to be contingent on national socioeconomic and cultural conditions. For example, while some publics are more supportive of CCU projects, others are likely to be less supportive if they are concerned about mitigation.[164] Fourth, publics are more supportive if the captured CO_2 is perceived to be of natural origin. Indeed, communities have often been more supportive of BECCS than fossil-CCS because the biomass feedstock is perceived as more sustainable.[165] Similarly, DAC plants are likely to elicit public support if they incorporate renewable energy sources to capture CO_2. Cox *et al.*'s recent study of public attitudes in the UK and US suggests that BECCS has the upper hand in this regard as it is perceived to be more 'natural' than DAC (and EW).[166] Fifth, public attitudes are influenced by potential side-effects and competition with other priorities. While the high-energy needs of DAC touches on the highly sensitive issue of energy security, the stakes are especially high in developing countries which are likely to prioritise development over climate protection. For BECCS, the socioeconomic impact of expanding biomass production raises several challenges that are likely to draw public scrutiny, such as, for example, increasing global food prices, displacement of local populations, land and water scarcities and loss of native forests.

Since EW and OF are still nascent technologies, public awareness is understandably low. Yet there is an expectation among academics and NGOs that these options will encounter the strongest public resistance because, out of all GGR approaches, they are perceived to be the least natural options and pose a range of unfamiliar results for environmental and human health. EW is likely to cross a red line for many publics due to its association with new mining.[167] However, the few existing studies on this subject suggest that more positive attitudes towards EW could result if provided with sufficient, balanced information about the need for negative emissions and nature of GGR technologies.[168,169] OF and EW (and BECCS and DAC) projects that involve ocean storage are also likely to encounter strong opposition because people often have strong emotional connections with oceans.[170] While EW is associated with potential benefits for soil productivity, there are many unknowns about the effects of OF on marine health, which has been shown to elicit opposition from communities that rely on income from fisheries such as, for example, the First Nations village in the Haida Gwaii Islands in western Canada.[171]

13.3.4.2 Public Engagement Strategies

Given the importance of public engagement for the viability of GGR technologies, academics and stakeholders have devoted a large amount of attention to exploring the underlying drivers of public approval. One frequent criticism is that developers often perceive the public as an obstacle rather than a constructive stakeholder, which could positively affect the decision-making process if project developers genuinely address issues of public concern.[172] Thus an obvious strategy for engaging publics is to ensure procedural justice by including local interests in the decision-making process to increase the legitimacy of GGR projects. Deliberative decision making is associated with greater efforts to address potential adverse impacts (*e.g.*, on health, housing, utilities, transport, air pollution and storage risks) and benefits (*e.g.*, employment and training opportunities and infrastructure investment) and, therefore, could be an effective way for pre-empting major opposition from arising further along in the process.[173]

Second, public attitudes are also influenced by the way that GGR technologies are framed. Most obviously, publics are more likely to be supportive when project developers convey that efforts have been made to mitigate potential risks and adverse side-effects. Of course, the nature of perceived risks varies across the different GGR approaches; for AR, SCS and biochar, the major concerns are about impacts on small farmers, resource-scarcity, food prices and accounting, for BECCS and DAC; storage, transport and resource-inputs and, for EW and OF; the sources of material for weathering and effects on marine health. More broadly, GGR technologies are more likely to elicit support if they are presented as temporary measures that are complementary to, rather than in conflict with, mitigation and the energy transition.[174] Such framing is important for dispelling moral hazard concerns and counteracting some of the negative imagery that is associated with the most visible projects to date – EOR/EGR and offsets. Research on public attitudes towards climate policy suggests that, in democratic contexts, GGR technologies are likely to elicit broader support if they are perceived to be autonomous from partisanship. In the US, for example, because Republicans have been stronger advocates of geoengineering than Democrats, it is likely that GGR technologies will increasingly come to be affiliated with the political right.[175]

Third, and relatedly, publics are more likely to be supportive when they trust the project developers. Since awareness of GGR is generally low and the technologies quite technical, lay people use their sense of trust in stakeholders as a cognitive shortcut to reach informed opinions about projects.[176] Consequently, attitudes are influenced by factors that go beyond the specific properties of the technology itself such as the perceived honesty, fairness and competence of the project developer. Climate sceptics, for example, are less likely to support GGR technologies, at least for the 'right' (mitigative) reasons, and have been found to believe that GGR are driven more by commercial rather than public interest.[177] Similarly, drawing on the partisan framing example above, individuals in democratic contexts are likely to support GGR projects if they are championed by their own political party, and oppose them if they are affiliated with the opposition.[178]

13.4 Governance

The ability to ramp up GGR to a climatically relevant scale depends on the emergence of a comprehensive governance system that can enable the technology in a market context and coordinate the interests of the diverse stakeholders involved.[179] On the surface, governance appears to be a technical exercise, involving issues such as financial incentivises, accounting, monitoring and verification. Yet, as we elaborate below, the regulation of these issues raises important sociocultural and political questions which will ultimately determine the viability of GGR technologies. For example, whether the government or another regulatory authority regulates individual lifestyle decisions or markets are given free rein to incentivise CDR depends on the sociocultural preferences of the nation.[180] Similarly, the ability of accounting, monitoring and verification systems to generate timely and socially valuable rewards to host communities will have a major impact on the social viability of upscaling.[181] Moreover, in addition to regulating the deployment process, a comprehensive governance structure is needed to manage the broader side-effects and implications for other political priorities that are affected by GGR.[182]

13.4.1 Incentivising GGR

National government have at their disposal a range of policies that could be used to potentially incentivise the full menu of GGR options – though none are unproblematic. Perhaps the most obvious measure is to make deployment mandatory for new power plants, although technology mandates would need to be international to prevent investments from being diverted to other countries. There is also a risk that, without adequate incentives for

CDR, mandates would prolong the use of old, high-carbon technologies and new plants involving GGR would simply not get built. Another option is to provide grants or tax breaks to fund large-scale demonstrations, which could mitigate some of the risks associated with investment. For example, capital grants in the UK have been used to support low carbon technologies (*e.g.*, offshore wind and solar PV), although substantial funds would be needed to overcome the large capital costs associated with GGR technologies.[183] Subsidies for downstream feedstocks such as renewables for DAC or sourcing of sustainable material for EW could provide additional incentives. For GGR approaches that involve land use (*i.e.*, BECCS, SCS, biochar and EW), agricultural subsidies could potentially engage farmers, although policies would need to accommodate the entrenched interests of the agriculture sector, which vary substantially across different national contexts.

13.4.2 Carbon Pricing

To achieve the Paris targets, most climate models are premised on the notion that the price of carbon needs to be significantly more ambitious to drive upscaling.[184] The underlying rationale is that because the environmental externality (the effect of CO_2 on the climate system) is not fully internalised, the price of CO_2 is significantly below where it needs to be in order to attract the required investment. For example, most scenarios in the IPCC AR5 assume that a carbon tax of around \$100/t CO_2 is a precondition for delivering BECCS at a global scale, although the costs are hugely uncertain.[185] Existing practices in national jurisdictions suggest that prices are hugely variable and largely reflect the interests of key sectors of the national economy.[186] Depending on how it was operationalised, a tax applied to carbon emissions could become a subsidy for negative emissions, which, to the extent that upscaling is successful, could eventually require massive expenditures – as much as a third of general government expenditure in advanced economies according to one study.[187] There are also questions about the ability of carbon pricing to simultaneously induce the required global transition from fossil fuels and optimal GGR deployment as both outcomes are likely to materialise at different prices.[188] An additional obstacle is that democratic publics are likely to perceive interventions in individual lifestyle decisions, *e.g.*, consumption and transport, as intrusive and oppose carbon taxes for passing on costs to end-users.[189] Moreover, it is becoming increasingly evident that carbon prices alone are unlikely to be enough to deliver the rapid upscaling required. Fundamentally, this is because, carbon pricing, as currently formulated, does not reward CDR, which renders alternative fossil-based decarbonisation options significantly cheaper than GGR technologies.[190]

13.4.3 Accounting

In this regard, a number of experts have suggested that storage-based accounting could serve as effective incentives for upscaling.[190,191,196] There are currently several different types of carbon accounting systems in operation spanning multiple (international, regional, national, subnational) levels and projects.[191] Yet all of them count CO_2 *emissions* rather than storage, which disincentivises CO_2 emissions but does not adequately remunerate storage. For example, under cap-and-trade systems, if a certain year's emissions fall below the cap, then there are no penalties for positive emissions or incentives for CDR. Similarly, while project-based schemes such as the CDM lend well to accounting negative emissions as credits (which can be traded to offset positive emissions from other sectors), at present, AR is the only GGR option that is recognised under existing accounting rules.[192]

The lack of renumeration for storage is especially problematic for BECCS and DAC as CCS is a costly process. This rationale underlies various proposals for storage-based accounting in relation to these GGR technologies. Daggash and Mac Dowell, for example, suggest that the use of a 'negative emissions credit' – a payment for each ton of CO_2 removed – could ensure that CDR would be appropriately renumerated.[193] This approach would create the added advantage of lower overall end prices. Similarly, Jenkins *et al.* propose the concept of a 'carbon takeback obligation' (CTO) to shift accounting away from the polluter pays principle, which is based on counting emissions, to producer storage, which involves counting storage.[194] In this approach, all primary extractors and importers of fossil carbon are required to demonstrate permanent sequestration of a progressively increasing fraction of CO_2 generated by production, transport and use of their products.

While storage-based accounting has the potential to incentivise CCS-based GGR technologies, its applicability to other options appears limited. Jenkins *et al.*,[194] for example, are keen to limit CTOs to BECCS and DAC because of the difficulty of monitoring and verifying storage in biological sinks and concerns over the permanence of nature-based solutions. Even within existing emissions-based accounting, these issues have posed serious obstacles for land-based solutions. For example, although AR was included in the CDM, it failed to take off because verification of negative emissions usually fails or happens too late to deliver benefits to the host society,[195] which has made developing country governments reluctant to host new projects.[196] Such concerns are also relevant to more nascent options – EW and OF, for which storage capacity and permanence are less understood. More broadly, upstream (*i.e.*, crop choice for BECCS, AR, SCS and biochar, renewable energy sources for DAC, and transport and sourcing material for OF and EW) and downstream flows (*i.e.*, CCU for BECCS and DAC, forest fires for AR, storage leakage) complicate emissions-based accounting for all GGR technologies.

13.4.4 Risk Management

Regardless of which accounting method one prefers, the emergence of an effective monitoring and verification system will be an essential first step. Not only will it be important for ensuring that carbon dioxide (emissions or storage) is counted accurately and transparently, but also for assessing and responding to risks. Since risks are an important driver of public opinion on GGR technologies, effective monitoring and verification could play an important role in making them more acceptable by regulating liability and insurance for risks. The publics may also be concerned about moral hazard risks relating to the effectiveness of GGR at preventing dangerous climate change.[197,198] In order to address such concerns, the governance system would need to cover diverse scales from upstream feedstocks (*e.g.*, biomass production) to long-term liability for storage as well as downstream flows from CCU and electricity generation.[199] Taking a broader approach that encompasses other kinds of risks as addressed by the UN Social Development Goals (SDGs) such as environmental sustainability, socio-economic development and equity could go a long way towards engaging publics in developing countries.

13.4.5 Side-effects and Broader Implications

Upscaling most GGR approaches entails significant LUCs, which is expected to exacerbate land and water scarcities, limiting the resources available for other land use purposes such as food and energy crop production. A significant fraction of the burden may be borne by small farmers in the global South, where disruption of livelihoods and dispossession of land, higher food prices and growing inequity are plausible outcomes. At present, there are only international frameworks for regulating the LUCs relating to AR (REDD+ and LULUCF), but even these are plagued by serious accounting and monitoring issues, which renders them ineffective in ensuring that the benefits of AR projects reach host communities in full and on time.[200] A way forward in regulating the adverse effects of all land-based GGR technologies might be to set criteria for land-use by including clearer rules about (locally defined) marginal lands and land rights to ensure that LUCs are sustainable socially, economically and environmentally.[201] In this way, it would be possible to engage with the context-specific barriers to deployment which is likely to result in more realistic understanding about the potential of nature-based solutions.

13.4.6 Procedural Justice

Perhaps the most important governance challenge is ensuring that the entire process of upscaling takes account of procedural justice. Stakeholders have the potential to play a vital role in designing governance systems right from the initial stages of deciding what role GGR technologies should play alongside a country's mitigation efforts through to setting the goals of governance and designing the mechanisms for regulating the different aspects of upscaling.[202] While the involvement of different actors is likely to complicate matters by creating more opportunities for voicing reservations and disagreements, the end result is likely to be more robust as the deliberative process would have addressed more issues in the early design phase, leaving less grounds for opposition further on in the process. For example, whether a policy instrument such as a carbon tax, a negative emissions credit, technology mandate, or CTO is deemed acceptable largely depends on society's norms and values.[203] While democratic contexts are arguably more accommodating to deliberative decision-making, their publics are likely to oppose the comprehensive governance that is needed to address the wide range of side-effects that go beyond what is directly seen to be related to GGR.[204] By contrast, authoritarian states would be less inclined to facilitate procedural justice, although their top-down, strong governance practices may be more conducive to upscaling if not necessarily to climate action in the first place.

13.5 Socio-political Aspects of International Governance

Global variation in geophysical conditions means that some regions are better suited for different GGR technologies and, consequently, can implement certain technologies at lower costs. Therefore, effective international coordination has the potential to deliver upscaling at the lowest (global) cost by incentivising countries to specialise in the GGR approaches for which they are most suited. Without such incentives, some regions are likely to not host GGR despite being geophysically suited to them.[205] For GGR technologies that involve multiple stages, this could facilitate the different components to be carried out in countries that can deliver the lowest costs. For example, for BECCS, biomass could be produced in Brazil, transported to the UK for use in a power plant and the captured CO_2 stored in Norway, which would reduce costs by increasing the efficiency of the component parts.[206] Of course, transnational deployment would raise a number of important questions such as, for example, which nation gets the credit for a GGR project – the host, the funder or the one that captures or stores the carbon?[207] Clearly, such questions would also need to be regulated under an international framework. Similarly, international offset markets would allow countries that are unsuited to hosting GGR to contribute to global temperature targets by funding negative emission activities in more efficient locations abroad. Furthermore, since the early stages of technology deployment are associated with high capital investments and risky payoffs, international cooperation would facilitate more learning and efficient research globally.[208] In addition, some key high-carbon industries such as aviation and marine transport fall outside the realm of national jurisdiction and therefore need to be regulated under a global framework.

But what would such a framework look like? Fundamentally, an effective global governance system would need to be premised on an acceptable framework for distributing responsibility for GGR internationally. National costs and capacities to deliver negative emissions are determined by a combination of geophysical factors, energy trajectories and population trends. Therefore, an obvious solution might be to let the markets dictate the governance framework. Economics suggests that international trade in carbon would transfer costs between countries and eventually arrive at the most efficient solution and have the added benefit of doing this in a non-prescriptive manner. Yet, political factors are likely to continue suppressing carbon prices for the foreseeable future, suggesting that markets alone will be unable to deliver sufficient negative emissions in time for preventing dangerous climate change.[209] Moreover, even without time constraints, it is unlikely that a laissez-faire international division of labour would be politically and socially acceptable. For example, most IAMs assume that maximum negative emissions can be delivered by deploying large-scale BECCS in the power sector, which suggests that countries that consume the most electricity have the largest potential for negative emissions, and therefore, should host the bulk of BECCS plants.[210] However, supplying large volumes of energy to power CCS and integrating biomass electricity produced *via* BECCS into existing energy systems is a formidable task even for industrialised countries, let alone emerging economies such as India and Pakistan, whose power systems already struggle to meet rising demand. Countries that have the highest biomass production capacity would experience the largest impacts on land use, which, as demonstrated by previous experience of biofuel booms and schemes to ramp up biomass production, is likely to encounter opposition from the agriculture sector and publics. Similarly, countries in the tropics that are suited for AR and SCS would experience various side-effects such as loss of livelihoods, resource scarcity and displacement, which raises understandable ethical concerns given that these countries are associated with historically low emissions.[211] These examples illustrate that the feasibility of an optimal framework for delivering negative emissions depends on its ability to address important political, social and ethical obstacles to upscaling.

Some recent studies speak directly to this challenge by exploring how different ethical principles might distribute responsibility for upscaling GGR technologies internationally. A central focus has been mapping out the implications of three core equity principles – namely, responsibility, egalitarianism and capacity – on sharing out the burdens to deliver negative emissions.[212,213] The responsibility principle implies that responsibility for CDR should be proportional to cumulative emissions, shifting most responsibility to industrialised nations which have historical complicity for climate change. This approach is severely limited from a performance perspective as it exempts the largest polluting countries both at present and going forward. Egalitarianism suggests that all countries, regardless of their complicity or capacity, should have the equal right to pollute and be protected from climate change. The governance implication is that equal efforts should be enforced across all countries in accordance with their population, resulting in the most populous nations taking the lead. Populous developing countries which are rapidly industrialising such as the BASIC states (Brazil, South Africa, India and China) would see this as unfair as they would be required to provide GGR at a similar or larger scale to industrialised nations that have had a larger contribution to climate change. Moreover, without adequate financial and capital transfers, egalitarian formulas for distributing responsibility for negative emissions would be susceptible to broader arguments about the global South's right to develop GGR at a scale that could be viewed as impeding growth. In this regard, international offset markets or alternative international financing mechanisms might offer a way forward by creating mechanisms for delivering Northern finance for GGR technologies in developing countries. By contrast, the capacity principle assigns responsibility to countries in accordance with their ability to finance GGR technologies, suggesting, much like the complicity principle, that (current) economic powers should lead the way. Yet, as argued throughout this chapter, the capacity to deploy GGR technologies is also affected by non-financial factors. For example, knowledge and technical expertise about GGR is generally lacking in the global South, which, some argue, has prevented the region from shaping global accounting methods to take account of their regional interests.[214] Furthermore, political institutions, governance capacity, land rights and energy and transport infrastructure tend to be more fragile in the global South, creating additional regional obstacles.

The principle of common but differentialised responsibilities and respective capabilities (CBDR-RC) could offer an effective compromise between the performative limitations of the responsibility and capacity principles and the regional development concerns raised by egalitarianism.[215,216] Extending its original usage in relation to climate mitigation, the CBDR-RC principle suggests that rich countries should take the lead in providing the initial investments for GGR technologies, and transfer additional finances until mechanisms (*e.g.* international offset markets) develop that enable developing countries to make a progressively increasing contribution to global upscaling without infringing on their own developmental rights.[217]

13.6 Conclusion

Given that our ability to meet an ambitious temperature target increasingly depends on rapid, large-scale deployment of GGR technologies, upscaling should already be well underway. While there have been some promising developments such as recent commitments by some national governments and private stakeholders to invest in large-scale CCS-based projects and AR, more often than not, technologically feasible options and economic mechanisms for arriving at optimal outcomes are obstructed by socioeconomic and political factors.

For technologies that are in the early stages of development such as DAC, EW and OF, the major obstacles are largely technical – relating to high operating costs and unknown side-effects or raise basic questions on whether work should go ahead in the first place (for OF and other ocean-based approaches in particular). Yet, as technologies mature and become deployment ready, they begin to encounter more practical challenges such as economic disincentives, accounting problems and regulatory gaps. On the surface, these issues appear like technical obstacles that can be overcome with design improvements, economic tools and good management. Closer analysis, however, suggests that the major sticking points are inherently social, ethical and political. For example, while an economic perspective underlines the side-effects and co-benefits associated with GGR technologies, the importance of these technology traits in obstructing or aiding deployment fundamentally depends on how much society values the thing that is endangered by the side-effect and the potential co-benefits. Most of these GGR technologies do not necessarily engender opposition at the project level but given that the logic for their development in the first place is driven by the need for massive upscaling then it is not surprising that concerns arise over the implications of their deployment at scale.

As more countries and corporations are committing to net-zero and carbon neutrality, stakeholder interest in GGR is growing. Governments are working to reconcile the substantial contribution that GGR technologies could make towards meeting a country's climate obligations with the broader trade-offs with other national priorities such as sustainability, food and energy security and mitigation, while, at the same time, trying to strike a balance between the diverse interests of stakeholders, civil society and publics.

Private actors are also looking to GGR technologies to reduce their carbon footprints as consumers and publics are increasingly prioritising sustainability. High-carbon sectors have begun taking some important steps. Farmers, for example, have started to look more positively at biomass production (for powering BECCS) and appear open to SCS and other land-management activities. Yet, given the political power of the agricultural sector, farmers are only likely to be forthcoming under the presence of strong state support, for example, subsidies or perhaps low-interest loans. Moreover, farmers are well informed about the potential for adverse effects on local communities – including their own livelihoods – and therefore need assurances from project developers that broader effects of land-based GGR will be addressed. Energy and utilities, on the other hand, have been exploring CCS-based options to reduce their carbon footprint and, in the case of BECCS, for its ability to generate net negative electricity. Yet there are major infrastructural issues and questions about the integration with existing energy structures and the strain that CCS will put on energy networks and energy security more broadly. Aviation and transport have also focused on CCS-based options, including DAC, particularly for producing low-carbon fuels. Both agriculture and transport (and aviation) have made major investments in offsets from AR and SCS, but this runs the risk of avoiding or postponing decarbonisation within these key sectors. It remains to be seen whether public and consumer pressure for decarbonisation will drive these sectors to prioritise GGR within their own industries.

Public awareness of newer GGR technologies is generally low, but is rising in parallel with deployment. Like NGOs, publics are generally more accepting of familiar activities – AR and SCS, and cautious of the unknown risks and side effects of newer technologies due to their potential to compromise other priorities such as public health and sustainability. Yet even nature-based options have been known to spark strong public opposition due to the disruptive effects of previous projects on local communities.

There are various measures that governments could use to incentivise GGR in their national jurisdiction such as, for example, subsidies, procurement, loans, technology mandates and carbon pricing. Yet the social viability of these policy instruments ultimately depends on the nation's political economy, social values, norms and priorities. Effective accounting and monitoring also have an important role to place in ensuring accurate, transparent practices and that benefits are returned to the host community, which would go a long way towards demonstrating the legitimacy of GGR technologies. Ensuring that governance is carried out under a framework of procedural justice could offer a way forward by providing a platform for publics and host communities to hold governments and private stakeholders accountable and address issues of public concern at early stages of project development.

International coordination has the potential to vastly reduce the cost of upscaling, but can only succeed if the international community is able to agree on a fair and equitable framework for assigning responsibility for CDR. Fundamentally, a compromise needs to be found between the global North's higher historical responsibility and capacity to deploy GGR technologies, and the need to persuade the rapidly industrialising global South to partake in upscaling in time for avoiding dangerous climate change. Knowledge, technology and financial transfers are likely to feature, but a 'techno-fix' which does not address the underlying socio-ethical issues is unlikely to be a long-term solution.[218] In this regard, the principle of CBDR-RC may provide some guidance by obligating rich countries to take the lead and poor countries to assume increasing responsibility as they develop. Taking the national sphere as a microcosm of the international governance challenges, a mutually acceptable framework is only likely to emerge if developing countries, which have so far been rather silent on GGR technologies, are involved in the design of the governance system to make sure that issues of regional importance are addressed from the outset.

References

1. J. Rogelj, M. Elzen, N. Hohne, T. Fransen, H. Fekete and H. Winler, *et al.*, *Nature*, 2016, **534**, 631.
2. Intergovernmental Panel on Climate Change, Fifth Assessment Report (AR5) Synthesis Report, 2014.
3. Intergovernmental Panel on Climate Change, IPCC Special Report on Global Warming of 1.5'C, 2018.
4. J. Fuhrman, H. McJeon, S. Doney, W. Shobe and A. Clarens, *Fron. Clim.*, 2019, **1**, 11.
5. Global Climate Capture and Storage Institute, *Bioenergy and Carbon Capture and Storage*, 2019.
6. G. Nemet, M. Callaghan, F. Creutzig, S. Fuss, J. Hartmann and J. Hilaire, *et al.*, *Environ. Res. Lett.*, 2018, **13**, 063003.
7. *Ibid.*
8. J. Strefler, T. Amann, N. Bauer, E. Kriegler and J. Hartmann, *Environ. Res. Lett.*, 2018, **13**, 034010.

9. G. Nemet, M. Callaghan, F. Creutzig, S. Fuss, J. Hartmann and J. Hilaire, *et al.*, *Environ. Res. Lett.*, 2018, **13**, 063003.
10. P. Smith, *Global Change Biol.*, 2016, **22**, 1315.
11. N. E. Vaughan and T. M. Lenton, *Clim. Change*, 2011, **109**, 745.
12. P. Boyd and M. Bressac, *Philos. Trans. R. Soc., A*, 2016, **374**, 20150299.
13. P. Williamson, D. W. Wallace, C. S. Law, P. W. Boyd, Y. Collos, P. Croot, K. Denman, U. Riebesell, S. Takeda and C. Vivian, *Process Saf. Environ. Prot.*, 2012, **90**, 475.
14. R. Socolow, M. Desmong, R. Aines, J. Blackstock, O. Bolland and T. Kaarsberg, *et al.*, *Direct Air Capture of CO_2 with Chemicals*, 2011.
15. D. Beerling, E. Kantzas, M. Lomas, P. Wade, R. Eufrasio and P. Renforth, *et al.*, *Potential for Large-scale CO_2 Removal via Enhanced Rock Weathering with Croplands*, 2020, vol. 583, p. 243.
16. S. W. Chisholm, *Nature*, 2000, **407**, 685.
17. WWF, *Carbon Dioxide Removal, Including Carbon Sequestration in Natural Systems, Climate and Energy Position Paper*, 2018. https://wwfint.awsassets.panda.org/downloads/wwf_1_5c_position_paper__carbon_dioxide_removal_including_carbon_sequestration_in_natur.pdf.
18. J. Unruh, *Soc. Nat. Res.*, 2011, **24**, 185.
19. H. J. Buck, *Clim. Change*, 2016, **139**, 155.
20. P. Smith, *Global Change Biol.*, 2016, **22**, 1315.
21. 4p1000 Secretariat, The 4 Per 1000 Initiative, 2018. https://www.4p1000.org/.
22. J. Fuhrman, H. McJeon, S. Doney, W. Shobe and A. Clarens, *Front. Clim.*, 2019, **1**, 11.
23. S. Fuss, W. Lamb, M. Callaghan, J. Hilaire, F. Creutzig and T. Amann, *et al.*, *Environ. Res. Lett.*, 2018, **13**, 063002.
24. D. McLaren, *Process Saf. Environm. Prot.*, 2012, **90**, 489.
25. M. Fridahl and M. Lehtveer, *Energy Res. Soc. Sci.*, 2018, **42**, 155.
26. M. Jones and F. Albanito, *Global Change Biol.*, 2020, **26**, 5358.
27. J. Fuhrman, H. McJeon, P. Patel, S. Doney, W. Shobe and A. Clarens, *Nat. Clim. Change*, 2020, **10**, 920.
28. G. Nemet, M. Callaghan, F. Creutzig, S. Fuss, J. Hartmann and J. Hilaire, *et al.*, *Environ. Res. Lett.*, 2018, **13**, 063003.
29. J. Fuhrman, H. McJeon, S. Doney, W. Shobe and A. Clarens, *Front. Clim.*, 2019, **1**, 11.
30. J. Hilaire, J. Minx, M. Callaghan, J. Edmonds, G. Luderer and G. Nemet, *et al.*, *Clim. Change*, 2019, **157**, 189.
31. G. Nemet, M. Callaghan, F. Creutzig, S. Fuss, J. Hartmann and J. Hilaire, *et al.*, *Environ. Res. Lett.*, 2018, **13**, 063003.
32. S. Fuss, C. D. Jones, F. Krazner, G. P. Peters, P. Smith and M. Tavoni, *et al.*, *Environ. Res. Lett.*, 2016, **11**, 115007.
33. Information Technology and Innovation Foundation, Across the 'second valley of death': Designing successful energy demonstration projects, 2017.
34. G. Nemet, M. Callaghan, F. Creutzig, S. Fuss, J. Hartmann and J. Hilaire, *et al.*, *Environ. Res. Lett.*, 2018, **13**, 063003.

35. V. Scott, S. Gilfillan, N. Markusson, H Chalmers and R. S. Haszeldine, *Nat. Clim. Change*, 2012, **3**, 105.
36. J. Bednar, M. Obersteiner and F. Wagner, *Nat. Commun.*, 2019, **10**, 1783.
37. M. Honegger and D. Reiner, *Clim. Policy*, 2018, **18**, 306.
38. C. Beuttler, L. Charles and J. Wurzbacher, *Front. Clim.*, 2019, **1**, 00010.
39. J. Temple, *Technol. Rev.*, 2020, **123**, 1.
40. S. Fuss, W. Lamb, M. Callaghan, J. Hilaire, F. Creutzig and T. Amann, *et al.*, *Environ. Res. Lett.*, 2018, **13**, 063002.
41. C. D. Fuss, F. Jones, G. P. Krazner, P. Peters, Smith and M. Tavoni, *et al.*, *Environ. Res. Lett.*, 2016, **11**, 115007.
42. J. Fuhrman, H. McJeon, S. Doney, W. Shobe and A. Clarens, *Front. Clim.*, 2019, **1**, 11.
43. *Ibid.*
44. S. Fuss, W. Lamb, M. Callaghan, J. Hilaire, F. Creutzig and T. Amann, *et al.*, *Environ. Res. Lett.*, 2018, **13**, 063002.
45. D. McLaren, *Process Saf. Environ. Prot.*, 2012, **90**, 489.
46. M. Honegger and D. Reiner, *Clim. Policy*, 2018, **18**, 306.
47. *Ibid.*
48. J. Strefler, T. Amann, N. Bauer, E. Kriegler and J. Hartmann, *Environ. Res. Lett.*, 2018, **13**, 034010.
49. C. D. Fuss, F. Jones, G. P. Krazner, P. Peters, Smith and M. Tavoni, *et al.*, *Environ. Res. Lett.*, 2016, **11**, 115007.
50. J. Fuhrman, H. McJeon, S. Doney, W. Shobe and A. Clarens, *Front. Clim.*, 2019, **1**, 11.
51. S. Fuss, W. Lamb, M. Callaghan, J. Hilaire, F. Creutzig and T. Amann, *et al.*, *Environ. Res. Lett.*, 2018, **13**, 063002.
52. P. Smith, *Global Change Biol.*, 2016, **22**, 1315.
53. J. Strefler, T. Amann, N. Bauer, E. Kriegler and J. Hartmann, *Environ. Res. Lett.*, 2018, **13**, 034010.
54. C. Pozo, A. Galan-Martin, D. M. Reiner, N. Mac Dowell and G. Guillen-Gosalbez, *Nat. Clim. Change*, 2020, **10**, 640.
55. J. Canadell and D. Schulze, *Nat. Commun.*, 2014, **5**, 5282.
56. P. Smith, S. Davis, F. Creutzig, S. Fuss, J. Minx and B. Gabrielle, *et al.*, *Nat. Clim. Change*, 2015, **6**, 42.
57. J. Fuhrman, H. McJeon, P. Patel, S. Doney, W. Shobe and A. Clarens, *Nat. Clim. Change*, 2020, **10**, 920.
58. E. Cox and N. R. Edwards, *Clim. Policy*, 2019, **19**, 1144.
59. J. Fuhrman, H. McJeon, S. Doney, W. Shobe and A. Clarens, *Front. Clim.*, 2019, **1**, 11.
60. A.-M. Dowd, M. Rodriguez and T. Jeanneret, *Energies*, 2015, **9**, 4024.
61. A. Aggarwal and D. Brockington, *Land Use Policy*, 2020, **95**, 104608.
62. P. Smith, *Global Change Biol.*, 2016, **22**, 1315.
63. J. Fuhrman, H. McJeon, S. Doney, W. Shobe and A. Clarens, *Front. Clim.*, 2019, **1**, 11.
64. S. Fuss, W. Lamb, M. Callaghan, J. Hilaire, F. Creutzig and T. Amann, *et al.*, *Environ. Res. Lett.*, 2018, **13**, 063002.

65. M. Honegger and D. M. Reiner, *Clim. Policy*, 2018, **18**, 306.
66. J. Strefler, T. Amann, N. Bauer, E. Kriegler and J. Hartmann, *Environ. Res. Lett.*, 2018, **13**, 034010.
67. M. Sharmina, O. Y. Edelenbosch, C. Wilson, R. Freeman, D. Gernat and P. Gilbert *et al.*, *Climate Policy*, (forthcoming).
68. K. Hamrick and M. Gallant, *Voluntary Carbon Markets: Outlooks and Trends*, 2018.
69. Global Climate Capture and Storage Institute, Bioenergy and Carbon Capture and Storage, 2019.
70. D. Reiner, *Carbon Capture and Storage*, ed. M. Bui and N. Mac Dowell, Royal Society of Chemistry, 2020, p. 536.
71. E. Cox and N. R. Edwards, *Clim. Policy*, 2019, **19**, 1144.
72. N. Mac Dowell, P. S. Fennell, N. Shah and G. C. Maitland, *Nat. Clim. Change*, 2017, **7**, 243.
73. Department for Business, Energy and Industrial Strategy, Direct Air Capture and Greenhouse Gas Removal Programme, 2020.
74. Energy Futures Initiative, Carbon Removal: Comparing Historical Federal Research Initiatives with the National Academies' Recommended Future Funding Levels, 2019.
75. W. Carton, A. Asiyanbi, S. Beck, H. J. Buck and J. Fund, *Climate Change*, (forthcoming).
76. Energy Futures Initiative, Carbon Removal: Comparing Historical Federal Research Initiatives with the National Academies' Recommended Future Funding Levels, 2019.
77. M. Fridahl, *Energy Policy*, 2017, **104**, 89.
78. A.-M. Dowd, M. Rodriguez and T. Jeanneret, *Energies*, 2015, **9**, 4024.
79. C. Pozo, A. Galan-Martin, D. M. Reiner, N. Mac Dowell and G. Guillen-Gosalbez, *Nat. Clim. Change*, 2020, **10**, 640.
80. P. Burnell, *Democratization*, 2012, **19**, 813.
81. S. L'Orange Siego, J. Arval, S. Dohle and M. Siegrist, *Int. J. Greenhouse Gas Control*, 2014, **25**, 23.
82. C. Campbell-Arvai, P. Hart, K. Raimi and K. Wolske, *Clim. Change*, 2017, **143**, 321.
83. A.-M. Dowd and M. James, *Social Epistemol.*, 2014, **28**, 364.
84. H. J. Buck, *Curr. Sustainable Renewable Energy Rep.*, 2019, **6**, 124.
85. A. Aggarwal and D. Brockington, *Land Use Policy*, 2020, **95**, 104608.
86. D. M. Reiner, *Nat. Energy*, 2016, **1**, 15011.
87. W. Carton, A. Asiyanbi, S. Beck, H. Buck and J. Fund, *Climate Change*, (forthcoming).
88. Global Carbon Capture and Storage Institute, Global Status of Bioenergy and Carbon Capture and Storage Projects 2010, 2010.
89. H. J. Buck, *Curr. Sustainable Renewable Energy Rep.*, 2019, **6**, 124.
90. R. S. Haszeldine, S. Flude, G. Johnson and V. Scott, *Philos. Trans. R. Soc., A*, 2018, **376**, 20160447.
91. H. J. Buck, *Curr. Sustainable Renewable Energy Rep.*, 2019, **6**, 124.
92. J. Unruh, *Soc. Nat. Resour.*, 2011, **24**, 185.

93. A. Aggarwal and D. Brockington, *Land Use Policy*, 2020, **95**, 104608.
94. A. Dowd, M. Rodriguez and T. Jeanneret, *Energies*, 2015, **9**, 4024.
95. C. Gough and S. Mander, *Curr. Sustainable Energy Rep.*, 2019, **6**, 116.
96. D. Reiner, *Carbon Capture and Storage*, ed. M. Bui and N. Mac Dowell, Royal Society of Chemistry, 2020, p. 536.
97. J. Kemper, *Int. J. Greenhouse Gas Control*, 2015, **40**, 401.
98. E. Cox and N. R. Edwards, *Climate Policy*, 2019, **19**, 1144.
99. J. Strefler, T. Amann, N. Bauer, E. Kriegler and J. Hartmann, *Environ. Res. Lett.*, 2018, **13**, 034010.
100. P. Smith, S. Davis, F. Creutzig, S. Fuss, J. Minx and B. Gabrielle, *et al.*, *Nat. Clim. Change*, 2015, **6**, 42.
101. P. Smith, *Global Change Biol.*, 2016, **22**, 1315.
102. E. Cox and N. R. Edwards, *Clim. Policy*, 2019, **19**, 1144.
103. E. Cox, E. Spence and N. Pidgeon, *Nat. Clim. Change*, 2020, **10**, 744.
104. P. Turner, C. Field, D. Lobell, D. Sanchez and K. Mach, *Nat. Sustainability*, 2018, **1**, 420.
105. E. Cox, E. Spence and N. Pidgeon, *Nat. Clim. Change*, 2020, **10**, 744.
106. NFU, Achieving Net Zero: Farming's 2040 Goal, November 2019. https://www.nfuonline.com/nfu-online/business/regulation/achieving-net-zero-farmings-2040-goal/.
107. T. Amann and J. Hartmann, *Biogeosciences*, 2019, **16**, 2949.
108. R. Lal, *Geoderma*, 2020, **378**, 114613.
109. E. Cox and N. R. Edwards, *Climate Policy*, 2019, **19**, 1144.
110. A. Aggarwal and D. Brockington, *Land Use Policy*, 2020, **95**, 104608.
111. R. Lal, *Geoderma*, 2020, **378**, 114613.
112. Ecosystems Marketplace, Voluntary Carbon Markets Insights: 2018 Outlook and First Quarter Trends, 2018.
113. A. Aggarwal and D. Brockington, *Land Use Policy*, 2020, **95**, 104608.
114. D. M. Reiner, *Nat. Energy*, 2016, **1**, 15011.
115. R. McCulley, Upstream, 2020, 2 November. https://www.upstreamonline.com/environment/oil-and-gas-companies-have-direct-air-capture-technology-in-their-sights/2-1-902736.
116. J. Morieria, S. Pacca and J. Goldemberg, *Earth Environ. Sci.*, 2019, **354**, 12107.
117. C. Beuttler, L. Charles and J. Wurzbacher, *Front. Clim.*, 2019, **1**, 00010.
118. P. Turner, C. Field, D. Lobell, D. Sanchez and K. Mach, *Nat. Sustainability*, 2018, **1**, 420.
119. E. Cox and N. R. Edwards, *Climate Policy*, 2019, **19**, 1144.
120. Shell, Media Release, 2019, https://www.shell.com/media/news-and-media-releases/2019/shell-invests-in-nature-to-tackle-co2-emissions.html.
121. BP, Press Release, 2019, https://www.bp.com/en/global/corporate/news-and-insights/press-releases/bp-invests-in-forest-carbon-offsets-leader-finite-resources.html.
122. World Resources Institute, Climate Watch Tool, 2020, https://www.wri.org/blog/2019/10/everything-you-need-know-about-fastest-growing-source-global-emissions-transport.

123. Parliamentary Debate, House of Commons, 31 January 2020, Number CDP-2020-0018.
124. E. Terrenoire, D. Hauglustaine, T. Gasser and O. Penanhoat, *Environ. Res. Lett.*, 2019, **14**, 84019.
125. T. Rayner, *Earth System Governance*, 2021, **8**, 100100.
126. S. Greene and C. Façanha, *Nat. Sustainability*, 2019, **2**, 994.
127. International Energy Agency, *Energy Technol. Perspect.*, 2017.
128. Reuters, United Airlines invests in carbon-capture project to be 100% green by 2050, 10 December 2020, https://mobile.reuters.com/article/amp/idUSKBN28K1NE.
129. British Airways, Sustainability at British Airways, https://mediacentre.britishairways.com/factsheets/details/86/Factsheets-3/208.
130. A. Stukas and E. Stechel, *Direct Air Capture of CO_2 and recycling CO_2 into Sustainable Aviation Fuels*, 2019, 20.
131. C. Beuttler, L. Charles and J. Wurzbacher, *Front. Clim.*, 2019, **1**, 00010.
132. E. Cox and N. R. Edwards, *Clim. Policy*, 2019, **19**, 1144.
133. H. J. Buck, *Global Sustainability*, 2018, **1**, E2.
134. Friends of the Earth, Negatonnes – An initial assessment of the potential for negative emission techniques to contribute safely and fairly to meeting carbon budgets in the 21st century, 2011.
135. D. Lenzi, *Global Sustainability*, 2018, **1**, E7.
136. N. Markusson, D. McLaren and D. Tyfield, *Global Sustainability*, 2018, **1**, E10.
137. Friends of the Earth, Negatonnes – An initial assessment of the potential for negative emission techniques to contribute safely and fairly to meeting carbon budgets in the 21st century, 2011.
138. Climate Land Ambition and Rights Alliance, Missing Pathways to 1.5' C: The role of the land sector in ambitious climate action, 2018.
139. Climate Institute, Negative Emission Technologies: Silver Bullet or Ethically Ambiguous?, 2019.
140. Friends of the Earth, Negatonnes – An initial assessment of the potential for negative emission techniques to contribute safely and fairly to meeting carbon budgets in the 21st century, 2011.
141. Climate Land Ambition and Rights Alliance, Missing Pathways to *1.5' C: The role of the land sector in ambitious climate action*, 2018.
142. D. Lenzi, W. F. Lamb, J. Hilaire, M. Kowarsch and J. C. Minx, *Nature*, 2018, **561**, 303.
143. Friends of the Earth, Negatonnes – An initial assessment of the potential for negative emission techniques to contribute safely and fairly to meeting carbon budgets in the 21st century, 2011.
144. Vivid Economics for WWF Scotland, Delivering on Net Zero: Next Steps for Scotland, 2019. https://www.wwf.org.uk/sites/default/files/2019-10/WWF_Report_VIVID_Climate_2019_web.pdf.
145. One Earth, BECCS: No time for false saviours, 2020, https://oneearth.org/beccs-no-time-for-false-saviours/.

146. Climate Land Ambition and Rights Alliance, Missing Pathways to 1.5' C: The role of the land sector in ambitious climate action, 2018.
147. Global Forest Coalition, Global Forest Coalition Annual Report 2017, 2017.
148. Climate Institute, Negative Emission Technologies: Silver Bullet or Ethically Ambiguous?, 2019.
149. A.-M. Dowd and M. James, *Social Epistemol.*, 2014, **28**, 364.
150. E. Cox, E. Spence and N. Pidgeon, *Nat. Clim. Change*, 2020, **10**, 744.
151. Ibid.
152. H. Buck, *Global Sustainability*, 2018, **1**, E2.
153. R. Amundson and L. Biardeau, *Proc. Natl. Acad. Sci.*, 2018, **115**, 11652.
154. *Ibid.*
155. J. Unruh, *Soc. Nat. Resour.*, 2011, **24**, 185.
156. A.-M. Dowd, M. Rodriguez and T. Jeanneret, *Energies*, 2015, **9**, 4024.
157. P. Upham and T. Roberts, *Int. J. Greenhouse Gas Control*, 2011, **5**, 1359.
158. E. Cox, E. Spence and N. Pidgeon, *Nature Climate Change*, 2020, **10**, 744.
159. S. L'Orange Siego, J. Arvai, S. Dohle and M. Siegrist, *Int. J. Greenhouse Gas Control*, 2014, **25**, 23.
160. P. Ashworth, S. Wade, D. Reiner and X. Liang, *Int. J. Greenhouse Gas Control*, 2015, **40**, 449.
161. F. Karimi and A. Toikka, *Int. J. Greenhouse Gas Control*, 2018, **70**, 93.
162. L. Whitmarsh, D. Xenias and C. Jones, *Palgrave Communications*, 2019, 5, 17.
163. R. Bellamy, M. Fridahl, J. Lezaun, J. Palmer, E. Rodriguez and A. Lefvert, *et al.*, *Environ. Sci. Policy*, 2021, **116**, 47.
164. H. J. Buck, *Curr. SustainableRenewable Energy Rep.*, 2019, **6**, 124.
165. G. Thomas, N. Pidgeon and E. Roberts, *Energy Res. Soc. Sci.*, 2018, **46**, 1.
166. E. Cox, E. Spence and N. Pidgeon, *Nat. Clim. Change*, 2020, **10**, 744.
167. Climate Land Ambition and Rights Alliance, Missing Pathways to 1.5' C: The role of the land sector in ambitious climate action, 2018.
168. E. Cox, E. Spence and N. Pidgeon, *Nat. Climate Change*, 2020, **10**, 744.
169. N. Pidgeon and E. Spence, *Biol. Lett.*, 2017, **13**, 24.
170. E. Spence, N. Pidgeon and P. Pearson, *Mar. Policy*, 2019, **97**, 287.
171. K. E. Gannon and M. Hulme, *Geogr. Environ.*, 2018, **54**, e00054.
172. L. Waller, T. Rayner, J. Chilver, C. Gough, I. Lorenzoni and A. Jordan, *et al.*, *Clim. Change*, 2020, **11**, 1.
173. A. Dowd, M. Rodriguez and T. Jeanneret, *Energies*, 2015, **9**, 4024.
174. M. Ferguson, K. Witt, C. Nisa and P. Ashworth, Effects of message framing on the support for carbon capture and storage and alternative energy technologies: Final report, 2019.
175. R. Colvin, L. Kemp, A. Talberg, C. De Castella, C. Downie and S. Friedl, *et al.*, *Environ. Commun.*, 2020, **14**, 23.
176. B. Terwel, F. Harinck, N. Ellemers and D. Daamen, *Int. J. Greenhouse Gas Control*, 2011, **5**, 181.
177. H. J. Buck, *Curr. Sustainable Renewable Energy Rep.*, 2019, **6**, 124.

178. R. Colvin, L. Kemp, A. Talberg, C. De Castella, C. Downie and S. Friedl, *et al.*, *Environ. Commun.*, 2020, **14**, 23.
179. S. Fuss, J. Canadell, G. Peters, M. Tavoni, R. Andrew and P. Ciais, *et al.*, *Nat. Clim. Change*, 2014, **4**, 850.
180. G. Peters and O. Geden, *Nat. Clim. Change*, 2017, **7**, 619.
181. A. Aggarwal and D. Brockington, *Land Use Policy*, 2020, **95**, 104608.
182. J. Fuhrman, H. McJeon, S. Doney, W. Shobe and A. Clarens, *Front. Clim.*, 2019, **1**, 11.
183. C. von Stechow, J. Watson and B. Praetorius, *Global Environ. Change*, 2011, **21**, 346.
184. H. Daggash and N. Mac Dowell, *Joule*, 2019, **3**, 2120.
185. J. Kemper, *Int. J. Greenhouse Gas Control*, 2015, **40**, 401.
186. E. Cox and N. R. Edwards, *Clim. Policy*, 2019, **19**, 1144.
187. J. Bednar, M. Obersteiner and F. Wagner, *Nat. Commun.*, 2019, **10**, 1783.
188. J. Fuhrman, H. McJeon, S. Doney, W. Shobe and A. Clarens, *Front. Clim.*, 2019, **1**, 11.
189. G. Peters and O. Geden, *Nat. Clim. Change*, 2017, **7**, 619.
190. H. Daggash and N. Mac Dowell, *Joule*, 2019, **3**, 2120.
191. P. Zakkour, J. Kemper and T. Dixon, *Energy Procedia*, 2014, **63**, 6824.
192. A. Torvanger, *Climate Policy*, 2019, **19**, 329.
193. H. Daggash and N. Mac Dowell, *Joule*, 2019, **3**, 2120.
194. S. Jenkins, E. Mitchell-Larson, S. Haszeldine and M. Allen, *Sustainable financing of a permanent CO_2 disposal through a Carbon Takeback Obligation*, 2020.
195. A. Aggarwal and D. Brockington, *Land Use Policy*, 2020, **95**, 104608.
196. E. Moe and J. Rottereng, *Energy Res. Soc. Sci.*, 2018, **44**, 199.
197. R. Bellamy, M. Fridahl, J. Lezaun, J. Palmer, E. Rodriguez and A. Lefvert, *et al.*, *Environ. Sci. Policy*, 2021, **116**, 47.
198. E. Cox, E. Spence and N. Pidgeon, *Nat. Clim. Change*, 2020, **10**, 744.
199. G. Peters and O. Geden, *Nat. Clim. Change*, 2017, **7**, 619.
200. A. Aggarwal and D. Brockington, *Land Use Policy*, 2020, **95**, 104608.
201. H. J. Buck, *Clim. Change*, 2016, **139**, 155.
202. R. Bellamy, *Nat. Energy*, 2018, **3**, 532.
203. S. Fuss, J. Canadell, P. Ciais, R. Jackson, C. Jones and A. Lyngfelt, *et al.*, *One Earth*, 2020, **3**, 145.
204. H. J. Buck, *Clim. Change*, 2016, **139**, 155.
205. M. Honegger and D. Reiner, *Climate Policy*, 2018, **18**, 306.
206. C. Pozo, A. Galan-Martin, D. M. Reiner, N. Mac Dowell and G. Guillen-Gosalbez, *Nat. Clim. Change*, 2020, **10**, 640.
207. C. Gough, S. Garcia-Frietes, C. Jones, S. Mander, B. Moore and C. Pereira, *et al.*, *Global Sustainability*, 2018, **1**, 1.
208. D. M. Reiner, *Nat. Energy*, 2016, **1**, 15011.
209. H. Daggash and N. Mac Dowell, *Joule*, 2019, **3**, 2120.
210. G. Peters and O. Geden, *Nat. Clim. Change*, 2017, **7**, 619.
211. E. Cox, N. Pidgeon and G. Thomas, *Front. Environ. Sci.*, 2018, **6**, 1.

212. C. Fyson, S. Baur, M. Gidden and C. Schleussner, *Nat. Clim. Change*, 2020, **10**, 836.
213. C. Pozo, A. Galan-Martin, D. M. Reiner, N. Mac Dowell and G. Guillen-Gosalbez, *Nat. Clim. Change*, 2020, **10**, 640.
214. W. Carton, A. Asiyanbi, S. Beck, H. Buck and J. Fund, *Climate Change*, (forthcoming).
215. C. Fyson, S. Baur, M. Gidden and C. Schleussner, *Nat. Clim. Change*, 2020, **10**, 836.
216. C. Pozo, A. Galan-Martin, D. M. Reiner, N. Mac Dowell and G. Guillen-Gosalbez, *Nat. Clim. Change*, 2020, **10**, 640.
217. M. Honegger and D. Reiner, *Clim. Policy*, 2018, **18**, 306.
218. E. Cox, N. Pidgeon and G. Thomas, *Front. Environ. Sci.*, 2018, **6**, 1.
219. N. L. Harris, D. A. Gibbs, A. Baccini, R. A. Birdsey, S. de Bruin, M. Farina, L. Fatoyinbo, M. C. Hansen, M. Herold, R. A. Houghton, P. V. Potapov, D. R. Suarez, R. M. Roman-Cuesta, S. S. Saatchi, C. M. Slay, S. A. Turubanova and A. Tyukavina, Global maps of twenty-first century forest carbon fluxes, *Nat. Clim. Change*, 2021, **11**(3), 234–240.
220. P. Poutlon, J. Johnston, A. Macdonald, R. White and D. Powlson, Major limitations to achieving '4 per 1000' increases in soil organic carbon stock in temperature regions: Evidence from long-term experiments at Rothamsted Research, United Kingdom, *Global Change Biol.*, 2017, **24**(6), 2563–2584.

CHAPTER 14

Priorities for Policy Design

HABIBA AHUT DAGGASH, MAI BUI AND NIALL MAC DOWELL*

Imperial College London, Exhibition Road, London SW7 2AZ, United Kingdom
*Email: niall@imperial.ac.uk

Results from integrated assessment models (IAMs) have made it increasingly evident that atmospheric greenhouse gas removal (GGR), namely carbon dioxide removal (CDR), otherwise known as negative emissions, is critical to delivering the Paris Agreement. The high levels of deployment of negative emissions technologies (NETs) anticipated in IAMs are in stark contrast with their early stage of development and deployment globally. Some CDR methods are yet to be proven as technically feasible at scale owing to insufficient research, development and demonstration (RD&D). Others, such as BECCS and DACCS, have failed to achieve commercialisation because of high costs and the significant uncertainties surrounding the wider environmental, social and political implications of their large-scale deployment. Furthermore, the necessary policy and regulatory frameworks that encourage technological innovation in NETs have not been implemented. Failure to design appropriate policies to encourage the commercialisation of NETs risks foregoing a portfolio of technologies that appear critical to the climate change mitigation challenge. This chapter discusses the innovation ecosystem of GGR and NETs and proposes policy measures that facilitate different aspects of innovation including RD&D, financing, niche markets, regulation and governance, and public acceptability.

Energy and Environment Series No. 31
Greenhouse Gas Removal Technologies
Edited by Mai Bui and Niall Mac Dowell

Published by the Royal Society of Chemistry, www.rsc.org

14.1 Introduction

Recent IPCC reports have highlighted that stabilising atmospheric greenhouse gas (GHG) concentrations at levels that would avoid catastrophic climate change—defined as an average temperature rise of "well below 2 °C above pre-industrial levels" by the 2015 Paris Agreement—requires significant deployment on negative emissions, principally in the second half of the century.[1–3] Analysis of the most recent deep mitigation pathways under the framework of policy assumptions known as Shared Socioeconomic Pathways (SSP) reveals that 95% of the unconstrained scenarios that limit temperature rise to 2 °C by the end of the century rely on both negative emissions and net-negative global carbon emissions post-2050.[4,5] This figure is 100% for scenarios that achieve the more stringent 1.5 °C limit of warming (also sought by the Paris Agreement).[4] The alternative scenarios that did not require carbon removal need CO_2 reduction to occur at rates that are currently deemed implausible.[6] Integrated assessment models (IAMs) show that even if emissions reduction targets committed to under the current nationally determined contributions (NDCs) are achieved, rapid upscaling of negative emissions is required to deliver the 2 °C target. By 2030, 0.2–0.8 Gt CO_2 per year of additional negative emissions are required to deliver 3–14 Gt CO_2 per year of carbon removal globally by 2050.[7] For 1.5 °C scenarios, even quicker upscaling may be required to deliver up to 16 Gt CO_2 per year of negative emissions by mid-century. Such high deployment levels are at odds with the relatively early stage of research and development of the portfolio of greenhouse gas removal (GGR) and negative emissions technologies (NETs) that are required in this effort.[8] The rate at which GGR and NETs can be scaled up to achieve commercial viability at affordable costs, deliver climate and non-climate benefits with limited harmful side-effects will determine their utility in addressing climate change and ultimately delivering the Paris Agreement.[9]

Historical evidence suggests that the commercialisation and widespread adoption of novel technologies can take up to several decades, although exceptions with relatively rapid adoption exist, *e.g.*, transition of energy technologies.[9–11] However, the urgency of the climate change problem means that the typically lengthy innovation process needs to be accelerated. Accelerating NETs deployment will require coordination and collaboration between both the public sector, which can use policy and regulation to incentivise research, development and demonstration, and the private sector which can channel vast capital and expertise to achieve commercialisation and scale.[12] The bulk of academic literature has thus far focused on technoeconomic assessments of the deployment potentials, economic costs and environmental impacts of the different NETs.[8] This cost-centric approach ignores the regulatory and social barriers to NETs deployment that also need to be overcome through policy.[12–14] Furthermore, the understanding of market potential and revenue opportunities that is needed to drive investment in specific NETs is lacking—this information is more valuable to the investment community given the unpredictability of gigatonne-scale NETs costs.[12] This chapter highlights how current policy mechanisms hinder the

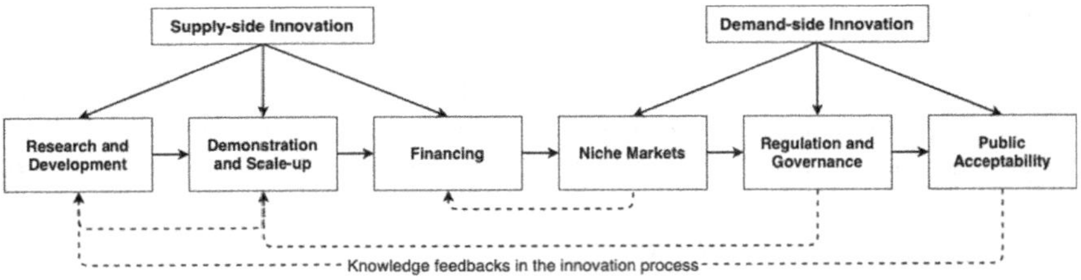

Figure 14.1 Supply- and demand-side innovation policies relevant to NETs development and deployment.

development of NETs and assesses the type of policies that are necessary to overcome existing barriers to innovation and upscaling of NETs.

Two distinct types of innovation are often highlighted: (1) supply-side measures that increase the supply of a technology, and (2) demand-side measures that increase the demand for a technology.[9,15] Supply-side measures typically reduce the costs and improve the performance of a technology through education, research and cheaper financing. Demand-side measures involve the creation of markets in which technologies can compete, consumer preferences for that technology, regulation and governance of technologies and their markets, and the public acceptability of technologies.[9,15] To investigate how policy can support the different stages of innovation of NETs to advance climate change mitigation, we define six aspects of NETs that need support: research and development, demonstration and scale-up, financing, niche markets, regulation and governance, and public acceptability (see Figure 14.1).

14.1.1 Characterising NETs for Policy Development

NETs typically consist of two steps: a capture process in which CO_2 is removed from the atmosphere, and the subsequent sequestration of the CO_2 in some form of permanent storage. Whilst broadly described as technologies, the NETs that have shown potential for scale-up (discussed in previous chapters) include capture and storage processes that directly interact with either the natural environment (as in afforestation and ocean fertilisation), technological systems (as in DACCS) or both (as in BECCS), or with the built environment (as with low-carbon concrete and building with biomass). The heterogeneity of NETs means that the new or adapted policy mechanisms needed to incentivise their deployment traverse different policy domains. For example, biomass-based mitigation measures such as BECCS and biochar need changes in agricultural and land policy, in addition to broader climate policy that governs carbon accounting frameworks. BECCS by virtue of its energy production also needs to be integrated into an evolving energy system so energy policy must transform to accommodate its participation. Similarly, NETs that create cross-border impacts such as ocean fertilisation will need changes to international law (in addition to coastal and marine policy). To understand the different policy domains that can be used to incentivise NETs deployment, we classify them by their capture method and carbon sequestration location (see Table 14.1). This will be used as

Table 14.1 Classification of NETs according to their CO_2 capture method and storage location. NETs with integrated CO_2 storage are indicated in *italic* while methods with separate storage mechanisms are in **bold**. Adapted from ref. 19, https://www.raeng.org.uk/publications/reports/greenhouse-gas-removal, under the terms of the CC BY 4.0 license https://creativecommons.org/licenses/by/4.0/.

		Greenhouse gas removal method		
		Increased biological uptake	Natural inorganic reactions	Engineered removal
Storage Location	Land vegetation (living)	*Afforestation, reforestation and forest management; Habitat restoration*		
	Soil and land vegetation (dead)	*Soil carbon sequestration; Biochar*	*Enhanced terrestrial weathering*	
	Geological reservoirs	**BECCS**	**Mineral carbonation at surface**	**DACCS with geological storage; DACCS with sub-surface mineral carbonation**
	Oceans	*Ocean fertilisation*	*Ocean alkalinity*	**DACCS with deep ocean storage**
	Built environment	**Building with biomass**		**Low-carbon concrete**

a framework to discuss the supply- and demand-side innovation measures that can facilitate the different archetypes of NETs.

14.2 Research and Development

Research and Development (R&D) is the process through which new knowledge is created for the purpose of developing new (or improving existing) technologies and processes. Within the NETs context, research involves scientific studies and computer modelling of NETs systems, whilst development involves conducting experiments and building prototypes to improve the technologies.[9,16] With the exception of afforestation and reforestation (AR), and soil carbon sequestration (SCS) that are available for large-scale deployment, most NETs require further R&D before they achieve technical and commercial viability, as set out in Table 14.2.[14]

Given the lack of policy certainty around NETs, the private sector has been reluctant to invest in R&D in the field.[9] Public R&D has historically been important in driving innovation in new technologies, when there are no incentives for the private sector.[17,18] It is therefore critical that governments ramp up investment in R&D to support the development of NETs in a timely fashion, whilst concurrently being cognisant of the various constraints that will be required to be met in order to effectively contract for these services.

14.2.1 Technological Feasibility

There remains significant uncertainty surrounding the ability of some NETs—most notably enhanced weathering, ocean fertilisation and biochar—to verifiably deliver accurately quantifiable CDR in practice, and to do so in a manner that does not pose other challenges to the environment, *e.g.*, the disruption of marine ecosystems. Reducing this uncertainty requires concerted efforts in basic scientific R&D; pilot- and demonstration-scale experiments to understand the viability of these technologies in field conditions; and social sciences research to understand how to improve the public acceptability of these technologies, in addition to the development of baselining protocols, as well as tools and techniques for the quantification of the amount of CO_2 permanently removed. NETs that have been proven to be technically feasible (such as BECCS and DACCS) also need further research to bring the technologies for full operational readiness and to determine the optimal conditions for their operation.[19]

14.2.2 Costs

A wide range of cost estimates have been proposed for NETs.[8,20,21] Uncertainty in costs often reflects uncertainty in the effectiveness of a technology,[19] and this limits investment. Further research on cost estimation of NETs and the technological (or other) innovations that can drive down initial costs is required.

Table 14.2 Side-effects and technology readiness levels of NETs. Adapted from ref. 8, https://doi.org/10.1088/1748-9326/aabf9f, under the terms of the CC BY 3.0 license https://creativecommons.org/licenses/by/3.0/.

	Side effects			
Technology	Positive	Negative	Development status of technology	Remaining barriers to development, deployment and upscaling
DACCS	Business opportunities; specific applications could improve indoor air quality	CO_2 penalty if high (thermal) energy demand is satisfied by fossil fuels; currently high upfront capital costs; mostly insufficiently studied; material/waste implications not known but cannot be excluded; some spatial requirements	Deployed in small niche markets; costs uncertain	Costs; mass production; leakage concerns; ensure governance that renders DACCS a CO_2 sequestration rather CO_2 utilisation technology
BECCS	Market opportunities; economic diversification; energy independence; technology development and transfer; GHG emissions substitution	Direct and indirect LUC; food security; biodiversity losses; deforestation and forest degradation; health impacts; impacts on soil and water; albedo change; CO_2 leakage	$\sim$1 full-scale demonstration plant in operation worldwide.	Costs; land-use competition; leakage concerns
Ocean Fertilisation	Potential increase in fish catches; enhanced biological production	Unknown impacts on marine biology and food web structure; changes to nutrient balance; anoxia in surface ocean; probable enhanced production of N_2O and CH_4	$\sim$10s of small-scale demonstrations	Efficiency of method; scale up; incentives for adoption; impacts on ocean and marine life
Afforestation and Reforestation	Implementation dependent benefits for soil carbon, soil quality, biodiversity, livelihoods, water retention, employment	Direct and indirect LUC; net-positive warming in high latitudes due to albedo effect; risks for biodiversity and food security.	Available at large scale	Lack of incentives for widespread adoption; competing land uses; concerns about offsetting temperature effect from albedo changes

Table 14.2 (*Continued*)

Technology	Side effects: Positive	Side effects: Negative	Development status of technology	Remaining barriers to development, deployment and upscaling
Soil Carbon Sequestration	Improved soil resilience and improved agricultural production; negative cost options; reduced pollution and improved soil quality; positive impacts on soil, water and air quality	Possible increase in N_2O emissions and N and P; need for addition of N and P to maintain stoichiometry of soil organic matter	Available at large scale	Incentives for widespread adoption; concerns about permanence
Biochar	Increased crop yields and reduced drought; reduced CH_4 and N_2O emissions from soils; improved soil carbon, nutrient and water cycling impacts	Competition for biomass resources; direct and indirect LUC; potential increase in plant vulnerability against insects, pathogens, and drought; albedo change partly offsetting mitigation effect	Pyrolysis capacity limited at present	Incentives for widespread adoption; costs of pyrolysis
Enhanced Weathering	Increase in crop yields; improved plant nutrition; improved soil fertility, nutrient and moisture; increase in soil pH; increasing cation exchange capacity in depleted soils	Human health risks associated with fine-grained material; ecological impacts of mineral extraction and transport; potential heavy metal release; changes in soil hydraulic properties	Tailored mineral production limited at present due to unknown optimal parameterisation of products	Incentives for widespread adoption; costs of mineral production and distribution

14.2.3 Scalability

The potential of NETs to deliver climate change mitigation at scale is still questioned because of the uncertainty in the availability of necessary inputs including suitable land, energy, water and CO_2 storage capacity. R&D is needed to quantify the potential scalability of NETs in different geographies based on the bio-geophysical resources that are available. Thus far, R&D has focused on NETs deployment in advanced economies, however the centres of population growth—and consequently economic growth and GHG emissions—are in developing countries.[22] Should developing countries pursue a carbon-intensive economic growth, there will be a greater need for NETs to achieve the GHG emissions reductions necessary to deliver the Paris Agreement. R&D funding needs to be diversified to focus on investigating the potential of NETs in different geographies and to determine the enabling factors for their deployment.

14.2.4 Measuring, Verification and Reporting Guidelines

The renumeration of negative emissions is critical to delivering NETs at scale. However, renumeration requires accurate measurement, verification and monitoring of the carbon sequestered *via* different methods. Widely accepted monitoring standards currently only exist for terrestrial biomass-based NETs such as afforestation and reforestation, and the accuracy of these approaches may not prove viable for afforestation/reforestation for CDR purposes. Research is needed to develop measurement, reporting and verification (MRV) guidelines for other NETs. These include biomass sustainability certification standards; CO_2 storage monitoring and certification; and cross-border carbon accounting frameworks owing to the transnational nature of NETs operation and impacts.

14.2.5 Environmental Impacts

Some NETs utilise significant natural resources and require the disruption of ecosystems—AR and BECCS require extensive land-use change, ocean fertilisation affects marine ecosystems, *etc.* R&D is needed to determine the extent of environmental disruption that NETs deployment will have in different geographies. This is especially important for ocean fertilisation as significant modifications to marine ecosystems are needed to achieve a small removal of CO_2.[19] Better ecosystem modelling capabilities are required before deployment.[19]

14.3 Demonstration and Scale-up

Although most NETs have been proven as technically feasible through pilot- and lab-scale experiments, they are yet to be progressed to a higher technology readiness level (TRL).[16,19,23] Demonstration projects are necessary to assess the performance and feasibility of NETs in field conditions. Thus far, both government and the private sector have been reluctant to fund such projects. The private sector is deterred by the free-riding problem; demonstrations of a technology by a firm will generate knowledge spill-overs that the firm's competitors can learn from without making any investment themselves.[24] Government is often hesitant, owing to the scale of investment needed, or to avoid the perceptions that it is "picking winners".[9] The weak incentives for the private sector mean that government efforts are needed to foster collaborations (either government–government or public–private) that encourage investment to demonstrate NETs feasibility. The innovation ecosystem relies on the feedback of knowledge between different innovation processes to refine research objectives and design choices (see Figure 14.1). Moderate-scale demonstration of NETs can reduce uncertainty in cost estimates and help to define technology cost reduction trajectories.

Scale-up involves the increase in the unit size of a technology. For NETs, the scaling of projects is crucial to achieve increased carbon dioxide removal from the atmosphere, and to assess the integration of NETs into existing infrastructure. BECCS and DACCS, for example, require access to CO_2 pipelines and may need to integrate with the electricity grid as they can produce and consume power, respectively. Enhanced weathering (EW) and ocean fertilisation (OF) require mining and transport of minerals. Initial large-scale projects will inform on the wider social and environmental impacts of NETs, and enable the process of cost discovery and reduction.[19] Given the scale of investment needed for commercial-scale projects, NETs cannot be solely funded by the government. Instead, the private sector will have to take a leading role in early demonstrators and large-scale deployment. Private investment is driven by risk-return prospects, thus financial incentives are needed for NETs to prove attractive. The following section discusses policy measures that can be implemented to encourage the financing of NETs projects.

14.4 Financing

Insufficient knowledge of the potential of NETs to deliver greenhouse gas removal at gigatonne-scale, their impacts on the natural environment, and their broader social and political acceptability—partially owing to the insufficient funding for research, development and demonstration projects—have contributed to uncertainty about their future utility in mitigation efforts. Consequently, they are perceived as high-risk technologies that should be a last resort for mitigation, if at all.[19,25,26] Studies have highlighted that even the NETs that have been proven to be technically feasible (soil carbon sequestration, BECCS and DACCS) have failed to achieve commercial viability and scale because their carbon abatement costs may be significantly greater than existing alternative mitigation options.[8,21,26–28] Furthermore, NETs do not qualify for the incentive mechanisms that have been implemented in many countries to promote climate change mitigation, such as carbon pricing and renewable energy subsidies.[29] The uncertainty in the future potential (and hence market) for negative emissions, and the absence of policy and regulatory frameworks to encourage deployment have led to a lack of financing for projects. An emerging body of literature is proposing a range of financing mechanisms and commercial delivery opportunities to drive investment in NETs. Broadly, these can be classified as: NET-based cost–benefit analysis, NET-based carbon markets, NET-based carbon taxes, and NET-based subsidies.

14.4.1 NET-based Cost–benefit Analysis

The social cost of carbon (SCC) is increasingly being used by governments in policy and regulatory design and evaluation, as it monetises the damage caused by GHGs, thereby allowing for a cost–benefit analysis (CBA) of policies that are designed to stimulate economic growth.[30,31] The SCC is the net economic cost or benefit that one tonne of carbon dioxide emissions poses to society.[32] The SCC has been found to be limited as a metric owing to the significant uncertainty and subjectivity (through assumed discount rates, for example) in the IAMs that produce the estimates.[33] A meta-analysis of 1214 SCC estimates found that they ranged from −10 to 105 000 $ per t_{CO2} (2010 USD) with a mean of US$317 per t_{CO2}.[34] The huge variation in estimates leaves them subject to bias in selecting a figure for policy design, and the large costs make their implementation politically infeasible in the near-term.

It has been suggested that the marginal cost of delivering one tonne of negative CO_2 emissions should replace the SCC in cost–benefit analyses, as the cost of removal serves as a practical ceiling for carbon prices. Snyder (2020)[33] proposes that a NET marginal abatement cost curve be used for CBAs such that a proposed project would need to generate a benefit to society greater than the costs to remove the resultant CO_2 emissions (see Figure 14.2).[33] Since the project brings a net benefit to society in the form of economic growth, society (and not the emitter) can bear the costs of removing the CO_2 emissions. This may be through government subsidies or other public financial support for NETs projects.

The use of the SCC for policy and regulatory analysis has been criticised for the transnational and intergenerational equity issues that it raises.[33,35] The 'polluter pays principle' that is widely accepted in environmental and climate policy mandates that a polluter pays for the damage they cause to the environment. However, policies based on the SCC concept such as carbon pricing or taxation mechanisms, which are principally implemented in developed countries, can increase government revenues, which in turn, are subsequently spent domestically. Since the burden of climate change impacts is felt in developing countries,[36] the affected party has not been compensated for the costs they bear—hence leading to transnational inequity. As previously highlighted, there is significant uncertainty in SCC estimates because of the uncertainty and subjective assumptions in the underlying models that are used. This leaves the possibility for policymakers

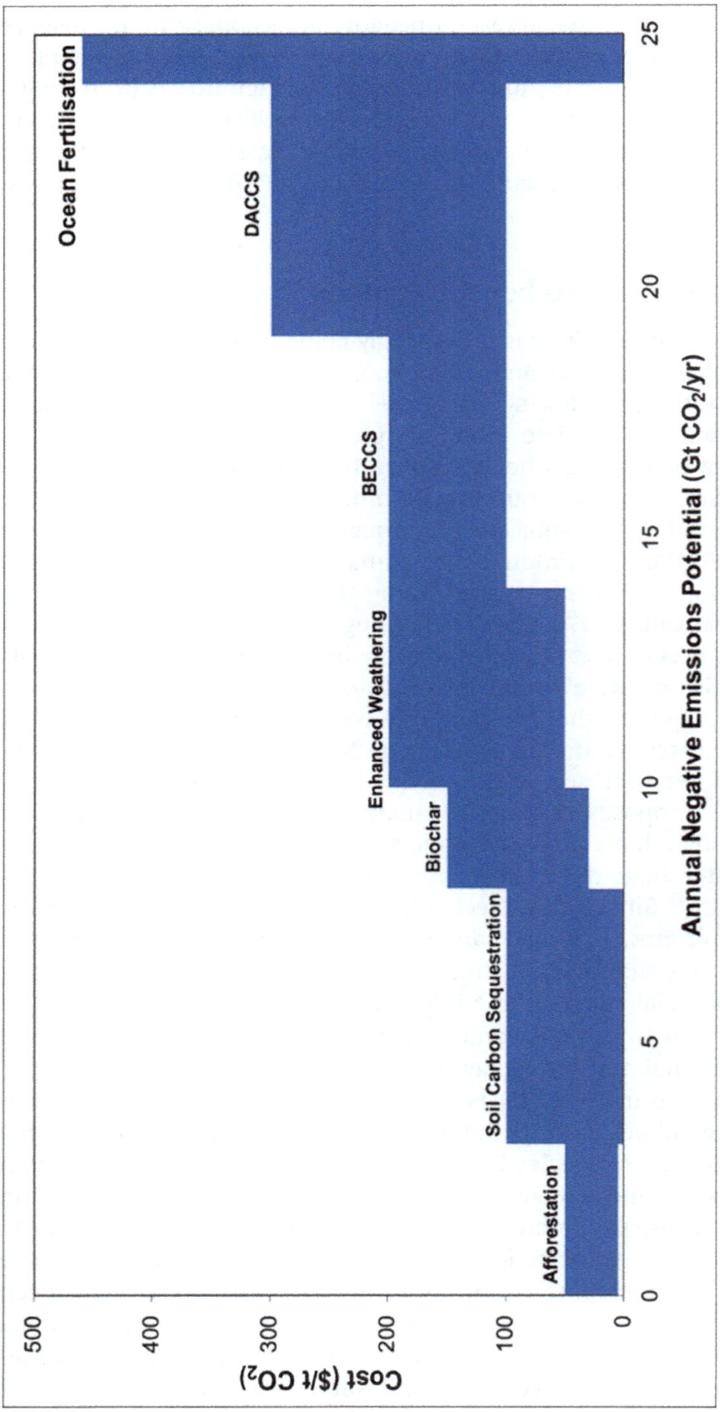

Figure 14.2 Marginal abatement cost curve for NETs. NETs CO_2 potentials and cost ranges obtained from expert estimates in Minx *et al.*[16]

to select SCCs that are either too low (so the present generation bears too little costs for future damage) or too high (so the present generation bears too much of the cost) thereby creating intergenerational inequity.

The use of the marginal abatement cost of NETs in CBAs avoids the highlighted issues as it ties the cost of carbon to a physical process which has a smaller variance in costs relative to the SCC. Additionally, it compares the current economic costs of a project to current CO_2 removal costs, not the future cost of climate change impacts as with the SCC. Finally, this kind of policy would also recognise the provision of negative emissions as a public good and thus make it eligible for public support.

14.4.2 NET-based Carbon Taxes

Economists often view climate change as a market failure to account for the negative externalities of economic activities that produce GHG emissions.[37] To address this failure, carbon pricing schemes that place a price on carbon and shift responsibility for the social costs of emissions to emitters have been proposed and adopted in many countries. Pricing carbon should drive down emissions and encourage investment in less carbon-intensive goods and services. Two types of carbon pricing schemes are predominantly implemented: (1) cap-and-trade programmes (sometimes known as an emissions trading scheme, ETS) which set a limit on GHG emissions for an industry or sector through the issuance of permits, and creates a market in which emitters can exchange permits, and (2) carbon taxes that penalise emitters for the GHGs they emit to the atmosphere at a pre-specified carbon price.[38]

Carbon taxation mechanisms have been limited in their effectiveness to deliver NETs for two reasons. First, they currently only penalise CO_2 emitters for the pollution they cause but fail to credit CO_2 removals.[22,29,39] This fails to recognise the value that CO_2 removals, and thus NETs, present to society. Adapting carbon taxation schemes to credit negative emissions would create a needed revenue stream for NETs project developers and encourage investment. A study found that such a reform could mean that lower—and therefore more politically-deliverable—carbon taxes are needed to deliver deep decarbonisation.[29] Secondly, carbon prices under taxation schemes have generally been well-below the marginal abatement cost of NETs. Thus, even if they credited negative emissions, they would be insufficient to incentivise investment. There are 30 carbon taxes currently implemented or scheduled globally; although prices range from US$1–119 per t_{CO2e}, more than half of emissions are taxed at less than US$10 per t_{CO2e}.[38] As observed in Figure 14.2, this is below the cost of delivering most NETs.

In addition to crediting NETs, higher carbon prices are needed to incentivise deployment. The social cost of carbon is often taken to the optimal carbon price and used to inform CO_2 pricing policy.[40] However, as previously described, there is considerable uncertainty and subjectivity in SCC estimates, which limits their effectiveness in policy design. The marginal abatement cost of NETs provides a better guide for price setting as it ties the tax to the physical removal of CO_2 from the atmosphere.[33] A NET-based tax could mean that revenues generated could be channelled into NETs development and deployment.

14.4.3 NET-based Carbon Markets

Several regional emissions trading and carbon offsetting schemes exist but they have failed to incentivise the verifiable, permanent sequestration of CO_2—crucial for NETs—by encouraging the short-term utilisation of CO_2 instead.[22,41] This limits the markets available for NETs developers. Nonetheless, some developers are already attempting to use voluntary carbon offset (VCO) markets to generate revenues for their projects. Climeworks, a Swiss company that has integrated a direct air capture process with the rapid mineralisation of CO_2, has begun to offer carbon credits for their provision of negative emissions on voluntary carbon offset markets.[42,43] However, VCO markets remain highly variable in the legitimacy and permanence of the offsets offered, and in the regulation that governs them.[42] The reform of these programmes to include NETs would create a market for the provision of negative emissions. Additionally, the establishment of a single framework to ensure the equivalence and validity of offsets will improve their effectiveness. The means to achieve this is discussed in Section 14.6.

Alternatively, government could create a separate market for NETs by mandating that emitters offset a share of their GHG emissions by purchasing negative emissions credits (NECs). The price of credits is then decided by supply-demand dynamics within this new market. In such a situation, the role of government would be as a regulator to ensure that NECs reflect actual CO_2 removal from the atmosphere; unlike in a carbon tax system where government specifies the carbon price, ensures collection of taxes and uses taxes to fund NETs deployment.[33] It has been suggested that the use of financial instruments called tradable put options (TPOs) and future contracts can be the basis of a carbon market that includes NETs.[42,44,45] Futures contracts (sometimes called futures) are contracts that enable the buying and selling parties in a transaction to fix a price for the provision of a commodity at a future date. Using them for NETs would allow developers to raise finance for current projects whilst offering price security to current emitters who want to offset their emissions in the future. Coffman and Lockley (2017)[46] highlight that a NET-based futures market needs to be regulated to ensure that the contracts are more reliable, and that moral hazard is minimised, and suggest that only governments can realistically support such a market in the long-term.[45] Put options are "contracts that give the holder the right, but not the obligation, to buy or sell an underlying asset at a set price, if it moves beyond that price within a set timeframe".[44] Government-backed TPOs would signal support for NETs development and deployment, and enable project developers to raise capital relatively quickly.[42,44]

14.4.4 NET-based Subsidies

Subsidies granted to low-carbon technologies, particularly wind and solar power, have been successful in driving down their costs and encouraging their widespread deployment.[46] Despite increasing consensus that NETs are critical to delivering the temperature goals set out in the Paris Agreement, they have yet to receive similar levels of support for their deployment at scale. The provision of tax credits or other forms of subsidy for NETs projects by government would lead to their recognition as a public good and encourage investment.[47] One example of such a subsidy is the 45Q tax credit in the USA which offers US$50 per t CO_2 for the permanent sequestration of CO_2 in geological reservoirs, and $35 per t CO_2 for the utilisation of CO_2.[12] Carbon storage payments have been shown to present the second most significant revenue capture potential for NETs delivery.[12] Thus, a policy allowing NETs to be paid for the CO_2 they sequester could be a route to deployment and commercialisation. The subsidy payments offered to renewable energy technologies were often determined by technoeconomic analysis by the government.[48] Extending this to NETs would require government to conduct assessments of the costs of CO_2 removals to determine appropriate subsidies which may prove difficult because of the variance in the literature.[8]

It has been shown that the regions that have the greatest afforestation and reforestation (AR) potential are in developing countries.[8] International incentive mechanisms such as the REDD+ scheme under the United Nations Framework Convention on Climate Change (UNFCCC) that offer financial incentives for improved forest management and the expansion of forests can be expanded to encourage CO_2 removal *via* AR. REDD+ has been included as a results-based payment mechanism within the Paris Agreement however the stringency of verification required to access financing has proved restrictive and needs to be addressed.[3,49] Other subsidies could also be directed at nature-based NET solutions. There is a growing interest in including natural capital in national accounts, *i.e.*, the stock of wealth of a country.[50] Should natural capital be increasingly valued, NETs such as afforestation, reforestation and habitat restoration that provide ecosystem services and mitigate economic damage from climate change (through flood regulation by woodland, for example) could be subsidised as a public good. An ecosystem services payment is thus an avenue for the realisation of some NETs, although this is likely to be a limited revenue stream.[12]

Given BECCS' ability to produce electricity in addition to the service of CO_2 removal, some subsidies for low-carbon energy sources (*e.g.*, feed-in tariffs, renewables obligation, *etc.*) can be directly extended to support its deployment. For example, the criteria for eligibility for the contracts-for-difference (CfD) mechanism in the UK which guarantees stable revenues for low-carbon generators can be extended to include BECCS. A study found that a commercial-scale BECCS power plant would need a strike price guarantee of £133–188/MWh to achieve a 4–15% return on investment, respectively, and this is similar to subsidy levels offered to nascent renewable energy technologies.[47]

Importantly, this price range would cover the provision of both the power and CDR services provided by this technology. In practice, these services could be readily separated, and the power product provided to the electricity market as normal, thus reducing the marginal revenue support required to finance the removal of CO_2 from the atmosphere. Finally, the availability of an appropriate infrastructure for the transport and storage of the CO_2 will further reduce costs.

14.5 Niche Markets

Some NETs (*e.g.*, BECCS, DACCS) have their CO_2 storage mechanism separate from the capture process such that a relatively pure stream of CO_2 can be obtained for commercial or other uses (see Figure 14.1). For them to be realised at gigatonne-scale, extensive CO_2 transport and storage (T&S) infrastructure needs to be in place, however, this is currently absent globally. To overcome this barrier to their deployment, the CO_2 that is extracted from the atmosphere could be used to satisfy CO_2 demand centres that are small, isolated or distributed which typically pay above-market prices for the chemical. Several industries have been shown to present niche market opportunities for NETs and present commercial delivery opportunities.[12,20]

- **Commercial agriculture:** Elevated CO_2 levels in greenhouses have been shown to improve crop yields. Currently, CO_2 for this purpose is sourced from fossil fuel combustion exhaust gases which are piped to the greenhouse or trucked with liquefied CO_2 tanks. Co-locating NETs with commercial greenhouses could represent a use for the extracted CO_2 in the absence of T&S infrastructure to geological reservoirs. The CO_2 can also be used for the cultivation of algae, a potential source of food, biofuel or bioenergy (biomass-derived heat or electricity).[20] Biochar which can be used as fertiliser can also serve agricultural markets.[19]
- **Merchant CO_2:** CO_2 used in a wide range of applications, including enhanced oil recovery (EOR), carbonation of beverages, food processing and chemical synthesis, is commonly referred to as merchant CO_2. The majority of merchant CO_2 is sourced from ammonia plants (which in turn, rely on natural gas), with selling prices typically around \$200 per t_{CO2} but with the potential to be higher as was the case during the 2018 supply shortage.[51] DACCS and BECCS may be able to deliver CO_2 at lower prices (see Figure 14.2) and thus be able to outcompete ammonia-derived CO_2 in these markets.
- **Synthetic renewable fuels:** CO_2 can be used as a chemical feedstock for the synthesis of some transport fuels.[51] NETs can serve this market.

Despite the limited scale, serving niche markets presents an opportunity to hedge against the policy uncertainty surrounding NETs in the near-term.[9] Additionally, for technologies that have the potential for significant cost reductions to be achieved through learning-by-doing (*e.g.*, standardised CCS equipment, or modular DACCS units), niche markets offer a valuable innovation prospect. Thus, policies that encourage the use of NETs and their by-products in the industries described above would contribute to their development and deployment.

14.6 Regulation and Governance

The potentially transnational environmental, social and geopolitical impacts of NETs deployment warrant an international governance regime.[52] No binding international laws have been enacted for the regulation of NETs or geoengineering methods in part because "no general and binding definition of climate engineering exists under international law".[53] However, several multilateral agreements contain binding and non-binding elements that are relevant to geoengineering which constitutes NETs (see Table 14.3). Talberg *et al.* have highlighted that the absence of purposive regulation of NETs (and more broadly geoengineering) by both national and international laws have been as a result of the failure of state actors to resolve the tension between two legal norms: 'precaution' and 'harm minimisation', with respect to NETs.[52,54]

Precautionary approaches to governance typically seek to limit the deployment of a technology or process until its effects are broadly understood and appreciated.[54] These approaches, formalised through ocean dumping and biodiversity regimes (see Table 14.2), have particularly stifled the deployment of ocean fertilisation, ocean mineral carbonation and marine CO_2 storage technologies as they require disruption of marine ecosystems with still poorly understood consequences. The London Protocol under the London Convention (LP/LC) was enacted to prevent ocean pollution through the dumping or incineration of wastes in the ocean. Ocean fertilisation (OF) is the only NET allowed under the LP, however, this is limited to research purposes only. Despite permitting ocean fertilisation activities, contracting parties to the LP have repeatedly expressed concerns about ocean fertilisation experiments.[55] CO_2 is also considered as a waste in the original LC/LP so the use of sub-seabed geological reservoirs for CO_2 storage was prohibited. This prohibits Carbon Capture and Storage (CCS) projects that store CO_2 in ocean reservoirs. In 2009, parties to the LP agreed on an amendment, which allows for cross-border transportation and sub-seabed sequestration of CO_2 for CCS purposes (provided that the protection standards of all other LP requirements have been met).[56,57] This amendment is yet to be ratified (formally accepted by two-thirds of the parties to the LP) so the consideration of marine-based climate change mitigation proposals has been limited.[57] The Convention on Biodiversity also cautions that climate geo-engineering activities (which constitute NETs) should not take place "in the absence of science-based, global, transparent and effective control and regulatory mechanisms", but "with the exception of small-scale scientific research studies".[58] However, with some NETs such as ocean fertilisation, it is often difficult to distinguish between meaningful field studies and deployment.[59] Consequently,

Table 14.3 International legal instruments relevant to NETs. Adapted from ref. 52 with permission from Springer Nature, Copyright 2018.

Instrument *(Membership)*	Treaty's main objective	Articles, protocols, or amendments of potential relevance to geoengineering	Affected technologies
UN Convention on the Law of the Sea (UNCLOS) *(168 parties—does not include the USA)*	Establishes the legal framework for allocating, protecting and preserving the marine environment	Countries with coastal areas have jurisdiction over their maritime zone and the airspace above it. To conduct an activity in another country's internal or territorial waters requires consent.[92] Environmental impact assessment is required for potentially risky projects.	Ocean fertilisation; CCS
Convention on the Prevention of Marine Pollution by Dumping of Wastes and Other Matter (London Convention) *(87 parties—includes the EU, the USA, Canada, Russia, and China)*	Regulates activities that may constitute ocean pollution	The London Protocol to the London Convention (LC/LP), which has 47 parties (and does not include the USA or Russia), has passed resolutions relating to ocean fertilisation. A 2008 resolution was passed stating that *'ocean fertilization activities other than legitimate scientific research...should be considered as contrary to the aims of the Convention and Protocol and not currently qualify for any exemption from the definition of dumping'*.[93] A 2010 resolution provides guidance to Parties on assessing ocean fertilisation research proposals and for completing an environmental assessment with risk management and monitoring.[94] An LC/LP amendment regulates marine geoengineering activities and adds two new annexes: a list of regulated marine geoengineering activities (currently only ocean fertilisation is listed) and an assessment framework to define legitimate scientific research.[95]	Ocean fertilisation; NETs that utilise sub-seabed CO_2 storage
International Convention for the Prevention of Pollution from Ships and Protocol (MARPOL73/78) *(155 parties)*	Aims to eliminate marine pollution and accidental spillage of oil and other harmful substances resulting from ships.	Within certain specially designated protected areas, the Convention and its Protocol limit discharge into the sea of hazardous substances.[3]	Ocean fertilisation
	Establishes a legal framework for managing a	The treaty's ultimate objective is stabilisation of atmospheric greenhouse gas concentrations at a level	All NETs

United Nations Framework Convention on Climate Change (UNFCCC) *(197 parties—includes the USA)*	global climate change response	that would prevent dangerous anthropogenic interference with the climate system. Article 4(d) urges all Parties to 'promote and cooperate in the conservation and enhancement, as appropriate, of sinks and reservoirs [. . .] including biomass, forests and oceans as well as other terrestrial, coastal and marine ecosystems'.[3] Also, the Paris Agreement to the UNFCCC establishes an aim 'to achieve a balance between anthropogenic emissions by sources and removals by sinks of greenhouse'.[96]	
Convention on Biological Diversity (CBD) *(196 parties—does not include the USA)*	Aims for the conservation and sustainable use of biological diversity with fair and equitable sharing of the benefits arising out of such resource use	The Convention itself cautions against activities that threaten biodiversity and promotes activities that conserve biodiversity. The Conference of the Parties (COP) has adopted three decisions that relate to geoengineering. In 2008, the COP cautioned precautionary behaviour with regard to ocean fertilisation.[58] In 2010, the COP invited Parties to ensure that no geoengineering activities with potential biodiversity impacts take place 'in the absence of science-based, global, transparent and effective control and regulatory mechanisms', but 'with the exception of small-scale scientific research studies'.[97] In 2012, the COP invited Parties to address climate engineering research gaps.[98,99] The scientific advisory body of the CBD COP (the Subsidiary Body on Scientific, Technical and Technological Advice (SBSTTA)) was instructed to undertake three studies relating to geoengineering. The SBSTTA produced three reviews: one on the science and impacts, one on the legal framework, and one on the view of indigenous and local communities.[100]	All NETs

Table 14.3 (*Continued*)

Instrument *(Membership)*	Treaty's main objective	Articles, protocols, or amendments of potential relevance to geoengineering	Affected technologies
United Nations Convention to Combat Desertification in Those Countries Experiencing Serious Drought and/or Desertification, particularly in Africa (UNCCD) *(196 parties)*	Aims to combat desertification and mitigate the effects of drought	The Convention encourages national programs and international cooperation targeting desertification. It is relevant to NETs to the extent that these may affect the hydrological cycle.	All NETs
Convention on the Prohibition of Military or Any Other Hostile Use of Environmental Modification Techniques (ENMOD) *(77 parties—includes, the EU, the USA, Russia, and China)*	Prohibits military or hostile usage of environmental modification	Environmental modification is defined as 'any technique for changing—through the deliberate manipulation of natural processes—the dynamics, composition or structure of the Earth, including its biota, lithosphere, hydrosphere and atmosphere, or of outer space' (ENMOD Convention art II). The Convention promotes research into and collaboration on such activities if they are for peaceful purposes.[101] Prohibition applies only if another contracting Party endures destruction, damage, or injury that is widespread, long-lasting, or severe.[101]	All NETs
Antarctic Treaty System (ATS) *(53 parties, only 29 with voting status—includes the USA, Russia, Australia, and some EU countries)*	Regulates the use of Antarctica	The Convention's Protocol on Environmental Protection to the Antarctic Treaty requires that activities be planned and conducted to avoid 'adverse effects on climate or weather patterns' (PEPAT art 3.2(b)(i)). Scientific research into geoengineering, such as ocean fertilisation, would be permitted with a requirement for 'due consideration [. . .] to avoid detrimental effects on dependent and associated ecosystems, outside the Antarctic Treaty area',[102] 'significant adverse effects on air or water quality'; or 'significant changes in the atmospheric, terrestrial (including aquatic), glacial or marine environments'.[102]	All NETs

research that is necessary to understand the effectiveness and limitations of NETs that potentially affect biodiversity is precluded.

The second legal norm that guides international earth system governance mechanisms is 'harm minimisation'.[52] This approach promotes the research and deployment of processes and technologies that provide a benefit to mankind through the preservation of its natural environment. Given the consensus that climate change mitigation is necessary to avoid harmful disruption of our physical environment and establishes socio-economic systems, 'harm minimisation' has led to international agreements that encourage the implementation of climate change mitigation and adaptation strategies. Notable examples of this approach are the Paris Agreement through which countries committed to limiting the catastrophic consequences of climate change, and amendments to the Convention on Biodiversity that allow for small-scale scientific research into climate change mitigation methods although they could potentially affect biodiversity.

Whereas a precautionary approach limits NETs deployment because of the uncertainties surrounding the broader effects of the proposed methods on the environment, seeking to minimise harm from climate change prompts their research and investigation. State actors have failed to reconcile both objectives and thus have left the NETs regulation agenda to be driven almost exclusively by the research community.[52] The heterogeneity of the community in terms of both organisation type (academia, non-governmental organisations, intergovernmental organisation) and specialisation (scientists, economists, legal scholars, philosophers)—each with their own biases—leads to varied opinions on what is possible and acceptable for NETs' research and deployment. Furthermore, the lack of governance from state actors has left researchers to almost exclusively determine the agenda for both NETs research and deployment.[52] Given academia's predominant role in research, that community has thus far determined which NETs warrant further investigation, which creates the possibility of lock-in to a subset of technologies that are being explored—this is evident in the overwhelming focus on BECCS in the scientific literature despite the complexities surrounding its effectiveness, and socio-political barriers to its deployment on a large-scale.[8,16,23,60]

Whilst increased research, development and demonstration (RD&D) is needed to reduce the uncertainties surrounding some NETs deployment and allow policymakers to design more effective regulation, better regulatory frameworks are needed to encourage the deployment of NETs that have been demonstrated to be technically feasible.

14.6.1 Recognising NETs

The international laws that contain elements (mostly non-binding) that govern some NETs have been designed for other purposes: prevention of environmental pollution, protection of biodiversity, and discouragement of environmental warfare, amongst others. The absence of explicit governance frameworks for NETs is the outcome of the precautionary approach that national and international state actors have taken towards their regulation and this has compounded the uncertainty surrounding their role in delivering deep decarbonisation targets. Regulatory uncertainty has in turn limited NETs development and scale-up. It is vital that the importance of NETs in minimising the harm from climate change and delivering the Paris Agreement—which has repeatedly been highlighted in the scientific literature[61,62]—translates to the domain of policymaking. As a first step, some existing international treaties can be readily altered to highlight that NETs deployment is beneficial and should be a priority given the scale of the climate change mitigation challenge.

Given their mandate to protect marine ecosystems from pollution (see Table 14.2), the London Convention (LC) and London Protocol (LP) can serve as a basis for a governance mechanism of marine-based NETs which include ocean fertilisation (OF), iron fertilisation and CCS technologies that utilise sub-seabed CO_2 storage. However, the London Convention can no longer be amended so only the LP can be the basis of a "global, transparent and effective control and regulatory mechanism for ocean fertilisation activities and other activities that fall within the scope of the London Convention and London Protocol and have the potential to cause harm to the marine environment".[63] Between 1999 and 2013, the International Maritime Organization (IMO) commissioned several working groups to develop an "Assessment Framework for Scientific Research Involving Ocean Fertilization", with the possibility of extending this to include other marine geoengineering activities.[55,63] Concerns surrounding the effects of marine geoengineering led to their prohibition—with the exception of OF for scientific research purposes only—*via* amendments of the LP in 2013.[55] Given the difficulty in distinguishing between meaningful field-scale experiments and deployment, obtaining the permits to carry out OF has been difficult.[59] As previously mentioned, marine CO_2 storage is also still prohibited (as it is considered a form of dumping in the original LP) because the 2009 amendment that allows for cross-border transportation and sub-seabed sequestration of CO_2 for CCS purposes (provided that the protection standards of all other LP requirements have been met) is yet to come into force.[56,57] Ratification of the amendment requires formal acceptance by two-thirds of the parties to the LP which has still not occured.[57]

Whilst caution is necessary in the consideration of NETs because of their potentially harmful side effects to the natural environment, it is important that caution does not preclude the opportunity to demonstrate NETs, which will serve to further reduce the uncertainties surrounding them. Adapting the London Protocol to recognise the role that marine-based NETs (discussed in Chapter 10) can play in minimising climate change and the urgency of the mitigation challenge will allow for easier demonstration of technologies and will facilitate their scale-up. Furthermore, as sub-seabed geological reservoirs present a significant CO_2 storage capacity, the 2009 amendment to the LP that allows for transnational export of CO_2 and offshore geological storage needs to be ratified (by the contracting parties to the agreement) to allow for commercial-scale projects to be developed.[55] Delivering an LP that encourages the demonstration and deployment of NETs therefore requires significant diplomacy between contracting parties. A resolution that allows the transboundary export of CO_2 for CCS under certain conditions was adopted in 2019 as an interim solution before the 2009 amendment comes into force.[64]

Non-marine-based NETs do not have a similar international law such as the London Protocol that can serve as a basis for a global regulatory framework (see Table 14.2). They require national and international state actors to develop new governance mechanisms for them. The foundation of any such mechanism needs to be the transparent verification and reporting of the carbon removal that is being delivered by proposed NETs.[22]

14.6.2 Measurement, Reporting and Verification

The commercialisation of NETs is dependent on the recognition of CO_2 removal as a public good and the incentivisation of its provision (see Section 14.1.4). For negative emissions to be renumerated, they need to be verifiable, thus CO_2 removal needs to be tracked in both space and time to ensure accountability. Different archetypes of NETs require different policies to enable easy measurement, reporting and verification (MRV). Moreover, all of these elements are key prerequisites for the safe abandonment of a GGR project.

14.6.2.1 Biomass Certification Standards

Land-based NETs are complex because land is both a source of GHG emissions (*via* forest burning and clearing and land-use change for agriculture, among other activities), and a sink (*via* photosynthesis).[49,65] Land-use change (LUC)—the altering of the natural landscape for commercial or other purposes—can therefore be a significant source of GHG emissions if unmanaged. High initial LUC emissions, sometimes referred to as a carbon debt,[66] therefore result in delayed *net carbon removal* from land-based NETs.[22] In addition to LUC emissions, monitoring biomass-based NETs also requires the consideration of the fuel supply chain emissions, which are the GHG emissions arising from the processing and transport of biomass for large-scale utilisation. Consequently, the NETs' CO_2 negativity (the amount of CO_2 they remove from the atmosphere when life cycle emissions are accounted for) or their CO_2 efficiency (the ratio of CO_2 removed to CO_2 absorbed by the biomass) are not uniform over time.[67,68] If lifecycle emissions are sufficiently high, then there may be no net negative emissions achieved over the lifetime of the NET; this has been shown to be the case for BECCS power plants that use biomass with high carbon footprints.[67] Consequently, it is critical to establish biomass sustainability standards for NETs projects. Such standards exist at national levels as in the UK.[69] However, the lack of globally accepted definitions of what should be considered sustainable biomass presents challenges for the establishment of an international regulatory framework. The absence of a common reference(s) for the determination of LUC emissions compounds the challenges. Thus, the establishment and implementation of sustainability criteria and supply chain certification frameworks for biomass should be a priority for international policymakers. After the Kyoto Protocol, the UNFCCC's Subsidiary Body for Science and Technological Advice (SBSTA) sought to develop accounting guidelines for the land use, land use change and forestry (LULUCF) sector.[49] However, difficult negotiations resulted in political compromise; the rules agreed upon only included specific land activities and have been criticised for having significant loopholes that undermine their effectiveness.[49] These LULUCF accounting guidelines can be the basis of setting sustainability standards for land-based mitigation activities, but they must first be adapted to improve transparency, and extended to include relevant NETs.

To ensure that negative emissions are achieved *via* afforestation and reforestation (AR), forest management is necessary as burning and clearing activities can result in GHG emissions.[65] Thus, forest certification—the recognition of the management of forests in line with challenging environmental, social and economic requirements—is required for the long-term MRV of AR practices.[70] The absence of universal certification guidelines and accreditation frameworks has resulted in large discrepancies between the claims of different projects.[70] This, in turn, limits the possibility of establishing an international regulatory mechanism for forestry-based biomass sources which are used in AR and BECCS projects.[69]

14.6.2.2 Reporting Metrics

As highlighted earlier, tracking CO_2 in space and time is important for the accurate accounting of negative emissions. This is also the case for remuneration, as it is important to have a metric that represents the actual amount of physical CO_2 removal that has been delivered by a NET project, and therefore needs to be remunerated. The heterogeneity of NETs—some offer immediate negative emissions (*e.g.*, DACCS) whilst others provide delayed carbon removal owing to the potential for initial carbon debt, as in the case of BECCS, or a non-negligible establishment phase, as in the case of afforestation—means that different metrics are needed to quantify their CO_2 removal capabilities. The *rate of CO_2 removal* has been proposed as a metric for biological and mineral sinks including BECCS, AR, EW and biochar, however significant uncertainty remains for removal rates through EW.[22,68] Standardised metrics need to be developed by policymakers for the appraisal of NETs projects.

14.6.2.3 CO_2 Storage Monitoring

For CDR to be meaningful, it must translate into permanent sequestration of CO_2 from the atmosphere. The NETs that have been proposed sequester

CO_2 in both biological sinks (trees, land and oceans) and geological reservoirs. Whereas biological sinks reach their maximum capacity within decades, underground geological storage usually take up to centuries to reach sink saturation.[22] In the case of AR, the net rate of CDR obtained from forests falls to zero when they reach maturity.[71] Thus, forests need to be managed and monitored to ensure optimal CO_2 uptake over time, as any activities that lead to the degradation of the carbon stock (*e.g.*, land use change, forest fires) will threaten the permanence of the negative emissions initially achieved.[22] Forest certification schemes have been widely practiced.[72] Under the UNFCCC, the REDD+ programme recommended the establishment of national forest monitoring systems (NFMS). REDD+ guidelines can serve as the basis of a standardised forest certification methodology but diplomacy and policy action by all parties to the UNFCCC is necessary. Moreover, once a given stand of trees reaches maturity, that carbon stock, *i.e.*, the forest, needs to be preserved in perpetuity - this will inevitably add significant cost to forest-based CDR relative to the cost of conventional afforestation/reforestation for the purposes of addressing deforestation.

Geological reservoirs have been shown to be provide reliable CO_2 storage with near-zero leakage over geological timescales (~10 000 years).[73] However, monitoring remains necessary for reasons of liability and insurance.[41,60,73] Standards for the performance of geological reservoirs need to be developed and agreed by countries. Furthermore, CCS projects have stalled because of a failure to resolve the commercial risk associated with the storage part of the value chain, amongst other challenges.[60,74] Governments must develop business models that share project risk such that private sector actors will not be discouraged from undertaking CCS projects. Taken together, all of this will be key for the safe abandonment of CCS projects.

14.6.3 Cross-border Carbon Accounting

NETs require bio-geophysical (land, sustainable biomass, water, CO_2 storage), technical (industry expertise) and economic (labour, financing) resources—all of which are unequally distributed geographically—for their deployment at scale. The quantity and quality of these resources possessed by a country will determine the amount and cost of negative emissions that it can deliver. Achieving both a cost- and resource-optimal deployment of negative emissions at gigatonne-scale will therefore require cross-border trade or exchange of resource inputs for NETs (*e.g.*, sustainable biomass, energy, investment), and thus cooperation between countries.[75,76] For example, a BECCS power plant in the UK may use cheaper sustainable biomass from South America and transport captured CO_2 to geological reservoirs in Norway. This poses a challenge for GHG accounting frameworks—cross-border, multi-stakeholder supply chains raise the issue of which country and sector gets credited for the provision of negative emissions.[25] The need for international governance mechanisms to align MRV standards for NETs has been previously highlighted. Two further actions by policymakers will also enable international collaboration for NETs deployment. Firstly, the plethora of GHG accounting standards that currently exist need to be harmonised to establish a widely accepted, credible and transparent standard that covers all relevant sectors of the economy.[8] Lack of such a standard has deterred climate investors.[77] Secondly, within a new framework, guidelines need to be agreed for the allocation of removal credits for NETs that include cross-border supply chains.

14.6.4 The Role of Diplomacy

The deployment of NETs has largely been discussed within the context of advanced economies.[8,9,16] However, given the scale of negative emissions required to meet the Paris Agreement targets, NETs would be required in developing countries as well, many of which have abundant resources required for cost-optimal NETs deployment (*e.g.*, land, renewable energy sources).[8] Furthermore, the centres of population (and thus energy demand) growth in the coming decades are in developing countries.[78] Should these countries pursue carbon-intensive growth models, there would be an increased need for NETs in the future to compensate for the resulting GHG emissions. The technical expertise and financial resources needed to establish the NETs deployment potential in lower-income countries and to encourage investment are lacking because of weak institutions and budgetary constraints. The Paris Agreement stated that the costs of addressing climate change will be distributed according to the "principle of equity and common but differentiated responsibilities and respective capabilities, in the light of different national circumstances."[79] Thus, finance, technology transfer and capacity-building from developed countries is needed to further NETs development globally. Under the UNFCCC, special funds such as the Green Climate Fund and Global Environment Facility have been created to serve this purpose, but investment pledges by high-income countries are yet to be realised, and the financing that has been committed has thus far eluded NETs projects and focused on alternative mitigation/adaptation solutions.[22] Financing must be inclusive of NETs if the necessary expertise is to be developed, so that the technologies are sufficiently developed when their deployment becomes necessary.

Widespread development and deployment of NETs will require the international community to scale-up climate finance, establish global certification standards for sustainable biomass resources, harmonise GHG accounting guidelines, and implement global CO_2 storage monitoring practices. Diplomacy—through which bilateral or multilateral partnerships are formed—is therefore critical to ensure NETs delivery at the lowest cost and equitable distribution of deployment costs. Policymakers should therefore seek out collaborations in developing NETs strategies as planning and deployment of these technologies in isolation will likely result in increased costs.[75,76]

14.7 Public Acceptability

The scientific literature on NETs has often framed costs as the principal barrier to NETs deployment and commercialisation.[8,19,23] History however shows that public and social acceptability of certain technologies could have a major impact on the feasibility of technologies, even when they are commercially viable.[25] The adopters of a technology include actors that have both direct and indirect agency in determining whether they adopt a technology or not.[9] Governments that issue permits or licences for projects, and public or private sectors that decide on investment in a technology have direct agency. Individuals and communities are not directly involved in the adoption process, but they may exert their agency through democratic processes or civil action (*e.g.*, protests, campaigns). Thus, the widespread adoption of a technology often involves addressing a range of economic, social, political and cultural factors. NETs are already facing several public acceptance issues.

14.7.1 Moral Hazard

Concerns have been raised that the development of NETs will deter more immediate climate change mitigation action, thus raising a moral hazard.[80–82] A failure to realise NETs at the scale suggested by IAMs will therefore result in a greater burden of climate change impacts or mitigation costs being borne by later generations. This raises the issue of intergenerational equity. It has been found that "the images or analogues that the public and key stakeholders find most evocative with regard to each of the NETs will have an impact on their likely acceptability and prospects for wide-scale deployment."[25] Thus, should connotations of 'moral hazard' or 'injustice' persevere with regards to NETs, the adoption of the technologies at the necessary scale will be limited. The resulting unavailability of NETs will greatly reduce the flexibility of the mitigation toolbox.

14.7.2 Food–Fuel Debate

Land-based, and especially biomass-based mitigation activities raise concerns over competition for land, deforestation, uncertain impacts of monocultures on land, land-use change emissions, conservation of natural habitats for dangerous species and food price inflation.[83–89] The recent history of biofuels in Europe highlights how some of these concerns can limit the success of land-based mitigation. Environmentalists fear that the bioenergy industry will encourage deforestation and unsustainable biomass use, and political actors have highlighted food security challenges that may emerge from aggressive targets for the share of biofuels in the transport sector.[25] Consequently, the European Parliament voted for a reduced share of conventional land-based biofuels in the transport fuel mix to 6% (from 10%), and to report emissions arising from indirect land-use change for biofuel production.[25] Land-based NETs (particularly BECCS and AR) are likely to face similar resistance to their adoption on a large scale.

14.7.3 CCS

The majority of the public is unfamiliar with CCS technology, which is needed for the deployment of BECCS and DACCS.[90] Surveys have revealed a "not in my backyard (NIMBY)" attitude towards both CO_2 transport (pipelines) and storage.[90,91] After the whereabouts of 408 potential CCS sites in Germany were released in 2011, a study found that would-be host communities of CCS projects had significantly lower acceptance rates than those that were not.[91] Living in tourist or coal-mining regions also markedly decreases acceptance, despite the economic benefits that CCS may provide for the latter. Another study showed that, after receiving basic information about CCS, people were more averse to the presence of CO_2 pipelines near their property than CO_2 storage.[90] For BECCS and DACCS projects, it is crucial to improve the acceptability of the technologies within the regions and communities that they are to be deployed. Policymakers should engage all stakeholders to identify the reasons for a lack of public acceptance of CCS projects, and design interventions (and projects) to counter the negative attitudes.

14.8 Conclusion

There is a growing consensus that deep decarbonisation consistent with the Paris Agreement targets cannot be met without negative emissions technologies. Integrated assessment models estimate that carbon removals of 10–20 Gt CO_2 per year are required by 2100 to keep the average global temperature rise below 2 °C. Such high deployment levels are in stark contrast with the early stage of development of many NETs methods. With the exception of afforestation and reforestation, no NET has previously been deployed at scale. This chapter has identified several areas that need urgent attention from policymakers to facilitate the technological innovation, commercialisation and adoption of NETs globally: research and development, demonstration and scale-up, financing, niche markets, regulation and governance, and public acceptability. Dedicated policy measures are needed within each of these domains to resolve uncertainties surrounding environmental, economic and socio-political implications of NETs deployment. Of greatest importance is the need for financial incentives for negative emissions provisions. The high costs of many NETs and the limited markets for carbon removals have deterred investment. Designing carbon markets, subsidies or taxes that remunerate NETs developers for the removals they provide will attract investors. Investment in research, development and demonstration is also needed to further constrain the uncertainties surrounding the effects of NETs deployment, to drive down technology costs, and to attract investment by de-risking projects.

The transnational nature of NETs necessitates international regulatory and governance frameworks. Where cross-border supply chains (for biomass, for example) are necessary to develop a NETs project, cross-border carbon accounting frameworks are required to apportion GHG emission reduction credits equitably. Additionally, the heterogenous nature of NETs—some NETs do not offer immediate negative emissions owing to carbon debts in their construction—means that spatial and temporal monitoring is necessary. It is crucial to develop measurement, reporting and verification methodologies that coherently account for the life-cycle emissions of NETs, and ensure the permanence of the mitigation achieved through storage certification and monitoring. The Kyoto Protocol established a largely production-based carbon accounting framework which fails to account for leakages and allows for non-disclosure of emissions from some sectors.[21] The current overhaul of existing carbon accounting guidelines and methodologies should be used to more accurately cover the LULUCF sector, and to incorporate NETs into GHG accounting frameworks. Their inclusion in such mechanisms which have been globally accepted (through the UNFCCC) will serve as a strong foundation for future NETs policy integration.

Finally, the scientific literature has paid little attention to the public acceptability of NETs. Historically, acceptability has proven to be a barrier to technology deployment even when the technology presents economic and environmental benefits (*e.g.*, nuclear power and CCS in Germany). Initial findings have shown that NETs are perceived to pose a moral hazard because, it is thought, that their development will prevent more urgent and immediate climate action. Furthermore, CCS which is critical for the realisation of NETs faces "not in my backyard" attitudes. It is important that early demonstrators of NETs engage civil society to understand how to overcome negative perceptions, and to stimulate public support for wider NETs deployment.

Abbreviations

ATS	Antarctic Treaty System
AR	Afforestation and Reforestation
BECCS	Bioenergy with Carbon Capture and Storage
CBD	Convention on Biological Diversity
CCS	Carbon Capture and Storage
CDR	Carbon dioxide removal
COP	Conference of the Parties
ENOD	Convention on the Prohibition of Military or Any Other Hostile Use of Environmental Modification Techniques
DACCS	Direct Air Carbon Capture and Storage
EW	Enhanced Weathering of Minerals
IAM	Integrated Assessment Model
IMO	International Maritime Organization
IPCC	Intergovernmental Panel on Climate Change

LC	London Convention
LP	London Protocol
LUC	Land-use change
LULUCF	Land use, land use change and forestry
MRV	Measurement, reporting and verification
NDC	Nationally Determined Contribution
NET	Negative Emissions Technology
NFMS	National Forest Monitoring System
NIMBY	Not in my backyard
OF	Ocean Fertilisation
PEPAT	Protocol on Environmental Protection to the Antarctic Treaty
RD&D	Research, Development and Demonstration
REDD+	Reducing emissions from deforestation and forest degradation and the role of conservation, sustainable management of forests and enhancement of forest carbon stocks in developing countries
SBSTA	Subsidiary Body for Science and Technical Advice
SCS	Soil Carbon Sequestration
T&S	Transport and storage
UNCCD	United Nations Convention to Combat Desertification
UNFCCC	United Nations Framework Convention on Climate Change

References

1. IPCC, Climate Change, *Mitigation of Climate Change. Working Group III Contribution to the Fifth Assessment Report of the Intergovernmental Panel on Climate Change*, Cambridge University Press, Cambridge, United Kingdom and New York, NY, USA, 2014.
2. IPCC, in *Global Warming of 1.5 °C. An IPCC Special Report on the Impacts of Global Warming of 1.5 °C above Pre-industrial Levels and Related Global Greenhouse Gas Emission Pathways, in the Context of Strengthening the Global Response to the Threat of Climate Change*, ed. V. Masson-Delmotte, P. Zhai, H.-O. Pörtner, D. Roberts, J. Skea, P. R. Shukla, A. Pirani, W. Moufouma-Okia, C. Péan, R. Pidcock, S. Connors, J. B. R. Matthews, Y. Chen, X. Zhou, M. I. Gomis, E. Lonnoy, T. Maycock, M. Tignor and T. Waterfield, World Meteorological Organization, Geneva, Switzerland, 2018, p. 32.
3. United Nations Framework Convention on Climate Change, *21st Conference of the Parties*, 2015.
4. M. Workman, K. Dooley, G. Lomax, J. Maltby and G. Darch, *Environ. Sci. Policy*, 2020, **103**, 77–84.
5. J. Rogelj, A. Popp, K. V Calvin, G. Luderer, J. Emmerling, D. Gernaat, S. Fujimori, J. Strefler, T. Hasegawa, G. Marangoni, V. Krey, E. Kriegler, K. Riahi, D. P. van Vuuren, J. Doelman, L. Drouet, J. Edmonds, O. Fricko, M. Harmsen, P. Havlík, F. Humpenöder, E. Stehfest and M. Tavoni, *Nat. Clim. Change*, 2018, **8**, 325–332.

6. C. Holz, L. S. Siegel, E. Johnston, A. P. Jones and J. Sterman, *Environ. Res. Lett.*, 2018, **13**, 064028.
7. J. C. Minx, W. F. Lamb, M. W. Callaghan, S. Fuss, J. Hilaire, F. Creutzig, T. Amann, T. Beringer, W. de Oliveira Garcia, J. Hartmann, T. Khanna, D. Lenzi, G. Luderer, G. F. Nemet, J. Rogelj, P. Smith, J. L. Vicente Vicente, J. Wilcox and M. del Mar Zamora Dominguez, *Environ. Res. Lett.*, 2018, **13**, 063001.
8. S. Fuss, W. F. Lamb, M. W. Callaghan, J. Hilaire, F. Creutzig, T. Amann, T. Beringer, W. de Oliveira Garcia, J. Hartmann, T. Khanna, G. Luderer, G. F. Nemet, J. Rogelj, P. Smith, J. L. V. Vicente, J. Wilcox, M. del Mar Zamora Dominguez and J. C. Minx, *Environ. Res. Lett.*, 2018, **13**, 063002.
9. G. F. Nemet, M. W. Callaghan, F. Creutzig, S. Fuss, J. Hartmann, J. Hilaire, W. F. Lamb, J. C. Minx, S. Rogers and P. Smith, *Environ. Res. Lett.*, 2018, **13**, 063003.
10. C. Wilson, *Energy Policy*, 2012, **50**, 81–94.
11. A. Grübler and C. Wilson, *Energy Technology Innovation*, Cambridge University Press, Cambridge, 2013.
12. D. Platt, M. Workman and S. Hall, *J. Cleaner Prod.*, 2018, **203**, 1003–1018.
13. G. Lomax, M. Workman, T. M. Lenton and N. Shah, *Energy Policy*, 2015, **78**, 125–136.
14. G. Lomax, T. M. Lenton, A. Adeosun and M. Workman, *Nat. Clim. Change*, 2015, **5**, 498–500.
15. J. Edler and L. Georghiou, *Res. Policy*, 2007, **36**, 949–963.
16. J. C. Minx, W. F. Lamb, M. W. Callaghan, S. Fuss, J. Hilaire, F. Creutzig, T. Amann, T. Beringer, W. de Oliveira Garcia, J. Hartmann, T. Khanna, D. Lenzi, G. Luderer, G. F. Nemet, J. Rogelj, P. Smith, J. L. Vicente Vicente, J. Wilcox and M. del Mar Zamora Dominguez, *Environ. Res. Lett.*, 2018, **13**, 063001.
17. C. I. Jones and J. C. Williams, *Q. J. Econ.*, 1998, **113**, 1119–1135.
18. W. M. Cohen, A. Goto, A. Nagata, R. R. Nelson and J. P. Walsh, *Res. Policy*, 2002, **31**, 1349–1367.
19. A. Azapagic, D. Beerling, C. Cheeseman, G. Henderson, C. Hepburn, J. HouseC. Le Quéré, N. Markusson, N. Shah, J. Shepherd and P. Smith, *Greenhouse Gas Removal*, Royal Society and Royal Academy of Engineering, 2018. https://www.raeng.org.uk/publications/reports/greenhouse-gas-removal.
20. H. A. Daggash, M. Fajardy, P. Heptonstall, N. Mac Dowell and R. Gross, *Bioenergy with Carbon Capture and Storage, and Direct Air Carbon Capture and Storage: Examining the Evidence on Deployment Potential and Costs in the UK*, 2019.
21. H. A. Daggash, M. Fajardy and N. Mac Dowell, *Carbon Capture and Storage (Energy and Environment Series)*, Royal Society of Chemistry, 2019, pp. 447–511.
22. M. Fajardy, P. Patrizio, H. A. Daggash and N. Mac Dowell, *Front. Clim.*, 2019, **1**, 1–6.

23. P. Smith, S. J. Davis, F. Creutzig, S. Fuss, J. Minx, B. Gabrielle, E. Kato, R. B. Jackson, A. Cowie, E. Kriegler, D. P. van Vuuren, J. Rogelj, P. Ciais, J. Milne, J. G. Canadell, D. McCollum, G. Peters, R. M. Andrew, V. Krey, G. Shrestha, P. Friedlingstein, T. Gasser, A. Grübler, W. K. Heidug, M. Jonas, C. D. Jones, F. Kraxner, E. Littleton, J. Lowe, J. R. Moreira, N. Nakicenovic, M. Obersteiner, A. Patwardhan, M. Rogner, E. Rubin, A. Sharifi, A. Torvanger, Y. Yamagata, J. Edmonds and C. Yongsung, *Nat. Clim. Change*, 2016, **6**, 42–50.
24. D. J. Teece, *Res. Policy*, 1986, **15**, 285–305.
25. M. Honegger and D. Reiner, *Clim. Policy*, 2018, **18**, 306–321.
26. K. Anderson and G. Peters, *Science*, 2016, **354**, 182–183.
27. N. McGlashan, N. Shah, B. Caldecott and M. Workman, *Process Saf. Environ. Prot.*, 2012, **90**, 501–510.
28. P. Smith, S. J. Davis, F. Creutzig, S. Fuss, J. Minx, B. Gabrielle, E. Kato, R. B. Jackson, A. Cowie, E. Kriegler, D. P. van Vuuren, J. Rogelj, P. Ciais, J. Milne, J. G. Canadell, D. McCollum, G. Peters, R. Andrew, V. Krey, G. Shrestha, P. Friedlingstein, T. Gasser, A. Grübler, W. K. Heidug, M. Jonas, C. D. Jones, F. Kraxner, E. Littleton, J. Lowe, J. R. Moreira, N. Nakicenovic, M. Obersteiner, A. Patwardhan, M. Rogner, E. Rubin, A. Sharifi, A. Torvanger, Y. Yamagata, J. Edmonds and C. Yongsung, *Nat. Clim. Change*, 2016, **6**, 42–50.
29. H. A. Daggash and N. Mac Dowell, *Joule*, 2019, **3**, 2120–2133.
30. G. E. Metcalf and J. H. Stock, *Rev. Environ. Econ. Policy*, 2017, **11**, 80–99.
31. Her Majesty's Treasury, *The Green Book: Central Government Guidance on Appraisal and Evaluation*, 2018.
32. K. Ricke, L. Drouet, K. Caldeira and M. Tavoni, *Nat. Clim. Change*, 2018, **8**, 895–900.
33. B. F. Snyder, *Ambio*, 2020, **49**, 1567–1580.
34. R. S. J. Tol, *The Impact of Climate Change and the Social Cost of Carbon The Impact of Climate Change and the Social Cost of Carbon*, Brighton, 2018.
35. M. C. Hänsel and M. F. Quaas, *Ecol. Econ.*, 2018, **146**, 520–535.
36. IPCC, *Climate Change Impacts, Adaptation, and Vulnerability. Part A: Global and Sectoral Aspects. Contribution of Working Group II to the Fifth Assessment Report of the Intergovernmental Panel on Climate Change*, Cambridge University Press, Cambridge, United Kingdom and New York, NY, USA, 2014.
37. R. N. Cooper, P. Cramton, O. Edenhofer, C. Gollier, É. Laurent, D. J. C. MacKay, W. Nordhaus, A. Ockenfels, I. Parry, J. E. Stiglitz, S. Stoft, J. Tirole and M. L. Weitzman, *Global Carbon Pricing*, MIT Press Ltd, Cambridge, MA, 2017.
38. World Bank, *State and Trends of Carbon Pricing 2020*, 2020.
39. M. Torres and P. Pinho, *Cities*, 2011, **28**, 576–582.
40. N. Kaufman, A. R. Barron, W. Krawczyk, P. Marsters and H. McJeon, *Nat. Clim. Change*, 2020, 1–8.

41. R. S. Haszeldine, S. Flude, G. Johnson and V. Scott, *Philos. Trans. R. Soc., A*, 2018, **376**, 20160447.
42. A. Lockley, Z. Mi and D. Coffman, *Front. Eng. Manage.*, 2019, **6**, 38–51.
43. Climeworks AG, Capturing CO_2 from the Air: Our Technology, https://climeworks.com, (accessed September 14, 2020).
44. A. Lockley and D. Coffman, *Environ. Res. Lett.*, 2018, **13**, 1–11.
45. D. Coffman and A. Lockley, *Environ. Res. Lett.*, 2017, **12**, 1–9.
46. R. Wall, S. Grafakos, A. Gianoli and S. Stavropoulos, *Clim. Policy*, 2019, **19**, 59–72.
47. H. A. Daggash and N. Mac Dowell, *Joule*, 2019, **3**, 1239–1251.
48. Department for Business Energy & Industrial Strategy, *Electricity Generation Costs 2020*, 2020.
49. K. Dooley and A. Gupta, International Environmental Agreements: Politics, *Law Econ.*, 2017, **17**, 483–500.
50. R. Smithers, O. Korkeala, G. Whiteley, S. Brace and B. Holmes, *Valuing Flood-regulation Services for Inclusion in the UK Ecosystem Accounts*, 2016.
51. H. A. Daggash, C. F. Patzschke, C. F. Heuberger, L. Zhu, K. Hellgardt, P. S. Fennell, A. N. Bhave, A. Bardow and N. Mac Dowell, *Sustainable Energy Fuels*, 2018, **2**, 1153–1169.
52. A. Talberg, P. Christoff, S. Thomas and D. Karoly, *Int. Environ. Agreements: Politics, Law Econ.*, 2018, **18**, 229–253.
53. W. Rickels, G. Klepper, J. Dovern, G. Betz, N. Brachatzek, S. Cacean, K. Güssow, J. Heintzenberg, S. Hiller, C. Hoose, T. Leisner, A. Oschlies, U. Platt, A. Proelß, O. Renn, S. Schäfer and M. Zürn, *Large-Scale Intentional Interventions into the Climate System? Assessing the Climate Engineering Debate. Scoping report conducted on behalf of the German Federal Ministry of Education and Research (BMBF)*, Kiel, 2011.
54. J. McGee, K. Brent and W. Burns, *Aust. J. Marit. Ocean Affairs*, 2018, **10**, 67–80.
55. T. Dixon, J. Garrett and E. Kleverlaan, *Energy Procedia*, 2014, **63**, 6623–6628.
56. J. Garrett and S. McCoy, *Energy Procedia*, 2013, **37**, 7747–7755.
57. International Maritime Organization, *The London Convention and the London Protocol: How Global Regulation Can Deal Responsibly With Climate Change Mitigation Technologies to Protect the Marine Environment*, 2019.
58. Convention on Biodiversity (Conference of Parties), Decision adopted by the Conference of the Parties to the Convention on Biological Diversity at its tenth meeting. UNEP/CBD/COP/DEC/X/33, Nagoya, 2010.
59. D. W. Keith, E. Parson and M. G. Morgan, *Nature*, 2010, **463**, 426–427.
60. M. Bui, C. S. Adjiman, A. Bardow, E. J. Anthony, A. Boston, S. Brown, P. S. Fennell, S. Fuss, A. Galindo, L. A. Hackett, J. P. Hallett, H. J. Herzog, G. Jackson, J. Kemper, S. Krevor, G. C. Maitland, M. Matuszewski, I. S. Metcalfe, C. Petit, G. Puxty, J. Reimer,

D. M. Reiner, E. S. Rubin, S. A. Scott, N. Shah, B. Smit, J. P. M. Trusler, P. Webley, J. Wilcox and N. Mac Dowell, *Energy Environ. Sci.*, 2018, **11**, 1062–1176.

61. IPCC, Global warming of 1.5 °C. An IPCC Special Report on the impacts of global warming of 1.5 °C above pre-industrial levels and related global greenhouse gas emission pathways, in the context of strengthening the global response to the threat of climate change, sustainable development, and efforts to eradicate poverty, In Press, 2018.
62. IPCC, *Climate Change 2014: Mitigation of Climate Change. Contribution of Working Group III to the Fifth Assessment Report of the Intergovernmental Panel on Climate Change*, Cambridge University Press, Cambridge, United Kingdom and New York, NY, USA, 2014.
63. International Maritime Organization, Ocean Fertilization under the LC/LP, https://tinyurl.com/y3vsah3c, (accessed September 28, 2020).
64. International Maritime Organization, Addressing barriers to transboundary carbon capture and storage, https://tinyurl.com/yxbtc8hr, (accessed September 29, 2020).
65. IPCC, Summary for Policymakers, in *Climate Change and Land: an IPCC Special Report on Climate Change, Desertification, Land Degradation, Sustainable Land Management, Food Security, and Greenhouse Gas Fluxes in Terrestrial Ecosystems*, 2019, https://www.ipcc.ch/srccl/.
66. J. Fargione, J. Hill, D. Tilman, S. Polasky and P. Hawthorne, *Science*, 2008, **319**, 1235–1238.
67. M. Fajardy and N. Mac Dowell, *Energy Environ. Sci.*, 2017, **10**, 1389–1426.
68. P. Renforth, *Int. J. Greenhouse Gas Control*, 2012, **10**, 229–243.
69. T. Searchinger, *Sound Principles and an Important Inconsistency in the 2012 UK Bioenergy Strategy*, 2012.
70. CIFOR, Evaluasi dampak sertifikasi Forest Stewardship Council (FSC) terhadap pengelolaan hutan alam di daerah tropis, 2013, vol. 19.
71. P. Smith, R. S. Haszeldine and S. M. Smith, *Environ. Sci.: Processes Impacts*, 2016, **18**, 1400–1405.
72. A. Agostini, J. Giuntoli and A. Boulamanti, 2014.
73. J. M. Miocic, S. M. V. Gilfillan, J. J. Roberts, K. Edlmann, C. I. McDermott and R. S. Haszeldine, *Int. J. Greenhouse Gas Control*, 2016, **51**, 118–125.
74. S. Budinis, S. Krevor, N. Mac Dowell, N. Brandon and A. Hawkes, *Energy Strategy Rev.*, 2018, **22**, 61–81.
75. M. Fajardy and N. Mac Dowell, *One Earth*, 2020, **3**, 214–225.
76. C. Pozo, Á. Galán-Martín, D. M. Reiner, N. Mac Dowell and G. Guillén-Gosálbez, *Nat. Clim. Change*, 2020, **10**, 640–646.
77. African Development Bank, Agence Française de Développement, Asian Development Bank, European Bank for Reconstruction and Development, European Investment Bank, Global Environment Facility, Green Investment Bank, Inter-American Development Bank, KfW, Nordic Development Fund, Nordic Environment Finance Corporation, Nordic Investment

Bank and World Bank Group, International Financial Institution Framework for a Harmonised Approach to Greenhouse Gas Accounting, 2015.
78. UN DESA, World Population Prospects 2019, Volume II: Demographic Profiles (ST/ESA/SER.A/427), New York, 2019.
79. UNFCCC, in Report of the Conference of the Parties on its twenty-first session, held in Paris from 30 November to 13 December 2015, 2015.
80. H. Shue, *J. Hum. Rights Environ.*, 2017, **8**, 203–216.
81. H. A. Daggash and N. Mac Dowell, *Joule*, 2019, **3**, 2120–2133.
82. K. Anderson and G. Peters, *Science*, 2016, **354**, 182–183.
83. R. J. Plevin, A. D. Jones, M. S. Torn and H. K. Gibbs, *Environ. Sci. Technol.*, 2010, **44**, 8015–8021.
84. T. Searchinger, R. Heimlich, R. A. Houghton, F. Dong, A. Elobeid, J. Fabiosa, S. Tokgoz, D. Hayes and T.-H. Yu, *Science*, 2008, **319**, 1238–1240.
85. V. Heck, D. Gerten, W. Lucht and A. Popp, *Nat. Clim. Change*, 2018, **8**, 151–155.
86. J. Popp, Z. Lakner, M. Harangi-Rákos and M. Fári, *Renewable Sustainable Energy Rev.*, 2014, **32**, 559–578.
87. U. Kreidenweis, F. Humpenöder, M. Stevanović, B. L. Bodirsky, E. Kriegler, H. Lotze-Campen and A. Popp, *Environ. Res. Lett.*, 2016, **11**, 085001.
88. M. Wise, K. Calvin, A. Thomson, L. Clarke, B. Bond-Lamberty, R. Sands, S. J. Smith, A. Janetos and J. Edmonds, *Science*, 2009, **324**, 1183–1186.
89. L. R. Boysen, W. Lucht and D. Gerten, *Global Change Biol.*, 2017, **23**, 4303–4317.
90. L. Wallquist, S. L. Seigo, V. H. M. Visschers and M. Siegrist, *Int. J. Greenhouse Gas Control*, 2012, **6**, 77–83.
91. C. Braun, *Risk Anal.*, 2017, **37**, 2264–2275.
92. LC/LP, Resolution LC-LP.1 (2008) on the regulation of ocean fertilization, 2008.
93. LC/LP, Resolution LC-LP.2 (2010) on the assessment framework for scientific research involving ocean fertilization, 2010.
94. LC/LP, Annex 4 Resolution LP.4(8) on the amendment to the London Protocol to regulate the placement of matter for ocean fertilization and other marine geoengineering activities, 2013.
95. R. Bodle, S. Oberthür, L. Donat, G. Homann, S. Sina and E. Tedsen, *Options and Proposals for the International Governance of Geoengineering*, Berlin, 2014.
96. Convention on Biodiversity (Conference of Parties), Decision adopted by the Conference of the Parties to the Convention on Biological Diversity at its ninth meeting. UNEP/CBD/COP/9/29, Bonn, 2008.
97. Convention on Biodiversity (Conference of Parties), Decision adopted by the Conference of the Parties to the Convention on Biological Diversity at its eleventh meeting. UNEP/CBD/COP/11/35, Hyderabad, 2012.

98. Secretariat of the Convention on Biological Diversity, *Geoengineering in relation to the Convention on biological diversity: technical and regulatory matters*, Montreal, 2012.
99. Convention on Biodiversity (Conference of Parties), *Impacts of Climate-Related Geoengineering on Biodiversity: Views and Experiences of Indigenous and Local Communities and Stakeholders UNEP/CBD/SBSTTA/16/INF/30*, Montreal, 2012.
100. G. Winter, *Rev. Eur. Commun. Int. Environ. Law*, 2011, **20**, 277–289.
101. A. C. Lin, *Issues Legal Scholarship*, 2009, **8**, 1–24.
102. K. N. Scott, *Michigan J. Int. Law*, 2013, **34**, 309–358.

CHAPTER 15

Socially Equitable Energy Transitions: Analytical Challenges and Policy Implications

PIERA PATRIZIO*[a] AND NIALL MAC DOWELL[a,b]

[a] Centre for Environmental Policy, Imperial College London, London SW7 1NE, United Kingdom; [b] Centre for Process Systems Engineering, Imperial College London, London SW7 2AZ, United Kingdom
*Email: p.patrizio@imperial.ac.uk

Driven by the urgency of mitigating climate change and on the global efforts to reach net-zero emissions, the transition to low-carbon energy systems will require unprecedent rates of decarbonisation, whilst being socially equitable and technically feasible. Quantitative modelling and analysis techniques are critical components of political and industrial decision making for the energy transition at the global, national, and sub-national levels. However, being near-exclusively cost oriented, current approaches do not adequately reflect the socio-economic impacts of decarbonisation strategies. This tendency has resulted in one size fits all emission mitigation pathways and in copycat policy strategies that may well exacerbate existing regional imbalances. Here, we present three key societal challenges associated with transitioning to low-carbon energy systems and we discuss current approaches in both energy modelling and policy making. We then suggest potential actions that can contribute to tackling these challenges and, in so

Energy and Environment Series No. 31
Greenhouse Gas Removal Technologies
Edited by Mai Bui and Niall Mac Dowell

Published by the Royal Society of Chemistry, www.rsc.org

doing, identify viable pathways for socially equitable energy systems transitions.

15.1 Introduction

Driven by global efforts to reach net-zero emissions, the transition to a sustainable energy system needs to occur at an unprecedented rate. In contrast with past energy transitions that have been mainly driven by energy sources and technology cost competitiveness,[1] the efforts required to reach national net-zero emission targets will entail a wide range of technical, political and societal transformations.[2]

In the scientific literature, the discussion around energy systems transition is predominantly facilitated *via* various energy systems modelling exercises, typically using cost metrics as the exclusive means of comparing low carbon scenarios. As pointed out by different energy transition scholars,[2–4] these costs do not fully address the socio-economic feasibility of climate mitigation.[5] Drivers such as socio-economic development and the capabilities of individual member states, are judged to be particularly influential for implementing effective decarbonisation strategies.[6] As such, detailed consideration of the country-specific natural resources, economic strengths, geopolitical perspectives, and levels of socio-economic and technological readiness for energy transition is an imperative.[4,7]

The interplay of costs and socio-economic readiness for energy transition, *i.e.* the availability of resources to support the transition or to compensate for its socio-economic consequences, is illustrated by Jewell *et al.* (2019)[8] with an analysis of worldwide pledges to phase out coal power. Their study shows that whether a country adopts such a pledge is explained by both the costs of coal phase-out (countries with older power plants and lower domestic extraction and use) and the capabilities of a country to bear these costs. These capabilities can be signalled by a range of socio-economic indicators, including income per capita,[8] access to low carbon energy technologies,[9] government's policy transparency and the level of regional stakeholder engagement in designing decarbonation strategies.[10]

The role of national governments in energy system decarbonisation is particularly relevant in the case of current energy transition, which is being driven by the need to remove atmospheric emissions *via* large-scale carbon dioxide removal (CDR) technologies rather than by technology cost competitiveness.[1] Apart from contributing to the achievement of emission reduction goals, a green policy can also have a strong economic impact. In fact, accelerating the transition towards a low-carbon future will decrease the costs of the transition and reduce future risks, including those related to climate change.[11] In addition, green stimulus will translate in strong fiscal multipliers and job creation.[7,12,13]

The economic stimulus packages and recovery plans that governments have being putting in place following the outbreak of COVID-19, offer a clear example of the role of public policy in promoting a green recovery. According to preliminary analysis conducted by the OECD Secretariat,[14] more than 30 OECD countries have already included measures directed at supporting the transition to greener economies as part of their recovery program. To promote an economic growth that is both low-carbon and inclusive, recovery packages should aim at long term structural goals, including addressing social inequalities and levelling up regional disparities, whilst decarbonising key sectors of the economy.[12,15,48]

In recent years, arguments around the efficiency of existing policies have dominated the energy transition debate at the national level. In Germany, for example, critics have been arguing that plans to phase out existing fossil fuels are extremely costly and have the potential to damage Germany's economy. Similar discussions have also been happening in the US[13] and Poland,[16] among other countries. While discussion about the efficiency of energy transition policies is important, the results of such analysis depend on how the socio-economic costs and benefits associated with the transition are accounted for.

Quantitative modelling and analysis techniques are critical components of political and industrial decision making for the energy transition at global, national, and sub-national levels. Incorporating socio-economic impacts in energy systems modelling would lead to the development and implementation of policy strategies that are both cost-effective and socio-politically feasible.[2] Similarly, accounting for the societal impacts in transition policies design, would contribute to tackle key regional issues such as regional economic imbalances and distributional effects.[17,49]

Here we review current approaches in climate change modelling and energy policy design, and we discuss their potential for addressing the socio-economic dimension of energy system decarbonisation. We then make some specific recommendations for how these approaches could be improved both in energy modelling and in policy making.

15.2 Energy Transition Modelling

15.2.1 Existing Approaches

There is a wide variety of energy transition models, at different scales, supply and/or demand, with different focuses on temporal, geographical, technological, and economic detail. Models at these various scales serve different purposes. Here, we present a review of the existing approaches in energy modelling, following the analytical framework developed by Li *et al.*[18] and we discuss their potential for capturing the socio-economic dimension of energy systems transition.[18]

Integrated Assessment Models (IAMs) are widely used in energy transition modelling as a framework for analysing the physical and socio-economic effects of climate change. These models are generally used in exploring the feasibility of achieving climate targets[19] and therefore setting the agenda for global environmental negotiations under the United Nations Framework

Convention on Climate Change (UNFCCC). IAMs are mainly composed of an economic and a physical block. The economic component is represented by a general or partial equilibrium model (CGE, PE), describing an aggregate economic agent deciding how much economic output to consume immediately and how much capital to invest into clean and dirty energy.[19] The physical component accounts for the emissions accumulating in the atmosphere as a result of economic and human activities which can be endogenously (*e.g.*, energy consumption) or exogenously (*e.g.*, population growth) determined.

Over recent years, IAMs have been subject to growing criticism particularly from the climate economics and the transition modelling communities. Such criticism points mainly at the lack of transparency in documenting the modelling structure and input requirements,[20] their near exclusive focus on a single or restricted set of CDRs options, and the inadequate representation of uncertainty.[19,21] For the purpose of our analysis, the most relevant shortcoming of IAMs stems from their aggregate nature, not only at a geographical level, since IAMs generally represent the world in a few global regions, but also at a sectoral level, with the industrial activities generally grouped in a few macro-sectors. Such aggregation levels mean that IAMs cannot capture the societal impacts of energy systems transition, which typically occur at the national or regional scale.

Bottom-up models are instead used to inform national energy and climate policy activities, such as exploring sector-specific policies or designing new market mechanisms. These approaches, referred to as energy systems optimisation models (ESOMs) are characterised by a detailed representation of technology options, and are based on data grounded in engineering processes. Their level of technological details makes them also particularly suitable for exploring critical features of CDR technologies, for instance biomass resource competition,[22] their integration into existing energy systems,[23] as well as the effect of technological innovation on competing CDR options.[24] As they can accommodate long time frames and different geographic scales,[25–27] they are often adopted to explore energy system configurations that might meet desired emissions or resource targets under constrained pathways.[23,24] This notwithstanding, ESOMs typically follow a least cost approach, and the socio-economic component of energy transition is generally excluded from the optimisation procedure.

Agent Based Models (ABMs) are simulations of the economy based on interactions of many heterogeneous agents according to specified rules that aim to allow more flexible and realistic characterisations of socio-economic systems. In addition to the effort required to empirically derive appropriate parameters values, ABM's flexibility to incorporate multiple agents can make their communication and interpretation extremely difficult. Such challenges have likely limited their use in energy and climate change applications: as noted by Farmer *et al.*,[19] ABMs have been successfully used in several areas of economics, including financial markets, technology adoption and labour networks. But they have not yet seen serious application to energy systems transition.[19]

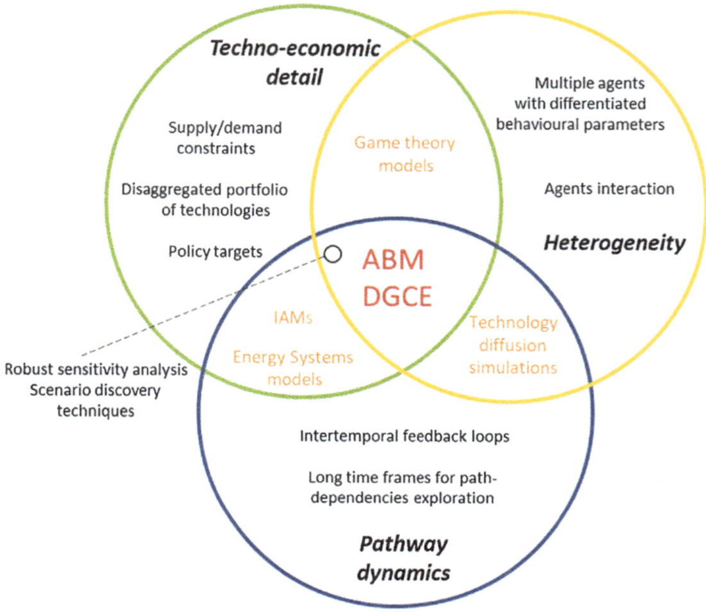

Figure 15.1 Classification of the main energy models based on key energy system transition features.

15.2.2 Key Features of Socially Equitable Energy Transition

There are important features of energy models that are key for capturing the socio-economic dimension of energy systems transitions, and thus for identifying socially equitable carbon removal pathways. Figure 15.1 classifies the main modelling approaches discussed above on the basis of key elements of energy systems transition – spatial and social heterogeneity, technological innovation and uncertainty. Promising approaches that can account for these features are identified in this section and summarised in Table 15.1.

15.2.2.1 Technological Innovation

The unprecedented rates of emissions reduction required to achieve country-wide net-zero targets, mandate the rapid deployment of CDR technologies. The speed at which these technologies can be scaled up so that they are commercially available at affordable costs to deliver climate benefits with reasonably tolerable adverse impacts, will determine their utility for addressing climate change.[28] Innovation in CDR technologies will be central to this scale-up process, and thus crucial to their outcomes on the climate and society.

Future technological progress is one of the most uncertain factors in modelling the economics of energy transition, since technologies often evolve stochastically over time through a process of imitation and mutation.[19] In addition, CDRs are expected to progress *via* very different rates of technological innovation, based on their limitations and supporting environment,[28] among other factors.

Table 15.1 Critical modelling features and socio-economic implications of energy systems transition.

Area	Main features	Current modelling approaches	Socio-economic implications	Suggested improvements	Modelling examples
Technological innovation	Collective process, specific for each technology	Global and technologically aggregated learning curves	Lack of interaction between actors driving technological progress; disruptive innovation (*e.g.*, automation) cannot be accurately represented.	Explicit representation of the R&D sector and of the interactions between actors driving innovation	QUEST; RHOMOLO[46]
Spatial and sectoral heterogeneity	Presence of intertwined dimensions shaping energy transition (*e.g.*, technology, economy, energy systems and society)	Cost-focused approaches (ESOMs); aggregate representative agents (CGE); static interactions between economic activities (I/O models)	One size fits all emission mitigation pathways; lack of understanding of distributional issues associated with policy measures (*e.g.*, carbon tax); sector specific decarbonisation challenges not addressed	Dynamic representation of sectoral interdependencies; explicit representation of regional assets (*e.g.*, industrial specialisation, labour market composition, energy resources)	ENGAGE; BLUE model[18,19,47]
Uncertainty	Inertia of energy systems; uncertainty of input parameters	Possibilistic rather than probabilistic range of decarbonisation pathways	Limited understanding of the socio-economic impacts of decarbonisation	What-if scenario-based modelling; robust sensitivity of modelling outcomes.	ESO; ENGAGE[24,47]

Learning curves are the most common way to deal with technological progress in IAMs and ESOMs, which imposes strong assumptions about perfect diffusion and spillovers, two important features of innovation. In their work, Wilson *et al.*[29] conducted a comprehensive analysis of technology innovation, highlighting the main drivers and barriers of energy technology development. Whilst recognising learning as the most important driver of the technological progress, the analysis emphasises the collective nature of innovation, involving many dimensions or actors, such as established and research organisations, institutions and end users.[29] Accounting for these actors and their dynamic interactions in shaping technological progress is therefore crucial when forecasting the development, scale-up and adoption of energy technologies.

Within this context, a useful approach is offered by Dynamic Stochastic General Equilibrium (DSGE) models, a class of energy models widely adopted in national monetary policy. DSGE models follow the rigorous market equilibrium concept of CGE models but model the R&D activities explicitly, by accounting for the dynamic interaction of key actors, such as institutions, firms and end users. Importantly, DSGE models derive agents' decision rules explicitly from inter-temporal optimisation under technological and budgetary constraints. This represents a distinctive feature of DSGE which overcomes the deterministic initialisation process of many ABMs.

In addition, the inclusion of inter-temporal considerations from economic agents makes DSGE particularly valuable to model disruptive innovation, which is generally neglected in energy models. A typical example of disruptive innovation is the advent of automation, which will have strong socio-economic impacts in the labour market, by inducing job losses in certain sectors or by triggering labour shifts between sectors. DSGE models can quantify these impacts as they are able to differentiate between labour categories whilst modelling sector interactions. They hence become particularly useful for: (i) quantifying sector specific innovation potentials, represented by the share of highly skilled workers in that sector, (ii) forecasting the employment disruption associated with automation within different sectors, and (iii) investigating different innovation scenarios at the technology level.

15.2.2.2 Spatial and Sectoral Heterogeneity

Energy transition can be defined as evolution in the energy systems driven by climate mitigation goals that entail social, technical, and economic changes and point to the inevitability of conflicts among actors. The transition to low carbon energy systems is hence a *multi-dimensional* challenge, requiring simultaneous analysis across four intertwined levels: energy technologies, economy, energy system, and ultimately society. Accounting for the complex relationship between such dimensions and how their interactions might change over time is key for identifying socially equitable decarbonisation futures.

To account for the spatial and sectoral heterogeneity of energy systems, Patrizio *et al.*[7] proposed a framework, presented in Figure 15.2, combining

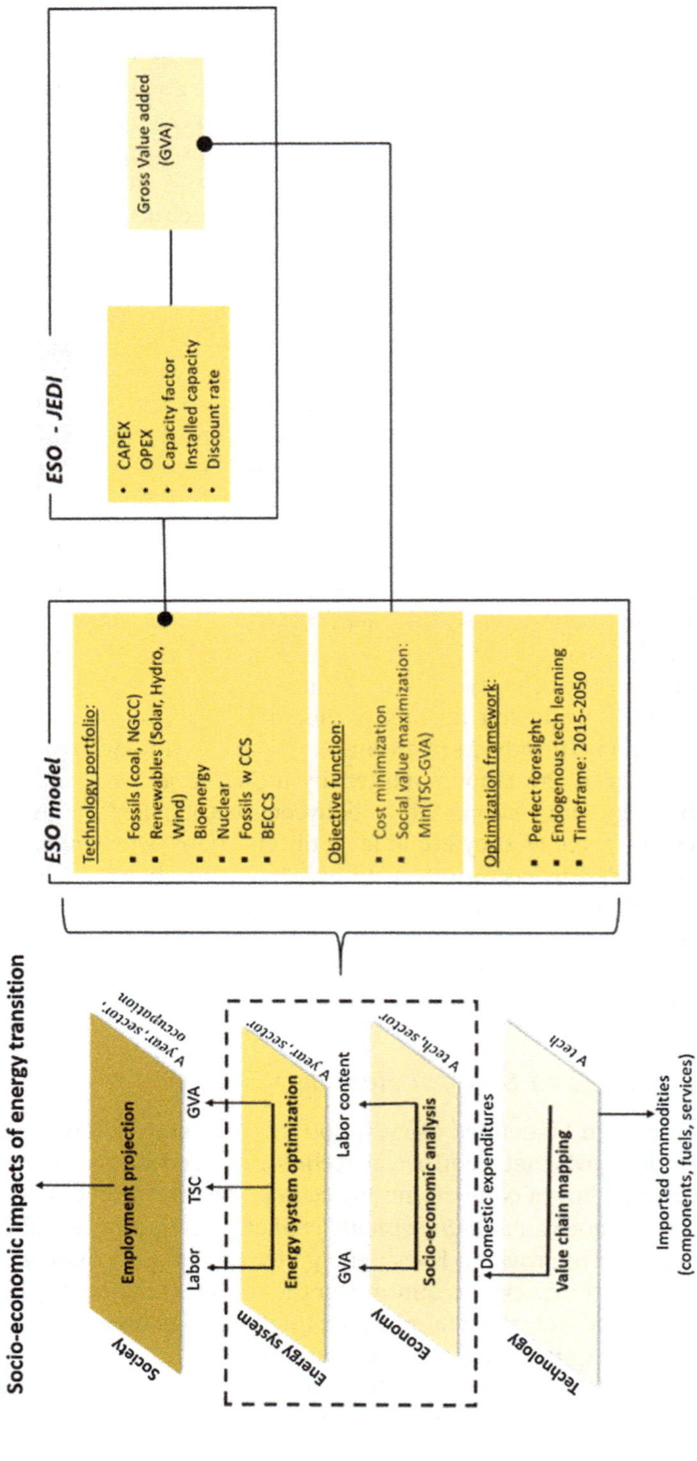

Figure 15.2 ESO-JEDI modelling framework accounting for the spatial and sectoral heterogeneity of energy systems transition. Reproduced from rref. 7 with permission from Elsevier, Copyright 2020.

an energy system optimisation model (ESO)[24,30] with a country-specific socio-economic tool. The latter is based on input-output (IO) tables,[31] based on the Leontief's production functions, that calculate *ex post* the economic gains, in terms of gross value added (GVA) deriving from specific industrial activities.

Whilst being able to take fully into account country-specific industrial inter-dependencies, the static nature of IO models means that these inter-dependencies are kept constant over time. In other words, IO models do not present a complete picture of the supply of the economy, in that they do not envision optimising behaviour of economic agents faced with alternative courses of action. Importantly, choices about the proportions in which inputs, *e.g.*, energy technologies, are to be combined in the production of a given output, *e.g.*, power supply, are excluded from the analysis. By modelling sectoral inter-dependencies as simulated interactions between agents, both DGCE and ABM models can capture the social and spatial heterogeneity of energy systems transition and its evolution over time. For instance, the BLUE model is a bottom-up hybrid energy system model simulating the interactions of a population of representative agents, each acting to minimise their own (imperfectly) perceived discounted system costs over time in a myopic fashion. Importantly, the model represents multiple economic sectors as separate actors whose interactions can affect the course of energy transitions, and features a probabilistic formulation for exploring uncertainties in future pathways.[18]

15.2.2.3 Uncertainty

Energy systems are characterised by inertia[32] since energy demand is driven by long-lasting patterns of infrastructure and technology interdependencies. Since time is needed to remove political, financial and social barriers for CDRs deployment, inertia increases the uncertainty of future energy system configurations. As such, energy transition modellers are often confronted with a range of possible decarbonisation futures, with assumptions around uncertain parameters, *e.g.*, energy prices, technology performances and costs, dictating the pace of the transition.

Robust scenarios investigation, to test the sensitivity of modelling outcomes to different assumptions, is a widely adopted practice in energy modelling to restrict the range of model outcomes. While scenarios are usually determined exogenously and fed into the model, ABMs are also able to undertake a relatively new 'scenario-discovery' technique, which instead allows modellers to identify scenarios that emerge endogenously from a set of model simulations.[19] In the case of DGSE, their stochastic nature makes them particularly sensible to uncertainty, especially regarding potential economic shocks and change in energy consumption patterns.

15.3 Energy Transition Policies

As already mentioned above, the current energy transition is policy driven, conducted by policy rather than by cost competitiveness.[1] National decarbonisation policies will hence play a key role in shaping the pace and rate of energy systems decarbonisation. Quantitative modelling and analysis techniques are critical components of national policy design, and cost focused approaches, such as cost benefits analysis (CBA), are generally adopted to evaluate the efficiency of such policies in terms of resource allocation. Here we present an overview of current approaches in policy strategies design and we discuss some key socio-economic indicators that could complement such approaches.

15.3.1 Cost-benefit Analysis (CBA)

The public sector usually evaluates the convenience of policy strategies and investment projects using cost–benefit analysis (CBA) or multicriteria analysis (MCA). The CBA methodology attempts to assess the economic efficiency of policy in terms of the efficient use of resources for the whole of society. In this context, the market benefit/cost ratio represents a key decision criterion, although benefits refer almost exclusively to market benefits.[33] Where there is evidence available, socio-economic costs and benefits, measured in monetary terms, can be compared to each other in order to discover if a policy is efficient in a CBA sense. Synthetic indicators such as the economic net present value (NPV), are generally adopted to compare the results arising from the CBA analysis. As frequently noted,[17,33,34] the exclusive reliance on cost metrics to guide public investment decisions, implies three main pitfalls, hereby presented.

15.3.1.1 Misleading Project Appraisals

A comparative longitudinal study reviewing cost data for transportation infrastructure projects across 20 nations, has found great differences between estimated and actual project outcomes, with 90% of the reviewed projects experiencing significant costs overrun.[35] According to the analysis, which spanned various project types including building, power plants and oil and gas extraction projects, substantial cost underestimates are often combined with equally significant overestimates of benefit, rendering cost–benefit analyses of projects not only inaccurate but biased. The cause of biased analyses lies on the foundation of the CBA methodology itself, which encourage proponents to underestimate costs and overestimate benefits, resulting in the 'survival of the unfittest' option. A recent investigation conducted by the Institute for the Government in the UK noticed similar overoptimistic behaviours in public project evaluations. The analysis concluded that the exclusive reliance on cost indicators lead public bodies to underestimate the deliverable challenge of multidimensional, uncertain and complex projects.[36]

15.3.1.2 Vicious Public Spending Cycles

CBA is rooted in welfare economics, intended to be a quantified, monetary assessment of the total impacts of a project on the lives of those affected over the future lifetime of the investment. However, in policy usage, it is almost always presented as a means of using scarce public funds as efficiently as possible to deliver higher productivity and economic growth, or in other words as a value for money exercise. This interpretation has led to the emergence of vicious cycles in public spending, contributing to wider existing economic imbalances, and to favour existing high-performing regions.[37] The procedures set out in the Treasury's Green Book[38] for transport project appraisal across the UK, offer an unequivocal example of this issue, with major funding allocation decisions almost wholly centralised. According to Coyle & Sensier,[17] the methodology favours productive regions of London and South-England as it includes the 'wider' benefits of agglomeration impacts, dependent development and increases in more productive jobs.[17] This implies that projects in areas that are already the densest in population, with the highest land values, and the most productive jobs will continue to score higher in appraisals of benefits. As noted by Patrizio, Sunny and Mac Dowell (2022), investment decisions exclusively motivated by economic efficiency considerations are likely to create adverse social equity implications and exacerbate existing regional imbalances.

15.3.1.3 Discounting Issues

Both the discounting mechanism and the discount rate are core elements of welfare economics. In CBA, they both make up the discounting function that is used to compare investments' costs and benefits that are expected to occur over the years. As already mentioned, energy transition policies are expected to produce fundamental technical, social and economic changes. In this sense, a major barrier to the reliance on efficiency analyses (CBA) framed in the very long term, with costs and benefits subject to multiple risks and great uncertainty, is that it is hard to accept that a simple discount function can adequately represent the vision of future generations. To overcome this issue, some national governments have recently proposed the adoption of a declining discount rate.[38] However, determining the point in time at which the discount rate should be scaled down remains an open question.[39]

15.3.2 Economic Impact Assessment (EIA)

Budget-constrained, CBA-type analyses are concerned with allocative or distributive efficiency, which involves making the best use of fixed resources *at a fixed point* in time. Ambitious projects such as decarbonizing national economies involves making the best use of resources to achieve changes *over time* including innovation, investment, and, most importantly, the deployment of new technologies.[40] The dynamic and disruptive nature of energy transition has two main implications for public investment appraisal. Firstly, instead of being evaluated against static metrics (such as those targeted by CBA) investments in decarbonization need to be dynamically monitored and evaluated throughout the process. Secondly, since energy transition is inherently transformative across the entire value chain, low-carbon investments need to be evaluated for their *socio-economic value* rather than for their *economic efficiency*.

Over recent decades, several contributions have introduced the notion of socio-economic value in public investment appraisal, particularly when these investments are being associated with the decarbonization of regional and national economies. Recent examples include the Fuel Cells and Hydrogen Joint Undertaking (FCH JU) report[41] and the Hy impact study,[42] focusing on quantifying the socioeconomic impacts of the hydrogen economy respectively in Europe and in the UK. Whilst these assessments provide preliminary insights on the economic opportunities of the energy transition, the underlying analysis is subject to specific baseline assumptions and market deployment forecasts. Employment and GVA figures are based on constant macro-economic assumptions in that wages, sectors interdependencies and price elasticities, are kept constant throughout the evaluation period. Because of that, the validity of these analysis is significantly reduced as the time-horizon lengthens.

There are different ways to reduce the uncertainties arising from these analyses. Firstly, rather than proposing absolute values, the socio-economic impacts of low-carbon investments should be evaluated against a counterfactual scenario, in which the same service, *i.e.* the supply of energy, is provided by conventional fuels. In this way, employment and GVA figures can be presented in relative terms or as differences between the proposed scenario and the counterfactual. Secondly, the analysis should detail the type of skills associated with certain economic activities, as this will have important socio-economic implications. This is because, the net economic impact of low-carbon investments within a certain sector is influenced, among other factors, by the capacity of the labour market to absorb variation in economic output. Labour skills imbalances, such as shortages or surpluses can increase labour turnover, reduce productivity and ultimately slow adoption of new technologies.[11,43] Hence, increasing the granularity of these analyses, by distinguishing job categories and wages, will provide a more detailed and realistic outlook of the socio-economic impacts of low-carbon investments.

15.3.3 Key Socio-economic Metrics

This section proposes a range of key socio-economic indicators that could assist decarbonisation strategies design. These include metrics aimed at capturing the existing structure of the economy including regional specialisations, as well as indicators that characterise the socio-economic impacts of energy policies across different sectors and regions. This list is not intended to be exhaustive, rather our aim is to present quantitative indicators, derived from national accounting databases, and therefore easily derivable, that can assist the identification of effective decarbonisation strategies.

15.3.3.1 Industrial Specialisation

Accounting for the regional concentration of certain economic activities is key in energy transition strategies, especially for those aiming at tackling industrial decarbonisation. A recent OECD study addressing productivity issues across the UK,[44] identifies a strong correlation between regional productivity[†] and the presence of knowledge intensive industrial sectors, for instance the chemical and the advanced machinery manufacturing sectors. In particular, the study found that the most productive regions specialise predominantly in knowledge intensive manufacturing and services, while the regional concentration of low-tech industries is larger in the less productive regions of Wales, North-East England and Northern Ireland.

Directing investments in innovation and R&D in these regions, through the deployment of CDR technologies in key industrial sectors, could unlock productivity benefits while decarbonising the economy.

15.3.3.2 Trade in Manufacturing Value Added

The production structure of energy technologies varies significantly. Conventional thermal power capacities such as coal-fired plants and combined cycle gas turbines (CCGT) as well as utility-scale bioenergy projects, are based on a simple production value chain configuration and consider mostly local sourcing. Value chains in the more dynamic sectors of renewable energy (RE) are far more complex and geographically extended: 50% of the global wind industry value-chain is in South-East Asia, while China alone controls 70% of the global solar PV industry. Hence, in assessing the socio-economic impacts of energy projects, it is critical to quantify the domestic value in service and manufacturing products associated with a certain energy project. Patrizio *et al.*[7] found that the GVA and associated employment derived from national decarbonisation strategies are inherently dependent on national strategic assets, with the presence of local industries for RE manufacturing being one of them.[7]

The trade in Manufacturing Value-Added indicators[‡] address this issue by considering the value added by each country in the production of goods and services that are consumed worldwide. Combining this data with the supply chain analysis of energy technologies, it is hence possible to provide realistic insights on national and regional manufacturing assets and on the economic multipliers associated with a certain decarbonisation strategy.

[†]In the industrial context, productivity refers to the amount of value created, in GVA terms, by each worker for a certain economic activity.

[‡]More information on the trade in value added indicators on the OECD database: http://www.oecd.org/industry/ind/measuring-trade-in-value-added.htm

15.3.3.3 Carbon Productivity

Carbon productivity indicators measure the CO_2 intensity of the economy, from both the production and the demand perspective.[45] Whilst consumption-based CO_2 productivity accounts for the carbon embodied in domestic consumption, industrial carbon productivity can be defined as the economic value generated, in terms of GDP, per unit of CO_2 emitted within a national/regional territory. Figure 15.3 depicts the carbon productivity of main EU economies, both from a supply and demand perspective. It can be noticed that, aside from a few exceptions represented by the yellow circles in Figure 15.2, most EU countries are net importers of CO_2, since emissions from domestic final demand exceed emissions from domestic production.

The decoupling of demand and production-based CO_2 is a useful indicator for assessing the socio-economic impacts of decarbonisation strategies, in that it allows the measuring of how much carbon productivity improvements are associated with low-carbon policies, and how much these are related to contraction in domestic demand (distributional effects).

Removal and Mitigation Value. Further indications on the socio-economic impacts of decarbonization strategies can be derived by quantifying the mitigation and/or removal value CDR technologies. As discussed in detail in Patrizio *et al.* (2021), the first step is to quantify the amount of CO_2 emissions that is effectively avoided and removed with each unit of energy produced with a certain technology. The *mitigation service* provided by CDR technologies, in $kg_{CO_2 avoided}$ per GJ, depends on the CO_2 intensity of the energy being displaced, *e.g.*, electricity from a low-carbon or fossil-based grid or gasoline. Conversely, the *removal service*, in $kg_{CO_2 removed}$ per GJ, is associated with the net CO_2 emissions that can be removed from the atmosphere within a certain CDR pathway. Hence, the mitigation and removal value of CDR pathways can be calculated as the ratio between the CO_2 avoided (and removed) within a specific CDR pathway and the GVA created across its value chain. In this sense, these indicators can also be interpreted as the avoidance and removal efficiency of public spending, since each unit of value created within the CDR supply chains will results in the avoidance and removal of a certain amount of carbon.

Figure 15.4 provides an example of the removal and mitigation value of selected biomass-based CDR pathways for the UK economy. These pathways are: BECCS in cement and iron and steel production (Bio-CaL and BF-BECCS respectively); BECCS to biofuels (Fischer–Tropsch diesel, biohydrogen and second-generation ethanol); BECCS for power production, and biochar.

Figure 15.4 shows that each of these pathways has a distinctive removal and avoidance value, depending on the process characterization, *e.g.*, capture efficiency and biomass conversion efficiency, biophysical factors, *e.g.*, carbon footprint of biomass, and region-specific factors. For instance, among BECCS to biofuel pathways, the production of FT diesel (FTL-CCS) has the highest avoidance value, since higher biofuels yields, are achievable compared to

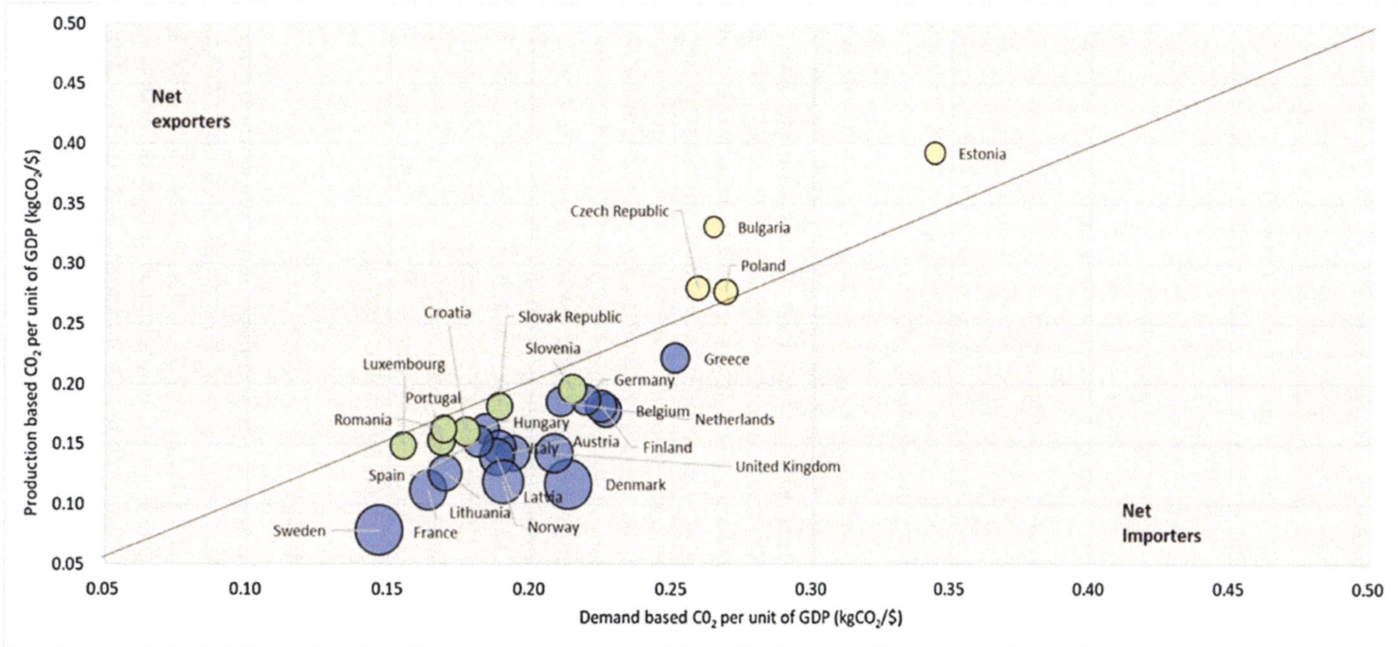

Figure 15.3 Consumption and production-based carbon productivity with the EU countries in 2018. The diagonal line in the figure allows distinguishing between countries which are net CO_2 importers and countries that are CO_2 exporters.

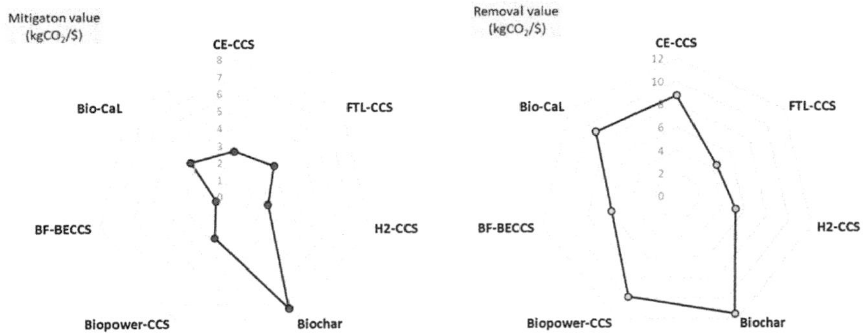

Figure 15.4 Mitigation and removal value of selected biomass-based CDR pathways: BF-BECCS: Integrated ironmaking route; Bio-CaL: Cement kiln with Tail End Calcium Looping; CE-CCS: Second-generation bioethanol with CCS; FTL-CCS: Fischer Tropsch diesel with CCS; H2-CCS: Hydrogen from biomass gasification with CCS; Biopower-CCS: large-scale biomass combustion plant with CCS.

second generation bioethanol (CE-CCS) hence higher amounts of fossil gasoline are being displaced. The cost of one GJ of FT diesel is also lower than 1 GJ of biohydrogen produced *via* biomass gasification (H_2-CCS), therefore higher amounts of CO_2 can be avoided with the same level of investment for their production. However, when it comes to the value of carbon removal, producing bioethanol represent the most efficient investment since 9 kg CO_2 are being removed from the atmosphere for each $ retained in the UK economy. This is because the conversion efficiency of fermentation-based processes is relatively low, $\sim$23%$_{LHV}$ compared to 37%$_{LHV}$ achievable in H_2-CCS pathways. Since larger quantities of biomass are required to produce one unit of biofuel, larger amounts of biogenic carbon can be sequestered within the conversion process. Note also that the deployment of large combustion BECCS power plants (Biopower-CCS) represent a valid investment both in terms of mitigation and removal services provided, since each unit of GVA created within the UK economy removes 10 kg and avoids 3 kg of CO_2 emissions otherwise emitted by fossil electricity.

15.4 Conclusions

Two decades of modelling efforts and developing low-carbon energy policies have shown analysts and policymakers the complexity and multi-dimensional nature of energy system transitions. This notwithstanding, energy models continue to be quasi exclusively cost-oriented, spatial and/or sectorial aggregated and unable to represent uncertainty. Similarly, the political and financial decisions around energy transition policies is often framed in terms of a cost benefit analysis, which lead to a disconnection between decarbonisation strategies and the regional context in which they are applied.

One of the premises of our analysis is that the current energy transition is policy driven and this has relevant consequences in terms of the role played by national government in designing effective decarbonization strategies. Achieving a socially equitable energy transition, requires a mission-oriented approach[40] where national governments and policymakers shape the direction of growth by making strategic investments across key industrial sectors, so that the socio-economic value, rather than the economic efficiency, is maximized.

The challenge for energy modellers and energy strategy analysts is hence to develop sectoral and spatial disaggregated analytical tools that can capture the complex socio-economic dimension alongside the technical and environmental one.[18] In this context agent-based approaches, such as ABMs and DSGE might offer a potential way forward, as they allow representation of the intertemporal interactions of the different dimensions shaping the transition. It remains an open question, however, as to what extent it is possible to represent all of the relevant societal elements of transitions in formal models.

References

1. G. Blazquez and R. Fuentez-Brancamontes, *A road map to navigate the energy transition*, 2019.
2. F. W. Geels, B. K. Sovacool, T. Schwanen and S. Sorrell, *Science*, 2017, **357**, 1242–1244.
3. J. Markard, *Nat. Energy*, 2018, **3**, 628–633.
4. J. Markard, R. Raven and B. Truffer, *Res. Policy*, 2012, **41**, 955–967.
5. J. Jewell and A. Cherp, *Wiley Interdiscip. Rev. Clim. Change*, 2020, 11.
6. K. J. Mach, C. M. Kraan, W. N. Adger, H. Buhaug, M. Burke, J. D. Fearon, C. B. Field, C. S. Hendrix, J. F. Maystadt, J. O'Loughlin, P. Roessler, J. Scheffran, K. A. Schultz and N. von Uexkull, *Nature*, 2019, **571**, 193–197.
7. P. Patrizio, Y. W. Pratama and N. Mac Dowell, *Joule*, 2020, **4**, 1700–1713.
8. J. Jewell, V. Vinichenko, L. Nacke and A. Cherp, *Nat. Clim. Change*, 2019, **9**, 592–597.
9. A. Goldthau, K. Westphal, M. Bazilian and M. Bradshaw, *Nature*, 2019, **569**, 29–31.
10. E3G, *A role model for European coal phase out? Five lessons from the German Coal Commission*, https://www.e3g.org/publications/a-role-model-for-european-coal-phase-out-five-lessons-from-the-german-coal/.
11. OECD, *Putting faces to the jobs at risk of automation*, 2018.
12. J. Allan, C. Donovan, P. Ekins, A. Gambhir, C. Hepburn, D. Reay, N. Robins, E. Shuckburgh and D. Zenghelis, *Smith Sch. Work. Pap. No. 20-01*.
13. P. Patrizio, S. Leduc, F. Kraxner, S. Fuss, G. Kindermann, S. Mesfun, K. Spokas, A. Mendoza, N. Mac Dowell, E. Wetterlund, J. Lundgren, E. Dotzauer, P. Yowargana and M. Obersteiner, *Joule*, 2018, **2**, 2633–2648.

14. OECD, *Making the Green Recovery Work for Jobs, Income and Growth*, 2020.
15. R. Hanna, Y. Xu and D. G. Victor, *Nature*, 2020, **582**, 178–180.
16. H. Brauers and P. Y. Oei, *Energy Policy*, 2020, **144**, 120–138.
17. D. Coyle and M. Sensier, *Reg. Stud.*, 2020, **54**, 283–295.
18. F. G. N. Li and N. Strachan, *Energy Res. Soc. Sci.*, 2019, **51**, 67–81.
19. J. D. Farmer, C. Hepburn, P. Mealy and A. Teytelboym, *Environ. Resour. Econ.*, 2015, **62**, 329–357.
20. A. Gambhir, I. Butnar, P. H. Li, P. Smith and N. Strachan, *Energies*, 2019, **12**(9), 1747.
21. J. Bednar, M. Obersteiner and F. Wagner, *Nat. Commun.*, 2019, 10.
22. M. Fajardy, S. Chiquier and N. Mac Dowell, *Energy Environ. Sci.*, 2018, **11**, 3408–3430.
23. H. A. Daggash and N. Mac Dowell, *Int. J. Greenh. Gas Control.*, 2019, **85**, 174–181.
24. C. F. Heuberger, I. Staffell, N. Shah and N. Mac Dowell, *Nat. Energy*, 2018, **3**, 1–7.
25. M. Xylia, S. Leduc, P. Patrizio, S. Silveira and F. Kraxner, in *Transportation Research Procedia*, 2017, vol. 27.
26. S. Mesfun, S. Leduc, P. Patrizio, E. Wetterlund, A. Mendoza-Ponce, T. Lammens, I. Staritsky, B. Elbersen, J. Lundgren and F. Kraxner, *Energy*, 2018, **164**, 676–693.
27. C. F. Heuberger, P. K. Bains and N. Mac Dowell, *Appl. Energy*, 2020, **257**, 113715.
28. G. F. Nemet, M. W. Callaghan, F. Creutzig, S. Fuss, J. Hartmann, J. Hilaire, W. F. Lamb, J. C. Minx, S. Rogers and P. Smith, *Environ. Res. Lett.*, 2018, 13.
29. C. Wilson, A. Grubler, K. S. Gallagher and G. F. Nemet, *Nat. Clim. Change*, 2012, **2**, 780–788.
30. C. F. Heuberger, I. Staffell, N. Shah and N. Mac Dowell, *Comput. Chem. Eng.*, 2017, **107**, 247–256.
31. OECD, STructural ANalysis (STAN) Database., http://www.oecd.org/sti/ind/stanstructuralanalysisdatabase.htm.
32. M. Ha-Duong, M. J. Grubb and J. C. Hourcade, *Nature*, 1997, **390**, 270–273.
33. S. Dietz and C. Hepburn, *Energy Econ.*, 2013, **40**, 61–71.
34. C. Hepburn and G. Gosnell, in *Handbook of Sustainable Development: Second Edition*, 2014.
35. B. Flyvbjerg, *Oxf. Rev. Econ. Policy*, 2009, **25**, 344–367.
36. Institute for Government, *How to value infrastructure*, 2016.
37. M. Sensier, G. Bristow and A. Healy, *Spat. Econ. Anal.*, 2016, **11**, 128–151.
38. H. M Treasury, *Central Government Guidance on Appraisal and Evaluation*, 2018.
39. R. Poudineh and D. Penyalver, *Social Discount Rate and The Energy Transition Policy*, 2020.
40. F. Lamperti, M. Mazzucato, A. Roventini and G. Semieniuk, *Vierteljahrshefte zur Wirtschaftsforsch.*, 2019, **88**, 73–88.
41. E4tech, *Value Added of the Hydrogen and Fuel Cell Sector in Europe Supporting European growth and competitiveness*, 2019.

42. Element Energy, *Hydrogen for economic growth: Unlocking jobs and GVA whilst reducing emissions in the UK*, 2019.
43. OECD, *Future of work and skills*, 2017.
44. OECD, *The UK productivity puzzle through the magnifying glass: A sectoral perspective*, 2018.
45. OECD, *Towards Green Growth: Monitoring Progress: OECD Indicators*, 2011.
46. F. Di Comite, A. Kancs, D. Kancs and A. Kancs, *Macro-Economic Models for R&D and Innovation Policies*, 2015.
47. M. D. Gerst, P. Wang, A. Roventini, G. Fagiolo, G. Dosi, R. B. Howarth and M. E. Borsuk, *Environ. Model. Softw.*, 2013, **44**, 62–75.
48. P. Patrizio, N. Sunny and N. Mac Dowell, Inefficient investments as a key to narrowing regional economic imbalances, *iScience*, 2022, **25**(3).
49. P. Patrizio, M. Fajardy, M. Bui and N. Mac Dowell, CO_2 mitigation or removal: The optimal uses of biomass in energy system decarbonization, *iScience*, 2021, **24**(7).

Subject Index

Note: Page numbers followed by "*fn*" indicate footnotes.

ABMs. *See* Agent Based Models (ABMs)
Above-ground biomass, 251–253
Absolute global warming potential (AGWP), 69
Accelerated weathering of limestone (AWL), 173
Accounting, 414–415
  for CO_2 sequestration in AR projects, 245–248
  time horizon, 60
Afforestation, 12, 19, 40–41, 365–366
Afforestation and land use management (AFOLU), 81
Afforestation and reforestation (AR), 3, 244, 390–391, 434, 443, 452
  accounting for CO_2 sequestration in AR projects, 245–248
  biogenic C sequestration model, 261–266
  CO_2 and N_2O balance model, 271–275
  cost balance model, 275–280
  forest growth model, 248–261
  forestry operations model, 269–271
  multi-dimensions assessment, 247–248
  risk-adjusted C sequestration model, 266–269
  scope of study, 244–245
Afforested above-ground biomass, 253
  FMC, 254–256
  forest cycle growth phases, 254
  initialisation year, 256–258
  loop, 258–259
AFOLU. *See* Afforestation and land use management (AFOLU)
Agent Based Models (ABMs), 468
Agitation, 159
Agricultural soils, 195
Agriculture, 404–405
Agroforestry systems, 205–206
Air handling, 118–120
Air Pollution Control Residue (APCR), 150
Albedo, 19*fn*, 201*fn*
  impact of biochar on, 226
  impact of SCS on, 209
Alkaline materials, 362–363
Alkaline minerals, 178
Alkaline paper mill wastes (APMW), 150
Alkaline-functionalised adsorbents, 125
Alkalinity, 157, 178
Allam Cycle process, 323
Amine degradation, 126
Ammonium sulphate (AS), 158
Amorphous mineral, 178
Annual forestry operations, 271
Annual growth (AG), 254

Anthropogenic alkaline materials, 147–150
carbonation, 172–173
resources, 154
APCR. *See* Air Pollution Control Residue (APCR)
APMW. *See* Alkaline paper mill wastes (APMW)
Aqueous processes, 158
carbonate precipitation, 161–162
CO_2 dissolution, 160–161
mineral dissolution, 159–160
Aquifer, 178
AR. *See* Afforestation and reforestation (AR)
AR1. *See* First Assessment Report (AR1)
ARPA-E. *See* US Department of Energy Advanced Research Projects Agency-Energy (ARPA-E)
Artificial substances, 148
Atmospheric ocean iron fertilisation, 303. *See also* Direct ocean iron fertilisation
cost-effectiveness and carbon accounting, 304
environmental consequences and context, 304
governance and acceptability, 304–305
scientific understanding of key processes, 303–304
Attrition, 158
Autotrophic respiration (R_A), 246
Aviation, 407–408
Avoided costs, 90
AWL. *See* Accelerated weathering of limestone (AWL)

Basalt, 153, 178
Basic oxygen furnace slag (BOF slag), 66*fn*
Basic rocks, 153, 178
Bauxite residue. *See* Red mud
Bauxite tailings, 149
BCAM. *See* Border carbon adjustment mechanisms (BCAM)
BECCS. *See* Bioenergy with Carbon Capture and Storage (BECCS)
Below-ground biomass, 259
Benthic organisms, 178
Bio-feedstocks, 91
Bio-geophysical constraints, 103
Bio-oil, 18
Biochar (BC), 3, 12, 18–19, 194, 196, 216–217, 363–365
application of biochar in grassland systems, 222
costs, 227
function as GGR, 222–228
as greenhouse gas removal technology, 218
impact on GHG emissions, 220–221
impact on soil carbon sequestration, 221
land required for biochar application, 221–222
traditional use, 217–218
Bioenergy, 83, 360
potential converted to removal potential, 103–104
Bioenergy with Carbon Capture and Storage (BECCS), 1, 3, 10, 14, 29, 40–41, 80, 83, 116, 291, 358–360, 381, 391
carbon metrics, 86–89
deployment opportunities and constraints, 103–108
energy metrics, 89–90
environmental metrics, 90–91
feedstocks, 91–93
financial metrics, 90
global sustainable potential and impacts, 103–105
for hydrogen, 98–101
in industry, 101
to liquid fuel, 95–98
low hanging fruit, 105–106

Bioenergy with Carbon Capture and Storage (BECCS) (*continued*)
mitigation *vs.* removal potential and system integration effects, 102
NET co-deployment and synergies, 107–108
net energy balance, 89
optimal BECCS deployment at global level, 106–107
pathways, 94–102
to power, 94–95
principle, 80–81
value chain, 83–86
value in decarbonisation pathways, 81–83
Biofuel crops, 43
Biogenic C sequestration model, 261
biogenic C pools, 261–262
biomass C pool, 262–264
dead organic matter C pool, 264
harvested wood product C pool, 264–266
Biogenic volatile compounds (BVOCs), 43
Biogeochemical models, 297
Biological carbon pump, 295
Biologically-based CO_2 removal
in coastal seas, 311–326
in Open Ocean, 295
Biomass, 81, 101
C pool, 262–264
certification standards, 451–452
conversion, 86
plantations, 222
processing and transport, 85
production, 83–85
Biomineralisation, 178
Blue carbon, 367
BLUE model, 473
Blue water footprint, 282
BOF slag. *See* Basic oxygen furnace slag (BOF slag)
Border carbon adjustment mechanisms (BCAM), 2
Bottom-up methods, 245–246, 468
Brine, 178
Bromoform ($CHBr_3$), 325
BVOCs. *See* Biogenic volatile compounds (BVOCs)

Calcifying, 178
Calcination, 178
Calcine, 179
Calciner, 128
Calcite, 162, 178
Calcium (Ca), 139
Calcium carbonate ($CaCO_3$), 178
precipitation, 161
Cap-and-trade programmes, 441
Capacity, 418–419
Capital recovery factor (CRF), 277
CarbFix project, 124, 172, 392
CarbFix2, 15, 124
Carbon (C)
accounting, 59–60
avoided, 87–90
breakeven time, 89
budget, 58
credit, 178
cycle feedbacks, 29
debt, 58
efficiency, 87
intensity of energy, 123
negativity, 89
neutrality, 101
price, 90
pricing, 414, 439, 441
productivity, 478–480
sequestration, 60
sink saturation, 215
storage, 60
taxes, 441
Carbon Capture and Storage (CCS), 357–358, 385, 392, 445, 456
water footprint, 91
Carbon Capture and Utilisation (CCU), 370, 399
Carbon dioxide (CO_2), 28, 115, 139, 179, 195, 292
avoidance, 71–73

balance model, 271–275
capture, 90
collector, 125
dissolution, 160–161
emission contributions, 271–273
emissions, 6
removal efficiency, 273–275
sequestration potential, 268–269
storage, 21, 71–73
monitoring, 452–453
Carbon dioxide capture and storage technologies (CCS technologies), 2–3, 86, 116
Carbon dioxide removal (CDR), 7, 29, 81, 179, 196, 244, 352, 378–379, 392, 430, 466
comparison of CDR technologies and practices, 358–368
cooperation, 384
effective policies to remove barriers, 385–386
fair CDR quotas, 380–381
pathways, 381–384
readiness framework for, 357–358
synthesis, 368–369
technologies, 58
Carbon dioxide removal model intercomparison project (CDRMIP), 34
Carbon engineering, 127–129, 408
Carbon Offsetting and Reduction Scheme for International Aviation (CORSIA), 407
Carbon takeback obligation (CTO), 415
Carbonate, 160
compensation depth, 157
minerals, 154
precipitation, 161–162
rocks, 178
Carbonation, 140
Carbonic acid (H_2CO_3), 139
Carbonic anhydrase, 161
Caustic, 179
CBA. *See* Cost benefits analysis (CBA)
CBCEs. *See* Coastal blue carbon ecosystems (CBCEs)
CBD. *See* Convention on Biological Diversity (CBD)
CBDR-RC. *See* Common but differentialised responsibilities and respective capabilities (CBDR-RC)
CCAMLR. *See* Convention on the Conservation of Antarctic Marine Living Resources (CCAMLR)
CCGT. *See* Combined cycle gas turbines (CCGT)
CCS. *See* Carbon Capture and Storage (CCS)
CCU. *See* Carbon Capture and Utilisation (CCU)
CDM. *See* Clean Development Mechanism (CDM)
CDR. *See* Carbon dioxide removal (CDR)
CDRMIP. *See* Carbon dioxide removal model intercomparison project (CDRMIP)
Cement, 179
Cement kiln dust, 148
CFB. *See* Circulating fluidised bed reactor (CFB)
CfD. *See* Contracts-for-difference (CfD)
Charcoal, 216
Chemical looping combustion plants (CLC plants), 94
Chemical weathering, 179
Chemicals, 67
CHP plants. *See* Combined heat and power plants (CHP plants)
Circular Economy Action Plan, 385
Circulating fluidised bed reactor (CFB), 128
Citrate, 160
Clays, 155, 179
formation, 155
CLC plants. *See* Chemical looping combustion plants (CLC plants)

Clean Development Mechanism (CDM), 321–322, 385
Climate models, 414
Climate Pledge, 58
Climate system, 27
 co-benefits and unintended consequences, 42–43
 Earth system implications for demand for negative emissions, 33–39
 Earth system implications for supply of negative emissions, 39–42
 Earth system in NETs research, 29–30
 need for negative emissions to achieve climate goals, 28–29
 non-CO_2 activities, 43–44
 supply and demand for negative emissions, 30–33
Clime Works, 124–126, 131, 172, 408
 in Iceland, 15
Clinopyroxene ($CaMgSi_2O_6$–$CaFeSi_2O_6$), 146
CMIP. *See* Coupled Model Intercomparison Project (CMIP)
Coal-fired plants, 477
Coastal blue carbon, 12, 20–21
Coastal blue carbon ecosystems (CBCEs), 311
 cost-effectiveness, 320–321
 environmental consequences and context, 316–320
 governance and carbon accounting, 321–322
 scientific understanding of key processes, 311–316
Coastal methods, 291
Coastal seas
 biologically-based CO_2 removal in, 311
 coastal blue carbon ecosystems, 311–322
 marine BECCS using seaweed, 322–326
Coefficient of performance (COP), 126
Columbia River basalt, 153
Combined cycle gas turbines (CCGT), 477
Combined heat and power plants (CHP plants), 94
Commercial agriculture, 444
Comminution, 179
Common but differentialised responsibilities and respective capabilities (CBDR-RC), 419
Concrete, 65–66, 179
Construction and demolition waste in soils (C&D waste in soils), 148
Constructive stakeholder, 412
Contactor design, 118–120
Contracts-for-difference (CfD), 443
Convention on Biological Diversity (CBD), 329
Convention on the Conservation of Antarctic Marine Living Resources (CCAMLR), 302
Cooperative approach, 379, 381–384
COP. *See* Coefficient of performance (COP)
Corn stover, 93
CORSIA. *See* Carbon Offsetting and Reduction Scheme for International Aviation (CORSIA)
Cost benefits analysis (CBA), 439, 474
 discounting issues, 475
 misleading project appraisals, 474
 vicious public spending cycles, 475
Costs, 434
 balance model, 275
 over 100 years, 277–278
 over 1000 years, 278–280
  cost contributions, 275–277
 of CO_2 removal, 90
Coupled Model Intercomparison Project (CMIP), 44–45

Cover crops, 201
CRF. *See* Capital recovery factor (CRF)
Crop growth cycle, 85
Crop rotations, 201
Cross-border carbon accounting, 453–454
CTO. *See* Carbon takeback obligation (CTO)

DAC. *See* Direct air capture (DAC)
DACCS. *See* Direct air CO_2 capture and storage (DACCS)
DCF. *See* Dynamic characterisation factors (DCF)
Dead organic matter C pool, 264
Decarbonisation pathways, BECCS value in, 81–83
Degraded land restoration, 206
Degraded soils, 206
Delayed emissions, 68
  dynamic life cycle assessment, 70–71
  tonne-year approaches, 69–70
Demand-side measures, 432
Demonstration, 437–438
Desorption, 121
Diatom growth, 175
DIC. *See* Dissolved inorganic carbon (DIC)
Dimethyl sulphide (DMS), 293
Diplomacy, 454
Direct air capture (DAC), 1, 10, 14–15, 40, 71, 107, 115, 144, 171, 179, 390, 392
  assessment, 135–136
  costs, 130–134
  generic sorption/desorption DAC processes, 117
  minimum work and energy requirements, 120–123
  process design, 123–129
  process fundamentals, 117–123
  processes, 116
Direct air CO_2 capture and storage (DACCS), 3, 360–362, 381–382
Direct aqueous mineral carbonation, 144
Direct capture of CO_2, 29
Direct carbonation, 140, 144, 179
Direct gas–solid carbonation, 144
Direct land use change, 87
Direct ocean iron fertilisation, 295. *See also* Atmospheric ocean iron fertilisation
  cost-effectiveness and carbon accounting, 298–301
  environmental consequences and context, 297–298
  governance and acceptability, 301–303
  scientific understanding of key processes, 295–297
Disruptive innovation, 471
Dissolution minerals, 151
  anthropogenic alkaline material resources, 154
  natural silicate resources, 151–154
Dissolved inorganic carbon (DIC), 157, 179
Dissolved organic carbon (DOC), 322
DMS. *See* Dimethyl sulphide (DMS)
DOC. *See* Dissolved organic carbon (DOC)
Dolomite, 179
DSGE models. *See* Dynamic Stochastic General Equilibrium models (DSGE models)
Dunite, 153
Dynamic characterisation factors (DCF), 70
Dynamic life cycle assessment, 70–71
Dynamic Stochastic General Equilibrium models (DSGE models), 471

Earth system
  implications for demand for negative emissions, 33
    complex models, 34–39

Earth system (*continued*)
implications for supply of negative emissions, 39
BECCS and afforestation, 40–41
land management and restoration, 41
in NETs research, 29–30
Earth system models (ESMs), 39, 45
Ecological water deficit, 282–283
Ecological zone, 248*fn*
Economic impact assessment (EIA), 475–476
EEZ. *See* Exclusive economic zone (EEZ)
Egalitarianism, 418–419
EGR. *See* Enhanced gas recovery (EGR)
EIA. *See* Economic impact assessment (EIA)
Electrochemical processes, 117
Emissions reduction, 410
Emissions trading schemes (ETS), 4, 441
Energy
impact of biochar on, 226–227
crops and residues, 91–93
efficiency, 89
requirements, 120–123
impact of SCS on, 209
and utilities, 405–407
Energy Return on Investment (EROI), 90–91
Energy system decarbonisation (ESD), 273, 275
Energy system optimisation model (ESO), 473
Energy systems optimisation models (ESOMs), 468
Energy transition
modelling, 467
existing approaches, 467–468
key features of socially equitable energy transition, 469–473
policies, 474
CBA, 474–475
EIA, 475–476
key socio-economic metrics, 476–480
Enhanced gas recovery (EGR), 399
Enhanced Oil Recovery (EOR), 83, 129, 361, 399, 444
Enhanced rock weathering (ERW), 146–147
Enhanced upwelling to increase productivity, 309–311
Enhanced weathering (EW), 3, 40, 179, 363, 390, 392, 438
Enhancement factor, 161
Environmental–economic constraints, 104
EOR. *See* Enhanced Oil Recovery (EOR)
EROI. *See* Energy Return on Investment (EROI)
ERW. *See* Enhanced rock weathering (ERW)
ESD. *See* Energy system decarbonisation (ESD)
ESMs. *See* Earth system models (ESMs)
ESO. *See* Energy system optimisation model (ESO)
ESOMs. *See* Energy systems optimisation models (ESOMs)
ETS. *See* Emissions trading schemes (ETS); European trade system (ETS)
EU Emission Trading System (EU ETS), 385
EU ETS. *See* EU Emission Trading System (EU ETS)
European trade system (ETS), 58
EVT dataset. *See* Existing vegetation type dataset (EVT dataset)
EW. *See* Enhanced weathering (EW)
Ex situ
mineral carbonation, 15–16, 140, 163–171
multi-step carbonation, 145
single-step carbonation, 140–144
mineralisation, 179

Exclusive economic zone (EEZ), 326
Existing vegetation type dataset (EVT dataset), 268

Fair CDR quotas, 380–381
FBC plants. *See* Fluidised bed combustion plants (FBC plants)
FCH JU report. *See* Fuel Cells and Hydrogen Joint Undertaking report (FCH JU report)
Feasibility, 357
  of using biologically-based processes, 291
    biologically-based CO_2 removal in coastal seas, 311–326
    biologically-based CO_2 removal in open ocean, 295–311
Felsic minerals, 179
Felsic rocks, 153, 179
Fermentation, 98
Financing, 438
  NET-based carbon markets, 442
  NET-based carbon taxes, 441–442
  NET-based cost–benefit analysis, 439–441
  NET-based subsidies, 443–444
First Assessment Report (AR1), 7
Fischer–Tropsch synthesis (FT synthesis), 98
Flood basalt, 179
Flue gas, 179
Fluidised bed combustion plants (FBC plants), 94
Fly ash, 180
FMC. *See* Forest management cycle (FMC)
Food–fuel debate, 455–456
Forest cycle growth phases, 254
Forest growth model, 248
  above-ground biomass, 251–253
  afforested above-ground biomass, 253–259
  below-ground biomass, 259
  model scope & characterisation, 248–251
  total biomass, 259
Forest management cycle (FMC), 254–256
Forest road construction and maintenance, 270–271
Forestry operations model, 269
  annual forestry operations, 271
  forest road construction and maintenance, 270–271
  site establishment, 270
Formiguer, 217*fn*
Forsterite (Mg_2SiO_4), 159
"4 per 1000" initiative, 228–229
45Q tax credit, 443
FT synthesis. *See* Fischer–Tropsch synthesis (FT synthesis)
Fuel Cells and Hydrogen Joint Undertaking report (FCH JU report), 476
Fuels, 67
  supply chain emissions, 451
Future contracts, 442
FW. *See* Weighting factor (FW)

G4M. *See* Global Forest Model (G4M)
Gabbro, 153
Gasification, 98
Gas–solid processes, 157–158
Geochemical negative emission technologies, 138
  anthropogenic alkaline material carbonation, 172–173
  ex-situ mineral carbonation, 163–171
  historical context, 139–150
  hybrid approaches, 174
  in situ mineral carbonation, 172
  kinetics, 157–163
  proposals for increasing ocean alkalinity, 173–174
  resources, 150
    chemistry, 150–151
    dissolution minerals, 151–154

Geochemical negative emission technologies (*continued*)
- ocean alkalinity, 155–157
- secondary minerals, 154–155
- risks and co-benefits, 174–177

Geological formation, 180
Geological reservoirs, 453
Geological storage, 180
GESAMP. *See* Joint Group of Experts on the Scientific Aspects of Marine Environmental Protection (GESAMP)
GEZ dataset. *See* Global ecological zone dataset (GEZ dataset)
GGR. *See* Greenhouse gas removal (GGR)
GHG emissions. *See* Greenhouse gas emissions (GHG emissions)
Gigatonnes of CO_2 (Gt CO_2), 10
Global ecological zone dataset (GEZ dataset), 268
Global Forest Coalition (2017), 409
Global Forest Model (G4M), 247
Global Warming Potential (GWP), 61, 317
Governance, 413
- accounting, 414–415
- carbon pricing, 414
- incentivising GGR, 413–414
- procedural justice, 416–417
- risk management, 415–416
- side-effects and implications, 416

Governments, 400–402
- revenues, 439

GPP. *See* Gross Primary Production (GPP)
Grasslands, 203
- application of biochar in grassland systems, 222

Green water footprint, 282
Greenhouse gas emissions (GHG emissions), 7, 41, 58, 195, 358, 391, 431
- balance/CO_2 removal potential, 275
- biochar impact on, 220–221
- monetary valuation, 59
- impact of SCS on, 208

Greenhouse gas removal (GGR), 1, 3, 6, 27, 194–195, 292, 390–391, 430–431
- challenges and opportunities, 10–21
- co-benefits and societal considerations, 21–22
- CO_2 storage, 21
- economic challenges, 395
  - compatibility with other priorities, 397–398
  - cost structures, 395–397
  - incentives, 398–399
- governance, 413–417
- historical CO_2 emissions, 9
- removing CO_2 from air, 9–10
- socio-political aspects of international governance, 417–419
- stakeholder views, 399–413
- technologies, 7–8

Grey water footprint, 283–284
Gross Primary Production (GPP), 246
Gross value added (GVA), 473
GVA. *See* Gross value added (GVA)
GWP. *See* Global Warming Potential (GWP)

Halomethanes, 325
Hard water, 140
Harm minimisation, 449
Harvested wood and residues from forestry, 91
Harvested wood products (HWPs), 64
- C pool, 264–266

Harzburgite, 153
Henry's coefficient, 160
Heterotropic soil respiration (R_H), 247
Host rock, 180
Hot gases, 18

HWPs. *See* Harvested wood products (HWPs)
Hy impact study, 476
Hybrid approaches, 174
Hydrochloric acid, 145
Hydrogen, 102
BECCS for, 98–101

IAMs. *See* Integrated assessment models (IAMs)
ICAO. *See* International Civil Aviation Organization (ICAO)
Igneous rock, 180
ILCD. *See* International Reference Life Cycle Data (ILCD)
iLUC. *See* Indirect Land Use Change (iLUC)
IMO. *See* International Maritime Organization (IMO)
In situ
carbonation, 145–146
mineral carbonation, 15, 172
mineralisation, 180
Incentives, 398–399
Incentivising GGR, 413–414
Indirect carbonation. *See* Multi-step carbonation
Indirect Land Use Change (iLUC), 87, 359
Industrial specialisation, 477
Initialisation year, 256–258
Inorganic ligands, 160
Input–output tables (IO tables), 473
Integrated assessment models (IAMs), 39, 81, 391, 430–431, 467–468
Inter-generational justice, 60
Intergovernmental Panel on Climate Change (IPCC), 7, 292, 391
International Civil Aviation Organization (ICAO), 407
International governance, 105
socio-political aspects of, 417–419
International law, 386
International Maritime Organization (IMO), 450
International Panel for Climate Change (IPCC), 81
International Reference Life Cycle Data (ILCD), 61
IO tables. *See* Input–output tables (IO tables)
IPCC. *See* Intergovernmental Panel on Climate Change (IPCC); International Panel for Climate Change (IPCC)
Iron salt aerosol (ISA), 303
ISA. *See* Iron salt aerosol (ISA)
Isolationist approach, 379, 381–384

Joint Group of Experts on the Scientific Aspects of Marine Environmental Protection (GESAMP), 292

Kaolinite, 155
Kimberlite, 153
Kraft pulping, 150
Kyoto Protocol, 452

Land management and restoration, 41
Land use, impact of SCS on, 208
Land use, land use change and forestry (LULUCF), 452
Land use changes (LUCs), 401, 416, 451
Land-based GGR, 292
Land-based NETs, 451
Landscape Fire and Resource Management Planning Tools, 267
Larnite (Ca_2SiO_4), 159
Lashof method, 70
Lawrence Livermore National Laboratory (LLNL), 93
LC. *See* London Convention (LC)
LC-LP. *See* London Convention and London Protocol (LC-LP)
Leontief's production functions, 473
Lherzolite, 153
LHV. *See* Lower heating value (LHV)

Lime (CaO), 17, 148, 173, 180
Limestone, 180
Liquid fuel, BECCS to, 95–98
LLNL. *See* Lawrence Livermore National Laboratory (LLNL)
London Convention (LC), 450
London Convention and London Protocol (LC-LP), 302, 329
London Protocol (LP), 450–451
London Protocol under the London Convention (LP/LC), 445
Loop, 258–259
Looping process, 16
Low hanging fruit, 105–106
Lower heating value (LHV), 98*fn*
LP. *See* London Protocol (LP)
LP/LC. *See* London Protocol under the London Convention (LP/LC)
LUCs. *See* Land use changes (LUCs)
LULUCF. *See* Land use, land use change and forestry (LULUCF)

Macroalgae Research Inspiring Novel Energy Resources program (MARINER program), 323
Macronutrient addition, 305–309
Mafic mineral, 180
Mafic rocks, 151–154, 180
MAG. *See* Mean Annual Growth (MAG)
Magnesite ($MgCO_3$), 139
Magnesium (Mg), 139
MAI. *See* Mean Annual Increment (MAI)
Manure, 93
  application, 202
Marine BECCS using seaweed, 322–326
Marine biologically-based CDR, 366–368
Marine macroalgae, 322
MARINER program. *See* Macroalgae Research Inspiring Novel Energy Resources program (MARINER program)
Maximum Mean Annual Growth (MMAG), 254
Maximum Mean Annual Increment (MMAI), 254
MCA. *See* Multicriteria analysis (MCA)
MEA. *See* Monoethanolamine (MEA)
Mean Annual Growth (MAG), 254–255
Mean Annual Increment (MAI), 254
Mean fire return interval (mFRI), 267–268
Measurement, reporting and verification (MRV), 451
  biomass certification standards, 451–452
  CO_2 storage monitoring, 452–453
  guidelines, 437
  reporting metrics, 452
Merchant CO_2, 444
Metamorphic rock, 180
Methane (CH_4), 195, 293, 298, 313
mFRI. *See* Mean fire return interval (mFRI)
Milling, 159
Mine tailings, 149
Mineral(s), 180
  carbonation, 12, 15–16, 66, 181, 362–363
  dissolution, 150*fn*, 159–160, 180
  for geochemical NETs, 153
  trap, 181
  weathering, 140
Mineralisation, 171
Minimum work, 120–123
Mining industry, 149
Mitigation *vs.* removal potential and system integration effects, 102
Mixing, 159
MMAG. *See* Maximum Mean Annual Growth (MMAG)
MMAI. *See* Maximum Mean Annual Increment (MMAI)

Monitoring, reporting and verification (MRV), 1
  platforms, 216
  processes, 317
Monoculture tree plantations, 409
Monoethanolamine (MEA), 15
Moral hazard, 455
Moura-Costa method, 69–70
MRV. *See* Measurement, reporting and verification (MRV)
MSW. *See* Municipal solid wastes (MSW)
Multi-sector energy systems modelling, 102
Multi-step carbonation, 145, 180
Multicriteria analysis (MCA), 474
Municipal solid wastes (MSW), 91, 93

NASA Land Use and Land Cover Program, 267
National decarbonisation policies, 474
National forest monitoring systems (NFM), 453
Nationally Determined Contributions (NDCs), 321, 380, 391, 431
Natural alkalinity function in ocean, 155–156
Natural capital of soils, 211
Natural silicate resources, 151–154
Natural sinks, 9–10
Nature-based solutions (NBS), 3, 39
Nature's Contribution to People (NCP), 196, 207, 223
  impact of biochar on, 227–228
  impact of SCS on, 209–211
NBS. *See* Nature-based solutions (NBS)
NCP. *See* Nature's Contribution to People (NCP)
NDCs. *See* Nationally Determined Contributions (NDCs)
NECs. *See* Negative emissions credits (NECs)
Negative CO_2 emissions, 181
Negative emission technologies (NETs), 292, 391
Negative emissions, 29, 57–58, 430
  carbon accounting, 59–60
  CO_2 storage and CO_2 avoidance, 71–73
  storage time, 60–67
  timing of emissions and removals, 67–71
Negative emissions credits (NECs), 90, 95, 275, 442
Negative emissions technologies (NETs), 7, 27, 29–30, 81, 181, 196, 294, 430–431
  characterising NETs for policy development, 432–434
  co-deployment and synergies, 107–108
  NET-based carbon markets, 442
  NET-based carbon taxes, 441–442
  NET-based cost–benefit analysis, 439–441
  NET-based subsidies, 443–444
  recognising, 449–451
NEP. *See* Net Ecosystem Productivity (NEP)
Net cost of removal, 90
Net Ecosystem Productivity (NEP), 247
Net present value (NPV), 474
Net Primary Production (NPP), 246
Net-zero emissions, 466
NETs. *See* Negative emission technologies (NETs); Negative emissions technologies (NETs)
NFM. *See* National forest monitoring systems (NFM)
NGOs, 408–409
Niche markets, 444–445
Nitric acid, 145
Nitrogen, phosphorus and potassium fertilisers (NPK fertilisers), 202
Nitrogen (N), 208
  fertiliser, 87

Nitrous oxide (N_2O), 195, 293, 298, 313
balance model, 271–275
emissions, 18
contributions, 271–273
Non-CO_2 activities, 43–44
Non-marine-based NETs, 451
Not in my backyard (NIMBY) attitude, 456
NPK fertilisers. *See* Nitrogen, phosphorus and potassium fertilisers (NPK fertilisers)
NPP. *See* Net Primary Production (NPP)
NPV. *See* Net present value (NPV)
Nutrients
impact of biochar on, 223–224
management, 202
impact of SCS on, 208

Occidental Petroleum (Oxy), 129
Ocean acidification, 157
Ocean alkalinity, 155
enhancement, 12, 16–17
natural alkalinity function in ocean, 155–156
proposals for increasing, 173–174
stability of increased alkalinity, 157
weathering and alkalinity, 156–157
Ocean fertilisation (OF), 12, 20, 390, 392, 438, 445
Ocean iron fertilisation (OIF), 294–295
Ocean Thermal Energy Conversion (OTEC), 310
OF. *See* Ocean fertilisation (OF)
OIF. *See* Ocean iron fertilisation (OIF)
Olivine (Mg_2SiO_4), 139, 146
1Point5 project, 408
Open ocean
atmospheric ocean iron fertilisation, 303–305
biologically-based CO_2 removal in, 295
direct ocean iron fertilisation, 295–303
enhanced upwelling to increase productivity, 309–311
macronutrient addition, 305–309
Optimal BECCS deployment at global level, 106–107
Organic amendments, 202–203
Organic biomass, 216
Organic ligands, 160
OSPAR Convention, 309
OTEC. *See* Ocean Thermal Energy Conversion (OTEC)
"Overshoot" behaviour, 34
Oxalate, 160

Paris Agreement, 1, 27, 44, 229, 294, 316, 321, 328, 380, 391, 431, 443, 454
Parts per million (ppm), 9
PAS. *See* Publicly Available Specification (PAS)
Passivating product layers, 181
Passive carbonation, 66*fn*, 181
Payments for Ecosystems Services (PES), 366
PC plants. *See* Pulverised combustion plants (PC plants)
PE. *See* Polyethylene (PE)
Pedogenic process, 181
PEF. *See* Product Environmental Footprint (PEF)
Pelleting, 85
Perennials, 203–205
Peridotite, 181
Permanence, 60
of biogenic CO_2 sequestration, 266–267
Permanent carbon storage, 60
Permanent-pyrogas, 222*fn*
PES. *See* Payments for Ecosystems Services (PES)
PET. *See* Polyethylene terephthalate (PET)

Peta-grammes, 28*fn*
pH swing, 145
Phosphate, 160
Phosphorus (P), 208
Physical Earth system, 27–28
Physical removals, 73
Plaggen soil cultivation, 217
Plastics, 63–64
Policy design, 431
 characterising NETs for policy development, 432–434
 demonstration and scale-up, 437–438
 financing, 438–444
 niche markets, 444–445
 public acceptability, 455–456
 regulation and governance, 445–454
 research and development, 434–437
Polluter pays principle, 415, 439
Polyester, polyamide, and acrylic fibres (PP&A fibres), 63
Polyethylene (PE), 63*fn*
Polyethylene terephthalate (PET), 63*fn*
Polypropylene (PP), 63*fn*
Polystyrene (PS), 63*fn*
Polyvinylchloride (PVC), 63*fn*
Positively charged cations, 156
Post-combustion CCS, 360
Potassium (K), 208
Potassium hydroxide, 127
Power, BECCS to, 94–95
PP. *See* Polypropylene (PP)
PP&A fibres. *See* Polyester, polyamide, and acrylic fibres (PP&A fibres)
ppm. *See* Parts per million (ppm)
PPP. *See* Purchasing power parity (PPP)
Precipitation kinetics, 162
Primary minerals, 181
Private sector views, 402
 agriculture, 404–405
 energy and utilities, 405–407
 transport and aviation, 407–408
Procedural justice, 416–417
Product Environmental Footprint (PEF), 61
Productivity, 477*fn*
 impact of biochar on, 224–225
Protocol on Environmental Protection to the Antarctic Treaty (Madrid Protocol), 302
PS. *See* Polystyrene (PS)
Public acceptability, 455–456
Public engagement, 410
 public attitudes, 410–412
 strategies, 412–413
Public sector, 431
Publicly Available Specification (PAS), 61, 70
 PAS 2050-method, 70
Pulverised combustion plants (PC plants), 94
Purchasing power parity (PPP), 277
PVC. *See* Polyvinylchloride (PVC)
PyCCS. *See* Pyrogenic carbon capture and storage (PyCCS)
Pyrogenic carbon capture and storage (PyCCS), 222
Pyrolysis, 216–217
Pyrolytic liquid, 222*fn*

Quantitative modelling and analysis techniques, 465, 467

R&D. *See* Research and Development (R&D)
Radiative forcing, 45
RCPs. *See* Representative Concentration Pathways (RCPs)
RD&D. *See* Research, development and demonstration (RD&D)
RE. *See* Renewable energy (RE)
Rebound effect, 34
Red mud, 149, 181
REDD+. *See* Reducing Emissions from Deforestation and Degradation (REDD+)

Reducing Emissions from Deforestation and Degradation (REDD+), 321–322, 366–367
Reforestation, 12, 19
Regional energy system modelling studies, 81
Remote sensing, 300
Renewable energy (RE), 477
    subsidies, 439
Representative Concentration Pathways (RCPs), 45
Research, development and demonstration (RD&D), 430
Research and development (R&D), 434
    costs, 434
    environmental impacts, 437
    MRV guidelines, 437
    scalability, 437
    technological feasibility, 434
Reservoir, 181
Residual emissions, 360
Residue management, 200–201
Responsibility, 418–419
Restoration of degraded land, 206
Reverse weathering reactions, 155
Reykjavik Energy, 124
Risk management, 415–416
Risk-adjusted C sequestration model, 266
    CO_2 sequestration potential, 268–269
    permanence of biogenic CO_2 sequestration, 266–267
    wildfire-penalty coefficient, 267–268
Rocks, 181
    extraction and processing, 362
    for geochemical NETs, 153

Saline formation, 181
Saline groundwater, 181
Salinity, 160
Saturation
    of SOC, 215
    state, 162, 181
SBSTA. *See* Subsidiary Body for Science and Technological Advice (SBSTA)
Scalability, 437
Scale-up, 437–438
SCC. *See* Social cost of carbon (SCC)
Scenario-discovery technique, 473
SCS. *See* Soil carbon sequestration (SCS)
SDG. *See* United Nation's Sustainable Development Goals (SDG)
Seaweeds. *See* Marine macroalgae
Secondary minerals, 154–155, 182
Sedimentary rock, 181
Sepiolite, 155
Sequestration, 182
Serpentine ($(Mg,Fe)_3Si_2O_5(OH)_4$), 146
Serpentine dissolution, 145
Shared Socioeconomic Pathways (SSPs), 45, 431
Short-term CDR approach, 363
Siberian Traps, 153
Siderite ($FeCO_3$), 154
Silica, 140
Silicic acid (H_4SiO_4), 159
Single-step carbonation, 140, 144
Slag, 148–149, 181
Slaker, 128
SOC. *See* Soil organic carbon (SOC)
Social acceptability, 104
Social and policy constraints, 104–105
Social cost of carbon (SCC), 439
Social license to operate. *See* Social acceptability
Socially equitable energy transitions, 466
    energy transition modelling, 467–473
    energy transition policies, 474–480
Soil carbon sequestration (SCS), 3, 12, 17–18, 182, 194–195, 197, 363–365, 390, 392, 434
    biochar impact on, 221

cost, 209
as greenhouse gas removal technology, 212
limitations and uncertainties, 215–216
potential and implementation, 212–214
impacts of SCS managements, 207
management, 197–207
Soil Conservation Service Method, 282
Soil organic carbon (SOC), 17, 182, 194–195
Soil organic matter (SOM), 195, 229
Soils, 210*fn*, 215*fn*
contamination, impact of biochar on, 225
degradation, 215*fn*
erosion, 215*fn*
fertility, impact of biochar on, 223–224
functions, impact of SCS on, 207
Solar radiation management (SRM), 43
SOLAS. *See* Surface Ocean-Lower Atmosphere Study (SOLAS)
Solubility trapping, 182
Solvent-based DAC, 15
SOM. *See* Soil organic matter (SOM)
Sonication, 159
Sorbent regeneration, 128
Sorption, 121
Spatial and sectoral heterogeneity, 471–473
Special Report on Global Warming of 1.5 1C (SR1.5), 57
Sponge effect, 66
SR1.5. *See* Special Report on Global Warming of 1.5 1C (SR1.5)
SRM. *See* Solar radiation management (SRM)
SS slag. *See* Stainless steel slag (SS slag)
SSPs. *See* Shared Socioeconomic Pathways (SSPs)
Stabilisation pathways, 28
Stability of increased alkalinity, 157
Stable geological formation, 182
Stainless steel slag (SS slag), 66*fn*
Stakeholder views of GGR, 399–413
Starch crops, 93
Steel, 65–66
making, 148
Stoichiometric solubility product, 162
Storage time of negative emissions, 60
time horizon and lifetime of carbon storage in products, 61–67
Stranded heat, 134
Subsidiary Body for Science and Technological Advice (SBSTA), 452
Substitution, 73
Sugar crops, 93
Sulphate, 160
Sulphuric acid, 145
Supersaturated, 181
Supply and demand for negative emissions, 30–33
Supply-side measures, 432
Surface Ocean-Lower Atmosphere Study (SOLAS), 302
Surficial, 146–147, 182
mineral carbonation, 15
Synthetic renewable fuels, 444

TA. *See* Total alkalinity (TA)
Tailings, 149, 181
Taxation mechanisms, 439
TCRE. *See* Transient Response to cumulative Carbon Emissions (TCRE)
TEA. *See* Total ecosystem aquaculture (TEA)
Technological feasibility, 434
Technological innovation, 469–471

Technological Readiness Levels (TRLs), 329, 437
Technology, 294
Temporal carbon storage, 60
Terra Preta (do Indio) soils, 217
Terrestrial sequestration, 182
TH. *See* Time horizon (TH)
Thermal energy, 131
Tillage management, 197–200
Timber production, 254
Time, 60
Time horizon (TH), 60, 69
    and lifetime of carbon storage in products, 61–67
Timing of emissions and removals, 67
    delayed emissions, 68–71
Tonne-year approaches, 69–70
Top-down methods, 246–247
Torrefaction, 85, 221*fn*
Total alkalinity (TA), 156, 182
Total biomass, 259
Total ecosystem aquaculture (TEA), 323
Toxic substances, impact of biochar on, 225
TPOs. *See* Tradable put options (TPOs)
Tradable put options (TPOs), 442
Trade in manufacturing value added, 477
Traditional kiln technologies, 225*n*
Transient Response to cumulative Carbon Emissions (TCRE), 28
Transport, 407–408
    costs, 85
Trapping
    geochemical, 182
    mineral, 182
    physical, 182
    residual, 182
    solubility, 182–183
TRLs. *See* Technological Readiness Levels (TRLs)

Ultrabasic rocks, 153, 182
Ultramafic rocks, 151–154, 182
Uncertainty, 473
UNFCCC. *See* United Nations Framework Convention on Climate Change (UNFCCC)
United Nation's Sustainable Development Goals (SDG), 33, 196, 207, 223, 416
    impact of biochar on, 228
    impact of SCS on, 211–212
United Nations Framework Convention on Climate Change (UNFCCC), 196, 379, 443, 467–468
Upland, 20*fn*
US Department of Agriculture (USDA), 282
US Department of Energy Advanced Research Projects Agency-Energy (ARPA-E), 323
US Ocean Dumping Ban Act (1988), 309
USDA. *See* US Department of Agriculture (USDA)

Valuable products, 174–175
Valuable services, 175
VCO markets. *See* Voluntary carbon offset markets (VCO markets)
VCS. *See* Verified Carbon Standard (VCS)
VDep. *See* Vegetation departure index (VDep)
Vegetation departure index (VDep), 267–268
Vegetation evapotranspiration, 281
Verified Carbon Standard (VCS), 322
Vertically-profiling Argo floats, 300
Voluntary carbon offset markets (VCO markets), 442

Wastes, 93
    impact of biochar on, 225
    wood, 91, 93
Water
    impact of biochar on, 226
    management, 206–207
    impact of SCS on, 208

Water footprint model (WF model), 280
  water balance, 284–285
  water contributions, 280–284
Weathering, 183
  of alkaline rocks, 138
  and alkalinity, 156–157
  of silicate rocks, 140
Wehrlite, 153
Weighting factor (FW), 70
Wet biomass, 91
Wetland ecosystems, 20–21
WF model. *See* Water footprint model (WF model)
Wheat straw, 93
Wildfire-penalty coefficient, 267–268
Wood products, 64–65
Woody biomass, 92

Yes in my back yard effect (YIMBY effect), 411
Yield, impact of biochar on, 224–225
YIMBY effect. *See* Yes in my back yard effect (YIMBY effect)